Bispecific Antibodies

Roland E. Kontermann
Editor

Bispecific Antibodies

Springer

Editor
Dr. Roland E. Kontermann
Universität Stuttgart
Inst. für Zellbiologie und Immunologie
Allmandring 31
70569 Stuttgart
Germany
roland.kontermann@izi.uni-stuttgart.de

ISBN 978-3-642-20909-3 e-ISBN 978-3-642-20910-9
DOI 10.1007/978-3-642-20910-9
Springer Heidelberg Dordrecht London New York

Library of Congress Control Number: 2011933913

Printed on acid-free paper

Springer is part of Springer Science+Business Media (www.springer.com)

Preface

The first description of bispecific antibodies dates back 50 years, when Nisonoff and Rivers described the recombination of a mixture of univalent antibody fragments of different specificity. However, it took another 20 years before bispecific antibodies were proposed for therapeutic applications, mainly for the retargeting of effector T cells to tumor cells. These early studies implied already the use of bispecific antibodies to extend the functions beyond that normally executed by antibodies. Limited by the availability of monoclonal antibodies obtained from animal sources, bispecific antibodies of this early phase were generated by somatic hybridization of two hybridomas or by chemical conjugation of two IgG molecules. The high expectations on these bispecific antibodies were, however, not fulfilled, mainly because of low efficacy, immunogenicity, and severe adverse effects seen in clinical trials. This resulted in a loss of interest in this kind of molecules during the last decade of the last century. However, with advancements in antibody engineering and the establishment of novel applications, bispecific antibodies experienced a revival at the beginning of this century. Besides effector cell retargeting for cancer immunotherapy, applications include, among others, pre-targeting strategies in radioimmunotherapy and more recently dual targeting strategies simultaneously attacking two disease-relevant targets. Genetic engineering allows nowadays to generate recombinant bispecific antibodies of defined composition, as well as with improved stability and producibility. Hence, bispecific antibodies have regained interest by the pharmaceutical industry, and many companies have meanwhile established their own bispecific antibody program. The first bispecific antibody for the retargeting of effector cells to EpCAM-positive tumor cells was approved in 2009 for the treatment of malignant ascites, and an increasing number of bispecific antibodies is currently in preclinical and clinical development.

Today, approaches to generate bispecific antibodies cover a broad spectrum including chemical conjugation, somatic hybridization, and genetic engineering. The latter has resulted in a multitude of recombinant IgG-like but also small-size bispecific antibody formats. Importantly, applications have been extended in the same manner. With many bispecific antibodies, especially of the second generation, in development, this class of molecules is rapidly advancing. This book is intended to provide a comprehensive overview of the various techniques and formats to generate bispecific antibodies and to give insights into the various applications which have emerged during the last two decades and which are actively explored for therapeutic and diagnostic purposes.

Stuttgart, Germany Roland E. Kontermann

Foreword

The development of monoclonal antibodies (mAbs) as therapeutics is an evolving discipline that relies heavily on innovation in biological engineering for its advancement. Therapeutic mAbs first entered clinical study in the early 1980s, soon after the hybridoma technology used to generate them was first described. Substantial improvements in the safety and efficacy of mAbs were a direct result of the advances in biological engineering that led to production of chimeric, humanized, and human antibody therapeutics. Advancement of technologies to design, engineer, and manufacture antibodies in the 1990s, as well as increases in the global sales of therapeutic mAbs, led to substantial commercial interest and investment in the 2000s. One measure of this investment is the number of novel mAbs that entered clinical study each year, which averaged approximately 20 candidates per year during the late 1990s and early 2000s but rose to nearly 50 per year in 2009. As of 2010, over 30 mAbs have been approved for marketing. Limitations inherent in the canonical monospecific IgG antibody prompted exploration of alternative molecular formats, including numerous bispecific versions of antibodies. This avenue of research recently yielded the first bispecific antibody to be approved, catumaxomab (Removab®), which is marketed in Europe for treatment of malignant ascites in epithelial cell adhesion molecule (EpCAM)-positive cancer patients.

The development of bispecific antibodies is a microcosm that has mimicked the past trends observed for overall antibody development and seems likely to share in a bright future. Historically, about half of all antibodies that entered clinical study sponsored by commercial firms were developed for cancer indications. Bispecific antibodies, with their inherent ability to bring two different targets into proximity, have been clinically studied almost exclusively in cancer patients. In order of frequency (first to fourth), the tumor-associated antigens most frequently targeted by anticancer antibodies evaluated in clinical studies sponsored by commercial firms to date are EpCAM, CD20, epithelial growth factor receptor (EGFR), and HER2/neu. Following these, insulin-like growth factor-1 receptor and carcinoembryonic antigen (CEA) were tied for fifth; CD30 and MUC1 were sixth;

and CD33, prostate-specific membrane antigen, tumor-related apoptosis-inducing ligand-receptor 2, and CD19 were seventh in frequency.

The tumor-associated antigens targeted by bispecific antibodies have mirrored these selections, with HER2/neu as the most frequent target followed by EpCAM, CEA, and CD30. A design challenge of the bispecific antibodies has been in the selection of the second specificity, which enables the cell-killing functionality of the anticancer molecules. Bispecific antibodies began entering clinical study in the early 1990s; the focus then was on CD64 (also known as FcγR1; found on macrophages and monocytes) and CD16 (T-cell co-receptor) as the second target of the molecules. At that time, Medarex had a robust bispecific antibody development program that focused on antigen-binding fragments (Fab) targeting CD64 that could be combined with Fab molecules targeting tumor-associated antigens such as HER2/neu, CD30, EGFR, or TAG72. However, all the bispecific molecules that entered clinical study in that period were ultimately terminated. The complexity of the biology, production issues, and competition from full-size IgG1 antibodies targeting the same tumor-associated antigens that were also in development at the same time were contributing factors in the decisions to discontinue development of these early bispecific candidates. The competitive IgG1 molecules proved to be superior in two cases – the humanized anti-HER2/neu trastuzumab (Herceptin®) was first approved in 1998 and the chimeric anti-EGFR cetuximab (Erbitux®) was first approved in 2003.

Although no therapeutic products resulted from the early work on bispecific antibodies, the ideas underlying it were sound and the research provided a foundation for advancement of the technology and improvement of the candidates. The combination of past experience with the recent increase in investments in antibody development has led to a notable revival of interest in bispecific formats. A path to approval for therapeutic bispecific antibodies has already been established by the work of Fresenius, the commercial sponsor of the approved bispecific antibody, catumaxomab. Numerous targets have now been clinically validated and bispecific molecules that might show enhanced efficacy compared with canonical full-size therapeutic antibodies are now entering clinical study in increasing numbers. The bispecific T-cell engager (BiTE) antibodies from Micromet are an excellent example of the potential of bispecific therapeutics. These molecules target CD3 on T cells, as well as tumor-associated antigens. Encouraging Phase 2 clinical results have been reported for blinatumomab, which targets CD3 and CD19 and is undergoing evaluation in patients with B-precursor acute lymphoblastic leukemia.

Considering the recent advances in bispecific antibody engineering and the obvious promise of the molecules as therapeutics, the publication of "Bispecific Antibodies," edited by Roland Kontermann, is timely. The book provides comprehensive coverage of the past, present, and probable future of bispecific antibody research and development. The available formats of these molecules have expanded rapidly and many of these, e.g., diabodies, dual variable domains, two-in-one antibodies, BiTEs, are discussed in detail. The potential of bispecific antibodies in gene therapy and their use as diagnostics is also explained. "Bispecific

Antibodies" contains a vast trove of in-depth knowledge about these versatile molecules and will thus be an invaluable resource for both experts and those new to the field.

Janice M. Reichert
Research Assistant Professor
Tufts Center for the Study of Drug Development
Tufts University School of Medicine
and
Editor-in-Chief, mAbs

janice.reichert@tufts.edu
janice.reichert@landesbioscience.com

Contents

Contributors

David Azria IRCM-Institut de Recherche en Cancérologie de Montpellier, Université MontpellierI, CRLC Val d'Aurelle-Paul Lamarque, 208 Rue des Apothicaires, 34298 Montpellier, Cedex 5, France

Patrick A. Baeuerle Micromet, Inc., 6707 Democracy Blvd, Bethesda, MD 20817, USA; Staffelseestr. 2, 81477 Munich, Germany, Patrick.baeuerle@micromet.com

Daniel Baty INSERM U624, 163 avenue de Luminy – case 915, 13288 Marseille, Cedex 09, France

Patrick Chames INSERM U624, 163 avenue de Luminy – case 915, 13288 Marseille, Cedex 09, France, patrick.chames@inserm.fr

Chien-Hsing Chang Immunomedics, Inc., 300 The American Road, Morris Plains, NJ 07950, USA, kchang@immunomedics.com

Diego Ellerman Department of Protein Chemistry, Research and Early Development, Genentech, 1 DNA Way, South San Francisco, CA 94080, USA

Georg H. Fey Microbiology, Department of Biology, University Erlangen-Nuremberg, Staudtstrasse 5, 91058 Erlangen, Germany

Wolfgang Fraunhofer Global Pharmaceutical Sciences, Abbott Laboratories, Abbott Park, IL 60064, USA

Fredrik Y. Frejd Affibody AB, Lindhagensgatan 133, 112 51 Stockholm, Sweden; Unit of Biomedical Radiations Sciences, Rudbeck Laboratory, Uppsala University, 751 85 Uppsala, Sweden, fredrik.frejd@affibody.se

Germaine Fuh Department of Antibody Engineering, Genentech, Inc, 1 DNA Way, South San Francisco, CA 94080, USA, gml@gene.com

Advaita Ganguly Faculty of Pharmacy and Pharmaceutical Sciences, University of Alberta, Edmonton, AB, Canada T6G 2N82

Tariq Ghayur Department of Biologics Abbott Bioresearch Center, 100 Research Drive, Worcester, MA 01605, USA

David M. Goldenberg Garden State Cancer Center, Center for Molecular Medicine and Immunology, 300 The American Road, 07950 Morris Plains, NJ, USA

Jijie Gu Department of Biologics Abbott Bioresearch Center, 100 Research Drive, Worcester, MA 01605, USA, jijie.gu@abbott.com

Juergen Hess TRION Pharma GmbH, Frankfurter Ring 193a, 80807 Munich, Germany

Pei Jin Protein Sciences and Design, Novartis Biologics, Novartis Pharma AG, 200 Technology Square, Cambridge, MA 02139, USA; Kadmon Pharmaceuticals LLC, 400 Madison Avenue, 11th Floor, New York, NY 10017, USA

Christian Kellner Division of Stem Cell Transplantation and Immunotherapy, 2nd Department of Medicine, Christian-Albrechts-University Kiel, Schittenhelmstr. 12, 24105 Kiel, Germany

Patrick Koenig Department of Antibody Engineering, Genentech, Inc, 1 DNA Way, South San Francisco, CA 94080, USA

Roland E. Kontermann Institut für Zellbiologie und Immunologie, Universität Stuttgart, Allmandring 31, 70569 Stuttgart, Germany, roland.kontermann@izi.uni-stuttgart.de

Christel Larbouret IRCM-Institut de Recherche en Cancérologie de Montpellier, Université MontpellierI, CRLC Val d'Aurelle-Paul Lamarque, 208 Rue des Apothicaires, 34298 Montpellier, Cedex 5, France

Horst Lindhofer TRION Pharma GmbH, Frankfurter Ring 193a, 80807 Munich, Germany; TRION Research GmbH, Am Klopferspitz 19, 82152 Martinsried, Germany, horst.lindhofer@trionpharma.de

John Löfblom Department of Molecular Biotechnology, School of Biotechnology, Royal Institute of Technology (KTH), AlbaNova University Center, 106 91 Stockholm, Sweden

Lawrence G. Lum BMT and Immunotherapy Program, Departments of Oncology, Medicine, and Immunology and Microbiology, Wayne State University, Detroit, MI, USA; Barbara Ann Karmanos Cancer Institute, Rm 740.1 HWCRC, 4100 John R, Detroit, MI 48201, USA, luml@karmanos.org

Jean-Pierre Mach Department of Biochemistry, University of Lausanne, Chemin des Boveresses 155, CH-1066 Epalinges, Switzerland

Nico Mertens Biotecnol SA, Lagoas Park, Edificio 7, 2741–901 Oeiras (Lisbon), Portugal, nm@biotecnol.com

Gerhard Moldenhauer Translational Immunology Unit (D015), Tumor Immunology Program, German Cancer Research Center, National Center for Tumor Diseases, Im Neuenheimer Feld 460, 69120 Heidelberg, Germany, g.moldenhauer@dkfz.de

Dafne Müller Institut für Zellbiologie und Immunologie, Universität Stuttgart, Allmandring 31, 70569 Stuttgart, Germany, dafne.mueller@izi.uni-stuttgart.de

Dirk M. Nettelbeck Department of Dermatology, German Cancer Research Center (DKFZ) and Heidelberg University Hospital, Helmholtz-University Group Oncolytic Adenoviruses, Im Neuenheimer Feld 242, 69120 Heidelberg, Germany, d.nettelbeck@dkfz.de

Archana Parashar Faculty of Pharmacy and Pharmaceutical Sciences, University of Alberta, Edmonton, AB, Canada T6G 2N82

André Pèlegrin IRCM-Institut de Recherche en Cancérologie de Montpellier, Université MontpellierI, CRLC Val d'Aurelle-Paul Lamarque, 208 Rue des Apothicaires, 34298 Montpellier, Cedex 5, France

Matthias Peipp Division of Stem Cell Transplantation and Immunotherapy, 2nd Department of Medicine, Christian-Albrechts-University Kiel, Schittenhelmstr. 12, 24105 Kiel, Germany

Bruno Robert IRCM-Institut de Recherche en Cancérologie de Montpellier, Université MontpellierI, CRLC Val d'Aurelle-Paul Lamarque, 208 Rue des Apothicaires, 34298 Montpellier, Cedex 5, France, bruno.robert@inserm.fr

Edmund A. Rossi Immunomedics, Inc., 300 The American Road, Morris Plains, NJ 07950, USA

Dominik Rüttinger Micromet, Inc., 6707 Democracy Blvd, Bethesda, MD 20817, USA; Staffelseestr. 2, 81477 Munich, Germany

Peter Ruf TRION Research GmbH, Am Klopferspitz 19, 82152 Martinsried, Germany

Jochen Salfeld Department of Biologics Abbott Bioresearch Center, 100 Research Drive, Worcester, MA 01605, USA

Susmita Sarkar Faculty of Pharmacy and Pharmaceutical Sciences, University of Alberta, Edmonton, AB, Canada T6G 2N82

Justin M. Scheer Department of Protein Chemistry, Research and Early Development, Genentech, 1 DNA Way, South San Francisco, CA 94080, USA, scheer.justin@gene.com

Ingo Schubert Microbiology, Department of Biology, University Erlangen-Nuremberg, Staudtstrasse 5, 91058 Erlangen, Germany

Robert M. Sharkey Garden State Cancer Center, Center for Molecular Medicine and Immunology, 300 The American Road, 07950 Morris Plains, NJ, USA

Sai Kiran Sharma Faculty of Pharmacy and Pharmaceutical Sciences, University of Alberta, Edmonton, AB, Canada T6G 2N82

Christoph Stein Microbiology, Department of Biology, University Erlangen-Nuremberg, Staudtstrasse 5, 91058 Erlangen, Germany, cstein@biologie.uni-erlangen.de

Mavanur R. Suresh Faculty of Pharmacy and Pharmaceutical Sciences, University of Alberta, Edmonton, AB, CanadaT6G 2N82, msuresh@pharmacy.ualberta.ca

Edit Tarcsa Department of Drug Metabolism and Pharmacokinetics, Abbott Bioresearch Center, 100 Research Drive, Worcester, MA 01605, USA

Archana Thakur BMT and Immunotherapy Program, Departments of Oncology, Wayne State University, Detroit, MI, USA

Thomas Valerius Division of Stem Cell Transplantation and Immunotherapy, 2nd Department of Medicine, Christian-Albrechts-University Kiel, Schittenhelmstr. 12, 24105 Kiel, Germany, t.valerius@med2.uni-kiel.de

Zhenping Zhu Protein Sciences and Design, Novartis Biologics, Novartis Pharma AG, 200 Technology Square, Cambridge, MA 02139, USA; Kadmon

Pharmaceuticals LLC, 400 Madison Avenue, 11th Floor, New York, NY 10017, USA, zzhu.mab@gmail.com

Gerhard Zugmaier Micromet, Inc., 6707 Democracy Blvd, Bethesda, MD 20817, USA; Staffelseestr. 2, 81477 Munich, Germany

Chapter 1
Bispecific Antibodies: Developments and Current Perspectives

Roland E. Kontermann

1.1 Introduction

All naturally occurring antibodies are multifunctional molecules, combining antigen-binding activity and Fc-mediated effector functions within a single molecule (Schroeder and Cavacini 2010). Antigen binding can lead to direct neutralization of the antigen, i.e., being antagonistic, but can also have agonistic activities, e.g., through activation of receptors (Fig. 1.1). The Fc region is capable of mediating further effector functions, which include antibody-dependent cellular cytotoxicity (ADCC), complement-dependent cytolysis (CDC) and opsonization, antibody-dependent cellular phagocytosis (ADCP), degranulation, antigen presentation, plasma recycling processes, and regulation of cell activation and proliferation (Fig. 1.1). Consequently, natural antibodies are well suited for a variety of therapeutic applications utilizing one or more of these effector functions. These effector functions can be further improved by protein and/or glyco-engineering (Carter 2006; Kubota et al. 2009). However, natural antibodies also face certain limitations. For example, antibodies are not capable of recruiting T cells because these cells do not express Fc receptors. Also, monoclonal antibodies are monospecific, thus, cannot address different target. Hence, strategies have been developed to extend the effector functions to novel activities normally not associated with antibodies, for example, by direct conjugation or fusion of therapeutic compounds or by indirect means using bispecific antibodies.

R.E. Kontermann (✉)
Institut für Zellbiologie und Immunologie, Universität Stuttgart, Allmandring 31, 70569 Stuttgart, Germany
e-mail: roland.kontermann@izi.uni-stuttgart.de

R.E. Kontermann (ed.), *Bispecific Antibodies*,
DOI 10.1007/978-3-642-20910-9_1, © Springer-Verlag Berlin Heidelberg 2011

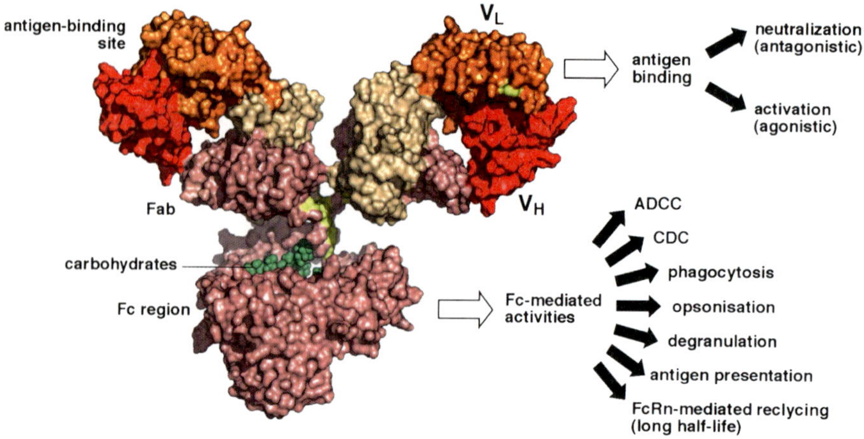

Fig. 1.1 The natural functions of antibodies. The structure of a mouse monoclonal IgG1 antibody (MAB61.1.3 directed against phenobarbital; pdb entry 1IGY) with the heavy chain shown in *light red* and the light chain shown in *light orange*. The variable heavy chain domain (V_H) is shown in *red* and the variable light chain domain (V_L) in *orange*. Carbohydrates are depicted in *green* and cysteine residues in *yellow*

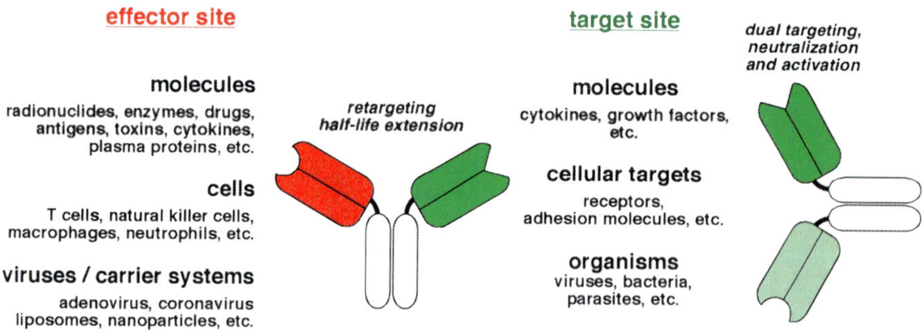

Fig. 1.2 Applications of bispecific antibodies (adapted from Müller and Kontermann 2007)

1.2 The Applications of Bispecific Antibodies

Bispecific antibodies are molecules with the ability to bind to two different epitopes on the same or on different antigens. Hence, bispecific antibodies are capable of bridging two different molecules, thus allowing a recruitment of effector molecules, effector cells, viruses, and drug carrier systems to target structures (Fig. 1.2). Recent approaches also utilize bispecific molecules binding with one site to a plasma protein such as albumin in order to prolong the half-life of a therapeutic antibody molecule. Furthermore, the simultaneous recognition of two different molecules, e.g., two growth factors or two different receptors on a target cell, allows for dual targeting strategies, thereby increasing selectivity and functional affinity, which can also lead to improved antagonistic or agonistic activities (Fanger et al. 1992;

Cao and Suresh 1998; van Spriel et al. 2000; Cao and Lam 2003; Thakur and Lum 2010). Applications of bispecific antibodies cover therefore a wide spectrum, for example, cancer therapy through improved binding, cellular cancer therapy through retargeting of effector cells, cancer radioimmunotherapy (RIT) through delivery of radionuclides, treatment of inflammatory diseases by targeting different proinflammatory cytokines, and many more including also diagnostic applications.

1.3 Developments

Bispecific antibodies with defined specificities are artificial molecules, i.e., are not produced by normal B cells. Hence, bispecific antibodies have to be produced by chemical or biological means, for example, by chemical conjugation of two antibodies, fusion of two antibody-producing cell lines, or genetic approaches resulting in recombinant bispecific antibody molecules. Genetic engineering has the advantage that the molecular design can be adapted to the desired activity, e.g., through modulation of valency and size. The first bispecific antibodies were described in the early sixties, produced by oxidative recombination of two univalent antibody preparations (Nisonoff and Rivers 1961). However, this was rather of academic interest and it took another two decades before bispecific antibodies with potential therapeutic use were described. An important step was the establishment of the hybridoma technology in 1975 by Köhler and Milstein, which allows the generation of monoclonal antibodies with predefined specificity (Köhler and Milstein 1975). In the early eighties, the group of Milstein also demonstrated that the fusion (somatic hybridization) of two different hybridomas results in hybrid hybridoma (quadroma) producing, amongst others, bispecific molecules (Fig. 1.3)

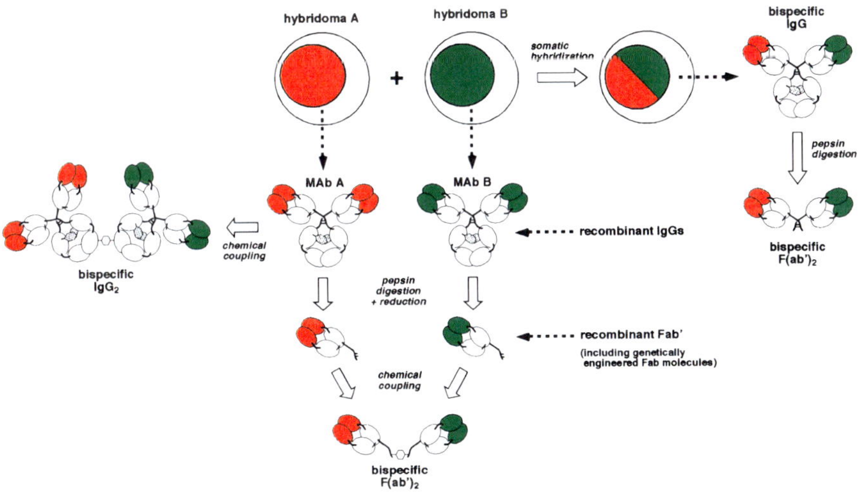

Fig. 1.3 Generation of bispecific antibodies by somatic hybridization or chemical conjugation

(Milstein and Cuello 1983). Several antibodies generated by somatic hybridization were developed for cancer immunotherapy, e.g., for the retargeting of effector T cells to tumor cells (Fanger et al. 1991; van Spriel et al. 2000). However, the high expectations were not fulfilled, mainly because of low antitumor activities, severe side effects caused by the presence of an Fc region, and immunogenicity of the murine bispecific antibodies leading to the production of neutralizing human antimouse antibodies (HAMA). Subsequent strategies focussed on the use of bispecific antibody fragments, e.g., F(ab′)$_2$ molecules, and the implementation of recombinant antibodies employing humanized and fully human antibodies for the generation of bispecific antibodies by chemical conjugation but also genetic engineering. Using the latter approach, a multitude of recombinant bispecific antibody formats were established during the past two decades including small bispecific antibody molecules lacking any constant domains as well as IgG-like bispecific antibodies comprising also immunoglobulin constant domains (Fig. 1.4). Several of these genetically engineered recombinant bispecific antibodies have meanwhile entered clinical trials (see Table 1.1, see also Chap. 15) and have opened the path to new and improved applications. Interestingly, however, the

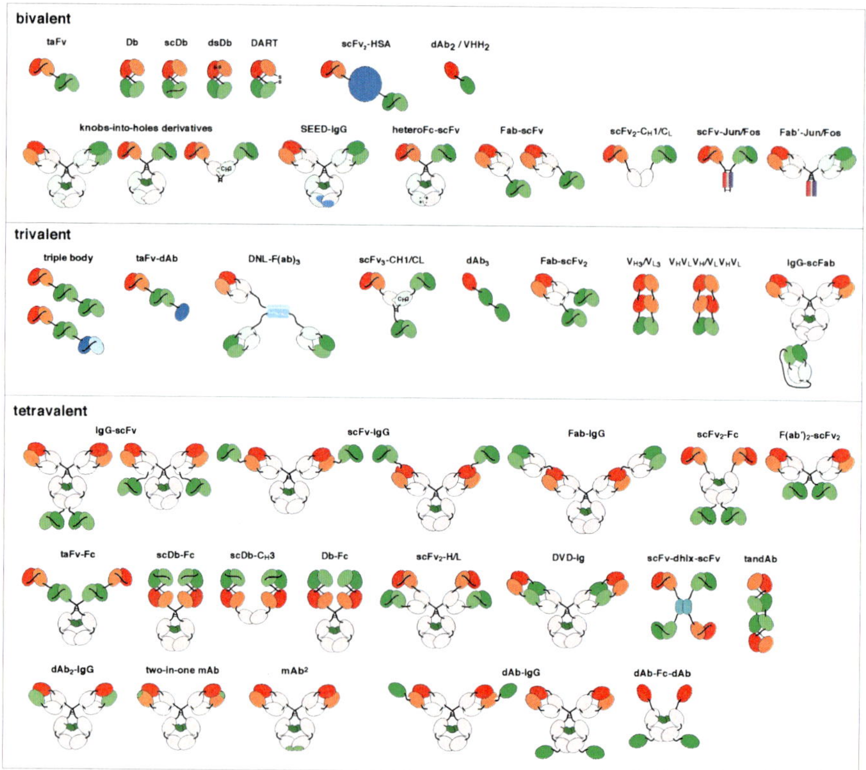

Fig. 1.4 Examples of recombinant bispecific antibody formats according to valency

Table 1.1 Recombinant bispecific antibodies in clinical development

Antibody	Specificities	Format	Origin	Indication	Status
BIS-1	EpCAM × CD3	IgG, F(ab′)$_2$	Mouse	Lung Ca	I $^{(t)}$
SHR-1	CD19 × CD3	IgG	Mouse	NHL	I $^{(t)}$
2B1	HER2 × CD16	IgG	Mouse	Various solid tumors	II $^{(t)}$
HEA125 × OKT3	EpCAM × CD3	IgG	Mouse	Ovarian Ca	I $^{(t)}$
H22 × Ki-4	CD30 × CD64	F(ab′)$_2$	Mouse/humanized	HL	I $^{(t)}$
HRS-3/A9	CD30 × CD16	IgG	Mouse	HL	I/II $^{(t)}$
OC/TR	FR × CD3	F(ab′)$_2$	Mouse	Ovarian Ca	I/II $^{(t)}$
MDX-H210	HER2 × C64	F(ab′)$_2$	Humanized	Various solid tumors	III $^{(t)}$
MDX-447	EGFR × CD64	F(ab′)$_2$	Humanized	Various solid tumors	II
CD20Bi	CD20 × CD3	IgG-IgG	Mouse/chimeric	MM, NHL	I
MM-111	HER2 × HER3	scFv$_2$-HSA	not known	Breast Ca	I/II
DT2219ARL	CD19 × CD22	scFv$_2$-DT	not known	B cell lymphoma	I
TF2	CEA × IMP228	Tri-Fab	Humanized	Colorectal Ca (radioimmunotherapy)	I
4G7 × H22	CD22 × CD64	IgG	Mouse	NHL, CLL	I
MT110	EpCAM × CD3	BiTE	Human	GI Ca, lung Ca	I
Blinatumomab	CD19 × CD3	BiTE	Mouse (deimmunized)	NHL, ALL	I/II
Lymphomun	CD20 × CD3	Triomab	Mouse/rat	B cell lymphoma	I/II
Ertumaxomab	HER2 × CD3	Triomab	Mouse/rat	Metastatic breast Ca	II
Catumaxomab	EpCAM × CD3	Triomab	Mouse/rat	Ovarian and gastric Ca, malignant ascites	II, approved 2009 in the EU

ALL acute lymphoplastic leukemia, *EpCAM* epithelial cell adhesion molecule, *FR* folate receptor, *HL* Hodgkin's lymphoma, *MM* multiple myeloma, *CLL* chronic lymphocytic leukemia, *Ca* Carcinoma, *NHL* non-Hodkgin's lymphoma, *DT* diphtheria toxin, *HSA* human serum albumin, (*t*) terminated

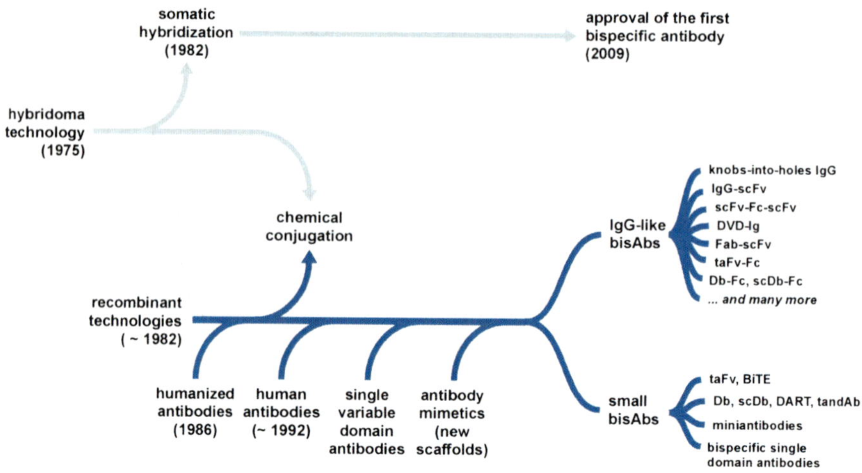

Fig. 1.5 The long path of bispecific antibodies

first bispecific antibody, which was approved by the EMEA in 2009, is a bispecific IgG molecule generated by somatic hybridization of a rat and a mouse hybridoma (Fig. 1.5). This antibody, developed by Trion Pharma and Fresenius Biotech, belongs to a family of trifunctional molecules (Triomabs) designed to retarget effector T cells to tumor cells (see Chap. 16).

1.4 Generation of Bispecific Antibodies by Somatic Hybridization

Hybridomas are generated by fusion of B cells with myeloma cells resulting in immortal cell lines producing monoclonal antibodies with predefined specificity. Further somatic hybridization of two hybridoma cells results in a quadroma cell line (hybrid hybridoma) producing within the same cell two different heavy and light chains (Milstein and Cuello 1983) (Fig. 1.3) (see also Chap. 2). Within in the cell, the heavy and light chains will randomly associate leading to a mixture of nonfunctional, monospecific but also bispecific molecules. On average only one in ten antibodies will be bispecific (De Lau et al. 1991; Smith et al. 1992). Therefore, in order to obtain homogenous bispecific antibodies, elaborate purification steps are normally required, e.g., employing two subsequent affinity chromatography columns. Interestingly, a species-restricted pairing was observed between heavy and light chains but not between heavy chains of different isotype or from different species (Corvalan and Smith 1987; Koolwijk et al. 1989; Link and Weiner 1993; Lindhofer et al. 1995). Importantly, bispecific antibodies composed of heavy chains with different isotypes are still able to exert Fc-mediated effector functions such as

ADCC and CMC (Koolwijk et al. 1991). However, this might be an undesirable feature for certain applications leading to Fc-mediated side effects such as cytokine-release syndrome, thrombocytopenia, and leukopenia. This can be avoided by proteolytical cleavage of bispecific antibodies into bispecific $F(ab')_2$ molecules, e.g., by pepsin digestion (Warnaar et al. 1994; Tutt et al. 1995) (Fig. 1.3). Interestingly, and as described in more detail in Chap. 16, bispecific antibodies produced from mouse/rat hybrid hybridoma and directed against a tumor antigen and CD3 have shown improved therapeutic activity. This has been attributed to Fc-mediated activation of immune cells of these trifunctional antibodies possessing a chimeric Fc region (Ruf and Lindhofer 2001).

1.5 Chemically Conjugated Bispecific Antibodies

An alternative way for the production of bispecific antibodies is provided by chemical coupling of two antibodies or antibody fragments obtained, e.g., from two hybridomas or using recombinant immunoglobulins (Fig. 1.3) (see also Chaps. 3 and 14). While initially oxidative reassociation strategies were used, current protocols utilize homo- or heterobifunctional cross-linking reagents (Graziano and Guptill 2004). Heterobifunctional cross-linkers exhibit reactivity toward two different reactive groups, e.g., amino and thiol-groups. For example, SPDP (*N*-succinimidyl-3-(2-pyridyldithio)propionate) reacts first with an amino group of the first molecule and introduces a thiol-reactive group, which then can react with a free thiol group in the second molecule. Similarly, EDC first reacts with a carboxyl group and subsequently with an amino group. Problems in the use of these heterobifunctional cross-linkers arise from the fact that antibodies or antibody fragments possess a large number of free amino or carboxyl groups, which leads to heterogeneous populations of cross-linked antibodies (Paulus 1985). Thiol-reactive homobifunctional cross-linking reagents can circumvent these obstacles. Two such reagents, 5,5′-dithiobis(2-nitrobenzoic acid) (DTNB) and *o*-phenylenedimaleimide (*o*-PDM), have been extensively applied for the generation of bispecific antibodies to crosslink two Fab′ fragments derived from different antibodies (Fig. 1.3). Fab′ molecules are produced by pepsin cleavage of intact antibodies and subsequent mild reduction of disulfide bonds of the remaining hinge region (Brennan et al. 1985; Glennie et al. 1987). The Fab′ fragments are then incubated with the cross-linking reagents in a two-step process leading to covalently linked bispecific $F(ab')_2$ molecules. Differences exist in the nature of crosslinking produced by these two reagents. DTNB produces disulfide bonds, while *o*-PDM results in highly stable thioether bonds, however, leaving an *o*-PDM moiety attached to the bispecific molecule (Graziano and Guptill 2004). Several bispecific antibodies generated by chemical crosslinking have entered clinical trials (Table 1.1) (Thakur and Lum 2010).

1.6 Recombinant Bispecific Antibody Molecules

Bispecific antibodies can be also produced by genetic means. Recombinant bispecific antibody formats allow for the production of structurally and functionally defined molecules. Furthermore, the implementation of humanized or fully human antibody molecules reduces immunogenicity (Shalaby et al. 1992), although the benefit of humanized and human antibodies has recently been questioned (Getts et al. 2010). A multitude of different antibody molecules have been developed over the past one to two decades (Fig. 1.4). These formats can be classified according to their structural properties. One classification can be made according to the valency of the bispecific molecules, i.e., whether they possess one antigen-binding site per antigen (bivalent) or more (tri-/tetravalent, multivalent). Furthermore, these molecules can be divided into (1) those based solely on the variable domains of two antibodies and (2) those using in addition constant immunoglobulin domains or nonimmunoglobulin domains for the assembly and formation of bispecific molecules (Fig. 1.4) (Kriangkum et al. 2001; Marvin and Zhu 2005; Plückthun and Pack 1997; Kontermann 2010; Chan and Carter 2010). Principally, all these strategies can also be utilized with single-domain antibodies and alternative scaffolds (antibody mimetics) (Saerens et al. 2008; Kontermann 2010; Skerra 2007).

Small recombinant bispecific antibodies, devoid of all constant immunoglobulin parts, are assembled from the variable light and heavy chain domains of two antibodies. The variable domains of an antibody can be obtained from existing hybridomas or isolated from antibody libraries, e.g., using phage display technology or transgenic animals. Usually, these antibodies are isolated in the form of single-chain Fv (scFv) or Fab fragments. ScFv fragments of different specificity can be easily converted into a bispecific molecule by genetically joining the two scFv moieties with a more or less flexible peptide linker. In these tandem scFv (taFv or $scFv_2$) molecules, each antigen-binding site forms a separate folding unit (see also Chap. 5). Consequently, tandem scFv molecules can be expressed in different configurations with scFv A and B having either the V_H–V_L or V_L–V_H arrangement of the variable domains. In a taFv molecule, the flanking linkers joining the variable domains of one folding unit have a length of approximately 15 amino acid residues, while the length of the middle linker connecting the two scFv molecules can be of varying length and composition. Thus, tandem scFv molecules were constructed possessing short Ala-Ala-Ala or Gly-Gly-Gly-Gly-Ser linkers (Brandão et al. 2003; Kufer et al. 1997; McCall et al. 2001), or longer linkers, which can adopt secondary structures (Hayden et al. 1994; Grosse-Hovest et al. 2004; Ren-Heidenreich et al. 2004; for an overview see: Kontermann 2005) (see also Chaps. 4 and 15). This strategy was extended for the generation of trivalent molecules, triple bodies, which can be either bispecific or trispecific (Kellner et al. 2008) (see also Chap. 4).

In contrast to tandem scFv molecules, bispecific diabodies Db are composed of two polypeptide chains V_HA–V_LB and V_HB–V_LA (V_H–V_L conformation), or V_LA–V_HB and V_LB–V_HA (V_L–V_H conformation) (Holliger et al. 1993) (see also Chap. 5). Reducing the linker sequence connecting the two variable domains within one chain from 15 residues usually used to produce scFv molecules to five residues forces heterodimerization of the two polypeptide chains within one cell. The head-to-tail pairing of the two chains produces a rigid molecule with two intact antigen-binding sites. Bispecific diabodies have been successfully expressed in different systems, including bacteria, yeast, or mammalian cells (overview in Kontermann 2005). However, the expression of the two different polypeptide chains leads also to the formation of nonfunctional homodimers possessing antigen-binding sites composed of V_H and V_L from two different antibodies. Attempts to improve formation of heterodimers include the introduction of disulfide bonds and knobs-into-holes structures between two adjacent variable domains, respectively (FitzGerald et al. 1997; Zhu et al. 1997) or adding cysteine residues at C-terminal extensions (DART, "dual affinity re-targeting"; Johnson et al. 2010). Another strategy is the introduction of an additional middle linker connecting the two polypeptide chains, thus, generating a single-chain diabody molecule scDb (Fig. 1.4). In scDb molecules, all four variable domains are present on this polypeptide chain. These molecules assemble into a diabody-like structure (Brüsselbach et al. 1999). The middle linker must have a length usually used for the generation of scFv fragments, i.e., approximately 15 residues (Völkel et al. 2001). This middle linker facilitates correct assembly of monomeric, bispecific molecules producing homogeneous preparations and improves stability compared to bispecific diabodies (Chap. 5). Reduction of the middle linker was shown to result in dimeric tetravalent and bispecific molecules, so-called tandAbs (Kipriyanov et al. 1999; Völkel et al. 2001).

ScFv or Fab fragments have been combined into bivalent and bispecific molecules using heterodimerizing peptides, such as jun-fos leucine zippers, fused to the C-terminus of the antibody molecules (Fig. 1.4). These heterodimers can be further stabilized by disulfide bond formation introducing additional cysteine residues or a natural hinge region (de Kruif and Logtenberg 1996; Kostelny et al. 1992; Pack and Plückthun 1992; Müller et al. 1998a).

A variety of additional formats are available using immunoglobulin constant regions for heterodimerization and the formation of bi- or multivalent, bispecific antibodies (Fig. 1.4) (see also Chap. 9). One strategy utilizes knobs-into-holes structures introduced into the C_H3 domain. This results in the heterodimerization of two C_H3 domains, which can be part of a complete heavy chain (Carter 2001; Shahied et al. 2004; Xie et al. 2005). C_H3 domains with knobs-into-holes are generated by replacing large amino acid side chains by small ones in one domain and introducing large side chain residues in the other domain (Ridgeway et al. 1996; Atwell et al. 1997). Problems arising from the fact that the light chains have no preference for one of the two knobs-into-holes heavy chains were solved by applying antibodies using the same light chain sequence (Merchant et al. 1998; Jackman et al. 2010). Alternatively, C_H3 knobs-into-holes domains can be part of an scFv-Fc fusion protein or can be directly fused to scFv or Fab fragments (Fig. 1.4). As an alternative to the knobs-into-holes approach, the Fc region was engineered to allow for a heterodimerization through an electrostatic steering effect (Gunasekaran et al. 2010). A further strategy to use constant immunoglobulin domains for heterodimerization is based on the natural occurring heterodimerization between the

C_L and C_H1 domain (Fig. 1.5). Thus, scFv fragments have been fused C-terminal of either the C_L or C_H1 domain of a Fab molecule (Schoonjans et al. 2000; Lu et al. 2002) or were fused N-terminal of both domains (Fig. 1.4) (Müller et al. 1998b) (see also Chap. 8).

Several bispecific antibody formats have been generated that possess more than one antigen-binding site for each antigen. Most of these formats are tetravalent, i.e., possess two antigen-binding sites for each antigen. Some of these formats contain an intact Fc region capable of mediating Fc effector functions. Recently developed dual-variable-domain antibodies DVD-Ig look like normal IgG antibodies with the exception that a second V_H and V_L domain is fused N-terminal of the V_H and V_L domain of the heavy and light chain, respectively (Fig. 1.4) (Wu et al. 2007, 2009) (see also Chap. 10). Thus, two variable domains are arranged in tandem on each chain. These molecules have the advantage that expression can be performed by established methods producing molecules of defined structure and function with the constant regions identical to natural antibodies. Another approach is the fusion of an scFv fragment (or a single domain antibody) to the N-terminus or C-terminus of either the heavy or light chain, respectively (Fig. 1.4) (Orcutt et al. 2010; Michaelson et al. 2009; Dimasi et al. 2009). More recently, the variable domains of an IgG were genetically modified to recognize two different antigens (Bostrom et al. 2009) (see also Chap. 11). These two-in-one antibodies are structurally indistinguishable form normal IgG molecules. Similarly, fusion of two different single-domain antibodies to the heavy or light chain, respectively, results in IgG-like bispecific molecules. A totally different approach is developed by f-star, engineering antigen-binding sites into the bottom loops of an Fc fragment (Wozniak-Knopp et al. 2010), which can then be used to generate bispecific IgG molecules (mAb^2).

Other approaches utilize fusion of a tandem scFv or a scDb to the Fc region of an antibody. The bispecific taFv-Fc and scDb-Fc fusion proteins are encoded by a single polypeptide and possess a similar molecular mass as IgG molecules (Fig. 1.4) (Connelly et al. 1998; Schneider et al. 2005; Lu et al. 2003, 2005; Alt et al. 1999; Zuo et al. 2000). For example, bispecific IgG-like molecules have been generated by fusing one chain of a bispecific diabody to the Fc part or two scFv molecules to the constant heavy and light chain region, respectively (Lu et al. 2005; Asano et al. 2007) (Chap. 9). The expression of two separate polypeptide chains is required for the production of these molecules. Other approaches employed fusion of an scFv fragment to the C-terminus of an intact heavy chain or a scFv-Fc fusion protein (Coloma and Morrison 1997). A trivalent molecule was produced fusing an scFv fragment C-terminal of an scFv-C_H3 fusion protein with a knobs-into-holes C_H3 domain coexpressed with a second scFv-C_H3 fusion protein. The dock-and-lock (DNL) method applies homo- and heterodimerizing domains obtained from cAMP-dependent protein kinase A and A-kinase anchor protein, to which two different Fab fragments are fused, for the generation of bispecific, trivalent molecules (Rossi et al. 2006) (see also Chap. 12). A tetravalent, bispecific molecule comprising the Fc region was generated by fusion of two different scFv molecules to the N- and C-terminus of an Fc region, combining bispecificity with Fc-mediated

activities. Bispecific and tri- or tetravalent molecules have also b
lack an intact Fc region, thus are not able to exert Fc-mediated e
for example, by fusing a bispecific scDb to a C_H3 domain. In an
scFv fragment was fused C-terminal of the hinge region generat
fusion protein. Bispecific and tetravalent molecules were also g
any immunoglobulin constant regions, e.g., fusing two scFv mo... dimerizing helix-loop-helix peptide (Müller et al. 1998a). Furthermore, the scDb format was modified to generate tetravalent bispecific molecules, for example, by reducing the middle linker to less than 12 amino acid residues, which leads to homodimerization of two scDb chains (Fig. 1.4) (Kipriyanov et al. 1999; Völkel et al. 2001).

1.7 Therapeutic Applications of Bispecific Antibodies

Certain classes and subclasses of natural antibodies (e.g., IgG1, IgG3, IgA) are capable of activating cellular effector functions of the immune system, e.g., ADCC and phagocytosis, through binding of their Fc region to Fc receptors on effector cells (Brekke and Sandlie 2003). These effector cells include macrophages expressing the high-affinity Fcγ receptor I (CD64) and natural killer cells (NK cells) expressing the medium affinity Fcγ receptor III (CD16). Both receptors preferentially bind to human IgG1 and IgG3. However, the induction of ADCC by therapeutic antibodies is often limited by binding to other nonactivating or inhibitory Fcg receptors (e.g., FcγRIIb, FcγRIIIb) expressed by macrophages, platelets, and B cells (Peipp and Valerius 2002). Furthermore, therapeutic antibodies have to compete for binding to the high-affinity receptor with serum IgG resulting in a poor recruitment of monocytes and macrophages (Valerius et al. 1997). Finally, cytotoxic T lymphocytes (CTL), the most potent effector cells of the immune system, do not express Fc receptors and cannot be recruited and activated by therapeutic antibodies. Hence, bispecific antibodies have been developed to selectively retarget certain effectors cells of the immune system by binding and triggering one of the Fc receptors or through binding to the T cell receptor (TCR) complex (de Gast et al. 1997; van Spriel et al. 2000; Müller and Kontermann 2010) (see Chap. 13).

One of the applications of bispecific antibodies is the retargeting of CTL for cellular cancer therapy (Fig. 1.6) (see also Chaps. 13–16). Under normal conditions, naive T cells are activated by binding of their TCR to MHC-displayed peptides (MHC class I on $CD8^+$ T cells, MHC class II on $CD4^+$ T cells) on antigen-presenting cells. However, proper activation of T cells requires a second signal through a costimulatory molecule, e.g., binding of B7 on antigen-presenting cells to CD28 on T cells. These activated $CD8^+$ T cells differentiate into CTLs capable of killing cells displaying the same antigenic peptide in the context of MHC class I. Killing is caused by release of cytotoxic molecules (perforin/granzyme pathway) and by death-receptor-mediated induction of apoptosis (FasL/CD95 pathway)

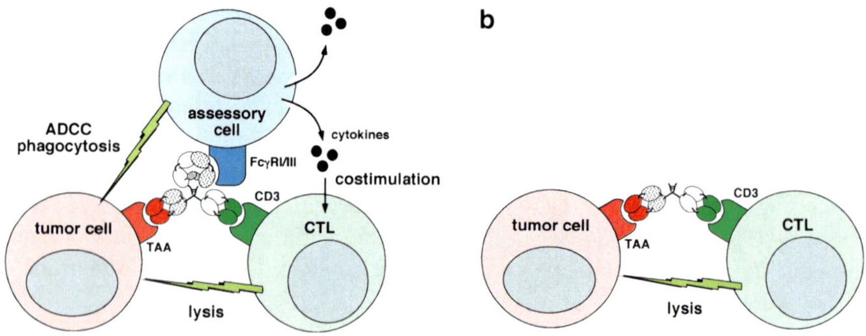

Fig. 1.6 Effector cell retargeting with bispecific IgG molecules (**a**) or bispecific F(ab′)$_2$ molecules (**b**)

(Russell and Ley 2002). Unfortunately, tumor cells are able to downregulate or lose MHC molecules, thus become invisible for T cells (Bubenik 2003).

Bispecific antibodies can bypass MHC-restricted target cell killing by T cells through binding to a trigger molecule that is part of the TCR complex and a cell surface-displayed antigen on the target cell, e.g., a tumor-associated antigen (TAA) (Lum and Davol 2005). Several trigger molecules (CD3, CD2, TCR) have been analyzed over the past two decades (Fanger et al. 1992). The most commonly trigger molecule used nowadays is CD3, a transmembrane protein involved in signal transduction by the TCR. Initial clinical studies employed bispecific antibodies [whole IgG, F(ab′)$_2$] directed against a trigger molecule and a TAA that were generated by somatic hybridization, i.e., were of mouse or rat origin. These molecules demonstrated potent tumor cell killing in vitro and in animal studies. However, despite some reported responses, clinical trials faced several problems. Amongst them was a severe toxicity due to release of inflammatory cytokines (cytokine storm), especially after i.v. administration, which was attributed to the presence of the Fc region in the bispecific IgG molecules. Furthermore, HAMA responses were observed in these studies. This led to the postulation of the following criteria of an ideal bispecific antibody for retargeting of effector cells: (1) strong selectivity for the target cell, (2) retargeting of relevant effector cells, (3) monovalent binding to the effector cells and activation only upon binding to target cells, (4) lack of an Fc region to avoid Fc-mediated side effects, (5) human or humanized to avoid a neutralizing antibody response, and (6) a size that allows penetration into tumor tissues and sufficiently long circulation (Segal et al. 1999).

Several new recombinant bispecific antibody formats developed in the past 10–15 years meet all or most of these criteria. These novel formats include tandem scFv molecules, diabodies, and single-chain diabodies. With these formats, bispecific antibodies of defined composition can be produced, which lack the Fc region. Furthermore, the implementation of humanized, deimmunized or fully human antibodies, e.g., isolated from human antibody phage libraries or transgenic animals, allows for the production of antibodies with reduced or absent immunogenicity in humans. Bispecific tandem scFv, diabodies, and single-chain diabodies with specificity for CD3 and TAA (CD19, CD20, EpCAM, erbB2, CEA) have shown efficacy in vitro as well as in animal studies (Kontermann 2005, Müller and Kontermann 2007). In these studies, preactivated T cells were used and in some cases included costimulatory signals provided by co-injection of anti-CD28 monoclonal antibodies (Cochlovius et al. 2000; Holliger et al. 1999). Several studies confirmed that costimulatory signals are necessary to obtain full activation of T cells and potent tumor cell killing (Reusch et al. 2004). Targeted delivery of such costimulatory signals was, for instance, achieved fusing the extracellular region of B7 or parts of it to scFv fragments, bivalent diabodies, or other antibody fragments directed against a TAA (Gerstmayer et al. 1997; Holliger et al. 1999; Rohrbach et al. 2000; Biburger et al. 2005; Müller et al. 2007). Several studies, especially those testing tandem scFv molecules of the BiTE (bispecific T cell engager) format, demonstrated potent tumor cell killing and antitumor activities using freshly isolated PBMCs from healthy donors without the need for preactivation (Löffler et al. 2000). Several BiTEs directed against different TAAs including CD19 and EpCAM have been developed and evaluated in vitro and in animal experiments (Dreier et al. 2003; Schlereth et al. 2005). A strictly target cell-dependent activation of T cells was demonstrated for these monomeric BiTE molecules possessing only one binding site for CD3, in contrast to molecules possessing two CD3 binding sites capable of activating T cells target cell-independent (Brischwein et al. 2007). BiTEs have been shown to mediate lysis of target cells at low antibody concentrations (pM–nM range) and low effector to target cell ratios (<10) (Wolf et al. 2005; Hoffmann et al. 2005). It was postulated that these molecules work costimulation independent and are highly efficient in recruiting and activating T cells (Wolf et al. 2005). One BiTE (MT103, MEDI-538, blinatumomab) directed against CD19 and CD3 has been investigated in a phase I dose-escalating trial for the treatment of non-Hodgkin's lymphoma (NHL). These patients received doses of 0.5–60 $\mu g/m^2$ per day over a period of 4–8 weeks. Dose-dependent complete and partial responses were observed in MCL, FL, and CLL patients. In addition, a complete removal or reduction of tumor load in bone marrow and other infiltrated organs was reported. The most common adverse effects were leukopenia, lymphopenia, and pyrexia, with the majority of adverse effects fully reversible in many cases. These are encouraging results obtained for this recombinant bispecific antibody molecule, which has been put forward for further clinical testing (see also Chap. 15).

Interestingly, several bispecific antibodies derived from mouse/rat hybrid-hybridomas are currently at various stages of clinical trials, despite the reported limitations of other whole bispecific IgG molecules found in early clinical trials. These bispecific antibodies are directed against CD3 and TAA and possess a hybrid Fc region composed of a mouse IgG2a and a rat IgG2b heavy chain that preferentially binds to the activating Fcγ receptors I and III but not to inhibitory FcγRII (Zeidler et al. 2000; Ruf and Lindhofer 2001). It was postulated that these trifunctional bispecific antibodies are capable of simultaneously activating T cells via binding to CD3 and accessory cells (NK cells, mononuclear blood cells) through interactions with their Fc region (Fig. 1.5). These activated accessory cells deliver necessary costimulatory signals to the T cells and further increase the immune response through phagocytosis of tumor material leading to long-lasting antitumor immunity (Zeidler et al. 2000, 2001). One of these trifunctional bispecific antibodies, removab (catumaxomab) directed against EpCAM and CD3 has been approved in 2009 by the EMEA for the treatment of malignant ascites and is further tested in clinical trials for treatment of other cancers, demonstrating that even whole bispecific antibodies can be safely applied i.p. and i.v. providing close monitoring of the patients (see also Chap. 16).

Bispecific antibodies have also been developed for retargeting other effector cells of the immune system (NK cells, macrophages, neutrophils) by binding to Fc receptors on these cells. These receptors include the medium-affinity FcγRIII (CD16) constitutively expressed by NK cells but also by monocytes and macrophages. Binding to FcγRIII triggers release of cytotoxic molecules (perforin, granzymes) and cell killing similar to T cell-mediated killing. The high-affinity IgG receptor FcγRI (CD64) is constitutively expressed by monocytes and macrophages. These cells also express FcγRIII and to some extent FcαRI (CD89). FcαRI is the main immunoglobulin receptor expressed by polymorphonuclear cells (PMNs), e.g., neutrophils, which represent the largest effector cell population in the blood (50–60% of all leukocytes). These cells also express FcγRI upon stimulation with IFN-γ and G-CSF (van Spriel et al. 2000). In contrast to T cells, these effector cells are constitutively active and do not require a second stimulus for efficient target cell killing and/or phagocytosis. Furthermore, because some of these receptors (e.g., FcγRI) are present on antigen-presenting cells, they are involved in antigen presentation and can therefore indirectly enhance antitumor activities. The first clinical trials with bispecific antibodies retargeting NK cells through binding to the CD16 revealed rather limited antitumor responses and were mainly characterized by toxicity (Weiner et al. 1995; Hartmann et al. 1997, 2001). As discussed before, this had been attributed to the presence of an intact Fc region leading to systemic leukocyte activation and massive cytokine release. Furthermore, most of the anti-CD16 used for construction of the bispecific antibodies did not discriminate between the activating FcγRIIIA on NK cells and the nonactivating FcγRIIIB on PMNs leading to a reduced activity of the bispecific antibodies unless a large excess is provided (Hombach et al. 1993; Amoroso et al. 1999). Subsequent studies were performed with bispecific F(ab$'$)$_2$ molecules, e.g., MDX-H210 (directed against erbB2 and CD64) and MDX-447 (directed against EGFR and CD64). These studies included application of growth factors and cytokines (e.g., GM-CSF, G-CSF, IFN-γ) in order to induce upregulation of Fc receptors and to increase the effector cell population. Although biological effects such as changes in circulating leukocyte subpopulation cell number and receptor expression, binding of the bispecific antibody to effector cells, enhanced ADCC/ phagocytic capacity in vitro, cytokine release and local infiltration of effector cells at the tumor site was observed, no objective tumor responses were achieved (Lewis et al. 2001; Wallace et al. 2001; Pullarkat et al. 1999; Repp et al. 2003; Curnow 1997; James et al. 2001; Fury et al. 2008). Nevertheless, these studies showed that bispecific F(ab$'$)$_2$ molecules directed against Fc receptors are well tolerated even at high doses, e.g., 30 mg/m^2 determined as recommended dose for further studies with MDX-447 (Fury et al. 2008).

Several studies also investigated FcαRI (CD89) as trigger molecule, because this receptor is constitutively expressed on neutrophils and the main trigger molecule to induce tumor cytolysis by these cells (Valerius et al. 1997; Deo et al. 1998). Direct comparison of various bispecific F(ab$'$)$_2$ molecules targeting CD20 or HER-2/neu and Fc receptors CD64, CD16, or CD89 in combination with G-CSF or GM-CSF revealed differences in the retargeting of effector cell cytotoxicity. This was dependent on Fc receptor expression, but also influenced by the growth factors and the tumor antigen involved (Stockmeyer et al. 2001).

The approach of retargeting effector cells through binding to Fcγ receptors was also extended to recombinant bispecific molecules. For example, bispecific tandem scFv molecules directed against CD16 and various TAAs (CD19, HER2/neu, HLA class II) have been developed for the retargeting of NK cells to tumor cells (McCall et al. 1999; Bruenke et al. 2004, 2005). Interestingly, in vitro cytotoxicity was enhanced by employing bispecific antibodies exhibiting increased binding to the TAA, either by increasing affinity or the number of TAA binding sites (McCall et al. 2001; Xie et al. 2003; Shahied et al. 2004). Furthermore, a synergistic antitumor activity in animal experiments of NHL was found using a combination of bispecific diabodies directed against CD19 × CD3 and CD19 × CD16 (Kipriyanov et al. 2002). These findings support the observation that retargeting and triggering of different effector cells to the same target cells results in increased cytotoxicity. In summary, these findings underline the complexity of this still challenging approach, but also highlight potentials and perspectives of bispecific and bifunctional antibodies for cellular cancer therapy.

1.8 Bispecific Antibodies and Retargeting of Effector Molecules

Bispecific antibodies have been extensively investigated for the recruitment of a large number of different effector molecules, including radionuclides, drugs, toxins, enzymes, cytokines, complement components, and immunoglobulins (Cao and Lam 2003). The use of bispecific antibodies circumvents chemical coupling of effector molecules, which is especially advantageous in cases where chemical modifications may lead to inactivation of the effector molecules or the antibody. Furthermore, the use of bispecific antibodies allows for an uncoupling of antibody-mediated targeting from delivery of effector molecules (Gruaz-Guyon et al. 2005). In this pretargeting strategy, the bispecific antibody is injected first and allowed to accumulate at the target site. After clearance of unbound antibodies from circulation and healthy tissues, the effector molecule is injected and is recovered by the second binding site of the bispecific antibody at the target site, while unbound effector molecules are rapidly eliminated. This approach reduces side effects associated with the effector molecule by reducing the exposure time to a minimum.

RIT is the main application of this pretargeting strategy combining antibodies and radionuclides of clinical interest, i.e., beta-emitters (^{131}I, ^{90}Y, ^{188}Rh) or alpha-emitters (^{211}At, ^{225}Ac) (Chang et al. 2002; Goldenberg 2003; Goldenberg et al. 2007; Goldenberg and Sharkey 2010). The efficacy of RIT is influenced by several factors including tumor location, size, morphology, physiology and radiosensitivity, physical/chemical properties of the radionuclides, and the nature of its radiation (low or high energy transfer), but also pharmacokinetic properties of the antibody (Goldenberg 2003). Clinical use of antibodies directly conjugated or complexes with radionuclides is often associated with toxicity in healthy tissues caused by an inappropriate tumor-to-normal tissue ratio, which limits the applicable dose and thus efficacy (Reilly 2006). This obstacle can be circumvented using a bispecific antibody-based pretargeting approach. Here, the bispecific antibody binds with one arm to a TAA and with the second arm to a radionuclide complexed with a chelating agent such as DTPA or DOTA. Alternatively, the radionuclide is complexed or conjugated with a molecule containing a short peptide sequence as hapten recognized by the bispecific antibody (Fig. 1.7). A widely used peptide sequence is HSG (histamine-succinyl-glycine), which has also been employed to generate bivalent molecules containing two HSG molecules exhibiting increased functional affinity. Similarly, an affinity enhancement system was developed by implementing two chelating molecules with the ability to bind two radionuclides per molecule (Le Doussal et al. 1990). Thus, these molecules exhibit an increased functional affinity for bispecific antibodies bound to target cells (Morandeau et al. 2005). Animal studies have shown that the use of the pretargeting approach in RIT results in reduced toxicity and allows for the administration of higher doses with enhanced antitumor effects (Sharkey et al. 2005a). The applied bispecific antibodies (bispecific $F(ab')_2$ molecules, bispecific diabodies) were directed against CEA, CD20, or the renal cell carcinoma marker G250 (Gautherot et al. 1997; Sharkey et al. 2005b; Kranenborg et al. 1998). One bispecific $F(ab')_2$ molecule (hMN14 × m734) directed against carcinoembryonic antigen (CEA) and a ^{131}I-labeled di-DTPA molecule was investigated in a phase I optimization clinical trial for the treatment of patients with various CEA-positive tumors (Kraeber-Bodéré et al. 2006). This bispecific antibody fragment is derived from a humanized anti-CEA antibody and a murine antihapten-chelate antibody and was generated by chemical

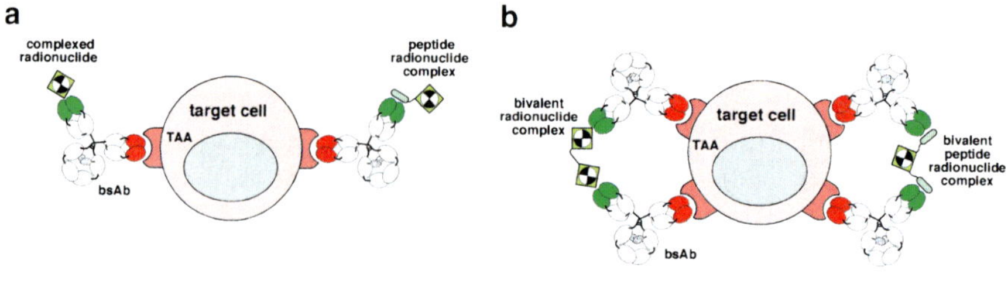

Fig. 1.7 Bispecific antibodies for radioimmunotherapy. (**a**) Monovalent radionuclide and radionuclide–peptide complexes. (**b**) Affinity enhancement systems using bivalent radionuclide complexes or bivalent peptide complexes

cross-linking with *o*-PDM. Patients suffering from nonmedullary thyroid carcinoma (non-MTC) or medullary thyroid carcinoma (MTC) received doses of 40 mg/m^2 and 75 mg/m^2 doses of the bispecific antibody and escalating activities of ^{131}I-di-DTPA-indium hapten 5 days later. A stabilization of disease was observed in 64% of patients receiving the 75 mg/m^2 of antibody. However, also higher toxicity was seen in these patients. Maximal tolerated activity was 3 GBq in MTC patients. The conclusion of this study was that a bispecific antibody dose of 40 mg/m^2 and a 5-day interval appeared to be a better dose/schedule regiment, with acceptable toxicity.

1.9 Dual Targeting Strategies

Like any other antibodies, bispecific antibodies are capable of acting as agonists or antagonists through binding to epitopes involved in activation or inhibition of a target molecule. The therapeutic use of antagonistic antibodies is best exemplified with antibodies directed against inflammatory cytokines, such as tumor necrosis factor (TNF). Inhibition of TNF binding to its two receptors (TNFR1, TNFR2) with antibodies (adalimumab, infliximab) or by soluble receptor-Fc fusion proteins (etanercept) has shown remarkable therapeutic efficacy in chronic inflammatory disease, e.g., rheumatoid arthritis (Kontermann et al. 2009). Conceptually, bispecific antibodies might be used to simultaneously bind to two different epitopes of the same cytokine or receptor, thus improving neutralization of one cytokine, or to simultaneously bind and neutralize two different cytokines or receptors, thus attacking the disease from two sites (Fig. 1.8). Several studies have demonstrated superiority of such bispecific antibodies as antagonists or agonists. For example, a bispecific diabody directed against two epitopes on vascular endothelial growth

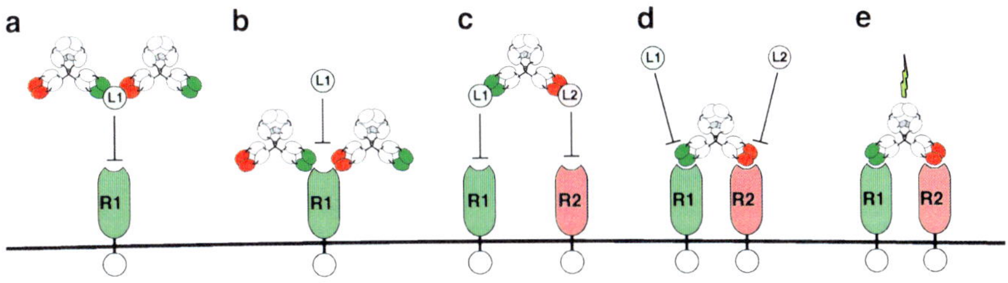

Fig. 1.8 Bispecific antibodies (exemplified by a bispecific IgG molecule) for dual targeting strategies. (**a**) Bispecific antibodies directed against different epitopes on one ligand (L1) inhibit binding of the ligand to its receptor (R1). (**b**) Bispecific antibodies directed against different epitopes on one receptor (R1) inhibit binding of the ligand (L1) to its receptor. (**c**) A bispecific antibody directed against two different ligands (L1, L2) simultaneously inhibits binding of the two ligands to its two receptors (R1, R2). (**d**) A bispecific antibody directed against two different receptors (R1, R2) simultaneously inhibits binding of two ligands (L1, L2). (**e**) Simultaneous antigen binding on a target cell can also lead to improved antibody-mediated activities such as ADCC

factor receptor 2 (VEGFR2) efficiently blocked binding of VEGF to its receptor and inhibited VEGF-induced activation of the receptor and mitogenesis of endothelial cells, while neither of the parental scFv fragments showed any inhibitory activity (Lu et al. 1999). An enhanced mitogenic activity was observed for a bispecific tandem scFv-Fc fusion protein directed against two epitopes of the T cell antigen CD2, compared to a combination of the two monoclonal antibodies (Connelly et al. 1998). A bispecific tandem scFv molecule directed against two epitopes on CTLA-4 was even able to convert CTLA-4 from an inhibitor to an activator of T cells (Madrenas et al. 2004). These T cell-activating bispecific antibodies may find clinical applications, e.g., in vaccination or cancer immunotherapy, by boosting immunity.

Enhanced inhibitory activity was observed for bispecific antibodies directed against two different target molecules (Marvin and Zhu 2006). For example, bispecific antibodies (diabody, diabody-Fc fusion protein) recognizing VEGF receptor 2 and 3 efficiently inhibited binding of VEGF and ligand-induced migration of endothelial cells (Lu et al. 2001, 2003; Jimenez et al. 2005). Similarly, IgG-like tetravalent and bispecific antibodies (diabody-Fc, $scFv_2$-H/L chain fusion proteins) with specificity for the EGF receptor and IGF receptor blocked binding of EGF and IGF and inhibited activation of signal transduction (Lu et al. 2004). These molecules were also capable of mediating ADCC through their Fc region. Animal experiments already demonstrated decreased growth of two different human tumor xenografts (Lu et al. 2005).

Recently, a novel bispecific, tetravalent IgG-like molecule [dual-variable-domain immunoglobulin (DVD-Ig)] was described for the neutralization of multiple disease mediators (Wu et al. 2007). These molecules were produced in mammalian cell lines (e.g., CHO) with similar yields as observed for normal IgG molecules. DVD-Ig directed against IL-12 and IL-18 was capable of binding both cytokines simultaneously and efficiently neutralized cytokine activity similar to the parental monoclonal antibodies. In a preclinical disease model of autoimmunity, this DVD-Ig was as potent as a mixture of the two parental antibodies and more potent than the two parental antibodies alone. Similar effects were observed for a DVD-Ig directed against IL-1α and IL-1β in a mouse collagen-induced arthritis model of rheumatoid arthritis. Besides improved activity, these and other types of bispecific molecules were described to facilitate production of drugs targeting two different disease-related factors while maintaining pharmacokinetic properties of whole IgG molecules.

Much work has also been done on dual targeting of immunotoxins. Here, two antibody moieties, e.g., in the form of tandem scFvs, are fused to a toxin. For example, fusion of two scFvs directed against CD19 and CD22 to the catalytic and translocation domains of Diptheria toxin resulted in an immunotoxin showing higher levels of binding to patient leukemia cells and efficient tumor cell killing in vitro and in vivo (Vallera et al. 2009). Similar results were obtained with a bispecific immunotoxin utilizing a fragment (P38) of Pseudomonas exotoxin (Vallera et al. 2010). Problems of immunogenicity of the bacterial toxin were overcome by engineering of a deimmunized version showing an approximately

80% reduction of neutralizing antitoxin antibodies, which will now allow multiple drug treatments (Vallera et al. 2010).

1.10 Bispecific Antibodies and Half-Life Extension

For most therapeutic applications, a long circulation time is desirable in order to maintain a therapeutically effective dose over an extended period of time (Mahmood and Green 2005). In humans, IgG molecules of the IgG1, IgG2, and IgG4 subclass exhibit a circulation half-life of approximately 3–4 weeks (Ternant and Paintaud 2005; Lobo et al. 2004). Thus, immunoglobulins have the longest half-life observed for plasma proteins. The only other plasma protein with a similar half-life of approximately 19 days in man is albumin. Responsible for the long circulation time of IgG antibodies and albumin is a reduced renal clearance and recycling processes mediated by the neonatal Fc receptor (FcRn, Brambell receptor) (Roopenian and Akilesh 2007).

Engineered small antibody molecules and molecules lacking an Fc region are rapidly cleared from circulation. For example, scFv fragments possess an initial half-life of a few minutes and a terminal half-life of only 1–4 h. Also, many of the small recombinant bispecific antibody molecules exhibit a fast clearance. Several strategies have therefore been developed to prolong circulation time of these molecules. These strategies can be divided into (1) those aiming to reduce renal clearance by increasing the apparent molecular mass and the hydrodynamic radius, e.g., through PEGylation and (2) those that in addition implement FcRn-mediated recycling processes (for review, see Kontermann 2009). The latter can be achieved through binding to long-circulating plasma proteins, such as albumin and immunoglobulins. Thus, small bispecific antibodies have been generated binding with one site to a therapeutic target, e.g., a cytokine or a cell surface receptor, and with the second binding site to albumin (Smith et al. 2001; Roovers et al. 2007).

1.11 Conclusion

Initial clinical trials with bispecific antibodies, developed mainly for cellular cancer immunotherapy, were not successful, which led to a declining interest in this type of molecules. However, during the past one to two decades, bispecific antibodies have experienced many new developments resulting from an increasing understanding of the molecular mechanisms of action as well as new developments in the field of antibody engineering. Thus, new bispecific formats were developed and new therapeutic approaches were established. While bispecific antibodies for cellular immunotherapy of cancer are still at the forefront of the developments, with one bispecific antibody approved for the treatment of malignant ascites and several small bispecific antibody molecules in clinical trials, new approaches especially in the field of dual targeting have emerged. Here, not only many new bispecific antibody formats have been established, but also new indications such as treatment of inflammatory diseases have been implemented. Therefore, one can expect that further bispecific antibodies will enter clinical trials in the near future.

References

Alt M, Müller R, Kontermann RE (1999) Novel tetravalent and bispecific IgG-like antibody molecules combining single-chain diabodies with the immunoglobulin g1 Fc or C_H3 region. FEBS Lett 454:90–94

Amoroso RA, Alpaugh RK, Barth MW, McCall AM, Weiner LM (1999) Production and characterization of mice transgenic for the A and B isoforms of human FcgRIII. Cancer Immunol Immunother 48:443–455

Asano R, Watanabe Y, Kawaguchi H, Fukazawa H, Nakanishi T, Umetsu M, Hayasi H, Katayose Y, Unno M, Kudo T, Kumagai I (2007) Highly effective recombinant format of a humanized IgG-like bispecific antibody for cancer immunotherapy with retargeting of lymphocytes to tumor cells. J Biol Chem 282:27659–27665

Atwell S, Ridgway JBB, Wells JA, Carter P (1997) Stable heterodimers from remodelling the domain interface of a homodimer using a phage display library. J Mol Biol 270:26–35

Biburger M, Weth R, Wels WS (2005) A novel bispecific tetravalent antibody fusion protein to target costimulatory activity for T-cell activation to tumor cells overexpressing ErbB2/HER2. J Mol Biol 346:1299–1311

Bostrom J, Yu SF, Kan D, Appleton BA, Lee CV, Billeci K, man W, Peale F, Ross S, Wiesmann C, Fuh G (2009) Variants of the antibody herceptin that interacts with HER2 and VEGF at the antigen binding site. Science 323:1610–1614

Brandão JG, Scheper RJ, Lougheed SM, Curiel DT, Tillman BW, Gerritsen WR, van den Eertwegh AJ, Pinedo HM, Haisma HJ, de Gruijl TD (2003) CD40-targeted adenoviral gene transfer to dendritic cells through the use of a novel bispecific single-chain Fv antibody enhances cytotoxic T cell activation. Vaccine 21:2268–2272

Brekke OH, Sandlie I (2003) Therapeutic antibodies for human diseases at the dawn of the twenty-first century. Nat Rev Drug Discov 22:52–62

Brennan M, Davison PF, Paulus H (1985) Preparation of bispecific antibodies by chemical recombination of monoclonal immunoglobulin G1 fragments. Science 229:81–83

Brischwein K, Parr L, Pflanz S, Volkland J, Lumsden J, Klinger M, Locher M, Hammond SA, Kiener P, Kufer P, Schlereth B, Baeuerle PA (2007) Strictly target cell-dependent activation of T cells by bispecific single-chain antibody constructs of the BiTE class. J Immunother 30:798–807

Bruenke J, Fischer B, Barbin K, Schreiter K, Wachter Y, Mahr K, Titgemeyer F, Niederweis M, Peipp M, Zunino S, Repp R, Valerius T, Fey GH (2004) A recombinant bispecific single-chain Fv antibody against HLA class II and FcγRIII (CD16) triggers effective lysis of lymphoma cells. Br J Haematol 125:167–179

Bruenke J, Barbin K, Kunert S, Lang P, Pfeiffer M, Stieglmaier K, Niethammer D, Stockmeyer B, Peipp M, Repp R, Valerius T, Fey GH (2005) Effective lysis of lymphoma cells with a stabilised bispecific single-chain Fv antibody against CD19 and FcgammaRIII (CD16). Br J Haematol 130:218–228

Brüsselbach S, Korn T, Völkel T, Müller R, Kontermann RE (1999) Enzyme recruitment and tumor cell killing in vitro by a secreted bispecific single-chain diabody. Tumor Target 4:115–123

Bubenik J (2003) Tumor MHC class I downregulation and immunotherapy. Oncol Rep 10: 2205–2208

Cao Y, Lam L (2003) Bispecific antibody conjugates in therapeutics. Adv Drug Deliv Rev 55: 171–197

Cao Y, Suresh MR (1998) Bispecific antibodies as novel bioconjugates. Bioconjug Chem 9:635–644

Carter P (2001) Bispecific human IgG by design. J Immunol Meth 248:7–15

Carter PJ (2006) Potent antibody therapeutics by design. Nat Rev Immunol 6:343–357

Chan AC, Carter PJ (2010) Therapeutic antibodies for autoimmunity and inflammation. Nat Rev Immunol 10:301–316

Chang CH, Sharkey RM, Rossi EA, Karacay H, McBride W, Hansen HJ, Chatal JF, Barbet J, Goldenberg DM (2002) Molecular advances in pretargeting radioimmunotherapy with bispecific antibodies. Mol Cancer 1:553–563

Cochlovius B, Kipriyanov SM, Stassar MJJG, Schuhmacher J, Benner A, Moldenhauer G, Little M (2000) Cure of Burkitt's lymphoma in severe combined immunodeficiency mice by T cells, tetravalent CD3 x CD19 tandem diabody, and CD28 costimulation. Cancer Res 60:4336–4341

Coloma MJ, Morrison SL (1997) Design and production of novel tetravalent bispecific antibodies. Nat Biotechnol 15:159–163

Connelly RJ, Hayden MS, Scholler JK, Tsu TT, Dupont B, Ledbetter JA, Kanner SB (1998) Mitogenic properties of a bispecific single-chain Fv-Ig fusion generated from CD2-specific mAb to distinct epitopes. Int Immunol 10:1863–1872

Corvalan JR, Smith W (1987) Construction and characterisation of a hybrid-hybrid monoclonal antibody recognising both carcinoembryonic antigen (CEA) and vinca alkaloids. Cancer Immunol Immunother 24:127–132

Curnow RT (1997) Clinical experience with CD64-directed immunotherapy. An overview. Cancer Immunol Immunother 45:210–215

de Gast GC, van de Winkel JGJ, Bast BEJEG (1997) Clinical perspectives of bispecific antibodies in cancer. Cancer Immunol Immunother 45:121–123

de Kruif J, Logtenberg T (1996) Leucine zipper dimerized bivalent and bispecific scFv antibodies from a semisynthetic antibody phage display library. J Biol Chem 271:7630–7634

de Lau WB, Keije K, Neefjes JJ, Oosterwegel M, Rozemüller E, Blast BJ (1991) Absence of preferential homologous H/L chain association in hybrid hybridomas. J Immunol 146: 906–914

Deo YM, Sundarapandiyan K, Keler T, Wallace PK, Graziano RF (1998) Bispecific molecules directed to the Fc receptor for IgA (FcαRI, CD89) and tumor antigens efficiently promote cell-mediated cytotoxicity of tumor targets in whole blood. J Immunol 160:1677–1686

Dimasi N, Gao C, Fleming R, Woods RM, Yao X, Shirinian L, Kiener PA, Wu H (2009) The design and characterization of oligospecific antibodies for simultaneous targeting of multiple disease mediators. J Mol Biol 393:672–692

Dreier T, Baeuerle PA, Fichtner I, Grün M, Schlereth B, Lorenczewski G, Kufer P, Lutterbüse R, Riethmüller G, Gjorstrup P, Bargou RC (2003) T cell costimulus-independent and very efficacious inhibition of tumor growth in mice bearing subcutaneous or leukemic human B cell lymphoma xenografts by a CD19-/CD3-bispecific single-chain antibody construct. J Immunol 170:4397–4402

Fanger MW, Segal DM, Romet-Lemonne J-L (1991) Bispecific antibodies and targeted cellular cytotoxicity. Immunol Today 12:51–54

Fanger MW, Morganelli PM, Guyre PM (1992) Bispecific antibodies. Crit Rev Immunol 12:101–124

FitzGerald K, Holliger P, Winter G (1997) Improved tumor targeting by disulphide stabilized diabodies expressed in Pichia pastoris. Protein Eng 10:1221–1225

Fury MG, Lipton A, Smith KM, Winston CB, Pfister DG (2008) A phase-I trial of epidermal growth factor receptor directed bispecific antibody MDX-447 without and with recombinant human granulocyte-colony stimulating factor in patients with advanced solid tumors. Cancer Immunol Immunother 57:155–163

Gautherot E, Bouhou J, Le Doussal JM, Manetti C, Martin M, Rouvier E, Barbet J (1997) Therapy of colon carcinoma xenografts with bispecific antibody-targeted, iodine-141-labeled bivalent hapten. Cancer 80(Suppl):2618–2623

Gerstmayer B, Hoffmann M, Altenschmidt U, Wels W (1997) Costimulation of T-cell proliferation by a chimeric B7-antibody fusion protein. Cancer Immunol Immunother 45:156–158

Getts DR, Getts MT, McCarthy DP, Chastain EML, Miller SD (2010) Have we overestimated the benefit of human(ized) antibodies? mAbs 2:682–694

Glennie MJ, McBride HM, Worth AT, Stevenson GT (1987) Preparation and performance of bispecific $F(ab'\gamma)_2$ antibody containing thioether-linked Fab'γ fragments. J Immunol 139:2367–2375

Goldenberg DM (2003) Advancing role of radiolabeled antibodies in the therapy of cancer. Cancer Immunol Immunother 52:281–296

Goldenberg DM, Sharkey RM (2010) Radioactive antibodies: a historical review of selective targeting and treatment of cancer. Hosp Pract 38:82–93

Goldenberg DM, Chatal JF, Barbet J, Boerman O, Sharkey RM (2007) Cancer imaging and therapy with bispecific antibody pretargeting. Update Cancer Ther 2:19–31

Graziano RF, Guptill P (2004) Chemical production of bispecific antibodies. Meth Mol Biol 283:71–85

Grosse-Hovest L, Müller S, Minoia R, Wolf E, Zakhartchenko V, Wenigerkind H, Lassnig C, Besenfelder U, Müller M, Lytton SD, Jung G, Brem G (2004) Cloned transgenic farm animals produce a bispecific antibody for T cell-mediated tumor cell killing. Proc Natl Acad Sci USA 101:6858–6863

Gruaz-Guyon A, Raguin O, Barbet J (2005) Recent advances in pretargeted radioimmunotherapy. Curr Med Chem 12:319–338

Gunasekaran K, Pentony M, Shen M, Garrett L, Forte C, Woodward A, Ng SB, Born T, Retter M, Mnachulenko K, Sweet H, Foltz IN, Wittekind M, Yan W (2010) Enhancing antibody Fc heterodimer formation through electrostatic steering effects: applications to bispecific molecules and monovalent IgG. J Biol Chem 18:19637–19646

Hartmann F, Renner C, Jung W, Deisting C, Juwana M, Eichentopf B, Kloft M, Pfreundschuh M (1997) Treatment of refractory Hodgkin's disease with an anti-CD16/CD30 bispecific antibody. Blood 89:2042–2047

Hartmann F, Renner C, Jung W, da Costa L, Tembrink S, Held G, Sek A, König J, Bauer S, Kloft M, Pfreundschuh M (2001) Anti-CD16/CD30 bispecific antibody treatment for Hodgkin's disease: role of infusion schedule and costimulation with cytokines. Clin Cancer Res 7:1873–1881

Hayden MS, Linsley PS, Gayle MA, Bajorath J, Brady WA, Norris NA, Fell HP, Ledbetter JA, Gilliland LK (1994) Single-chain mono- and bispecific antibody derivatives with novel biological properties and antitumor activity from a COS cell transient expression system. Ther Immunol 1:3–15

Hoffmann P, Hofmeister R, Brischwein K, Brandl C, Crommer S, Bargou R, Itin C, Prang N, Baeuerle PA (2005) Serial killing of tumor cells by cytotoxic T cells redirected with a CD19-/CD3-bispecific single-chain antibody construct. Int J Cancer 115:98–104

Holliger P, Prospero T, Winter G (1993) "Diabodies": small bivalent and bispecific antibody fragments. Proc Natl Acad Sci USA 90:6444–6448

Holliger P, Manzke O, Span M, Hawkins R, Fleischmann B, Qianghua L, Wolf J, Diehl V, Cochet O, Winter G, Bohlen H (1999) Carcinoembryonic antigen (CEA)-specific T-cell activation in colon carcinoma induced by anti-CD3xanti-CEA bispecific diabodies and B7xanti-CEA bispecific fusion proteins. Cancer Res 59:2909–2926

Hombach A, Jung W, Pohl C, Renner C, Sahin U, Schmits R, Wolf J, Kapp U, Diehl V, Pfreundschuh M (1993) A CD16/CD30 bispecific monoclonal antibody induces lysis of Hodgkin's cells by unstimulated natural killer cells in vitro and in vivo. Int J Cancer 55:830–836

Jackman J, Chen Y, Hhuang A, Moffat B, Scheer JM, Leong SR, Lee WP, Zhang J, Sharma N, Lu Y, Iyer S, Shields RL, Chiang N, Bauer MC, Wadley D, Roose-Girma M, Vandlen R, Yansura DG, Wu Y, Wu LC (2010) Development of two-part strategy to identify a therapeutic human bispecific antibody that inhibits IgE receptor signaling. J Biol Chem 285:20850–20859

James ND, Atherton PJ, Jones J, Howie AJ, Tchekmedyian S, Curnow RT (2001) A phase II study of the bispecific antibody MDX-H210 (anti-HER2xCD64) with GM-CSF in HER2^{+} advanced prostate cancer. B J Cancer 85:152–156

Jimenez X, Lu D, Brennan L, Persaud K, Liu M, Miao H, Witte L, Zhu Z (2005) A recombinant, fully human, bispecific antibody neutralizes the biological activities mediated by both vascular endothelial growth factor receptors 2 and 3. Mol Cancer Ther 4:427–434

Johnson S, Burke S, Huang L, Gorlatov S, Lim H, Wangm W, Zhangm W, Tuaillon N, Raineym J, Baratm B, Yangm Y, Jinm L, Ciccaronem V, Moorem PA, Koenigm S, Bonvini E (2010) Effector cell recruitment with novel Fv-based dual-affinity re-targeting protein leads to potent tumor cytolysis and in vivo B-cell depletion. J Mol Biol 399:436–449

Kellner C, Bruenke J, Stieglmaier J, Schwemmlein M, Schwenkert M, Singer H, Mentz K, Peipp M, Lang P, Oduncu F, Stockmeyer B, Fey GH (2008) A novel CD19-directed recombinant bispecific antibody derivative with enhanced immune effector functions for human leukemic cells. J Immunother 31:871–884

Kipriyanov SM, Moldenhauer G, Schuhmacher J, Cochlovius B, Von der Lieth CW, Matys ER, Little M (1999) Bispecific tandem diabody for tumor therapy with improved antigen binding and pharmacokinetics. J Mol Biol 293:41–56

Kipriyanov SM, Cochlovius B, Schäfer HJ, Moldenhauer G, Bähre A, Le Gall F, Knackmuss S, Little M (2002) Synergistic antitumor effect of bispecific CD19xCD3 and CD19xCD16 diabodies in a preclinical model of non-Hodgkin's lymphoma. J Immunol 169:137–144

Köhler G, Milstein C (1975) Continuous cultures of fused cells secreting antibody of predefined specificity. Nature 256:495–497

Kontermann RE (2005) Recombinant bispecific antibodies for cancer therapy. Acta Pharmacol Sin 26:1–9

Kontermann RE (2009) Strategies to extend plasma half-lives of recombinant antibodies. BioDrugs 23:93–109

Kontermann RE (2010) Alternative antibody formats. Curr Opin Mol Ther 12:176–183

Kontermann RE, Scheurich P, Pfizenmaier K (2009) Antagonists of TNF action: clinical experience and new developments. Expert Opin Drug Discov 4:279–292

Koolwijk P, Spierenbur GGT, Frasa H, Boot JH, van de Winkel JH, Bat BJ (1989) Interaction between hybrid mouse monoclonal antibodies and the human high-affinity IgG FcR, huFcgRI, on U937. Involvement of only one of the mIgG heavy chains in receptor binding. J Immunol 143:1656–1662

Koolwijk P, Boot JH, Griep R, Bast BJ (1991) Binding of human complement subcomponent C1q to hybrid mouse monoclonal antibodies. Mol Immunol 28:567–576

Kostelny SA, Cole MS, Tso JY (1992) Formation of a bispecific antibody by the use of leucine zippers. J Immunol 148:1547–1553

Kraeber-Bodéré F, Rousseau C, Bodet-Milin C, Ferrer L, Faivre-Chauvet A, Campion L, Vuillez JP, Devillers A, Chang CH, Goldenberg DM, Chatal JF, Barbet J (2006) Targeting, toxicity and efficacy of 2-step, pretargeted radioimmunotherapy using a chimaeric bispecific antibody and 131I-labelled bivalent hapten in a phase I optimization clinical trial. J Nucl Med 47:247–255

Kranenborg MH, Boeman Oc, Oosterwijk-Wakka JC, de Weijert MC, Corstens FH, Oosterwijk E (1998) Two-step radio-immunotargeting of renal-cell carcinoma xenografts in nude mice with anti-renal-cell-carcinoma X anti-DTPA bispecific monoclonal antibodies. Int J Cancer 75:74–80

Kriangkum J, Xu B, Nagata LP, Fulton RE, Suresh MR (2001) Bispecific and bifunctional single chain recombinant antibodies. Biomol Eng 18:31–40

Kubota T, Niwa R, Satoh M, Akinaga S, Shitara K, Hanai N (2009) Engineered therapeutic antibodies with improved effector functions. Cancer Sci 1000:1566–1572

Kufer P, Mack M, Gruber R, Lutterbuse R, Zettl F, Riethmuller G (1997) Construction and biological activity of a recombinant bispecific single-chain antibody designed for therapy of minimal residual colorectal cancer. Cancer Immunol Immunother 45:193–197

Le Doussal JM, Gruaz-Guyon A, Martin M, Gautherot E, Delaage M, Barbet J (1990) Targeting of indium 111-labeled bivalent hapten to human melanoma mediated by bispecific monoclonal antibody conjugates: imaging of tumors hosted in nude mice. Cancer Res 50:3445–3452

Lewis LD, Cole BF, Wallace PK, Fisher JL, Waugh M, Guyre PM, Fanger MW, Curnow RT, Kaufman PA, Ernstoff MS (2001) Pharmacokinetic-pharmacodynamic relationship of the bispecific antibody MDX-H210 when administered in combination with interferon gamma: a multiple-dose phase-I study in patients with advanced cancer which overexpresses HER-2/neu. J Immunol Meth 248:149–165

Lindhofer H, Mocika R, Steipe B, Thierfelder S (1995) Preferential species-restricted heavy/light chain pairing in rat/mouse quadromas. Implications for a single-step purification of bispecific antibodies. J Immunol 155:219–225

Link BK, Weiner GJ (1993) Production and characterization of a bispecific IgG capable of inducing T-cell-mediated lysis of malignant B cells. Blood 81:3343–3349

Lobo ED, Hansen RJ, Balthasar JP (2004) Antibody pharmacokinetics and pharmacodynamics. J Pharm Sci 93:2645–2668

Löffler A, Kufer P, Lutterbüse R, Zettl F, Daniel PT, Schwenkenbecher JM, Riethmüller G, Dörken B, Bargou RC (2000) A recombinant bispecific single-chain antibody, CD19 x CD3, induces rapid and high lymphoma-directed cytotoxicity by unstimulated T lymphoctes. Blood 95:2089–2103

Lu D, Kotanides H, Jimenez X, Zhou Q, Persaud K, Bohlen P, Witte L, Zhu Z (1999) Acquired antagonistic activity of a bispecific diabody directed against two different epitopes on vascular endothelial growth factor receptor 2. J Immunol Meth 230:159–171

Lu D, Jimenez X, Zhang H, Wu Y, Bohlen P, Witte L, Zhu Z (2001) Complete inhibition of vascular endothelial growth factor (VEGF) activities with a bifunctional diabody directed against both VEGF kinase receptors, fms-like tyrosine kinase receptor and kinase insert domain-containing receptor. Cancer Res 61:7002–7008

Lu D, Jimenez X, Zhang H, Bohlen P, Witte L, Zhu Z (2002) Fab-scFv fusion protein: an efficient approach to production of bispecific antibody fragments. J Immunol Meth 267: 213–226

Lu D, Jimenez X, Zhang H, Atkins A, Brennan L, Balderes P, Bohlen P, Witte L, Zhu Z (2003) Di-diabody: a novel tetravalent bispecific antibody molecule by design. J Immunol Meth 279:219–232

Lu D, Zhang H, Ludwig D, Persaud A, Jimenez X, Burtrum D, Balderes P, Liu M, Bohlen P, Witte L, Zhu Z (2004) Simultaneous blockade of both the epidermal growth factor receptor and the insulin-like growth factor receptor signaling pathways in cancer cells with a fully human recombinant bispecific antibody. J Biol Chem 279:2856–2865

Lu D, Zhang H, Koo H, Tonra J, Balderes P, Prewett M, Corcoran E, Mangalampalli V, Bassi R, Anselma D, Patel D, Kang X, Ludwig DL, Hicklin DJ, Bohlen P, Witte L, Zhu Z (2005) A fully human recombinant IgG-like bispecific antibody to both the epidermal growth factor receptor and the insulin-like growth factor receptor for enhanced antitumor activity. J Biol Chem 280:19665–19672

Lum LG, Davol PA (2005) Retargeting T cells and immune effector cells with bispecific antibodies. Cancer Chemother Biol Response Modif 22:273–291

Madrenas J, Chau LA, Teft WA, Wu PW, Jussif J, Kasaian M, Carreno BM, Ling V (2004) Conversion of CTLA-4 from inhibitor to activator of T cells with a bispecific tandem single-chain Fv ligand. J Immunol 172:5948–5956

Mahmood I, Green MD (2005) Pharmacokinetic and pharmacodynamic considerations in the development of therapeutic proteins. Clin Pharmacokinet 44:331–347

Marvin JS, Zhu Z (2005) Recombinant approaches to IgG-like bispecific antibodies. Acta Pharmacol Sin 26:649–658

Marvin JS, Zhu Z (2006) Bispecific antibodies for dual-modality cancer therapy: killing two signaling cascades with one stone. Curr Opin Drug Discov Develop 9:184–193

McCall AM, Adams GP, Amoroso AR, Nielsen UB, Zhang L, Horak E, Simmons H, Schier R, Marks JD, Weiner LM (1999) Isolation and characterization of an anti-CD16 single-chain Fv fragment and construction of an anti-HER2/neu/anti-CD16 bispecific scFv that triggers CD16-dependent tumor cytolysis. Mol Immunol 36:433–445

McCall AM, Shahied L, Amoroso AR, Horak EM, Simmons HH, Nielson U, Adams GP, Schier R, Marks JD, Weiner LM (2001) Increasing the affinity for tumor antigen enhances bispecific antibody cytotoxicity. J Immunol 166:6112–6117

Merchant AM, Zhu Z, Yuan JQ, Goddard A, Adams CW, Presta LG, Carter P (1998) An efficient route to human bispecific IgG. Nat Biotechn 16:677–681

Michaelson JS, Demarest SJ, Miller B, Amatucci A, William S, Wu X, Huang F, Pfan S, Gao S, Doern A et al (2009) Anti-tumor activity of stability-engineered IgG-like bispecific antibodies targeting TRAIL-R2 and LTβR. mAbs 1:128–141

Milstein C, Cuello AC (1983) Hybrid hybridomas and their use in immunohistochemistry. Nature 305:537–540

Morandeau L, Benoist E, Loussouarn A, Quadi A, Lesaec P, Mougin M, Faivre-Chauvet A, Le Boterff J, Chatal JF, Barbet J, Gestin JF (2005) Synthesis of new bivalent peptides for applications in the affinity enhancement system. Bioconjugat Chem 16:184–193

Müller D, Kontermann RE (2007) Recombinant bispecific antibodies for cellular cancer immunotherapy. Curr Opin Mol Ther 9:319–326

Müller D, Kontermann RE (2010) Bispecific antibodies for cancer immunotherapy: current perspectives. BioDrugs 24:89–98

Müller KM, Arndt KM, Plückthun A (1998a) A dimeric bispecific miniantibody combines two specificities with avidity. FEBS Lett 432:45–49

Müller KM, Arndt KM, Strittmatter W, Plückthun A (1998b) The first constant domain (C_H1 and C_L) of an antibody used as heterodimerisation domain for bispecific miniantibodies. FEBS Lett 422:259–264

Müller D, Karle A, Meißburger B, Höfig I, Stork R, Kontermann RE (2007) Improved pharmacokinetics of recombinant bispecific antibody molecules by fusion to human serum albumin. J Biol Chem 282:12650–12660

Nisonoff A, Rivers MM (1961) Recombination of a mixture of univalent antibody fragments of different specificity. Arch Biochem Biophys 93:460–462

Orcutt KD, Ackerman ME, Cieslewicz M, Quiroz E, Slusarczyk AL, Frangioni JV, Wittrup KD (2010) A modular IgG-scFv bispecific antibody topology. Protein Eng Des Sel 23:221–228

Pack P, Plückthun A (1992) Miniantibodies: use of amphipathic helices to produce functional, flexibly linked dimeric Fv fragments with high avidity in Escherichia coli. Biochemistry 31:1579–1584

Paulus H (1985) Preparation and biomedical applications of bispecific antibodies. Behring Inst Mitt 78:118–132

Peipp M, Valerius T (2002) Bispecific antibodies targeting cancer cells. Biochem Soc Trans 30:507–511

Plückthun A, Pack P (1997) New protein engineering approaches to multivalent and bispecific antibody fragments. Immunotechnol 3:83–105

Pullarkat V, Deo Y, Link J, Spears L, Marty V, Curnow R, Groshen S, Gee C, Weber JS (1999) A phase I study of a HER2/neu bispecific antibody with granulocyte-colony-stimulating factor in patients with metastatic breast cancer that overexpresses HER2/neu. Cancer Immunol Immunother 48:9–21

Reilly RM (2006) Radioimmunotherapy of solid tumors: the promise of pretargeting strategies using bispecific antibodies and radiolabelled haptens. J Nuc Med 47:196–199

Ren-Heidenreich L, Davol PA, Kouttab NM, Elfenbein GJ, Lum LG (2004) Redirected T-cell cytotoxicity to epithelial cell adhesion molecule-overexpressing adenocarcinomas by a novel recombinant antibody, E3Bi, in vitro and in an animal model. Cancer 100:1095–1103

Repp R, van Ojik HH, Valerius T, Groenewegen G, Wieland G, Oetzel C, Stockmeyer B, Becker W, Eisenhut M, Steininger H, Deo YM, Blijham GH, Kalden JR, van de Winkel JGJ, Gramatzki M (2003) Phase I clinical trial of the bispecific antibody MDX-H210 (anti-FcγRIxanti-HER-2/neu) in combination with Filgrastim (G-CSF) for treatment of advanced breast cancer. B J Cancer 89:2234–2243

Reusch U, Le Gall F, Hensel M, Moldenhauer G, Ho AD, Little M, Kipriyanov SM (2004) Effect of tetravalent bispecific CD19xCD3 recombinant antibody construct and CD28 costimulation on lysis of malignant B cells from patients with chronic lymphocytic leukaemia by autologous T cells. Int J Cancer 112:509

Ridgeway JB, Presta LG, Carter P (1996) "Knobs-into-holes" engineering of antibody C_H3 domains for heavy chain heterodimerization. Protein Eng 9:617–621

Rohrbach F, Gerstmayer B, Biburger M, Wels W (2000) Construction and characterization of bispecific costimulatory molecules containing a minimized CD86 (B7-2) domain and single-chain antibody fragments for tumor targeting. Clin Cancer Res 6:4314–4322

Roopenian DC, Akilesh S (2007) FcRn: the neonatal Fc receptor comes of age. Nat Rev Immunol 7:715–725

Roovers RC, Laeremans T, Huang L, De Taeye S, Verkleij AJ, Revets H, de Haard HJ, van Bergen en Henegouwen PMP (2007) Efficient inhibition of EGFR signalling and of tumor growth by antagonistic anti-EGFR nanobodies. Cancer Immunol Immunother 56:303–317

Rossi EA, Goldenberg DM, Cardillo TM, McBride WJ, Sharkey RM, Chang CH (2006) Stably tethered multifunctional structures of defined composition made by the dock and lock method for use in cancer targeting. Proc Natl Acad Sci USA 103:6841–6846

Ruf P, Lindhofer H (2001) Induction of a long-lasting antitumor immunity by a trifunctional bispecific antibody. Blood 98:2526–2534

Russell JH, Ley TJ (2002) Lymphocyte-mediated cytotoxicity. Annu Rev Immunol 20:323–370

Saerens D, Ghassabeh GH, Muyldersmans S (2008) Single-domain antibodies as building blocks for novel therapeutics. Curr Opin Pharmacol 8:600–608

Schlereth B, Fichtner I, Lorenczewski G, Kleindienst P, Brischwein K, da Silva A, Kufer P, Lutterbüse R, Juhnhahn I, Kasimir-Bauer S, Wimberger P, Kimmig R, Baeuerle PA (2005) Eradication of tumors from a human colon cancer cell line and from ovarian cancer metastases in immunodeficient mice by a single-chain Ep-CAM-/CD3-bispecific antibody construct. Cancer Res 65:2882–2889

Schneider MA, Bruhl H, Wechselberger A, Cihak J, Stangassinger M, Schlondorff D, Mack M (2005) In vitro and in vivo properties of a dimeric bispecific single-chain antibody IgG-fusion protein for depletion of CCR2(+) target cells in mice. Eur J Immunol 35:987–995

Schoonjans R, Willems a, Schoonooghe S, Fiers W, Grooten J, Mertens N (2000) Fab chains as an efficient heterodimerization scaffold for the production of recombinant bispecific and trispecific antibody derivatives. J Immunol 165:7050–7057

Schroeder HW, Cavacini L (2010) Structure and function of immunoglobulins. J Allergy Clin Immunol 125:S41–S52

Segal DM, Weiner GJ, Weiner LM (1999) Bispecific antibodies in cancer therapy. Curr Op Immunol 11:558–562

Shahied LS, Tang Y, Alpaugh RK, Somer R, Greenspon D, Weiner LM (2004) Bispecific minibodies targeting (HER2/neu and CD16 exhibit improved tumor lysis when placed in a divalent tumor antigen binding format. J Biol Chem 279:53907–53914

Shalaby MR, Shepard HM, Presta L, Rodrgues ML, Beverley PCL, Feldmann M, Carter P (1992) Development of humanised bispecific antibodies reactive with cytotoxic lymphocytes and tumor cells overexpressing the HER2 protoncogene. J Exp Med 175:217–225

Sharkey RM, Cardillo TM, Rossi EA, Chang CH, Karacay H, McBride WJ, Hansen HJ, Horak ID, Goldenberg DM (2005a) Signal amplification in molecular imaging by pretargeting a multivalent, bispecific antibody. Nat Med 11:1250–1255

Sharkey RM, Karacay H, Chang CH, McBride WJ, Horak ID, Goldenberg DM (2005b) Improved therapy of non-Hodgkin's lymphoma xenografts using radionuclides pretargeted with a new anti-CD20 bispecific antibody. Leukemia 19:1064–1069

Skerra A (2007) Alternative non-antibody scaffolds for molecular recognition. Curr Opin Biotechnol 18:295–304

Smith W, Jarrett AL, Beattie RE, Corvalan JR (1992) Immunoglobulins secreted by a hybrid-hybridoma: analysis of chain assemblies. Hybridoma 11:87–98

Smith BJ, Popplewell A, Athwal D, Chapman AP, Heywood S, West SM, Carrington B, Nesbitt A, Lawson AD, Antoniw P, Eddelston A, Suittler A (2001) Prolonged in vivo residence times of antibody fragments associated with albumin. Bioconjug Chem 12:7450–7756

Stockmeyer B, Elsässer D, Dechant M, Repp R, Gramatzki M, Glennie MJ, van de Winkel JGJ, Valerius T (2001) Mechanisms of G-CSF- or GM-CSF-stimulated tumor cell killing by Fc receptor-directed bispecific antibodies. J Immunol Meth 248:103–111

Ternant D, Paintaud G (2005) Pharmacokinetics and concentration-effect relationships of therapeutic monoclonal antibodies and fusion proteins. Expert Opin Biol Ther 5:S37–S47

Thakur A, Lum LG (2010) Cancer therapy with bispecific antibodies: clinical experience. Curr Opin Mol Ther 12:340–349

Tutt AL, Reid R, Wilkins BS, Glennie MJ (1995) Activation and preferential expansion of rat cytotoxic (CD8) T cells in vitro and in vivo with a bispecific (anti-TRC a/b x anti-CD2) $F(ab')_2$ antibody. J Immunol 155:2960–2971

Valerius T, Würflein D, Stockmeyer B, Repp R, Kalden JR, Gramatzki M (1997) Activated neutrophils as effector cells for bispecific antibodies. Cancer Immunol Immunother 45:142–145

Vallera DA, Chen H, Sichender AR, Panoskaltsis-Mortari A, Taras EP (2009) Genetic alteration of a bispecific ligand-directed toxin targeting human CD19 and CD22 receptors resulting in improved efficacy against systemic B cell malignancies. Leuk Res 33:1233–1242

Vallera DA, Oh S, Chen H, Shu Y, Frankel AE (2010) Bioengineering a unique deimmunized bispecific targeted toxin that simultaneously recognizes human CD22 and CD19 receptors in a mouse model of B-cell metastasis. Mol Cancer Ther 9:1872–1883

van Spriel AB, van Ojik HH, van de Winkel JGJ (2000) Immunotherapeutic perspective for bispecific antibodies. Immunol Today 21:391–396

Völkel T, Korn T, Bach M, Müller R, Kontermann RE (2001) Optimized linker sequences for the expression of monomeric and dimeric bispecific single-chain diabodies. Protein Eng 14:815–823

Wallace PK, Kaufman PA, Lewis LD, Keler T, Givan AL, Fisher JL, Waugh MG, Wahner AE, Guyre PM, Fanger MW, Ernstoff MS (2001) Bispecific antibody-targeted phagocytosis of HER-2/neu expressing tumor cells by myeloid cells activated in vivo. J Immunol Meth 248:167–182

Warnaar SO, de Paus V, Lardenoije R, Machielse BN, de Graaf J, Bregonje M, van Haarlem H (1994) Purification of bispecific $F(ab')_2$ from murine trinoma OC/TR with specificity for CD3 and ovarian cancer. Hybridoma 13:519–526

Weiner LM, Clark JI, Davey M, Li WS, Garcia de Palazzo I, Ring DB, Alpaugh RK (1995) Phase I trial of 2B1, a bispecific monoclonal antibody targeting c-erbB-2 and FcγRIII. Cancer Res 55:4586–4593

Wolf E, Hofmeister R, Kufer P, Schlereth B, Baeuerle PA (2005) BiTEs: bispecific antibody constructs with unique anti-tumor activity. Drug Disov Today 10:1237–1244

Wozniak-Knopp G, Bartl S, Bauer A, Mostageer M, Woisetschläger M, Antes B, Ettl K, Kainer M, Weberhofer G, Wiederkum S, Himmler G, Mudde GC, Rüker F (2010) Introducing antigen-binding sites in structural loops of immunoglobulin constant domains: Fc fragments with engineered Her2/neu-binding sites and antibody properties. Protein Eng Des Sel 23: 289–297

Wu C, Ying H, Grinnell C, Bryant S, Miller R, Clabbers A, Bose S, McCarty D, Zhu R-R, Santora L, Davis-Taber R, Kunes Y, Fung E, Schwartz A, Sakorafas P, Gu J, Tarcsa E, Murtaza A, Ghayur T (2007) Simultaneous targeting of multiple disease mediators by a dual-variable-domain immunoglobulin. Nat Biotechnol 25:1290–1297

Wu C, Ying H, Bose S, Miller R, Medina L, Santora L, Ghayur T (2009) Molecular construction and optimization of anti-human IL-1alpha/beta dual variable domain immunoglobulins (DVD-Ig) molecules. mAbs 1:339–347

Xie Z, Shi M, Feng J, Yu M, Sun Y, Shen B, Guo N (2003) A trivalent anti-erbB2/anti-CD16 bispecific antibody retargeting NK cells against human breast cancer cells. Biochem Biophys Res Commun 311:307–312

Xie Z, Guo N, Yu M, Hu M, Shen B (2005) A new format of bispecific antibody: highly efficient heterodimerization, expression and tumor cell lysis. J Immunol Meth 296:95–101

Zeidler R, Mysliwietz J, Csanady M, Walz A, Ziegler I, Schmitt B, Wollenberg B, Lindhofer H (2000) The Fc-region of a new class of intact bispecific antibody mediates activation of accessory cells and NK cells and induces direct phagocytosis of tumor cells. Br J Cancer 83:261–266

Zeidler R, Mayer A, Gires O, Schmitt B, Mack BM, Lindhofer H, Wollenberg B, Walz A (2001) TNFa contributes to the antitumor activity of a bispecific, trifunctional antibody. Anticancer Res 21:3499–3503

Zhu Z, Presta LG, Zapata G, Carter P (1997) Remodeling domain interfaces to enhance heterodimer formation. Protein Sci 6:781–788

Zuo Z, Jimenez X, Witte L, Zhu Z (2000) An efficient route to the production of an IgG-like bispecific antibody. Protein Eng 13:361–367

Chapter 2
Bispecific Antibodies from Hybrid Hybridoma

Gerhard Moldenhauer

2.1 Introduction

Hybrid hybridomas (also termed quadromas or tetradomas) are man-made cell lines that secrete bispecific antibodies (bsAb) with two different specificities being able to crosslink two distinct molecules. Such antibodies do not occur in nature and have been originally developed to improve immunohistochemical staining procedures and immunoassays (Milstein and Cuello 1983; Suresh et al. 1986). Interestingly, the fusion of two immunoglobulin-producing myeloma cells (Cotton and Milstein 1973) was described even before the seminal publication of monoclonal antibody technology (Köhler and Milstein 1975). This early experiment showing expression of both parental immunoglobulin genes in the hybrid cell was performed to better understand allelic exclusion, whereby under normal conditions each B lymphocyte produces antibodies encoded by only one of two possible alleles.

In the following years it became obvious that bsAb can be used to redirect immunological effector cells or molecules toward tumor cells. Targeting an immune response to the tumor site has evolved as an attractive concept since it recruits many effector cells and obviates several drawbacks connected with classical anti-tumor responses (reviewed by Fanger et al. 1992; Renner and Pfreundschuh 1995; van Spriel et al. 2000; Müller and Kontermann 2007a).

Basically, there are three methods by which bsAbs can be obtained. The first generation of bsAb was produced either by chemical coupling of different immunoglobulin Fab′ fragments at the hinge region (Glenie et al. 1987) or by cell fusion of two hybridoma cell lines, resulting in a quadroma cell which secretes among other immunoglobulin combinations also bsAb (Milstein and Cuello 1983).

G. Moldenhauer (✉)
Translational Immunology Unit (D015), Tumor Immunology Program, German Cancer Research Center, National Center for Tumor Diseases, Im Neuenheimer Feld 460, 69120 Heidelberg, Germany
e-mail: g.moldenhauer@dkfz.de

R.E. Kontermann (ed.), *Bispecific Antibodies*,
DOI 10.1007/978-3-642-20910-9_2, © Springer-Verlag Berlin Heidelberg 2011

Although some of these early reagents showed remarkable efficacy in the treatment of certain malignancies, the vast majority of them did not met the high initial expectations because of toxicity and low therapeutic potency. Advance in genetic engineering has facilitated the creation of second-generation bispecific molecules of different sizes and binding strengths. These recombinant bispecific constructs paved the way for a revival of bsAb and renewed the interest in this novel type of modified antibodies (Little et al. 2000; Kufer et al. 2004; Weiner 2007; Müller and Kontermann 2007b, 2010; Chames and Baty 2009a, b; Beck et al. 2010; Chan and Carter 2010).

In this chapter, I describe the principles of hybrid hybridoma creation, give an overview on screening and purification procedures of bsAb, and finally outline the mechanisms by which this special class of antibodies exerts its effector functions. For the sake of clarity, I have restricted my viewpoint mainly to the field of cancer research and oncology/hematology.

2.2 Production of Hybrid Hybridoma by Cell Fusion

The major goal of fusing two permanently growing hybridoma lines is the selection of resulting hybrids and the exclusion of non-fused parental cells and intra-hybridoma fused cells. This can essentially be achieved by chemical selection procedures or by cell sorting techniques (Segal and Bast 2001).

2.2.1 Introduction of Selection Markers

In case of chemical selection, a mutation of the gene coding for the enzyme hypoxanthine guanine phosphoribosyl transferase (HGPRT) can be induced in one hybridoma line (Chervonsky et al. 1988). HGPRT-negative mutants can be selected by culturing the cells in the presence of toxic purine analogs like 8-azaguanine or 6-thioguanine. Cells harboring the enzyme are killed after incorporation of the toxic nucleotide. Spontaneously arising mutants can simply be established because the enzyme is encoded on the X chromosome and only one gene locus has to be targeted. Proliferation of the other parental line can be irreversibly inhibited by treatment with iodoacetamide just before hybridization (Gilliland et al. 1988). If the fusion mixture is cultured in medium containing hypoxanthine, aminopterin, and thymidine (HAT medium) only hybrid hybridomas can actively grow. This is schematically depicted in Fig. 2.1. The combination of HAT sensitivity and resistance to the drug ouabain was also successfully applied (Link and Weiner 1993).

Alternatively, one hybridoma line is selected for two mutations, e.g., HAT sensitivity and resistance to neomycin (neo^r). For this, a HGPRT-deficient hybridoma is subsequently transfected by electroporation with a vector containing the neo^r

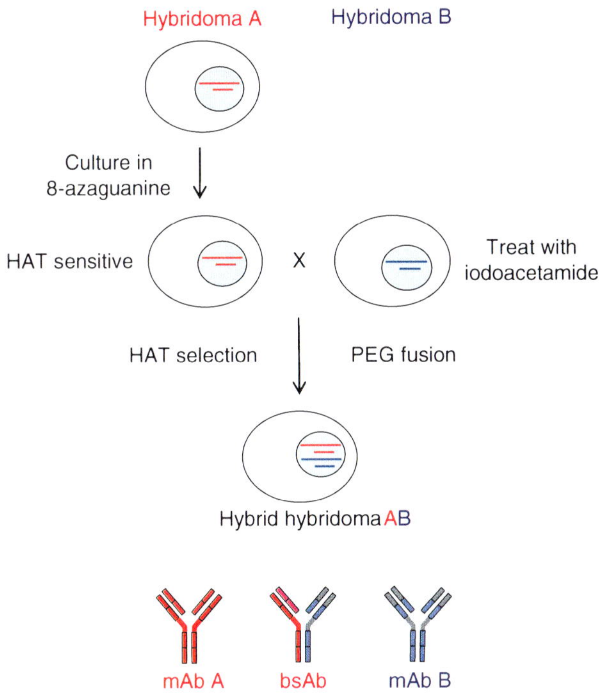

Fig. 2.1 Principle of hybrid hybridoma production by cell fusion. Sensitivity to HAT is induced in one hybridoma by selection with 8-azaguanine. Proliferation of the other hybridoma is inhibited by iodoacetamide treatment. After PEG fusion only hybrid hybridoma cells can actively grow in HAT medium

gene (De Lau et al. 1989). The advantage of such a double-mutant hybridoma line is that it can be fused with any unmodified second hybridoma line. As above, only hybrids can survive selection in medium supplemented with HAT and the neomycin analog G418.

2.2.2 *Selection by Flow Cytometry*

Fluorescence-activated cell sorting (FACS) offers another possibility to select hybrid hybridomas (Karawajew et al. 1987; Koolwijk et al. 1988). In this setting, one hybridoma cell line is labeled for instance with a green fluorochrome and the other with a red fluorescent marker. After cell fusion, double-labeled hybrid hybridomas are sorted and immediately cloned using a single-cell deposition device. Although this method is fast since it circumvents lengthy insertion of selectable markers there is a certain risk of contamination with parental cells.

2.2.3 Cell Fusion Procedures

For a long time polyethylene glycol is the fusion agent of choice (Pontecorvo 1975). It renders the membrane of cells to be fused gluey so that they will adhere together. Subsequently, plasma membrane fusion occurs giving rise to a cell with two (or more) nuclei. During cell division the nuclear membranes are degraded and the chromosomes are distributed into the daughter cell. These hybrid cells contain only one nucleus but the genetic material of both parents. Since hybridomas usually are hyperploid, a fusion of those will harbor a large excess of chromosomes that causes genetic instability during further mitoses leading to improper segregation or loss of chromosomes. To stabilize the chromosomal inventory of hybrid hybridoma repeated cloning by limiting dilution is inevitable. This also allows the selection of subclones producing high amounts of the desired bsAb.

As an alternative to the polyethylene glycol method, electrically induced cell fusion has been developed. It is based on the delivery of high-voltage electrical field pulses to physically fuse hybridoma cells (Cao et al. 1995; Kreutz et al. 1998).

2.2.4 Screening Hybrid Hybridoma Cultures for Specific Antibody

Establishment of a reliable, sensitive, and fast screening assay for the detection of desired bispecific antibody is the most important prerequisite for successful quadroma production. It is not recommended to start a fusion experiment before an appropriate screening assay has been set up. There are different test types available for the initial screening that all are based on the measurement of antigen–antibody binding. Enzyme-linked immunosorbent assay (ELISA), flow cytometry, and cytotoxicity are the most commonly used methods (Moldenhauer 2007). Choosing parental hybridoma cell lines of distinct isotype will facilitate the screening procedure tremendously.

Solid-phase ELISA (Engvall and Perlman 1971) where the antigen is immobilized on the well of a microtiter plate represents a universal test system that can easily be customized and allows rapid analysis of many samples in parallel. By ELISA one can simply detect bsAb with two different heavy chains. For this, the assay plate is coated with a catcher antibody specific for one isotype (e.g., mouse IgG1). Binding of bsAb is recognized by a detector antibody specific for the other isotype (e.g., mouse IgG2a). If one or both target antigens are available as proteins, they can be immobilized onto the assay plate. Again, reactivity with isotype-specific second step reagents is indicative for bsAb.

Bispecific antibodies developed for cancer therapy are usually directed against cell surface molecules on tumor and effector cells. Fluorescence-activated cell analyzer and cell sorter provide extremely valuable tools for the rapid, reliable, and quantitative screening of antibodies interacting with cell surface receptors.

Binding of bsAb to cells is monitored by isotype-specific fluorescently labeled secondary antibody. Thereby, the mean fluorescence intensity reflects the binding strengths of the respective antibody arm to its cellular target.

To test for bsAb-induced redirected cytotoxicity is a straight forward but technically demanding approach (Clark and Waldmann 1987). Target and effector cells are incubated in the presence of bsAb (and appropriate controls) leading to specific lysis of antigen-bearing targets. For the quantitative evaluation either the classical chromium release test (that requires initial labeling of target cells with radioactive chromium-51) or non-radioactive methods measuring the release of the endogenous cytosolic enzymes as for instance lactate dehydrogenase (Decker and Lohmann-Matthes 1988) or glyceraldehyde-3-phosphate dehydrogenase (Corey et al. 1997) may be applied. During the initial screening of culture supernatants (containing all immunoglobulin combinations) the signals obtained by flow cytometry and cytotoxicity test may be relatively weak which is caused by competition of parental monoclonal antibodies with bsAb for antigen binding.

2.2.5 Cloning

Rapid cloning of hybrid hybridoma cultures is mandatory to select for stable bsAb-secreting cell lines. The method of choice for single-cell cloning is limiting dilution. In principle, quadroma cells are distributed in 96-well plates so that one well will contain theoretically 0.5 or 1 cell. Usually the cloning efficiency of hybrid hybridomas is quite high and feeder cells are not required. Due to the genetic instability of hybrids it is possible to use repeated cloning steps to identify subclones with enhanced production of bsAb that can be verified by quantitative ELISA.

2.2.6 Purification of Bispecific Antibody

Once a stable hybrid hybridoma line is established, large quantities of antibody can be produced employing modern cell culture devices for long-term propagation. At least two systems are on the market meeting the demands of laboratory-scale production because they can simply be installed in a normal CO_2 incubator and do not require complicated pumping and other sophisticated equipment. Both, the miniPERM modular minifermenter (Falkenberg et al. 1995) and the two-chamber cell culture device CELLine 1000 (Trebak et al. 1999) are easy to handle and allow culturing of antibody-producing cells at high density (above 10^7 cells per ml). Harvest of the antibody-enriched product can be performed several times until productivity ceases.

Due to the huge variability in physicochemical properties of quadroma-derived IgG that largely depends (among other traits) from the chain composition no standard procedure for purification can be recommended. In fact, the method has to be established for every individual bsAb. Only in rare cases the respective target antigens are available as (recombinant) proteins which can be employed for affinity purification (Gupta and Suresh 2002). Anti-idiotypic antibodies specifically interacting with the parental antibody were also employed but are very difficult to obtain (Bruynck et al. 1993). In retrospect, most investigators have used sophisticated two-step methods to purify bsAb from a mixture of mono- and bispecific immunoglobulin species. Often protein A or protein G affinity chromatography was combined with ion-exchange or size-exclusion chromatography (Tarditi et al. 1992). More recently, a hydrophobic interaction chromatographic technique was described that resolves bsAb, monospecific immunoglobulins, and culture medium supplements in one single step from bioreactor harvest (Manzke et al. 1997). The purity of bsAb can be assessed by SDS-PAGE, isoelectric focusing, Western blotting, isotype-specific ELISA or functional assays like T-cell proliferation.

2.2.7 *Molecular Composition of Bispecific Antibody*

So far, mostly mouse–mouse, rat–rat, and mouse–rat hybrid hybridomas have been created. Besides fusion of two IgG-secreting hybridomas also IgA-producing cell lines and hybrids between IgA and IgM were successfully established (Urnovitz et al. 1988). One major drawback connected with the production of bsAb by hybrid hybridoma is the co-dominant expression of immunoglobulin genes of both parents. This leads to random association of both parental heavy and light chains resulting in ten different antibody species only one of which constitutes the desired bsAb (Fig. 2.2). When assuming that the rate of synthesis is the same for all four chains and pairing happens randomly one would expect 12.5% of the total immunoglobulin to be bispecific (Staerz and Bevan 1986). Ideally, for a high yield of bsAb heterologous H–H chain pairing together with homologous H–L chain association

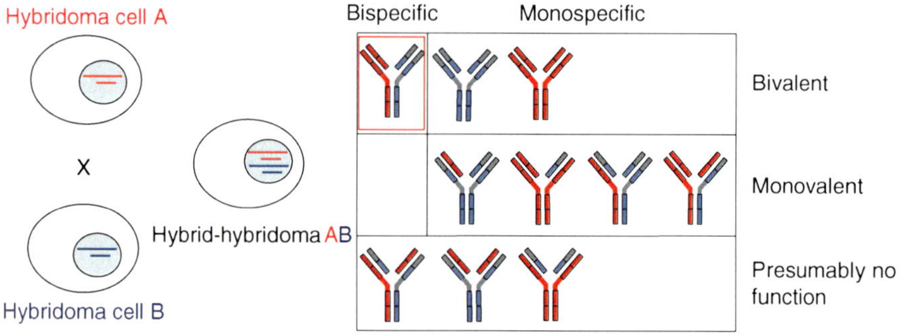

Fig. 2.2 Antibody species secreted by hybrid hybridoma. Immunoglobulin genes of both parental hybridomas are codominantly expressed in hybrid hybridoma leading to random association of heavy and light chains. Only one among the ten different antibody species is the desired bsAb

should occur. Unfortunately, this is not always achieved and preferential heterologous H–L chain association giving rise to antibodies with unknown specificity has been described (De Lau et al. 1991). The use of identical IgG isotype appears to favor mixed H–H chain combinations, whereas different isotypes lead to a reduced yield of bsAb (Milstein and Cuello 1984). Preferential species-restricted H–L chain pairing was observed in certain rat/mouse quadromas (Lindhofer et al. 1995). However, parental hybridomas of different IgG isotype offer great advantages with regard to purification since the resulting bsAb are bi-isotypic and can therefore be separated from parental immunoglobulins.

2.3 Functional Features of Bispecific Antibody

The key function of a bsAb for tumor therapy is of course the recruitment of immune effector cells to the tumor site. Depending on the type of effector cell, additional functional requirements have to be fulfilled. Under optimal conditions, the tumor cell will be finally destroyed by the common mechanisms of necrosis or apoptosis. A schematic representation of bsAb-mediated retargeting of various effector cells is given in Fig. 2.3.

BsAb from hybrid hybridoma display the same size (MW approximately 150 kDa) and high stability as conventional IgG antibodies. Thus, their serum half life and tissue penetration is comparable. To avoid interaction with Fc receptor-bearing cells and to increase tumor accessibility $F(ab')_2$ fragments were produced by enzymatic digestion (Warnaar et al. 1994). Fragmentation of bsAb is not an easy task and the protocol has to be adapted for each individual candidate to prevent complete degradation.

The avidity of both binding arms usually reflects the binding strengths of the parental antibodies at least under monovalent binding conditions in vitro.

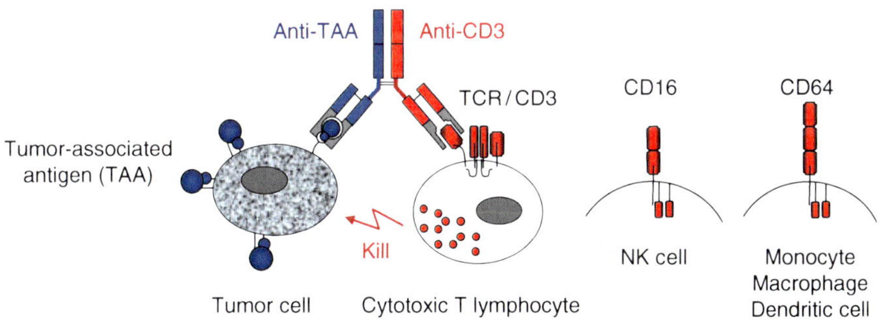

Fig. 2.3 BsAb-mediated retargeting of effector cells. BsAb binds with one arm to a triggering molecule on the effector cell and with the other arm to a tumor-associated antigen on the malignant cell. Conjugate formation of both cells results in lysis of the target cell. The main trigger molecules on cytotoxic T lymphocytes, NK cells, and myeloid cells are depicted

The actual avidity can be estimated by a flow cytometric binding competition assay. For this, increasing amounts of bsAb compete for binding to the antigen-bearing target cell with the fluorescently labeled parental antibody of known affinity. More accurate affinity measurements by surface plasmon resonance (BIACORE technology) are possible if the target antigens are available as (fusion) proteins (van Regenmortel 2003).

The innate and adaptive immune system contains distinct effector cell populations that can elicit cytotoxicity; they comprise T lymphocytes, natural killer cells, monocytes/macrophages, and polymorphonuclear neutrophils (PMN). These cell types carry specific trigger molecules on their cell surface that interact with natural ligands on the target cell. Binding of bsAb to the trigger molecule replaces the interaction with the ligand and leads to activation.

2.3.1 Recruitment of Cytotoxic T Lymphocytes

From the beginning, cytotoxic T lymphocytes were mostly used as effector cells for tumor attack as they represent the professional killers of the immune system (Liu et al. 1985; Perez et al. 1985; Staerz et al. 1985; Lanzavecchia and Scheidegger 1987). Under physiological conditions T cell specificity is determined by the T cell receptor (TCR) that recognizes peptides in conjunction with major histocompatibility complex (MHC) molecules. T cells need a second signal to become fully activated. This is delivered by the CD28 receptor on T cells that interact with costimulatory molecules of the B7 family exposed on antigen-presenting cells. Injected into the circulation, a bsAb binds only with one arm to a T lymphocyte that is not sufficient for triggering. Subsequently, if the bsAb-coated T cell finds and interacts with the target antigen expressing tumor cell functional crosslinking of the CD3/TCR complex occurs leading to a first step of activation. If costimulation is also provided at the tumor site either by CD28 ligation on the same T cell or secretion of lymphokines and cytokines by neighboring cells than cytotoxicity is fully established. This mechanism has been denoted as target cell-induced T cell activation (Jung et al. 1987; Jung and Müller-Eberhard 1988).

Retargeting of cytotoxic T cells versus tumor cells offers two fundamental advances. First, in contrast to conventional mono- or oligoclonal T cell activation by tumor-associated antigens bsAb containing an anti-CD3 moiety are able to induce a polyclonal T cell proliferation. Thus, a huge number of effector T cells are recruited to the tumor site. Secondly, bsAb-guided T cells also attack MHC-negative target cells because the specificity of the reaction is dictated by the antibody and not by the genetically determined TCR. This feature is of special importance in light of the fact that many tumors show reduced or even absent MHC expression as a consequence of dedifferentiation in the course of progression (Garrido and Algarra 2001). MHC loss represents an important tumor escape mechanism that potentially can be overcome by bsAb therapy. Antitumor cytotoxicity is mainly achieved by $CD8^+$ lymphocytes but in the human system also $CD4^+$ cells have the capacity to kill.

Tumor cell lysis following retargeting of T cells by bsAb engages the granule exocytosis system that relies on the directed release of granules containing perforin and granzymes from activated lymphocytes after specific recognition and conjugate formation with the target cell (Renner et al. 1997).

As mentioned above, full activation of effector T lymphocytes is an important prerequisite for the establishment of a robust cytotoxic reaction. This can be achieved via stimulation of the TCR/CD3 complex in combination with CD28 costimulation. Commonly, tumor cells are not able to provide costimulatory signals since they lack expression of members of the B7 family that interact with CD28 on the effector cell. Consequently, in many preclinical and clinical studies costimulation was made available by the additional application of conventional bivalent anti-CD28 monoclonal antibody (Manzke et al. 2001a; Manzke et al. 2001b) or by the use of two bsAb one of which is directed against CD28 (Bohlen et al. 1993; Pohl et al. 1993; Kroesen et al. 1995). In this setting both bsAb recognize the same target antigen but one reacts with CD3 and the other with CD28 to ensure activation (via CD3) and costimulation (via CD28) of T cells in a target cell restricted manner.

Depending on the epitope they bind on the CD3 coreceptor, some bsAb do not require further costimulation in order to induce a vigorous T cell activation. Although this strengthens the potency of therapeutic bsAb, at the same time it enhances the risk of adverse side effects caused by the release of lymphokines and cytokines. In the worst case a global T cell activation can cause a cytokine-release syndrome ("cytokine storm") that represents a life-threatening adverse event. Based on an incidence in 2006 following the application of a superagonistic anti-CD28 antibody in six human healthy volunteers with the development of unexpected severe adverse events including multiorgan failure (Suntharalingam et al. 2006) the German Paul-Ehrlich-Institut (regulatory authority for the approval of mAb) is classifying bsAb as high-risk drugs (Schneider et al. 2006).

To avoid unwanted side effects caused by excessive cytokine release, some investigators have administered bsAb in a locoregional fashion, e.g., by direct intratumoral or intralymphatic injection in case of low-grade B cell lymphomas (Manzke et al. 2001a; Manzke et al. 2001b). In a previous study we found that tumor-associated lymphocytes present in malignant ascites and pleural effusion from patients with ovarian and breast cancer are already in a preactivated state and able to efficiently lyse tumor cells in the presence of a bsAb of EpCAM × CD3 specificity without further costimulation (Strauss et al. 1999). These data provided the rationale for a pilot study of intraperitoneal bsAb therapy in patients with advanced ovarian cancer and malignant ascites (Marmé et al. 2002). Another possibility to minimize adverse reactions is to expand and to preload T lymphocytes from peripheral blood ex vivo with bsAb and to re-infuse them into the patient. This strategy representing a special form of adoptive immunotherapy was pursued in several studies of epithelial tumors leading to malignant effusions. BsAb-coated preactivated T lymphocytes were injected either alone or with additional soluble bsAb in the pleural or abdominal cavity of the patient (Bolhuis et al. 1992; Kroesen et al. 1993; Canevari et al. 1995). Lastly, the most demanding procedure is the systemic application of bsAb. A phase I study was conducted in patients with renal cell cancer receiving increasing doses of $F(ab')_2$ fragments of an EpCAM × CD3 bsAb together with subcutaneous IL-2 (Kroesen et al. 1994). Although high serum levels of pro-inflammatory cytokines were noted in the patients, the feasibility of this approach could be demonstrated. Further clinical experience was reported in a dose escalation study applying a CD19 × CD3 bsAb in non-Hodgkin's lymphoma patients (de Gast et al. 1995). Noteworthy, only limited toxicity (WHO grade II) occurred which was mainly attributed to tumor necrosis factor alpha release.

Intact bsAb carry a Fc portion that potentially can bind to Fc-receptors on a variety of accessory cells. Mouse bsAb composed solely of IgG1 or a combination of IgG1 and IgG2a heavy chains are poor mediators of ADCC in human effector cells due to their weak interaction with activating Fcγ receptors like CD16 and CD64. To improve binding to Fc-receptors, so-called trifunctional bsAb consisting of mouse IgG2a and rat IgG2b and directed against CD3 and a tumor-associated antigen were created. In addition to their dual specificity via the two binding domains, these reagents are able to efficiently recruit and activate accessory cells as macrophages, NK cells, and dendritic cells by their Fc region leading to secretion of various cytokines. This third function is able to provide the essential costimulatory signals to cytotoxic T-lymphocytes (Zeidler et al. 1999). Importantly, most likely by phagocytes ingesting tumor material a long-lasting anti-tumor immunity is achieved (Zeidler et al. 2000; Ruf and Lindhofer 2001). One trifunctional bsAb named catumaxomab (Removab®) with specificity for the epithelial cell adhesion molecule (EpCAM) and CD3 has proven to be especially effective for the intraperitoneal treatment of malignant ascites which develops during advanced stages of a variety of intra-abdominal malignancies as ovarian and stomach cancer (Burges et al. 2007; Heiss et al. 2010). As first bsAb worldwide catumaxomab was approved in April 2009 in the European Union (Seimetz et al. 2010) and will be discussed in a separate chapter of this book.

In addition to the TCR-associated CD3 molecule other T cell antigens as CD2 (Wild et al. 1999), CD5, and the TCR itself (Ferrini et al. 1989) have been engaged as trigger molecules of bsAb. Our group has recently reported on a new quadroma-derived bsAb of the specificity CD19 × CD5 that was employed to target an ex vivo expanded and activated T cell subset called cytokine-induced killer (CIK) cells expressing CD5 (Tita-Nwa et al. 2007). Importantly, CD5 targeting bsAb may be particularly useful in combination with adoptive T cell transfer, e.g., in the setting of allogeneic stem cell transplantation, as it neither activates nor induces proliferation of naïve T cells potentially directed against host antigens. Thus, the danger of a graft versus host reaction in the recipient might be diminished.

But what actually happens when a target cell and an effector cell encounter each other in the presence of bsAb? We have addressed this question for a bsAb with EpCAM × CD3 specificity in two experimental systems. In the first setting a collagen gel three-dimensional tumor reconstruct was used which closely resembled the tumor microenvironment. Dynamic tumor cell–lymphocyte interactions were recorded by time-lapse video microscopy. Contact duration was about three times longer in the presence of bsAb compared with control Ab, whereas lymphocyte velocity was not influenced (Salnikov et al. 2009). This finding indicates that bsAb facilitates and prolongs interactions between tumor cells and lymphocytes regardless of TCR specificity. In a second set of experiments the molecular binding characteristics of the EpCAM × CD3 bsAb were analyzed by Single Cell Force Spectroscopy using an atomic force microscope. The investigation reveals that bsAb-induced conjugate formation between T cells and tumor cells occurs in a biphasic process that finally leads to the development of a complete immune synapse. The early phase of cell adhesion is specific and reversible and consecutively adhesive forces increase in a time and contact force-dependent manner (Hoffmann et al. 2011). It appears likely that extended contact time and increasing binding forces between effector and target cell will ultimately initiate T cell activation and signaling as indicated by the formation of a mature synapse.

2.3.2 Recruitment of Fc Receptor-Bearing Effector Cells

Besides T lymphocytes, Fc receptor-carrying cells of the innate immune system have been extensively exploited as effectors. Natural killer cells, monocytes, macrophages, dendritic cells, and neutrophils expose (either constitutively or upon induction with cytokines) Fc receptors on the surface that can be engaged by bsAb. In general, two classes of Fc receptors can be distinguished according to their function: activating and inhibitory receptors that differ with respect to their cytosolic signal transduction pathway. For bsAb-mediated targeting strategies especially the activating receptors FcγRIA (CD64), FcγRIIA (CD32), FcγRIIIA (CD16), and FcαRI (CD89) are of interest. Nevertheless, one important characteristic of the Fc receptor system is the coexpression of activating and inhibitory Fc receptors on the same cell whereby inhibitory signaling dominates activation in order to prevent unspecific stimulation of the immune system (Nimmerjahn and Ravetch 2006; Nimmerjahn and Ravetch 2007). Natural killer cells represent an exception from this rule since they carry only the activating FcγRIIIA (CD16). Functionally, they are cytotoxic lymphocytes which without prior stimulation can kill a variety of target cells in a non-MHC-restricted fashion. They belong to the first line of cellular defense of the innate immune system. Ligation of CD16 by bsAb is able to activate resting NK cells and induce lysis of tumor cells. Under certain conditions additional application of IL-2 can even enhance the efficacy (Ferrini et al. 1992; Hombach et al. 1993). First clinical studies evaluating NK targeting via CD16-specific bsAb showed some minor responses that were achieved at the expense of severe adverse reactions.

Cells of the myeloid lineage constitutively express activating Fc receptors that have been employed as exquisite trigger molecules for the induction of cytotoxicity. Especially the high affinity receptor for IgG, termed FcγRIA or CD64, was identified as a well-suited target molecule for bsAb-based immunotherapy of cancer (Deo et al. 1997; van Spriel et al. 2000; Schweizer et al. 2002). CD64 constitutes an activating Fc receptor on monocytes, macrophages, and dendritic cells, whereas its expression can be induced on neutrophils and eosinophils by IFN-γ or G-CSF treatment. This is of special relevance since polymorphonuclear neutrophils (PMNs) are the most abundant circulating white blood cells and able to elicit strong cytolytic and phagocytic activities. They release soluble chemotactic factors that recruit further nonspecific and specific immune effector cells. Importantly, induction of FcγRIA on granulocytes and substantial upregulation on macrophages can be easily achieved in patients by systemic application of the respective cytokine or growth factor (Repp et al. 1995).

Several bsAb recognizing the CD64 molecule in conjunction with an epithelial tumor-associated antigen were established during the past years. The tyrosine kinase receptor HER2/neu, the EGF receptor, the EpCAM cell adhesion molecule, and the MUC1 antigen are prominent examples for the second binding specificity for those bsAb being prepared or having already entered clinical testing. So far, most bsAb trials taking advantage of FcγRIA targeting have demonstrated moderate to low toxicity but a sustained tumor regression was achieved only in few patients (Curnow 1997; James et al. 2001). Dendritic cells that also express Fc receptors were shown to efficiently take up dying antibody-coated tumor cells. This might lead to an enhanced cross presentation and generation of tumor-specific T cell responses and ultimately boost antitumor immunity (Weiner et al. 2009). Finally, in recent studies the Fc receptor for IgA, CD89, has attracted much attention for bsAb-guided therapy. Comparison of different trigger molecules expressed on neutrophils revealed that CD89 is the most potent Fc receptor for tumor cytolysis (Valerius et al. 1997).

2.3.3 Human Anti-Rodent Immunoglobulin Response Following Bispecific Antibody Therapy

One of the problems related to the therapeutic application of a murine or rat bsAb is the occurrence of a human anti-mouse antibody (HAMA) or human anti-rat antibody (HARA) response. It has been shown that HAMA/HARA production may lead to allergic reactions ranging from mild cutaneous eczema to anaphylactic shock (Khazaeli et al. 1994). Apart from inducing adverse events, HAMA was described to block the binding of bsAb to its targets and thereby inhibit cytotoxicity toward the tumor cells (Lamers et al. 1995). Whether HAMA directed against the constant part of the mouse immunoglobulin molecule or merely anti-idiotype antibodies is responsible for such an inhibitory effect in patients has to be further investigated. In a retrospective study data from three different clinical trials employing $F(ab')_2$ fragments of mouse bsAb of the specificity CD3 $\times$ folate receptor for the intraperitoneal treatment of ovarian cancer were analyzed. As one parameter, the influence of HAMA response on the therapeutic efficacy of the murine bispecific antibody was evaluated. Surprisingly, patients with high HAMA levels had a significantly longer median survival probability than patients with a minor HAMA response (Miotti et al. 1999). Similar results were reported from a study using conventional bivalent mouse monoclonal antibodies to treat B cell lymphoma patients (Azinovic et al. 2006).

There are at least two explanations for this unexpected phenomenon. It is well established that mouse antibodies can induce a significant anti-idiotypic response in humans especially after multiple applications. According to the network hypothesis, the anti-idiotype antibody is immunogenic itself and induces a humoral immune response by the formation of anti-anti-idiotypic antibodies. Since it is believed that the anti-idiotype mimics an epitope of the antigen the initial antibody binds to (the anti-idiotype represents an internal image of the antigen) this immune response is not only directed against the anti-idiotypic antibody but at the same time against the primary target. In addition, complexes containing the murine therapeutic antibody and the human anti-idiotype are internalized and presented by professional antigen presenting cells. Through the presentation of internal image peptides on MHC molecules T cells might be stimulated which subsequently elicit an anti-tumor response. A second possible mechanism may be the direct induction of a cellular immune response by the murine antibody through its MHC-complexed presentation by antigen presenting cells. In this case, the processed mouse antibody presented on the tumor cell would serve as a new artificial tumor-associated antigen that could be recognized by cytotoxic T cells. In a previous investigation we found that tumor patients treated with mouse mAb develop T cells that recognize processed murine immunoglobulin on autologous antigen presenting cells in a MHC class II-restricted fashion. Mouse mAb directed against various cell surface molecules can thus be used as antigens to focus these T cells against an MHC class II positive target of choice (Lanzavecchia et al. 1988). A cellular immune response would certainly explain the efficacy of an antibody therapy even after formation of neutralizing HAMA.

2.4 Conclusion and Outlook

The past 25 years of bsAb development have seen several excitements and drawbacks. Nevertheless, the idea to use redesigned antibody molecules for the recruitment and retargeting of immunological effector cells to the tumor site has retained its attraction. One recent milestone was the approval of the first bsAb catumaxomab in Europe for treatment of malignant ascites arising from EpCAM-expressing carcinomas (Heiss et al. 2010; Seimetz et al. 2010). Another highlight was the impressive clinical results obtained with the recombinant CD19 × CD3 bispecific T cell engaging (BiTE) construct blinatumomab in patients suffering from non-Hodgkin's lymphoma (Bargou et al. 2008; Baeuerle and Reinhardt 2009).

A multitude of bispecific therapeutics is currently under clinical evaluation. New applications of bsAb will be further investigated as targeting of two independent receptors on the surface of one cell or pre-targeting strategies to improve chemo- and radiotherapy (Müller and Kontermann 2010). The induction of specific and long-lasting antitumor T cell responses seems to be feasible by enhancing the capacity of bsAb to engage professional antigen-presenting cells via Fc receptors (Weiner et al. 2009). Moreover, bsAb hold great promise to guide gene-modified effector cell populations (e.g., TCR-modified T lymphocytes) in order to increase specificity and affinity in adoptive cellular therapies. After all, there are many good reasons to believe that now we shall see the light at the end of the tunnel of bsAb-based cancer therapy in near future (Chames and Baty 2009a).

Acknowledgments I would like to thank Dr. Reinhard Schwartz-Albiez for critical reading of the manuscript and Dr. Dorothée Deppe for help preparing the figures.

References

Azinovic I, DeNardo GL, Lamborn KR, Mirick G, Goldstein D, Bradt BM, DeNardo SJ (2006) Survival benefit associated with human anti-mouse antibody (HAMA) in patients with B-cell malignancies. Cancer Immunol Immunother 55:1451–1458

Baeuerle PA, Reinhardt C (2009) Bispecific T-cell engaging antibodies for cancer therapy. Cancer Res 69:4941–4944

Bargou R, Leo E, Zugmaier G, Klinger M, Goebeler M, Knop S, Noppeney R, Viardot A, Hess G, Schuler M, Einsele H, Brandl C, Wolf A, Kirchinger P, Klappers P, Schmidt M, Riethmüller G, Reinhard C, Baeuerle PA, Kufer P (2008) Tumor regression in cancer patients by very low doses of a T cell-engaging antibody. Science 321:974–977

Beck A, Wurch T, Bailly C, Corvaia N (2010) Strategies and challenges for the next generation of therapeutic antibodies. Nat Rev Immunol 10:345–352

Bohlen H, Manzke O, Patel B, Moldenhauer G, Dörken B, von Fliedner V, Diehl V, Tesch H (1993) Cytolysis of leukemic B-cells by T-cells activated via two bispecific antibodies. Cancer Res 53:4310–4314

Bolhuis RLH, Lamers CHJ, Goey SH, Eggermont AMM, Trimbos JBMZ, Stoter G, Lanzavecchia A, Di Re E, Miotti S, Raspagliesi F, Rivoltini L, Colnaghi MI (1992) Adoptive immunotherapy of ovarian carcinoma with Bs-MAb-targeted lymphocytes: a multicenter study. Int J Cancer 7:78–81

Bruynck A, Seemann G, Bosslet K (1993) Characterisation of a humanized bispecific monoclonal antibody for cancer therapy. Br J Cancer 67:436–440

Burges A, Wimberger P, Kümper C, Gorbounova V, Sommer H, Schmalfeldt B, Pfisterer J, Lichinitser M, Makhson A, Moiseyenko V, Lahr A, Schulze E, Jäger M, Ströhlein M, Heiss MM, Gottwald T, Lindhofer H, Kimmig R (2007) Effective relief of malignant ascites in patients with advanced ovarian cancer by a trifunctional anti-EpCAM x anti-CD3 antibody: a phase I/II study. Clin Cancer Res 13:3899–3905

Canevari S, Stoter G, Arienti F, Bolis G, Colnaghi MI, Di Re EM, Eggermont AMM, Goey SH, Gratama JW, Lamers CHJ, Nooy MA, Parmiani G, Raspagliesi F, Ravagnani F, Scarfone G, Trimbos JB, Warnaar SO, Bolhuis RLH (1995) Regression of advanced ovarian carcinoma by intraperinoneal treatment with autologous T lymphocytes retargeted by a bispecific monoclonal antibody. J Natl Cancer Inst 87:1463–1469

Cao Y, Vinayagamoorthy T, Noujaim A, Suresh MR (1995) A rapid non-selective method to generate quadromas by microelectrofusion. J Immunol methods 187:1–7

Chames P, Baty D (2009a) Bispecific antibodies for cancer therapy: the light at the end of the tunnel? MAbs 1:539–547

Chames P, Baty D (2009b) Bispecific antibodies for cancer therapy. Curr Opin Drug Discov Develop 12:276–283

Chan AC, Carter PJ (2010) Therapeutic antibodies for autoimmunity and inflammation. Nat Rev Immunol. May;10(5):301-16

Chervonsky AV, Faerman AI, Evdonina LV, Jazova AK, Kazarov AR, Gussev AI (1988) A simple metabolic system for selection of hybrid hybridomas (tetradomas) producing bispecific monoclonal antibodies. Mol Immunol 25:913–915

Clark MR, Waldmann H (1987) T-cell killing of target cells induced by hybrid antibodies: comparison of two bispecific monoclonal antibodies. J Natl Cancer Inst 79:1393–1401

Corey MJ, Kinders RJ, Brown LG, Vessella RL (1997) A very sensitive coupled luminescent assay for cytotoxicity and complement-mediated lysis. J Immunol Meth 207:43–51

Cotton RGH, Milstein C (1973) Fusion of two immunoglobulin-producing myeloma cells. Nature 244:42–43

Curnow R (1997) Clinical experience with CD64-directed immunotherapy. An overview. Cancer Immunol Immunother 45:210–215

De Gast GC, Van Houten AA, Haagen IA, Klein S, De Weger RA, Van Dijk A, Philips J, Clark M, Bast BJEG (1995) Clinical experience with CD3 x CD19 bispecific antibodies in patients with B cell malignancies. J Hematother 4:433–437

De Lau WBM, Van Loon AE, Heije K, Valerio D, Bast BJEG (1989) Production of hybrid hybridomas based on HAT(s)-neomycin(r) double mutants. J Immunol Meth 117:1–8

De Lau WBM, Heije K, Neefjes JJ, Oosterwegel M, Rozemuller E, Bast BJEG (1991) Absence of preferential homologous H/L chain association in hybrid hybridomas. J Immunol 146:906–914

Decker T, Lohmann-Matthes ML (1988) A quick and simple method for the quantitation of lactate dehydrogenase release in measurements of cellular cytotoxicity and tumor necrosis factor (TNF) activity. J Immunol Meth 115:61–69

Deo YM, Graziano RF, Repp R, van de Winkel JGJ (1997) Clinical significance of IgG Fc receptors and FcgammaR-directed immunotherapies. Immunol Today 18:127–135

Engvall E, Perlman P (1971) Enzyme-linked immunosorbent assay (ELISA): quantitative assay of immunoglobulin G. Immunochem 8:871–879

Fanger MW, Morganelli PM, Guyre PM (1992) Bispecific antibodies. Crit Rev Immunol 12: 101–124

Falkenberg FW, Weichert H, Krane M, Bartels I, Palme M, Nagels HO, Fiebig H (1995) In vitro production of monoclonal antibodies in high concentration in a new and easy to handle modular minifermenter. J Immunol Methods. Feb 13;179(1):13–29.

Ferrini S, Prigione I, Mammoliti S, Colnaghi MI, Ménard S, Moretta A, Moretta L (1989) Re-targeting of human lymphocytes expressing the T-cell receptor gamma/delta to ovarian carcinoma cells by the use of bispecific monoclonal antibodies. Int J Cancer 44:245–250

Ferrini F, Cambiaggi A, Cantoni C, Canevari S, Mezzanzanica D, Colaghi MI, Moretta L (1992) Targeting of T or NK lymphocytes against tumor cells by bispecific monoclonal antibodies: role of different triggering molecules. Int J Cancer 7:15–18

Garrido F, Algarra I (2001) MHC antigens and tumor escape from immune surveillance. Adv Cancer Res 83:117–158

Gilliland LK, Clark MR, Waldmann H (1988) Universal bispecific antibody for targeting tumor cells for destruction by cytotoxic T cells. Proc Natl Acad Sci USA 85:7719–7723

Glenie MJ, McBride HM, Worth AT, Stevenson GT (1987) Preparation and performance of bispecific F(ab' gamma)2 antibody containing thioether-linked Fab' gamma fragments. J Immunol 139:2367–2375

Gupta S, Suresh M (2002) Affinity chromatography and co-chromatography of bispecific monoclonal antibody immunoconjugates. J Biochem Biophys Meth 51:203–216

Heiss MM, Murawa P, Koralewski P, Kutarska E, Kolesnik OO, Ivanchenko VV, Dudnichenko AS, Aleknaviciene B, Razbadauskas A, Gore M, Ganea-Motan E, Ciuleanu T, Wimberger P, Schmittel A, Schmalfeld B, Burges A, Bokemeyer C, Lindhofer H, Lahr A, Parsons SL (2010) The trifunctional antibody catumaxomab for the treatment of malignant ascites due to epithelial cancer: results of a prospective randomized phase II/III trial. Int J Cancer 127:2209–2221

Hoffmann SC, Wabnitz GH, Samstag Y, Moldenhauer G, Ludwig T (2011) Functional analysis of bispecific antibody (EpCAMxCD3) mediated T-lymphocyte and cancer cell interaction by single cell force spectroscopy. Int J Cancer 128(9):2096–2104

Hombach A, Jung W, Pohl C, Renner C, Sahin U, Schmitts R, Wolf J, Kapp U, Diehl V, Pfreundschuh M (1993) A CD16/CD30 bispecific antibody induces lysis of Hodgkin cells by unstimulated natural killer cells in vitro and in vivo. Int J Cancer 55:830–837

James ND, Atherton PJ, Jones J, Howie AJ, Tchekmedyian S, Curnow RT (2001) A phase II study of the bispecific antibody MDX-H210 (anti-HER2 x CD64) with GM-CSF in HER2+ advanced prostate cancer. Br J Cancer 85:152–156

Jung G, Müller-Eberhard HJ (1988) An in-vitro model for tumor immunotherapy with antibody heteroconjugates. Immunol Today 9:257–260

Jung G, Ledbetter JA, Müller-Eberhardt HJ (1987) Induction of cytotoxicity in resting human T lymphocytes bound to tumor cells by antibody heteroconjugates. Proc Natl Acad Sci USA 84:4611–4615

Karawajew L, Micheel B, Behrsing O, Gaestel M (1987) Bispecific antibody-producing hybrid hybridomas selected by a fluorescence activated cell sorter. J Immunol Meth 96:265–270

Khazaeli MB, Conry RM, LoBuglio AF (1994) Human immune response to monoclonal antibodies. J Immunother 15:42–52

Köhler G, Milstein C (1975) Continuous cultures of fused cells secreting antibody of predefined specificity. Nature 256:495–497

Koolwijk P, Rozemuller E, Stad RK, De Lau WB, Bast BJ (1988) Enrichment and selection of hybrid hybridomas by Percoll density gradient centrifugation and fluorescent-activated cell sorting. Hybridoma 7:217–225

Kreutz FT, Xu DZ, Suresh MR (1998) A new method to generate quadromas by electrofusion and FACS sorting. Hybridoma 17:267–273

Kroesen BJ, ter Haar A, Spakman H, Willemse P, Sleijfer DT, de Vries EG, Mulder NH, Berendsen HH, Limburg PC, The TH et al (1993) Local antitumor treatment in carcinoma patients with bispecific-monoclonal-antibody-redirected T cells. Cancer Immunol Immunother 37:400–407

Kroesen BJ, Buter J, Sleijfer DT, Jansen RAJ, van der Graaf WTA, The TH, de Leij L, Mulder NH (1994) Phase I study of intravenously applied bispecific antibody in renal cell cancer patients receiving subcutaneous interleukin 2. Br J Cancer 70:652–661

Kroesen BJ, Bakker A, van Lier RAW, The HT, de Leij L (1995) Bispecific antibody-mediated target cell-specific costimulation of resting T cells via CD5 and CD28. Cancer Res 55:4409–4415

Kufer P, Lutterbüse R, Baeuerle PA (2004) A revival of bispecific antibodies. Trends Biotechnol 22:238–244

Lamers CHJ, Gratama JW, Warnaar SO, Stoter G, Bolhuis RLH (1995) Inhibition of bispecific monoclonal antibody (bsAb)-targeted cytolysis by human anti-mouse antibodies in ovarian carcinoma patients treated with bsAb-targeted activated T-lymphocytes. Int J Cancer 60:450–457

Lanzavecchia A, Scheidegger D (1987) The use of hybrid hybridomas to target human cytotoxic T lymphocytes. Eur J Immunol 17:105–111

Lanzavecchia A, Abrignani S, Scheidegger D, Obrist R, Dörken B, Moldenhauer G (1988) Antibodies as antigens. The use of mouse monoclonal antibodies to focus human T cells against selected targets. J Exp Med 167:345–352

Lindhofer H, Mocikat R, Steipe B, Thierfelder S (1995) Preferential species-restricted heavy/light chain pairing in rat/mouse quadromas. Implications for a single-step purification of bispecific antibodies. J Immunol. Jul 1;155(1):219–25

Link BK, Weiner GJ (1993) Production and characterization of a bispecific IgG capable of inducing T-cell-mediated lysis of malignant B cells. Blood 81:3343–3349

Little M, Kipriyanov SM, Le Gall F, Moldenhauer G (2000) Of mice and men: hybridoma and recombinant antibodies. Immunol Today 21:364–370

Liu MA, Kranz DM, Kurnick JT, Boyle LA, Levy R, Eisen HN (1985) Heteroantibody duplexes target cells for lysis by cytotoxic T lymphocytes. Proc Natl Acad Sci USA 82: 8648–8652

Manzke O, Tesch H, Diehl H, Bohlen H (1997) Single-step purification of bispecific monoclonal antibodies for immunotherapeutic use by hydrophobic interaction chromatography. J Immunol Meth 208:65–73

Manzke O, Tesch H, Borchmann P, Wolf J, Lackner K, Gossmann A, Diehl V, Bohlen H (2001a) Locoregional treatment of low-grade B-cell lymphoma with CD3xCD19 bispecific antibodies and CD28 costimulation: I. Clinical phase I evaluation. Int J Cancer 91:508–515

Manzke O, Tesch H, Lorenzen J, Diehl V, Bohlen H (2001b) Locoregional treatment of low-grade B-cell lymphoma with CD3xCD19 bispecific antibodies and CD28 costimulation: II. Assessment of cellular immune responses. Int J Cancer 91:516–522

Marmé A, Strauss G, Bastert G, Grischke E-M, Moldenhauer G (2002) Intraperitoneal bispecific antibody (HEA125xOKT3) therapy inhibits malignant ascites production in advanced ovarian carcinoma. Int J Cancer 101:183–189

Milstein C, Cuello AC (1983) Hybrid hybridomas and their use in immunohistochemistry. Nature 305:537–540

Milstein C, Cuello AC (1984) Hybrid hybridomas and the production of bi-specific monoclonal antibodies. Immunol Today 5:299–304

Miotti S, Negri DRM, Valota O, Calabrese M, Bolhuis RLH, Gratama JW, Colnaghi MI, Canevari S (1999) Level of anti-mouse-antibody response induced by bispecific monoclonal antibody OC/TR in ovarian carcinoma patients is associated with longer survival. Int J Cancer 84:62–68

Moldenhauer G (2007) Selection strategies I: monoclonal antibodies. In: Dübel S (ed) Handbook of therapeutic antibodies. Wiley-VCH, Weinheim, pp 19–44

Müller D, Kontermann RE (2007a) Bispecific antibodies. In: Dübel S (ed) Handbook of therapeutic antibodies. Wiley-VCH, Weinheim, pp 345–378

Müller D, Kontermann RE (2007b) Recombinant bispecific antibodies for cellular cancer immunotherapy. Curr Opin Mol Ther 9:319–326

Müller D, Kontermann RE (2010) Bispecific antibodies for cancer immunotherapy. Current perspectives. Biodrugs 24:89–98

Nimmerjahn F, Ravetch JV (2006) Fc-gamma receptors: old friends and new family members. Immunity 24:19–28

Nimmerjahn F, Ravetch JV (2007) Antibodies, Fc receptors and cancer. Curr Opin Immunol 19:239–245

Perez P, Hoffman RW, Shaw S, Bluestone JA, Segal DM (1985) Specific targeting of cytotoxic T cells by anti-T3 linked to anti-target cell antibody. Nature 316:354–356

Pohl C, Denfeld R, Renner C, Jung W, Bohlen H, Sahin U, Hombach A, van Lier R, Schwonzen M, Diehl V, Pfreundschuh M (1993) CD30-antigen-specific targeting and activation of T cells via murine bispecific monoclonal antibodies against CD3 and CD28: potential use for the treatment of Hodgkin's lymphoma. Int J Cancer 54:820–827

Pontecorvo G (1975) Production of mammalian somatic cell hybrids by means of polyethylene glycol treatment. Somat Cell Genet 1:397–400

Renner C, Pfreundschuh M (1995) Tumor therapy by immune recruitment with bispecific antibodies. Immunol Rev 145:179–209

Renner C, Held G, Ohnesorge S, Bauer S, Gerlach K, Pfitzenmeier JP, Pfreundschuh M (1997) Role of perforin, granzymes and the proliferative state of the target cells in apoptosis and necrosis mediated by bispecific-antibody-activated cytotoxic T cells. Cancer Immunol Immunother 44:70–76

Repp R, Valerius T, Wieland G, Becker W, Steininger H, Deo Y, Helm G, Gramatzki M, van de Winkel JG, Lang N, Kalden JR (1995) G-CSF-stimulated PMN in immunotherapy of breast cancer with a bispecific antibody to Fc gamma RI and to HER-2/neu (MDX-210). J Hematother 4:415–421

Ruf P, Lindhofer H (2001) Induction of a long-lasting antitumor immunity by a trifunctional bispecific antibody. Blood 98:2526–2534

Salnikov AV, Groth A, Apel A, Kalifatidis G, Beckermann BM, Khamidjanov A, Ryschich E, Buechler MW, Herr I, Moldenhauer G (2009) Targeting of cancer stem cell marker EpCAM by bispecific antibody EpCAMxCD3 inhibits pancreatic carcinoma. J Cell Mol Med 13:4023–4033

Schneider CK, Kalinke U, Löwer J (2006) TGN1412 -a regulator's perspective. Nat Biotechnol 24:493–496

Schweizer C, Strauss G, Lindner M, Marmé A, Deo YM, Moldenhauer G (2002) Efficient carcinoma cell killing by activated polymorphonuclear neutrophils targeted with an Ep-CAMxCD64 (HEA125x197) bispecific antibody. Cancer Immunol Immunother 51:621–629

Segal DM, Bast BJEB (2001) Production of bispecific antibodies. In: Coligan JE, Kruisbeek AM, Margulies DH, Shevach EM, Strober W (eds) Current protocols in immunology. Wiley, New York, 2.13.1–2.13.16

Seimetz D, Lindhofer H, Bokemeyer C (2010) Development and approval of the trifunctional antibody catumaxomab (anti-EpCAM x anti-CD3) as a targeted cancer immunotherapy. Cancer Treat Rev 36:458–467

Staerz UD, Bevan MJ (1986) Hybrid hybridoma producing a bispecific monoclonal antibody that can focus effector T-cell activity. Proc Natl Acad Sci USA 83:1453–1457

Staerz UD, Kanegawa O, Bevan MJ (1985) Hybrid antibodies can target sites for attack by T cells. Nature 314:628–631

Strauss G, Gueckel B, Wallwiener D, Moldenhauer G (1999) Without prior stimulation tumor-associated lymphocytes from malignant effusions lyse autologous tumor cells in the presence of bispecific antibody HEA125xOKT3. Clin Cancer Res 5:171–180

Suntharalingam G, Perry MR, Ward S, Brett SJ, Castello-cortes A, Brunner MD, Panoskaltsis N (2006) Cytokine storm in a phase 1 trial of the anti-CD28 monoclonal antibody TGN1412. New Engl J Med 355:1018–1028

Suresh MR, Cuello AC, Milstein C (1986) Advantages of bispecific hybridomas in one-step immunocytochemistry and immunoassays. Proc Natl Acad Sci USA 83:7989–7993

Tarditi L, Camagna M, Parisi A, Vassarotto C, De Monte LB, Letarte M, Malavasi F, Mariani M (1992) Selective high-performance liquid chromatographic purification of bispecific monoclonal antibodies. J Chromatogr 599:13–20

Trebak M, Chong JM, Herlyn D, Speicher DW (1999) Efficient laboratory-scale production of monoclonal antibodies using membrane-based high-density cell culture technology. J Immunol Methods. Nov 19;230(1-2):59–70

Tita-Nwa F, Moldenhauer G, Herbst M, Kleist C, Ho AD, Kornacker M (2007) Cytokine-induced killer cells targeted by the novel bispecific antibody CD19xCD5 (HD37xT5.16) efficiently lyse B-lymphoma cells. Cancer Immunol Immunother 56:1911–1920

Urnovitz HB, Chang Y, Scott M, Fleischman J, Lynch RG (1988) IgA:IgM and IgA:IgA hybrid hybridomas secrete heteropolymeric immunoglobulin's that are polyvalent and bispecific. J Immunol 140:558–563

Valerius T, Stockmeyer B, van Spriel AB, Graziano RF, van den Herik-Oudijk IE, Repp R, Deo YM, Lund J, Kalden JR, Gramatzki M, van de Winkel JG (1997) FcalphaRI (CD89) as a novel trigger molecule for bispecific antibody therapy. Blood 90:4485–4492

Van Regenmortel MH (2003) Improving the quality of BIACORE-based affinity measurements. Dev Biol 112:141–151

Van Spriel AB, van Ojik HH, van de Winkel JGJ (2000) Immunotherapeutic perspective for bispecific antibodies. Immunol Today 21:391–397

Warnaar SO, De Paus V, Lardenhoije R, Machielse BNM, De Graaf J, Bregonje M, Van Haarlem H (1994) Purification of bispecific F(ab')2 from murine trioma OC/TR with specificity for CD3 and ovarian cancer. Hybridoma 13:519–526

Weiner LM (2007) Building better magic bullets: improving unconjugated monoclonal antibody therapy for cancer. Nat Rev Cancer 7:701–706

Weiner LM, Dhodapkar MV, Ferrone S (2009) Monoclonal antibodies for cancer immunotherapy. Lancet 373:1033–1040

Wild MK, Strittmatter W, Matzku S, Schraven B, Meuer SC (1999) Tumor therapy with bispecific antibody: the targeting and triggering steps can be separated employing a CD2-based strategy. J Immunol 163:2064–2072

Zeidler R, Reisbach G, Wollenberg B, Lang S, Chaubal S, Schmitt B, Lindhofer H (1999) Simultaneous activatin of T cells and accessory cells by a new class if intact bispecific antibody results in efficient tumor cell killing. J Immunol 163:1246–1252

Zeidler R, Mysliwietz J, Csanady M, Walz A, Ziegler I, Schmitt B, Wollenberg B, Lindhofer H (2000) The Fc-region of a new class of intact bispecific antibody mediates activation of accessory cells and NK cells and induces direct phagocytosis of tumor cells. Br J Cancer 83:261–266

Chapter 3
Generation of Bispecific Antibodies by Chemical Conjugation

Diego Ellerman and Justin M. Scheer

3.1 Introduction

Originally reported as an academic achievement and later envisioned as diagnostic tools, bispecific antibodies (bsAbs) have now drawn substantial attention as therapeutic agents. Interestingly, the first bispecific polyclonal antibodies were produced by chemically coupling two different polyclonal antibody mixtures (Nisonoff and Rivers 1961). This pioneering work involved the dissociation of the two different antibodies and re-association of the two halves. A fraction of the resulting dimeric antibody was shown to be bispecific by immunoprecipitation. The advent of monoclonal antibody technology in 1975 (Köhler and Milstein 1975) allowed working with more defined antibody reagents and thus represented a significant advance toward the goal of producing a homogenous bispecific antibody preparation. Another breakthrough in the field was the development of recombinant DNA technology in the 1990s, which was soon adopted for the expression of antibodies. Since the first report of recombinant antibody expression (Songsivilai et al. 1989), a plethora of Ab fragments and whole Ab formats have been created. Although domain fusion by DNA engineering is the prevalent methodology for producing bsAbs, a significant number of studies are being performed with chemically crosslinked Abs derived from either hybridoma or recombinant methods. One longstanding limitation for the introduction of bsAbs into the clinic has been the difficulty in producing large quantities of homogenous material. In this chapter, we review the methods currently used for the generation of bsAbs by chemical conjugation and we discuss recent advances in the field that will likely improve the yields and quality of the final product as well as open new paths for bispecific antibodies.

D. Ellerman • J.M. Scheer (✉)
Department of Protein Chemistry, Research and Early Development, Genentech, 1 DNA Way, South San Francisco, CA 94080, USA
e-mail: scheer.justin@gene.com

R.E. Kontermann (ed.), *Bispecific Antibodies*,
DOI 10.1007/978-3-642-20910-9_3,

3.2 Current Methodologies

3.2.1 Overview of bsAbs Generated by Chemical Conjugation

Different antibody structures, or formats, can be produced by chemical conjugation (Fig. 3.1). By using different starting material and various conjugation strategies, bispecific fragments, full-length Abs, or combinations thereof can be generated. For example, two different full-length Abs can be crosslinked to generate a ~300 kDa bispecific molecule which is bivalent for each antigen (Fig. 3.1a). Alternatively, a full-length Ab (~150 kDa) can be made bispecific and bivalent for each antigen by the "chemical programming" approach (Fig. 3.1b). However, most cases of chemical conjugation utilize $F(ab')_2$ fragments generated by enzymatic digestion of parent antibodies, followed by crosslinking of the reduced Fab′ to produce a monovalent bispecific $F(ab')_2$ (Fig. 3.1c) or a bispecific $F(ab')_3$ bivalent for one of the two antigens (Fig. 3.1d).

The synthesis of bsAb with different formats is often associated with distinct chemistries: while the crosslinking of intact Igs usually utilizes amino-reactive reagents targeting lysine side-chains, the production of a bispecific $F(ab')_2$ generally relies on sulfhydryl-reactive crosslinkers. In addition to the differences in primary chemical reactivity, crosslinkers can also be hetero- or homo-bifunctional. Homo-bifunctional compounds contain the same reactive group at both ends of the molecule, whereas hetero-bifunctional reagents posses two different reactive groups. The amino-reactive crosslinkers used for producing bsAbs are in general hetero-bifunctional whereas the sulfhydryl-reactive crosslinkers are homo-bifunctional. The pros and cons of each method are described in the following sections.

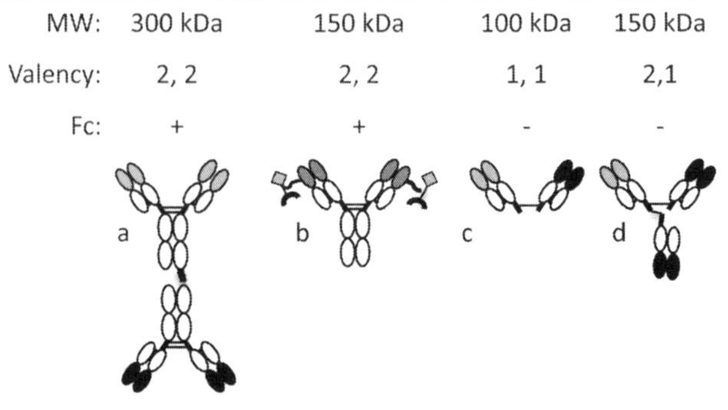

Fig. 3.1 *Formats can create different valencies*. Two full-length antibodies can be conjugated to produce a ~300 kDa construct bivalent for each antigen (**a**). "Chemical programming" covalently modifies an Ab incorporating a small molecule module targeting two different proteins (**b**). Crosslinking of Fab′s can generate bsAbs monovalent for each antigen (**c**) or bivalent for one of the two antigens (**d**)

3.2.2 Hetero-bifunctional Reagents

3.2.2.1 SPDP

Most of the hetero-bifunctional reagents used for producing bsAbs react with amino groups. An amino-reactive crosslinker, which has been extensively used for the conjugation of different proteins, including antibodies, is SPDP (succinimidyl-3 (2-pyridylthiol)propionate). The different steps involved in crosslinking Abs with SPDP are shown in Fig. 3.2a. First, both antibodies are incubated separately with SPDP to react the activated succinimidyl ester end of SPDP with the ε-amino groups on lysine residues, forming an amide linkage. The amount of SPDP used in the reaction influences the degree of crosslinking of the Ab, so higher concentrations of SPDP produce increased crosslinking rates. This, however, risks inactivation of the Ab due to reaction with residues involved in binding the antigen.

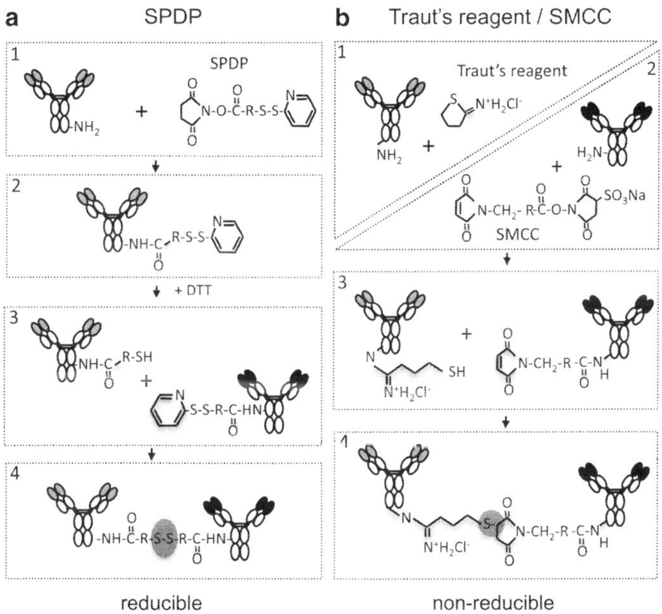

Fig. 3.2 *Hetero-bifunctional crosslinkers.* (**a**) Antibody conjugation using SPDP. Each antibody is incubated separately with SPDP (1) (R = CH_2CH_2). The ε-amino group of lysine react with the succinimidyl group of the crosslinker, generating the pyridyldithiol activated protein (2). Treatment with DTT at low pH cleaves the crosslinker's internal disulfide producing the sulfhydryl-activated protein (3). The sulfhydryl group in the first antibody reduces the disulfide present in the modified second antibody, generating the heteroconjugate linked by a reducible disulfide bond (highlighted in *gray*). (**b**) Crosslinking using Traut's reagent and sulpho SMCC. One antibody is treated with Traut's reagent (1) while the other is treated with sulpho-SMCC (2) (R=cyclohexane). Co-incubation of the two modified antibodies leads to the reaction of the sulfhydryl group introduced by the Traut's reagent with the maleimido group introduced by sulpho-SMCC (3) generating the bsAb linked by a non-reducible thioether bond (4)

A four molar excess of SPDP has been found to provide adequate crosslinking while modifying less than 10% of the total lysine residues in an Ig molecule (Segal and Bast 1995). After modification with SPDP, one of the two Abs is treated with DTT at low pH to reduce the antibody-bound pyridylthiol propionate (PDP) groups in preference to the intrinsic disulfide bonds of the antibody. Upon co-incubation with the other SPDP-modified Ab, the SH introduced in the first Ab reacts with the 2-pyridil thiol group generating the bispecific heteroconjugate linked by a disulfide bond. The resulting bsAb is a heterogenous sample conjugated at random sites. In addition to the bsAb, crosslinking produces a significant percentage of aggregates of antibodies of varying sizes from which the bsAb needs to be purified by methods such as gel filtration. Yields of the dimeric bsAb have been reported to be in the 30–40% range (Karpovsky et al. 1984).

3.2.2.2 Traut's Reagent/SMCC

Although used widely in protein conjugation, bsAbs crosslinked with SPDP have a reducible disulfide bond that may be unstable in the presence of reducing agents. A way to generate a more stable link is to use two different amino-reactive hetero-bifunctional crosslinkers for the two parental Abs. One of the crosslinkers introduces a SH group (like SPDP) while the other incorporates a maleimide group, which reacts with a free-thiol (SH) to generate a non-reducible thioether crosslink. A strategy that has been widely employed for generating bsAbs is to use Traut's reagent (2-iminothiolane) to introduce a SH in one of the antibodies and Sulpho-SMCC (sulpho-[succinimidyl-4-(*N*-maleimidomethyl)-4-cyclohexane-1-carboxylate]) as the maleimido-containing crosslinker used to modify the second Ab (Fig. 3.2b). The resulting product is a bsAb that is bivalent for each antigen, as well as some unwanted multimeric product. One of the major caveats of this method is that the reported yields for the dimeric product are low (~10–29%) (Reusch et al. 2006; Lee et al. 2007).

Hetero-bifunctional crosslinkers can be used on intact Abs or on Ab fragments generated by enzymatic digestion. However, the crosslinking of intact Abs is reported more frequently, probably because of the associated simplicity. One of the main advantages of the use of hetero-bifunctional reagents is that they rely on amino groups present in any Ig molecule, making it amenable to all Ig subclasses and Igs from different species. Thus the use of hetero-bifunctional crosslinkers provides a fairly quick and easy way of preparing bsAbs. On the other hand, one noted disadvantage is that because there is no control on the attachment site, reaction of residues involved in antigen binding could inactivate the Ab. In addition, because of the multiple reactive sites per molecule, a trade off between maximizing the crosslinker incorporation and reducing the formation of multimers reduces yields. Therefore, the main drawback of these methods is the heterogeneity of the bsAb and the consequent batch-to-batch variation. In spite of these limitations, bsAbs obtained by these approaches have been used for a variety of applications such as directing cytotoxic T cells to cancer cells (Reusch et al. 2006), targeting endothelial cells with adenoviruses for gene therapy (Miller et al. 2005), delivering toxins to target cells (Raso et al. 1997), immunosuppressive therapies (Vasu et al. 2003), in vivo imaging (Khaw et al. 2006), immunoassay development (Khaw et al. 2005), and now recently for directing the migration of stem cells to injured tissues (Lee et al. 2007; Langer et al. 2010).

3.2.3 Homo-bifunctional Reagents

As mentioned above, the abundance of amine groups in an Ig molecule capable of reacting with hetero-bifunctional reagents is the main source of the heterogeneity of the final bsAb product. In contrast, the frequency of free sulfhydryl occurrence in antibodies is usually low or nonexistent compared to amine groups. The use of sulfhydryl-reactive chemistries thus can restrict modification to only a limited number of sites in an antibody. Homo-bifunctional reagents used for producing bsAbs react with sulfhydryl residues, either generated by reduction of the hinge disulfide or by using engineered cysteines. While crosslinking using amino-reactive reagents has been more often carried out on intact antibodies, Ab conjugation based on SH groups is generally applied to $F(ab')_2$ fragments.

One shortcoming of the original method described by Nisonoff was the poor yields of the bispecific product because of the formation of homodimers and the formation of intra-chain disulfide bonds. Later on Brennan et al. (1985) reported the use of sodium arsenite to prevent the formation of intrachain disulfide bonds in one of the Abs and the activation of the thiol groups in the other Ab by reaction with DTNB [5,5′-dithiobis-(2-nitrobenzoic acid)]. The methods currently used are variations of the original method using the crosslinker *o*-phenylenedimaleimide (*o*-PDM) (Glennie et al. 1987). The first step consists in reducing the cysteine residues in the hinge region of the $F(ab')_2$ to generate Fab′ fragments. Then one of the Fab′ species is selected for alkylation with *o*-PDM, a reagent that contains two maleimide groups capable of reacting with vicinal thiol to form a cycle. In order to crosslink the two Fab′ species, one of the two maleimide groups needs to be available to react with the second Fab′, therefore this method is useful only for Ig molecules containing an odd number of cysteine residues in the hinge region (i.e., mouse IgG1, IgG3, IgG2a, rabbit IgG, rat IgG1). In Fig. 3.3a mouse IgG1 Fab′ with three cysteines in the hinge is illustrated: two of the cysteine residues react with a single *o*-PDM while the third is linked to a *o*-PDM molecule containing a nonreacted maleimide. The modified Fab′ (Fab-A-mal) is then reacted with the other reduced Fab′ (Fab-B). When equal amounts of the two parental Fabs are used, the major product is a bispecific $F(ab')_2$, resulting from the reaction of Fab-A-mal with the SH in Fab-B. Increasing the proportion of Fab-A-mal in the reaction produces a significant amount of $F(ab')_3$ generated by the reaction of two molecules of Fab-A-mal with two free SH groups in a single Fab-B molecule. The remaining SH groups in the Fab-B can be alkylated to prevent the formation of multimers and finally the different products are separated by size exclusion chromatography.

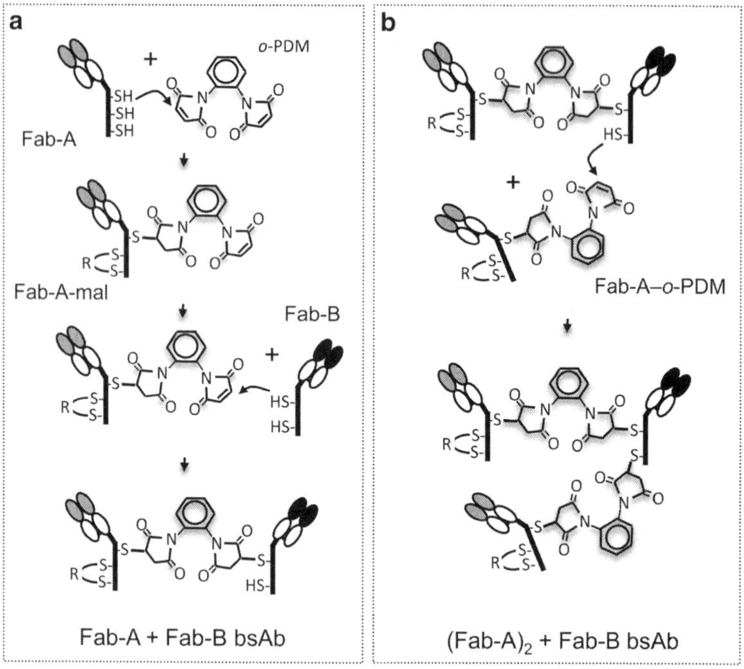

Fig. 3.3 *Homo-bifunctional crosslinker o-PDM*. (**a**) Generation of a bsAb monovalent for each antigen. Incubation of the Fab′ containing an odd number of cysteines in the hinge region (Fab-A) with *o*-PDM produces the crosslinker-mediated cyclization of two cysteines (R=*o*-PDM), while the third cysteine is conjugated to a crosslinker molecule containing an unreacted maleimide group. One of the SH groups in the second Fab′ (Fab-B) reacts with the maleimide group to generate the monovalent bsAb. (**b**) If Fab-A is present in excess, it reacts with free SH groups in the conjugated Fab-B generating a bsAb bivalent for Fab-A and monovalent for Fab-B

As mentioned above, the application of this methodology is restricted to IgGs with an even number of Cys residues in the hinge. This is a limitation for human or humanized antibodies often used for human therapy to avoid eliciting an anti-antibody immune response. A way to circumvent this problem is to express humanized Fab′ recombinantly with only one of the cysteines in the hinge region of the antibody (Shalaby et al. 1992). An alternative strategy we have developed uses *N*-ethylmaleimide (NEM) to partially block the SH groups. $F(ab')_2$ fragments generated by enzymatic digestion are reduced and then reacted with one equivalent of NEM so three different products are formed (a) unreacted Fab′, (b) Fab′ with one sulfhydryl alkylated, and (c) Fab′ with all two sulfhydryl groups alkylated. Only the first two products, containing either two or one free thiols respectively, are reactive in the subsequent incubation with a bis-maleimide crosslinker. This reaction generates a Fab′ with its two Cys linked to a single bis-malemide molecule, and a Fab′ with one NEM and one crosslinker molecule. This last species is the only one capable of reacting with a second Fab′.

Although aggregates are not significantly formed when homo-bifunctional crosslinkers are used, a measurable amount of heterotrimers is formed if the antibodies contain more than one cysteine residue in the hinge region. Typical yields are 65–75% of $F(ab')_2$ heterodimer, and 25–35% $F(ab')_3$ heterotrimer. Besides the improved yields compared to methods based on hetero-bifunctional reagents, methods using homo-bifunctional reagents are directed to specific sites (such as cysteines in the hinge region) and thus do not cause antibody inactivation.

3.2.4 Chemically Programmed Antibodies

This approach uses a specific antibody (Ab 38C2) as a scaffold onto which small drugs or peptides targeting a specific protein are covalently linked. The Ab 38C2 is a catalytic Ab with aldolase activity. The active site contains a lysine residue with an exceptional nucleophilic property, which can be selectively and covalently labeled with small molecules or peptides containing a diketone or a β lactam functional group. The chemical programming of Abs potentially combines the structural diversity and simple manufacturing features of small drugs with the tunable half-life and valencies of antibodies (Rader et al. 2003). As a proof of concept for this approach, mAb 38C2 was reprogrammed to target the endothelin receptor with a small molecule (Doppalapudi et al. 2007) and integrins $\alpha_v\beta_3$ and $\alpha_v\beta_5$ with a peptide (Rader et al. 2003). More recently, conjugation of RNA aptamers to the same antibody scaffold has been shown to significantly increase the half-life of the aptamer while retaining its biological activity (Wuellner et al. 2010). The chemical programming of antibodies can also be carried out in vivo. Mice actively immunized with KLH conjugated to a small molecule containing a diketone raise an immune response producing Abs that can covalently bind to the diketone-containing molecule (Popkov et al. 2009). Subsequent administration of a chemical module containing a diketone leads to the covalent modification of the Abs in vivo and can redirect the antibody toward the specified targets. Antibodies programmed in vivo by this approach showed the ability to reduce the growth of syngeneic tumors in mouse models (Popkov et al. 2009). If the diketone-containing module targets two proteins instead of one, a bivalent, bispecific, chemically programmed antibody can be generated. Using this approach, a bsAb targeting integrins $\alpha_v\beta_3$, $\alpha_v\beta_5$ and the luteinizing hormone releasing hormone receptor (LHRHR) was created (Gavrilyuk et al. 2009). The resulting bsAb was able to specifically bind the targeted proteins expressed on cancer cells as well as the isolated proteins (Gavrilyuk et al. 2009). Clinical trials using chemically programmed Abs are underway (http://www.clinicaltrials.gov).

3.2.5 Clinical Applications

Starting in the early 1990s, a number of bsAbs have been tested in clinical trials. The first bsAbs that were tested for their safety and efficacy in patients were generated either by the quadroma technique or by chemical crosslinking. Chemically conjugated bsAbs have been used mainly in oncologic clinical studies. Within the oncologic applications, different bsAbs have exploited distinct therapeutic strategies, which are briefly described below.

3.2.5.1 Retargeting Immune Effector Cells

The first bsAbs tested in clinical trails were a series of $F(ab')_2$s generated by crosslinking with *o*-PDM and were directed against the Fc receptor FcγRI (CD64) and a tumor antigen with the goal of recruiting effector cells to the tumor. The FcγRI receptor is constitutively present on monocytes, macrophages, and on activated polymorphonuclear granulocytes (PMN) and mediates cellular responses such as cytotoxicity, phagocytosis, and respiratory burst. Two of these bsAbs generated by chemical conjugation have been tested in the clinic: MDX-210 and MDX-447. BsAbs MDX-210 and its derivative, a partially humanized version, MDX-H210, targeted HER2, which is overexpressed in a variety of human carcinomas, whereas MDX-447 targeted the epidermal growth factor receptor (EGFR) overexpressed in some neck, kidney, bladder, and prostate cancers. Initial in vitro studies showed that MDX-210 and MDX-H210 could induce high cytotoxic effect on cell lines overexpressing HER2 in the presence of activated PMN cells (Repp et al. 2003). Similarly, MDX-447 effectively induced cytolysis in vitro of cells expressing high levels of EGFR. When administrated to patients the bsAbs were well tolerated, but in spite of the encouraging preclinical data, they did not produce consistent antitumor results (Pullarkat et al. 1999; Repp et al. 2003; Fury et al. 2008). This result was thought to be due in part to the suboptimal concentration of the bsAbs achieved within the tumor, the short half-life of the therapeutic proteins (2–8 h), and the high effector to target ratios required for lysis (Valone et al. 1995).

3.2.5.2 In Vitro Arming T Cells

An approach related to the recruitment of cells expressing Fc-receptors is the recruitment of cytotoxic T cells. Cytotoxic T cells are the most efficient cell-lysing effector cells of the body, but unlike macrophages, PMN and NK, they do not express Fcγ receptors. Thus, bsAbs aimed at recruiting cytotoxic T cells target instead the T-cell receptor (TCR) component CD3. However, for full activation of T cells, a secondary signal is required in addition to engaging the TCR. This limitation applies to most bsAb formats in the published literature to date with the exception of bispecific T-cell engagers (BiTE®), which are discussed in Chap. 15. One way to overcome the requirement for the administration of multiple drugs to achieve full T-cell activation in vivo is to activate and expand autologous T-cell in vitro, "arm" them with the bsAb and then transfer the cells back to the patient to redirect the lysis of tumor cells. Expansion and activation is achieved by using the anti-CD3 Ab OKT3 together with low doses of IL-2. One advantage of this ex vivo strategy over the in vivo redirecting of T cells is that the bsAbs are not administrated systemically, reducing the chances for toxic side effects. Three crosslinked intact antibodies are being tested for in vitro T-cell arming: CD20-Bi (OKT3 × Rituximab) for the treatment of multiple myeloma, HER2-Bi (OKT3 × Trastuzumab) for the treatment of metastatic breast cancer and hormone-refractory prostate cancer, and EGFR-Bi (OKT3 × Cetuximab). Ongoing phase I/II studies with HER2-Bi-armed T cells have shown the treatment is well-tolerated (with occasional chills and fever) and improved survival rates (Lum et al. 2007). Preclinical studies showed T cells armed in vitro with EGFR-Bi can delay the growth of tumors in animal models for pancreatic cancer (Reusch et al. 2006). Two Phase I clinical studies for this bsAb in advanced nonsmall cell lung cancer patients and in patients with advanced solid tumors are ongoing (http://www.clinicaltrials.gov). Application of this method is further described in Chap. 14.

3.2.5.3 Pretargeting

Radioimmunotherapy has long been used for the treatment of cancer. One problem originally associated with this technique was the high toxicity due to the presence in circulation of the radiolabeled antibody. To overcome this problem, a strategy called "pretargeting" was developed in the 1990s. A bsAb specific for a tumor antigen and for a short peptide is administered and allowed to bind to the tumor. After some time all the unbound Ab is cleared from circulation and the radionuclide complexed to the peptide is then injected. A fraction of the peptide binds to the tumor-bound antibody and the free radionuclide is cleared very rapidly from the body resulting in concentrations in the tumor ten times higher than in the blood or normal tissue after 1 h of the radionuclide injection. The separate delivery of antibody and radionuclide significantly reduces the radionuclide residence time in the blood, thus reducing dose-limiting toxicities. In preclinical studies, pretargeting showed reduced toxicity and improved efficacy in animal models. A crosslinked $F(ab')_2$ bsAb binding both the carcinoembryonic antigen (CEA) and a short peptide has been evaluated for the treatment of medullary thyroid cancer. Initial phase I/II studies showed that the dose-limiting toxicity was hematological and a tumor stabilization rate of 45% in a 1-year assessment was observed (Kraeber-Bodere et al. 1999, 2006). A retrospective study comparing the treated patients vs. the nontreated patients indicated that the treated patients had a longer survival than nontreated patients (10-year overall survival of 83% vs. 14%) (Chatal et al. 2006). More recently, a study aimed at evaluating the clinical response was carried out, and the preliminary analysis of the results showed efficacy rates, assessed by positron emission tomography and biomarker levels, ranging between 50 and 60% (Kraeber-Bodere et al. 2010).

The use of bsAbs for pretargeting has been improved recently by using Abs generated by the "dock and lock" method, discussed in Chap. 12. The Abs generated by this method are trifunctional molecules made of three Fab fragments, one targeting the peptide and two able to bind the tumor-specific antigen.

The presence of more than one arm targeting the tumor confers a higher avidity which leads to a superior accumulation and retention in the tumor.

3.3 Future Possibilities

The main hurdle for the transfer of chemically conjugated bsAbs to clinical applications has been the low yield and heterogeneity of the products. Several incipient technologies in the area of bioconjugation chemistry may help to improve both the yield and the homogeneity of bsAbs as well as the pharmacokinetics properties.

3.3.1 Bis-Fabs

Antibody-drug conjugation is a current therapeutic approach that also produces heterogeneous products due to the presence of multiple reactive sites. In an effort to improve yields and product quality, a recent study aimed at achieving site-specific conjugations scanned different positions in an antibody for the introduction of cysteine residues having high reactivity and without interfering with protein folding or antigen binding. Using Trastuzumab Fab as a model, serine, alanine and valine residues with high solvent accessibility were selected for mutagenesis and the proteins expressed in a phage-display-based assay for reactive cysteines (Junutula et al. 2008a). The assay revealed some positions that had high reactivities and no interference with protein folding or antigen binding. The identified high-reacting sites were used in full-length antibodies to create engineered molecules with single-site reactivity which were coined Thio-mAbs. Some of the engineered variants showed greater than 90% conjugation efficiencies along with near-equivalent stoichiometry. In addition, using several drug-conjugated Abs targeting tumor antigens, Thio-mAbs showed improved safety without loss of efficacy compared to Ab-drug conjugates generated by standard methods (Junutula et al. 2008b). The high conjugation rates and product homogeneity achieved with Thiomabs made them a very attractive platform to develop a more efficient process to produce bsAbs by chemical conjugation. Thus we adapted the Thiomab technology to develop a bsAb production process. We refer to the resulting bsAbs as "bis-Fabs" to include the name of the crosslinker employed, bis-maleimide (Fig. 3.4a). The two-step production method is a variation of the strategy using a SH reactive homo-bifunctional crosslinker.

After designing two Thio-Fabs with engineered cysteines, the first synthesis step converts more than 90% of the thio-Fab into the desired product. This product can be separated from the non-reacted crosslinker by gel filtration. In the second step, the modified Fab (thio-Fab-mal) is incubated with the unmodified Thio-Fab generating the bispecific bis-Fab. Usually the Thio-Fab present in limiting amounts

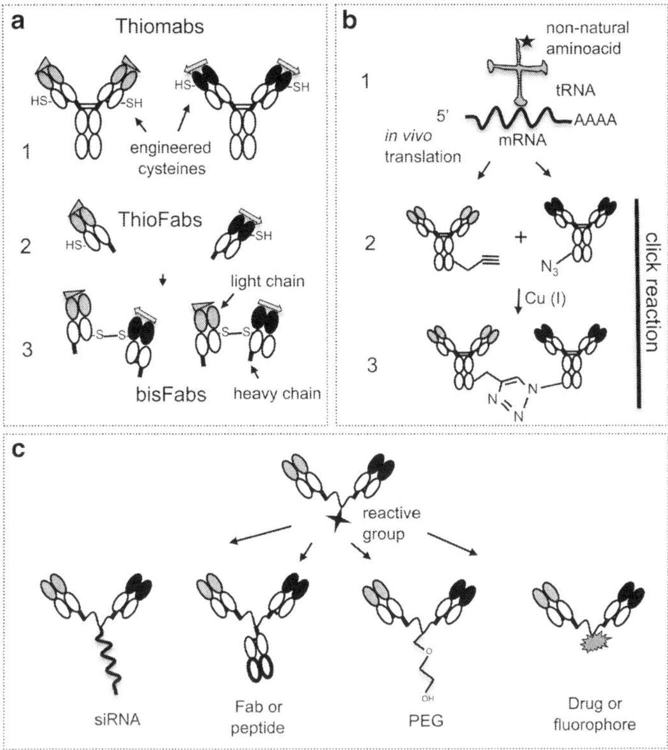

Fig. 3.4 *Prospective developments*. (**a**) Bis-Fabs production for engineering structural variants. Thiomabs are mAbs containing engineered cysteine residues at varying positions (1). ThioFabs are the Fabs derived from Thiomabs (2). Sulfhydryl crosslinking of ThioFabs generate bis-Fabs. Because the site of attachment is an engineered cysteine, the location of the covalent attachment can be moved to nearly any position in the heavy chain or light chain. Thus, structural variants of the bis-Fab can be generated by altering the engineered cysteine. Note the different positions of the crosslinking and the orientation of the binding sites with respect to each other (3). (**b**) Non-natural amino acids and click reactions. Non-natural amino acids are incorporated into proteins during in vivo translation using engineered tRNAs (1). Various non-natural amino acids can be added using different recombinant strains of *E. coli* (2). The non-natural amino acids allow using different chemistries such as the click reaction azide-alkyne [3+2] cycloaddition catalyzed by copper (I) (3). (**c**) Addition of other functionalities into the crosslinker. Fabs can be modified using a crosslinker carrying a third reactive group. This reactive group can be used to site-specifically conjugate other molecules, such as siRNA, peptides, PEG, toxic drugs, or fluorophores

reacts completely, while a fraction of the Thio-Fab in excess remains unreacted. Thus bis-Fab technology allows the production of bsAbs with rates of conversion greater than 90%, which is a significant improvement over the traditional crosslinking methods. Moreover, it generates conjugates with an exquisitely defined composition and batch-to-batch reproducibility.

To characterize the bis-Fab production process, we initially generated a series of monospecific bis-Fabs using Trastuzumab as a model. Different Trastuzumab Fabs with engineered cysteine residues at varying positions in either the light or heavy chain were conjugated to each other according to a matrix arrangement. Trastuzumab binds to HER2, inhibiting its tyrosine kinase activity that is required for cellular replication. Therefore, the functional tests consisted in assaying the effect of the bis-Fabs on the replication of HER2-positive cells. Although most of the conjugates showed an inhibitory effect on cell proliferation, unexpectedly other bis-Fab variants showed a stimulatory effect (unpublished observations). Additional analysis of the bis-Fabs promoting cell proliferation indicated they caused an increase in the tyrosine kinase activity of HER2 (unpublished observations). The agonist-like bis-Fabs differ from the parent antibody only in the spatial orientation of the two antigen binding sites with respect to each other. It is remarkable that by solely changing the three-dimensional configuration, an antibody can acquire the opposite biological effect to the parent Ab. However, this technology provides increases in (1) efficiency, (2) reproducibility, and (3) novel strategies for engineering activity. This observation opens important opportunities for antibody development.

3.3.2 Non-natural Amino Acids/Click Chemistry

As mentioned, the presence of multiple reactive sites in a protein leads to heterogeneity in the conjugation products. A way to overcome this problem is to incorporate into proteins non-natural amino acids with different functional groups. These non-natural amino acids have a wide range of structures and functions not present in the natural amino acid repertoire, and thus provide the opportunity to use other chemistry. In protein therapeutics, the incorporation of non-natural amino acids containing orthogonal reactive functional groups allows site-specific modification with many additional conjugates. The potential modifications include conjugation of toxins, radioisotopes, polyethylene glycol, and siRNA. Different strategies are being explored for the incorporation of non-natural amino acids. A semi-synthetic approach involves two steps: a short peptide (<50 aa) containing the non-natural amino acid is generated in vitro by solid phase synthesis, while the rest of the protein is produced by recombinant expression. The two pieces are modified at their termini with mutually reactive groups and linked to each other (de Graaf et al. 2009). An in vivo strategy for the incorporation of non-natural amino acids, which may be more versatile, is to use engineered or foreign pairs of tRNA and aminoacyl-tRNA synthetases (Fig. 3.4b). Using this approach, over 70 non-natural amino acids have been added to the genome of bacteria, yeasts, and mammalian cells (Liu and Schultz 2010).

An in vitro enzymatic approach to introduce non-natural amino acids into a protein is based on the activity of the formylglycine generating enzyme (FGE) (Wu et al. 2009). The enzyme recognizes a 12 aa sequence (LCTPSRAALLTGR) present in the active site of Type 1 sulfatases, and catalyzes the conversion of cysteine into formylglycine which possesses an aldehyde group. Heterologous proteins containing a minimized target sequence (6 aa) can be expressed in *Escherichia coli* and efficiently modified by a co-expressed FGE (Carrico et al. 2007). In addition to modifing proteins expressed in *E. coli*, the system has been shown to work also for the expression of recombinant proteins, including IgG, in mammalian cells. Unlike expression in bacteria, the initial rates of conversion of cysteine to formylglycine in mammalian cells are variable, but they can be increased to ~90% with some optimization (Wu et al. 2009). The aldehyde tag can be then reacted with hidrazides or aminooxy-functionalized molecules, thus allowing the site-specific modification of the protein.

The main advantage of non-natural amino acids is that they offer reactivities not found in natural amino acids, some of them pertaining to the "click chemistry." This term refers to chemical reactions with high rates and selectivity, requiring simple reaction conditions (Kolb et al. 2001). One such reaction that has been successfully used for the conjugation of biomolecules is the copper (I) catalyzed azide-alkyne [3+2] cycloaddition (CuAAC) (Fig. 3.3b). This reaction can be carried out at biologically compatible conditions (physiological pH and temperature) and in a variety of solvents achieving high conjugation efficiencies. A particularly relevant example of non-natural amino acids used for site-specific modification emerged in a study describing protein–protein crosslinking. Bundy and Swartz (2010) described the site-specific, linker-less, one-step crosslinking of two proteins containing non-natural amino acids and using a CuAAC reaction, showing that the strategy could also be used to improve the production of bsAbs. Given the robustness of click reactions and the fact that bio-orthogonal conjugation reactions can be carried out in the presence of other proteins that contain only natural amino acids, it is possible to envision the future development of a technology to crosslink antibodies in a crude extract, requiring only the downstream purification of the crosslinked product. This "one-pot" approach for the generation of chemically crosslinked bsAbs would translate in to reduced costs and complexity of the process. Because copper is toxic to cells, a disadvantage of copper-catalyzed reactions for the production of therapeutics is the potential presence of residual copper in the product. A series of compounds have been recently developed for the fast and selective cycloaddition reaction with azide-containing biomolecules in a copper-free manner (Jewett and Bertozzi 2010). Since copper-free cycloaddition reactions do not interfere with normal metabolism, they can be used without interfering with cell viability. The application of these reactions to the production of bsAbs would not only eliminate the risk of copper contamination, but in systems supporting the co-expression of both Ab arms, it could also allow integrating the crosslinking step to the protein expression process.

3.3.3 Addition of New Functionalities

The use of crosslinkers to produce bsAbs provides an opportunity to incorporate more functionality into the protein. In addition to the two reactive groups, some crosslinkers contain a third reactive group that can be used to incorporate other molecules such as siRNA, fluorophores, cytotoxic drugs, stimulating peptides, PEG, or a third protein (Fig. 3.4c). The addition of other molecules into the crosslinker has the advantage of being site-specific and thus reducing both the risk of inactivating the Ab and the heterogeneity of the product.

An area where this strategy is already being used is the PEGylation of proteins. Most of the bsAbs produced by chemical crosslinking are in the $F(ab')_2$ format. The smaller size of $F(ab')_2$ bsAbs with respect to an IgG endow them with some advantages, i.e., a better tumor penetrance. However, they have a significantly shorter half-life in serum, which could reduce the therapeutic potential for some applications. PEGylation is a well-established modification that extends the protein half-life and has been clinically proven in marketed products. Current PEGylation methods are inefficient because they yield heterogeneous products of varying potency. Although several methods have been recently developed to PEGylate proteins in a more controlled fashion (Deiters et al. 2004; Brocchini et al. 2008; Maullu et al. 2009), the use of a new generation of crosslinkers carrying PEG of various MW make easier the production of Ab fragments with half lives that can be adjusted to the needs of the intended application. In a similar way, the ability to design new crosslinkers already loaded with cytotoxic drugs or fluorophores will likely simplify the production of bsAbs for antibody-directed chemotherapy and imaging purposes.

3.4 Concluding Remarks

The concept of bispecific antibodies originated more than 40 years ago. However, only in recent years, have they been considered as viable therapeutic tools. A number of crosslinked bsAbs are at present being tested in clinical trials (Hollander 2009; Gu and Ghayur 2010). Although the treatment of cancer is probably the area where most of the efforts are currently made, other therapeutic areas include the treatment of infectious diseases, regenerative medicine, neuromedicine and immune diseases. Besides therapeutics, bsAbs are also used in fundamental research, imaging, and diagnostics. These diverse applications of bsAbs have specific and distinct requirements (i.e., protein size, pharmacokinetic properties, valency, presence or absence of the Fc). Thus there is the need for a wide variety of formats with different properties. Chemical conjugation is a versatile tool that allows the production in a modular fashion of several different bsAb formats. Although the main limitation of bsAbs produced by chemical conjugations for clinical applications has been the heterogeneity of the products and low yield of the process, the adoption of technologies discussed in this chapter will allow a more efficient and site-specific crosslinking, contributing to improve the yields and product quality. Moreover, the development of bis-Fabs technology with its unique ability of modifying the relative geometry of the two antigen binding sites further expands the general capabilities of bsAbs. In summary, advances in the field of bioconjugation will help make chemically conjugated bsAbs an affordable, attractive way of generating new biotherapeutics.

Acknowledgement We thank Dr. Byoung-Chul Lee for critically reading the manuscript.

References

Brennan M, Davison PF, Paulus H (1985) Preparation of bispecific antibodies by chemical recombination of monoclonal immunoglobulin G1 fragments. Science 229(4708):81–83

Brocchini S, Godwin A, Balan S, Choi JW, Zloh M, Shaunak S (2008) Disulfide bridge based PEGylation of proteins. Adv Drug Deliv Rev 60(1):3–12

Bundy BC, Swartz JR (2010) Site-specific incorporation of p-propargyloxyphenylalanine in a cell-free environment for direct protein-protein click conjugation. Bioconjug Chem 21(2): 255–263

Carrico IS, Carlson BL, Bertozzi CR (2007) Introducing genetically encoded aldehydes into proteins. Nat Chem Biol 3(6):321–322

Chatal JF, Campion L, Kraeber-Bodere F, Bardet S, Vuillez JP, Charbonnel B, Rohmer V, Chang CH, Sharkey RM, Goldenberg DM, Barbet J (2006) Survival improvement in patients with medullary thyroid carcinoma who undergo pretargeted anti-carcinoembryonic-antigen radioimmunotherapy: a collaborative study with the French Endocrine Tumor Group. J Clin Oncol 24(11):1705–1711

de Graaf AJ, Kooijman M, Hennink WE, Mastrobattista E (2009) Nonnatural amino acids for site-specific protein conjugation. Bioconjug Chem 20(7):1281–1295

Deiters A, Cropp TA, Summerer D, Mukherji M, Schultz PG (2004) Site-specific PEGylation of proteins containing unnatural amino acids. Bioorg Med Chem Lett 14(23):5743–5745

Doppalapudi VR, Tryder N, Li L, Aja T, Griffith D, Liao FF, Roxas G, Ramprasad MP, Bradshaw C, Barbas CF 3rd (2007) Chemically programmed antibodies: endothelin receptor targeting CovX-Bodies. Bioorg Med Chem Lett 17(2):501–506

Fury MG, Lipton A, Smith KM, Winston CB, Pfister DG (2008) A phase-I trial of the epidermal growth factor receptor directed bispecific antibody MDX-447 without and with recombinant human granulocyte-colony stimulating factor in patients with advanced solid tumors. Cancer Immunol Immunother 57(2):155–163

Gavrilyuk JI, Wuellner U, Salahuddin S, Goswami RK, Sinha SC, Barbas CF 3rd (2009) An efficient chemical approach to bispecific antibodies and antibodies of high valency. Bioorg Med Chem Lett 19(14):3716–3720

Glennie MJ, McBride HM, Worth AT, Stevenson GT (1987) Preparation and performance of bispecific F(ab' gamma)2 antibody containing thioether-linked Fab' gamma fragments. J Immunol 139(7):2367–2375

Gu J, Ghayur T (2010) Rationale and development of multispecifc antibody drugs. Expert reviews of clinical pharmacology 3(4):491–508

Hollander N (2009) Bispecific antibodies for cancer therapy. Immunotherapy 1(2):211–222

Jewett JC, Bertozzi CR (2010) Cu-free click cycloaddition reactions in chemical biology. Chem Soc Rev 39(4):1272–1279

Junutula JR, Bhakta S, Raab H, Ervin KE, Eigenbrot C, Vandlen R, Scheller RH, Lowman HB (2008a) Rapid identification of reactive cysteine residues for site-specific labeling of antibody-Fabs. J Immunol Methods 332(1–2):41–52

Junutula JR, Raab H, Clark S, Bhakta S, Leipold DD, Weir S, Chen Y, Simpson M, Tsai SP, Dennis MS, Lu Y, Meng YG, Ng C, Yang J, Lee CC, Duenas E, Gorrell J, Katta V, Kim A, McDorman K, Flagella K, Venook R, Ross S, Spencer SD, Lee Wong W, Lowman HB, Vandlen R, Sliwkowski MX, Scheller RH, Polakis P, Mallet W (2008b) Site-specific conjugation of a cytotoxic drug to an antibody improves the therapeutic index. Nat Biotechnol 26(8):925–932

Karpovsky B, Titus JA, Stephany DA, Segal DM (1984) Production of target-specific effector cells using hetero-cross-linked aggregates containing anti-target cell and anti-Fc gamma receptor antibodies. J Exp Med 160(6):1686–1701

Khaw BA, Rammohan R, Abu-Taha A (2005) Bispecific enzyme-linked signal-enhanced immunoassay with subattomole sensitivity. Assay Drug Dev Technol 3(3):319–327

Khaw BA, Tekabe Y, Johnson LL (2006) Imaging experimental atherosclerotic lesions in ApoE knockout mice: enhanced targeting with Z2D3-anti-DTPA bispecific antibody and 99mTc-labeled negatively charged polymers. J Nucl Med 47(5):868–876

Köhler G, Milstein C (1975) Continuous cultures of fused cells secreting antibody of predefined specificity. Nature 256(5517):495–497

Kolb HC, Finn MG, Sharpless KB (2001) Click chemistry: diverse chemical function from a few good reactions. Angew Chem Int Ed Engl 40(11):2004–2021

Kraeber-Bodere F, Bardet S, Hoefnagel CA, Vieira MR, Vuillez JP, Murat A, Ferreira TC, Bardies M, Ferrer L, Resche I, Gautherot E, Rouvier E, Barbet J, Chatal JF (1999) Radioimmunotherapy in medullary thyroid cancer using bispecific antibody and iodine 131-labeled bivalent hapten: preliminary results of a phase I/II clinical trial. Clin Cancer Res 5(Suppl 10):3190s–3198s

Kraeber-Bodere F, Rousseau C, Bodet-Milin C, Ferrer L, Faivre-Chauvet A, Campion L, Vuillez JP, Devillers A, Chang CH, Goldenberg DM, Chatal JF, Barbet J (2006) Targeting, toxicity, and efficacy of 2-step, pretargeted radioimmunotherapy using a chimeric bispecific antibody and 131I-labeled bivalent hapten in a phase I optimization clinical trial. J Nucl Med 47(2):247–255

Kraeber-Bodere F, Salaun PY, Oudoux A, Goldenberg DM, Chatal JF, Barbet J (2010) Pretargeted radioimmunotherapy in rapidly progressing, metastatic, medullary thyroid cancer. Cancer 116 (4 Suppl):1118–1125

Langer HF, von der Ruhr JW, Daub K, Schoenberger T, Stellos K, May AE, Schnell H, Gauss A, Hafner R, Lang P, Schumm M, Buhring HJ, Klingel K, Conrad S, Schaller M, van Zandvoort M, Jung G, Dimmeler S, Skutella T, Gawaz M (2010) Capture of endothelial progenitor cells by a bispecific protein/monoclonal antibody molecule induces reendothelialization of vascular lesions. J Mol Med 88(7):687–699

Lee RJ, Fang Q, Davol PA, Gu Y, Sievers RE, Grabert RC, Gall JM, Tsang E, Yee MS, Fok H, Huang NF, Padbury JF, Larrick JW, Lum LG (2007) Antibody targeting of stem cells to infarcted myocardium. Stem Cells 25(3):712–717

Liu CC, Schultz PG (2010) Adding new chemistries to the genetic code. Annu Rev Biochem 79:413–444

Lum LG, Rathore R, Dizon D, Wang D, Al-Kadhimi Z, Uberti J, Ratanathahathorn V (2007) Induction of immune response and improved survival after infusions of T-cells armed with anti-CD3 x anti-Her2/Neu bispecifc antibody in stage IV breast cancer patients (PhaseI). Blood 110:2747

Maullu C, Raimondo D, Caboi F, Giorgetti A, Sergi M, Valentini M, Tonon G, Tramontano A (2009) Site-directed enzymatic PEGylation of the human granulocyte colony-stimulating factor. FEBS J 276(22):6741–6750

Miller WH, Brosnan MJ, Graham D, Nicol CG, Morecroft I, Channon KM, Danilov SM, Reynolds PN, Baker AH, Dominiczak AF (2005) Targeting endothelial cells with adenovirus expressing nitric oxide synthase prevents elevation of blood pressure in stroke-prone spontaneously hypertensive rats. Mol Ther 12(2):321–327

Nisonoff A, Rivers MM (1961) Recombination of a mixture of univalent antibody fragments of different specificity. Arch Biochem Biophys 93:460–462

Popkov M, Gonzalez B, Sinha SC, Barbas CF 3rd (2009) Instant immunity through chemically programmable vaccination and covalent self-assembly. Proc Natl Acad Sci U S A 106 (11):4378–4383

Pullarkat V, Deo Y, Link J, Spears L, Marty V, Curnow R, Groshen S, Gee C, Weber JS (1999) A phase I study of a HER2/neu bispecific antibody with granulocyte-colony-stimulating factor in patients with metastatic breast cancer that overexpresses HER2/neu. Cancer Immunol Immunother 48(1):9–21

Rader C, Sinha SC, Popkov M, Lerner RA, Barbas CF 3rd (2003) Chemically programmed monoclonal antibodies for cancer therapy: adaptor immunotherapy based on a covalent antibody catalyst. Proc Natl Acad Sci U S A 100(9):5396–5400

Raso V, Brown M, McGrath J (1997) Intracellular targeting with low pH-triggered bispecific antibodies. J Biol Chem 272(44):27623–27628

Repp R, van Ojik HH, Valerius T, Groenewegen G, Wieland G, Oetzel C, Stockmeyer B, Becker W, Eisenhut M, Steininger H, Deo YM, Blijham GH, Kalden JR, van de Winkel JG, Gramatzki M (2003) Phase I clinical trial of the bispecific antibody MDX-H210 (anti-FcgammaRI x anti-HER-2/neu) in combination with Filgrastim (G-CSF) for treatment of advanced breast cancer. Br J Cancer 89(12):2234–2243

Reusch U, Sundaram M, Davol PA, Olson SD, Davis JB, Demel K, Nissim J, Rathore R, Liu PY, Lum LG (2006) Anti-CD3 x anti-epidermal growth factor receptor (EGFR) bispecific antibody redirects T-cell cytolytic activity to EGFR-positive cancers in vitro and in an animal model. Clin Cancer Res 12(1):183–190

Segal D. SD Bast B (1995) Production of Bispecific Antibodies. In: Current Protocols in Immunology, Supplement 14. John Wiley & Sons, New York

Shalaby MR, Shepard HM, Presta L, Rodrigues ML, Beverley PC, Feldmann M, Carter P (1992) Development of humanized bispecific antibodies reactive with cytotoxic lymphocytes and tumor cells overexpressing the HER2 protooncogene. J Exp Med 175(1):217–225

Songsivilai S, Clissold PM, Lachmann PJ (1989) A novel strategy for producing chimeric bispecific antibodies by gene transfection. Biochem Biophys Res Commun 164(1):271–276

Valone FH, Kaufman PA, Guyre PM, Lewis LD, Memoli V, Ernstoff MS, Wells W, Barth R, Deo Y, Fisher J et al (1995) Clinical trials of bispecific antibody MDX-210 in women with advanced breast or ovarian cancer that overexpresses HER-2/neu. J Hematother 4(5):471–475

Vasu C, Gorla SR, Prabhakar BS, Holterman MJ (2003) Targeted engagement of CTLA-4 prevents autoimmune thyroiditis. Int Immunol 15(5):641–654

Wu P, Shui W, Carlson BL, Hu N, Rabuka D, Lee J, Bertozzi CR (2009) Site-specific chemical modification of recombinant proteins produced in mammalian cells by using the genetically encoded aldehyde tag. Proc Natl Acad Sci U S A 106(9):3000–3005

Wuellner U, Gavrilyuk JI, Barbas CF 3rd (2010) Expanding the concept of chemically programmable antibodies to RNA aptamers: chemically programmed biotherapeutics. Angew Chem Int Ed Engl 49(34):5934–5937

Chapter 4
Trivalent and Trispecific Antibody Derivatives for Cancer Therapy

Christoph Stein, Ingo Schubert, and Georg H. Fey

4.1 Introduction

Antibodies carry two main functional domains, the antigen binding site and the Fc region, which recruit effector functions. The Fc-domain can be bound by a number of different types of Fc-receptors (FcRs) present on effector cells. This diversity has the advantage that multiple populations of effector cells can be recruited to carry out various effector functions, which overall benefit the host defense against infectious agents. However, the diversity carries a price, which becomes apparent when antibodies are used for cancer therapy. Here, a sizeable fraction of the injected dose never reaches the cancer cells, because it is absorbed by FcRs on a number of healthy cells. This often results not only in a net reduction of the effective dose, but also causes undesired side-effects, because some of the cells carrying FcRs modulate immune-responses. An example is FcγRIIb (CD32b), a low-affinity receptor for polymeric and aggregated immunoglobulin (Ig). Binding of antibodies by CD32b leads to a dampening of the B-cell response, and this can be one of the mentioned unfavorable effects. Therefore, investigators have tried to replace or to modify the Fc-domain, which permit them to address only the desired type of FcR on the effector cells, and thereby to reduce side effects. With this intent, bispecific antibody derivatives have been produced for the last two decades (Chames and Baty 2009; Hartmann et al. 1997; Muller and Kontermann 2007). However, the production of bispecific conjugates between full-length Igs or their Fab fragments presented major manufacturing problems, and therefore, this format has been abandoned. As a consequence, interest in bispecific antibodies was low for many years, and bispecific agents have only become more attractive again with the arrival of recombinant formats, which circumvent these manufacturing problems.

C. Stein • I. Schubert (✉) • G.H. Fey
Microbiology, Department of Biology, University Erlangen-Nuremberg, Staudtstrasse 5, 91058 Erlangen, Germany
e-mail: cstein@biologie.uni-erlangen.de; ischuber@biologie.uni-erlangen.de

R.E. Kontermann (ed.), *Bispecific Antibodies*,
DOI 10.1007/978-3-642-20910-9_4,

In the past 5 years, the interest in bispecific and trispecific agents has seen a great revival.

The renewed interest was largely spurred by the first successful application of a recombinant bispecific agent, Blinatumomab™, in clinical trials. This fusion protein between scFvs specific for CD19 on B-lymphoid cells and CD3 on T lymphocytes, respectively, was highly successful in clinical trials for B-cell lymphomas of adults (Bargou et al. 2008) and ongoing clinical studies for acute lymphoblastic leukemia (ALL) in children (Handgretinger et al. 2010; Topp et al. 2009).

Although bispecific agents of the Blinatumomab™ format present clear advantages, they still leave room for improvement. One of their shortcomings is a short plasma half-life, which in humans is on the order of 1 h. This requires application in multiple-repeat doses or continuous delivery, which can be inconvenient for the patient. The short plasma half-life of bsscFvs probably is due to a large extent to the excretion of the agents on first pass through the kidney. To improve the plasma half-life of bsscFvs, our team has designed sctbs, by extending bsscFvs through the addition of a third scFv, carried in the same polypeptide chain. This addition raised the mass to about 90 kDa, well above the kidney threshold, which should result in a prolonged plasma half-life. The expectation was born out, at least in mice, and the sctb format showed other favorable features. The main additional advantages were an increased avidity for the cancer cell, resulting in greatly reduced concentrations for cytotoxic activity, and the ability to design dual-targeting sctbs, carrying binding sites for two different antigens on the tumor cell. This latter feature generated attractive new applications of the format, which cannot be reached by bsscFvs, due to their monovalent binding to cancer cells. These new features have not yet been fully explored, and a rigorous proof of the benefit of dual targeting for cancer therapy is still missing. However, the initial results are encouraging and are therefore restated here.

4.2 The Single-Chain Triplebody Format

4.2.1 General Structure of Single-Chain Triplebodies

Sctbs (Fig. 4.1a) carry two specific binding sites for antigens on the cancer cell, and one for a trigger molecule on an effector cell, permitting to recruit them for the elimination of the cancer cell. In strict terminology, this cytotoxic effect is called "redirected lysis," although many authors also refer to it as antibody-dependent cellular cytotoxicity (ADCC). The linkers connecting the three scFvs are multiples of the (Gly4Ser) building blocks, mostly 20 amino acids in length. Computer-assisted modeling of the sctb with two linkers of 20 amino acids each predicted the distance between the two distal binding sites to be 20 nm in the fully extended configuration (Singer et al. 2010; Fig. 4.1b). However, we have also used linkers of

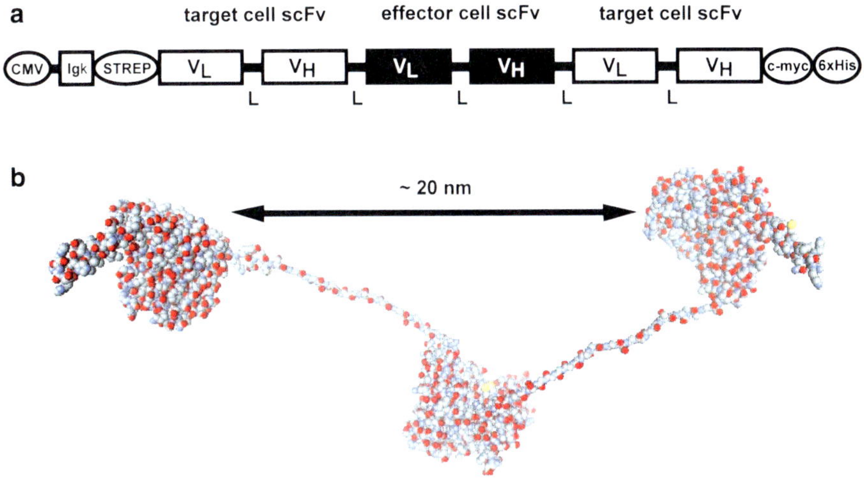

Fig. 4.1 Design of a recombinant single-chain triplebody. (**a**) Block-structure of the expression vector for sctbs. CMV, cytomegalovirus early promotor; Igk secretion leader sequence from the murine Ig kappa L chain; V_L,V_H, cDNA sequences coding for the V regions of Ig L- or H-chains; L, cDNA coding for a 20 amino acid flexible linker $(Gly_4Ser)_4$; Strep, c-myc, 6× His, cDNA coding for a strep, c-myc or a hexahistidine tag. (**b**) Graphic representation of the sctb 33-16-33 geometries as space-filling models. ScFv domains are geometry optimized homology models based on existing X-ray structures

10, 30 and 50 amino acids in length, and sctbs carrying these linkers were also stable and functionally active. Through the design we hoped to maximize the probability that one sctb molecule would simultaneously bind with both distal binding sites to the cancer cell, and at the same time to the effector cell and activate it for ADCC. In this manner, the central binding domain for the trigger molecule serves as a spacer between the two binding sites for tumor antigens. This arrangement should increase the probability for bivalent binding to the cancer cell, even for low-density tumor antigens. If the effector cell is a macrophage, we anticipate that sctbs will bind with their central binding site to a trigger molecule on the macrophage surface and activate it for phagocytosis, although this effect has not yet been directly studied. We have built sctbs carrying a scFv binding domain specific for CD64 (FcγRI), present on macrophages and cytokine-activated neutrophilic granulocytes. Using neutrophilic granulocytes from donors treated in vivo with cytokines to activate the granulocytes and to upregulate CD64, ADCC was observed in a small number of experiments. Similarly, one may try to place an scFv specific for CD89, the receptor for IgA (FcαR), which is constitutively expressed on neutrophilic granulocytes, and thus to recruit these cells as effectors. This has not yet been attempted, although we have worked with a bsscFv HLAII-CD89, carrying scFvs specific for HLA class II on malignant B-lymphoid cells and for CD89, and have observed strong ADCC effects using human granulocytes as effectors (Guettinger et al. 2010).

4.2.2 *All Three Binding Sites of a Single-Chain Triplebody are Functional*

The evidence supporting the claim that all three binding sites of a sctb are functional active is threefold. The first type of evidence comes from comparative measurements of the affinities and avidities of bsscFvs and sctbs for equilibrium binding to antigen-positive cells. The second comes from binding studies to all three antigens, in which one of the antigens was cell-bound, the other two in the fluid phase. The third type comes from ADCC experiments showing enhanced cytotoxic function of a sctb over the corresponding bsscFv controls. Examples of the first type of experiments were published for the sctbs 19-16-19 (Kellner et al. 2008), 33-16-33 (Singer et al. 2010), 33-16-19 (Schubert et al. 2011), and 123-16-33 (Kugler et al. 2010). For the prototype 19-16-19, the avidity of the sctb for CD19-positive cells was compared to the affinity of bsscFv 19-16. As a result, the affinity (K_D) of the sctb 19-16-19 for CD19-positive cells was 13.0 $\pm$ 1.2 nM, and the affinity of the bsscFv 19-16 for the same cell line was 42.4 $\pm$ 5.7 nM ($P = 0.0048$; Kellner et al. 2008). Thus, the sctb reached equilibrium binding to this cell line at approximately threefold lower molar concentrations than the bsscFv, and this is only possible, if both CD19 binding sites contributed to the overall binding. Similar data were obtained from a comparison between the sctb 33-16-33 and the bsscFv 33-16 (Singer et al. 2010). Here the difference in affinities was 3.5-fold, with K_D values for the sctb and the bsscFv for CD33 of 7.9 $\pm$ 1.1 nM and 28.9 $\pm$ 1.9 nM, respectively. This gain in avidity due to addition of the second binding site was not numerically large, but clearly existed and was of similar order of magnitude, approximately threefold, for both the sctbs 19-16-19 and 33-16-33 on different leukemia-derived cell lines. These data sets are consistent with each other and mutually reinforce confidence in the validity of the conclusion. From these data we conclude that, averaged over time, both distal binding sites of a sctb can simultaneously be engaged in binding to the same target cell. Similar data suggest that the same is also true for sctbs 123-16-33 (Kugler et al. 2010) and 33-16-19 (Schubert et al. 2011).

For the second type of experiment, recombinant fusion proteins were generated consisting of the extracellular domain of CD16 fused to a green fluorescent protein (CD16ex-GFP) and the extracellular domain of CD33, fused to a red-fluorescent protein (CD33ex-RFP). The CD19-positive and CD33-negative subclone of the SEM leukemia cell line was then incubated with the sctb 33-16-19 and subsequently with the two fluorescent reagents. Strong cell-bound red and green fluorescent signals were observed in a double-positive population of cells (Schubert et al. 2011). Similar data were obtained for the sctb 123-16-33 (Kugler et al. 2010). Here the fluorescent fusion protein CD123ex-RFP was used, consisting of the extracellular domain of CD123 fused to RFP. In this case, CD33-positive U937 leukemia cells were incubated with the sctb first, and then with CD123ex-RFP and CD16ex-GFP. Again a single population carrying both red and green fluorescent labels was observed. This result indicates that the sctb mediated binding CD123ex-RFP and CD16ex-GFP to the same target cell.

In cytotoxicity studies, the third type of experiments, the addition of a secon scFv specific for the target cell resulted in a shift of the dose–response curves toward lower concentrations in the low picomolar range. All investigated sctbs display not only an increase in avidity compared to the corresponding bsscFv but also a strong increase in cytotoxic activity. Currently, we are not able to explain this far greater-than-proportional gain in anti-tumor activity. From the aggregate of these data, we propose that all three binding sites of a single sctb can be simultaneously engaged and thereby contribute to increased ADCC function of the sctbs over the corresponding bsscFvs.

4.3 Trivalent, Bispecific Triplebodies: Mono-targeting Agents

4.3.1 Prototype 19-16-19 for the Treatment of B-Cell Malignancies

The best characterized mono-targeting sctb is the prototype 19-16-19, with two distal binding sites for CD19 and a central site for CD16 (Kellner et al. 2008), designed for use against CD19-positive B-lymphoid malignancies.

4.3.1.1 CD19 Is an Attractive Target for Antibody Therapy

CD19 is an attractive target for antibody therapy, because the antigen is present on a wide range of B-cell maturation stages, from early pro-B-cells to late mature B-cells, but is absent from plasma cells and hematopoietic stem cells (HSCs), and is restricted in its expression to the hematopoietic system. This latter point is important, because agents specific for CD19 would therefore spare HSCs and allow for hematopoietic reconstitution of the patients from their HSCs or from a stem cell transplant after the end of therapy. CD19 is expressed on a broad range of malignancies corresponding to different B-cell maturation stages, from early acute B-cell precursor leukemias (BCP-ALL), mostly of infants, children and young adults, over chronic B-cell leukemias (B-CLL) to B-cell lymphomas (NHL) of adults.

It is not yet clear whether leukemia stem cells (LSCs) of different types of B cell and BCP leukemias are CD19-positive, because LSCs for lymphoid leukemias are not yet as well defined as those for myeloid leukemias (Dick 2008). For some of these B-lineage malignancies the LSCs carry CD19 (le Viseur et al. 2008), in which case agents targeting CD19 would be very valuable. Elimination of the LSC compartment is an important objective in order to reach long-lasting therapeutic effects by eliminating minimal residual disease, which is probably due at least in part to LSCs. CD19 is a particularly useful target for the treatment of childhood for ~3/4 of these patients the blasts lack CD20 but are highly These patients therefore cannot benefit from CD20 antibodies, n CD19-specific agents (Handgretinger et al. 2010). In addition, on the surface of tumor cells is robust. The antigen is not shed d is also not down-modulated extensively after the cells are bodies.

4.3.1.2 Properties of CD16

CD16 is the low-affinity receptor for IgG, present on NK cells, macrophages, mast cells, neutrophilic granulocytes, and cytokine-stimulated dendritic cells (DCs, Nimmerjahn and Ravetch 2005). On human leukocytes, two forms of CD16 are present, CD16a and CD16b, products of two separate genes, which differ in their C-terminal portions. CD16a has a transmembrane domain and a cytoplasmic tail and can associate with either the γ chain of the FcεRI receptor or the T-cell antigen receptor ζ chain, or both (Nimmerjahn and Ravetch 2006). On human mast cells, CD16 is also associated with the FcεR β chain. These subunits carry cytoplasmic signaling domains and transmit the signals resulting from receptor occupation. CD16b carries no transmembrane domain and no cytoplasmic tail, but is attached to the membrane through a glycosyl-phosphatidylinositol (GPI) linker, this form does not assemble with the γ and ζ chains. The human b-form is expressed mainly on neutrophilic granulocytes. As granulocytes are approximately tenfold more abundant in human blood and tissues than NK cells and macrophages, they represent a sink for CD16 antibodies cross-reacting with both variants. Therefore, for therapeutic purposes, it would probably be an advantage, to use CD16a-specific agents only, as in the case of the tandem diabodies developed by the company Affimed, which are based on proprietary CD16a-specific scFvs. The scFvs used in our triplebodies are derived from the murine hybridoma 3G8 (Fleit 1991), and react with CD16a and b.

On human NK cells, the expression of CD16 is not constitutive but regulated. It is downregulated on aging NK cells and NK cells having exhausted their cytotoxic capacity by sequential degranulation events. Conversely, the CD16 density can be increased ex vivo by expansion of NK cells under optimized conditions (Sutlu and Alici 2009), and in vivo presumably by periods of recovery, cell division and by stimulation with appropriate cytokines and other signals.

4.3.1.3 Cytotoxic Effects Mediated by the Single-Chain Triplebody 19-16-19 for Leukemia-Derived Cell Lines

Sctb 19-16-19 was extensively tested in ADCC reactions in cell culture. The effector cells were mononuclear cells (MNCs), freshly prepared from unrelated healthy donors. Three different CD19-positive leukemia-derived cell lines were used as targets: SEM, BV-173, and ARH-77 (Fig. 4.2). SEM cells are derived from

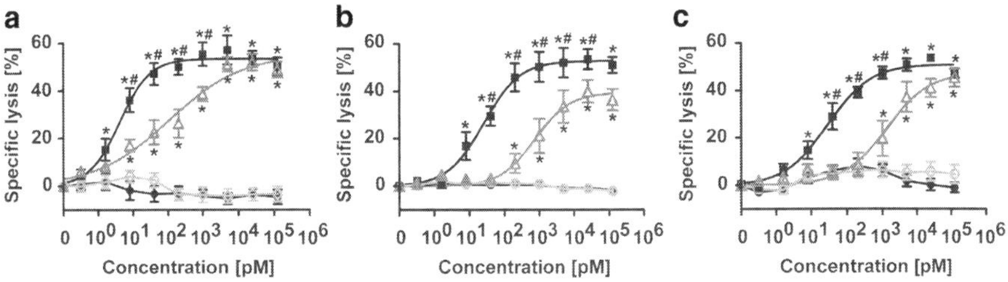

Fig. 4.2 Dose-dependent induction of ADCC of different tumor cell lines by the single-chain triplebody 19-16-19 and the bsscFv 19-16. The CD19-positive tumor cell lines SEM (**a**), BV-173 (**b**), and ARH-77 (**c**) were used as targets to compare efficacy of both antibody derivatives at a constant E:T cell ratio of 40:1. The sctb (*filled black square*) and the bsscFv (*open gray triangle*) triggered ADCC in a dose-dependent manner. Neither the nonrelevant control sctb (*filled black circle*) nor the nonrelevant control bsscFv (*open gray circle*) induced significant killing. Data points represent mean percentage of lysis ± SEM obtained with isolated MNCs from at least six different healthy donors. *Statistically significant differences in ADCC compared to the control without added protein. #Statistically significant differences between killing induced by sctb and bsscFv

a pediatric BCP-ALL with translocation t(4;11) to the MLL gene (Marschalek et al. 1995), BV-173 from a Philadelphia-positive, aggressive BCP-ALL with mixed-lineage phenotype, and ARH-77 is an Epstein–Barr virus positive B-lymphoblastoid cell line. These lines represent different disease entities and different maturation stages of B-lymphoid cells, the first two an immature pro-B-cell stage, the latter a mature B-cell stage. The effects of the sctb 19-16-19 were compared to those of the bsscFv 19-16 (Bruenke et al. 2005). On a mole-per-mole base, the sctb was far more effective than the bsscFv (Fig. 4.2a–c). The EC_{50} values for the sctb for the three lines derived from these curves were 4, 29 and 29 pM for SEM, BV-173 and ARH-77 cells, respectively. The corresponding values for the bsscFv were 113, 753 and 1,277 pM, respectively. Therefore, the sctb produced half-maximum cytotoxicity in these experiments at 28 , 26- and 44-fold lower concentrations than the bsscFv for these three target lines, respectively. These lines carry different densities of CD19 and probably differ in their spectrum of mutations contributing to the malignant phenotype, as well as in their cellular properties, reflecting their independent origin from different diseases. The fact that for all three targets the sctb achieved similar effectiveness as the bsscFv at 26- to 44-fold lower concentrations permits the expectation that the sctb may be generally more effective than the bsscFv for a broad range of B-cell malignancies.

4.3.1.4 Cytotoxic Effects of the Single-Chain Triplebody 19-16-19 for Primary Leukemic Cells

Sctb 19-16-19 was also tested for cytotoxic activity in similar experiments with primary cells from leukemia patients as targets (Fig. 4.3). Cells from peripheral blood and bone marrow of seven B-CLL patients, one patient with a mantle cell

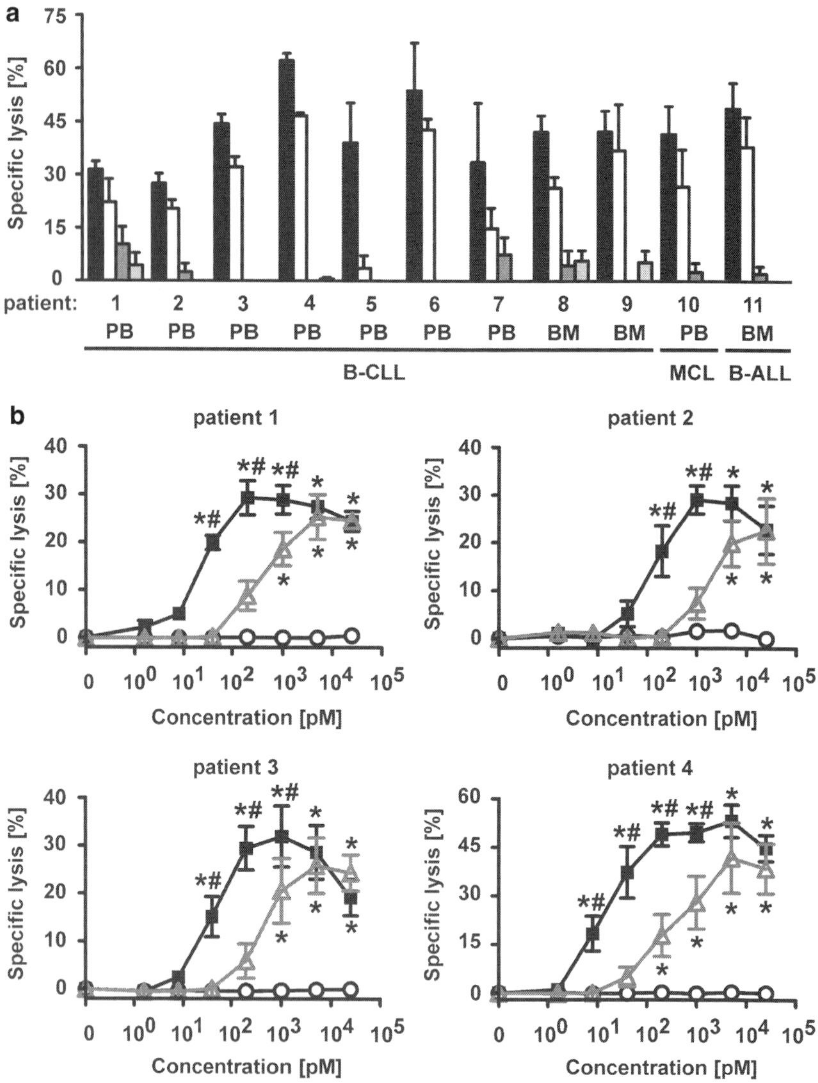

Fig. 4.3 Potent lysis of CD19-positive primary leukemia blasts and lymphoma cells by the single-chain triplebody 19-16-19 and the bsscFv 19-16. (**a**) Malignant cells from peripheral blood (PB) or bone marrow (BM) from nine B-CLL patients, and from one MCL and one B-ALL patient, were lysed by the sctb (*black bars*) and the bsscFv (*white bars*) at the concentration of 1 nM, using MNCs from a healthy donor at an E:T ratio of 40:1. No lysis was induced by a nonrelevant sctb (*dark gray bars*) or by MNCs alone (*light gray bars*). Data points are presented as mean values from triplicate determinations, error bars represent SEM. (**b**) Dose–response curves using cells from B-CLL patients 1–4 as targets, at an E:T ratio of 40:1. The sctb (*filled black square*) induced lysis with 10- to 20-fold lower EC_{50} values than the bsscFv (*open gray triangle*). No significant lysis was observed with a nonrelevant control sctb (*open circle*). Data points represent mean percentage of lysis ± SEM obtained with isolated MNCs from at least three different healthy donors. *Statistically significant differences in ADCC compared to the control without added protein. #Statistically significant differences between killing induced by the sctb and the bsscFv

lymphoma (MCL) and one B-ALL patient, were used. ADCC experiments were performed in comparison with the bsscFv 19-16. In all 11 samples tested, the sctb achieved a greater maximum lysis than the bsscFv (Fig. 4.3a). Dose–response curves were recorded for four B-CLL patients (Fig. 4.3b). For all four patients the dose–response curves were shifted toward lower effective concentrations by approximately one order of magnitude, and the EC_{50} values were about 11-fold lower for the sctb than for the bsscFv for one patient, and about 20-fold lower for the other three. These differentials are remarkably close to the values derived for leukemia-derived cell lines (Fig. 4.2), although those were monoclonal cell lines, whereas the patient cells used here were MNCs, including both healthy and malignant cells. The observation that the sctb scored so well for primary cells from B-CLL patients suggests that it may be useful to develop this agent for clinical use against both acute B-cell leukemias and chronic B-CLL.

4.3.1.5 Plasma Half-Life and Surface Retention of the Single-Chain Triplebody 19-16-19

One of the reasons to develop the triplebody format was the expectation that an improved plasma half-life may be reached by the addition of the extra scFv domain. Indeed, the bsscFv 19-16 had a plasma half-life in mice of 2 h, the sctb of 4 h (Fig. 4.4). The prolonged time is probably mainly due to the increased mass, with the mass of the bsscFv below and that of the sctb above the kidney exclusion limit.

Surface retention of the sctb and the bsscFv on CD19-positive leukemic cells in vitro were directly measured in separate experiments, and the sctb showed a sixfold increased surface retention relative to the bsscFv, but no difference was observed for retention of both molecules on CD16 transfected

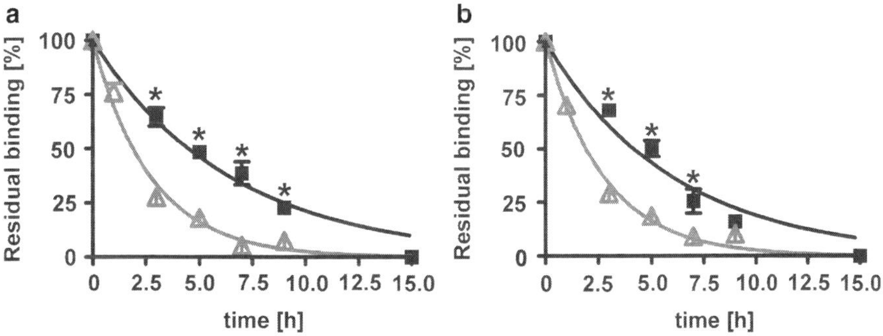

Fig. 4.4 In vivo plasma retention times. The sctb 19-16-19 (*filled black square*) and the bsscFv 19-16 (*open gray triangle*) were separately injected i.v. into NOD-SCID mice. The mice were bled at several time points and serum was analyzed for the presence of immunoreactive antibody derivatives by FACS using CD19-positive SEM (**a**) and CD16-transfected CHO cells (**b**). *Statistically significant differences between residual binding of both antibody derivatives

cells (Kellner et al. 2008). Analog results were obtained for sctbs with other specificity, independently if it was a mono- or dual-targeting sctb.

4.4 Dual-Targeting Single-Chain Triplebodies

4.4.1 Anticipated Benefits of Dual-Targeting: The Working Hypothesis

Our working hypothesis is that by virtue of dual-targeting, a preferential elimination of antigen double-positive cells over cells single-positive for only one of the two target antigens should be possible. In other words, dual-targeting is claimed to be able to produce a degree of selectivity for the elimination of double-positive cells. This would be important for cancer therapy, if it were possible to identify a pair of target antigens that is either uniquely present on the cancer cell, and only one of the pair is present on healthy normal cells of the patient, or for a pair, which is present in far greater surface density on the cancer cells than on normal cells. The first prototype of a dual-targeting sctb, 33-16-19, illustrates the principle. This agent is directed against the tumor antigens CD33 and CD19. Normal human hematopoietic cells carry only one of these antigens, not both simultaneously. By contrast, in a certain type of acute leukemias, the so-called "mixed lineage" or "biphenotypic leukemias," the blasts have both myeloid and lymphoid characteristics and simultaneously carry CD19 and CD33 on their surface. According to our hypothesis, a sctb 33-16-19 should therefore bind with higher avidity to the double-positive blasts than to single-positive healthy cells, to which it will only bind with monovalent affinity. It should therefore achieve a preferential elimination of the blasts over healthy cells. This claim sounds obvious, but this claimed selectivity is new. If the hypotheses were true, then dual-targeting would constitute important progress, because the cancer therapeutics currently available have no inbuilt preference for cancer cells over normal cells. The same is true for CD20 antibody Rituximab and Blinatumomab™. However, for many cancer cells it is often possible to identify a pair of antigens, which is either uniquely present on the cancer cells, such as CD19 and CD33 on mixed lineage leukemia blasts, or which is expressed on cancer cells in a greater combined density than on healthy normal cells. The probability of simultaneous binding of a sctb to such a pair is proportional to their combined densities, and the second prototype tries to take advantage of this property. This is the sctb 123-16-33. The target antigens CD123 and CD33 are both simultaneously present in normal myeloid cells and their progenitors, but CD123 is present in AML leukemia stem cells in five- to tenfold greater density than in normal human HSCs, normal myeloid cells, and the bulk of AML blasts (Jin et al. 2009; Jordan et al. 2000). Therefore, if the hypotheses were true, this sctb should permit a preferential elimination of AML leukemic stem cells (LSCs), an important objective for leukemia therapy. We have begun to test this hypothesis about the benefit of dual targeting, but the work is still in progress, and definitive proof of the hypothesis has still not been achieved. However, the initial data are encouraging and are summarized below.

4.4.2 *Prototype 123-16-33 Directed Against AML Cells*

One prototype of a dual-targeting sctb is 123-16-33 (Kugler et al. 2010), which was generated with the intent to preferentially target acute myeloid leukemia LSCs.

4.4.2.1 AML Stem Cells Are a Defined Subgroup with Distinct Immunophenotype

The intention underlying this design was to find a combination of two target antigens, which would be expressed on the bulk of AML cells, but which would be present in greater combined density on AML-LSCs than on the bulk and normal HSCs. If the working hypotheses were true, then a sctb directed against such a pair of target antigens should achieve a preferential elimination of the AML-LSCs, and possibly spare a few normal HSCs for hematopoietic reconstitution after the end of therapy. A number of antigens were considered for this purpose, including CD123, which is overexpressed on AML-LSCs (Jin et al. 2009; Jordan et al. 2000), CD96, which is exclusively expressed on the LSCs of a number of AML subtypes (Hosen et al. 2007), CLL-1, which was reported to be expressed exclusively on AML-LSCs but not on normal HSCs (van Rhenen et al. 2007), and CD33, which is present on a subset of AML-LSCs and a subset of normal HSCs (Taussig et al. 2005). The combination of CD33 and CD123 was chosen, because both of these antigens are clinically validated target antigens for the treatment of AML, and because we anticipated that due to this validation they presented the highest probability for the generation of a clinically useful agent. The glycoprotein CD33 is a member of the immunoglobulin superfamily and a useful target for antibody-derived therapeutics against AML (Sievers et al. 1999; Silla et al. 1995). CD33 is validated by the clinical results obtained with Mylotarg for AML (Zwaan et al. 2010). CD123 as a component of the interleukin-3 receptor is also a validated target, because a fusion protein between IL-3 and a fragment of diphtheria toxin (DT388) was successful in a phase I clinical study for AML and is currently tested in a phase II study (Feuring-Buske et al. 2002; Frankel et al. 2008). In addition, a CD123 antibody is currently tested in a phase I study for AML (Roberts et al. 2008).

4.4.2.2 Cytotoxic Effects of Single-Chain Triplebody 123-16-33 on Antigen Double-Positive Cell Lines

Sctb 123-16-33 was tested in ADCC assays in cell culture similar to those described above for sctb 19-16-19, except with CD123 and CD33 double-positive AML cells

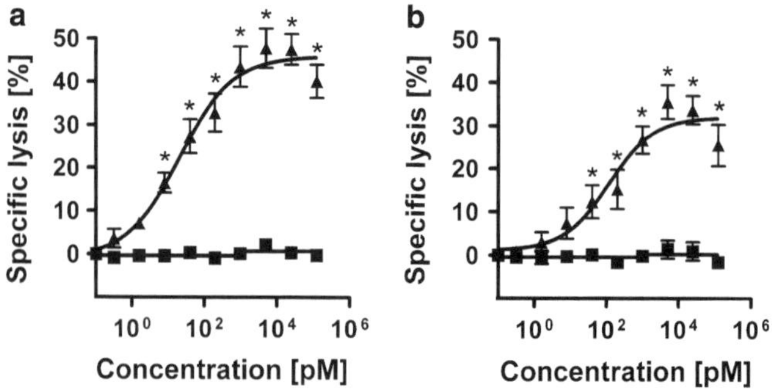

Fig. 4.5 Single-chain triplebody 123-16-33 induces ADCC of double-positive AML cell lines. Sctbs 123-16-33 (*closed triangle*) mediated dose-dependent ADCC of double-positive MOLM-13 (**a**) and THP-1 (**b**) cells, whereas a control sctb (*closed square*) did not induce cell lysis. Data points represent mean percentage of specific lysis obtained with isolated MNCs from six different healthy donors at an E:T ratio of 40:1. *Statistically significant differences ($P < 0.05$) in ADCC relative to the control sctb

as targets. The target cells were labeled with ^{51}Cr, and MNCs from unrelated healthy human donors were used as effectors. First, two established AML-derived cell lines were tested: MOLM-13 and THP-1 (Fig. 4.5). For both cell lines, the sctb produced strong ADCC. As a control, a sctb of identical format, but specific for CD7, was carried along in the same experiments in parallel. The EC_{50} values of the sctb were 21 and 118 pM for MOLM-13 (Fig. 4.5a) and THP-1 (Fig. 4.5b), respectively, and in the same range of other sctbs with similar format but different specificity. The data demonstrate that CD16-positive MNCs and purified NK cells can be used as effectors in vitro to combat AML cells by ADCC, and that the sctb was a highly effective molecule to mediate this effect (Kugler et al. 2010).

4.4.2.3 Cytotoxic Effects of Single-Chain Triplebody 123-16-33 on Primary Leukemic Cells

The sctb 123-16-33 has also been tested on primary cells from AML patients. Cells were prepared by density gradient centrifugation from primary material obtained from bone marrow and peripheral blood of seven untreated AML patients. The patients had different subtypes of AML according to FAB and WHO classification. MNCs from unrelated healthy donors were used as effectors at an E:T ratio of 40:1. No deliberate effort was made to avoid a histocompatibility match between effector and target cells. For direct comparison of the potencies of dual-targeting and mono-targeting sctbs in ADCC of primary leukemia cells, sctbs 123-16-123 and 33-16-33 were carried along. All three sctbs showed potent lysis of primary AML cells in a concentration-dependent manner (Fig. 4.6). Four representative ADCC reactions of samples from patients 1 and 2 (peripheral blood; Fig. 4.6a, b) and patients 7 and 6 (bone marrow; Fig. 4.6c, d) are depicted. At a concentration of 5 nM, all three recombinant proteins produced potent specific lysis (Fig. 4.6e). In four of eight samples, the dual-targeting sctb 123-16-33 reached the highest extent of lysis. This finding was confirmed when ADCC data from all six peripheral blood samples were

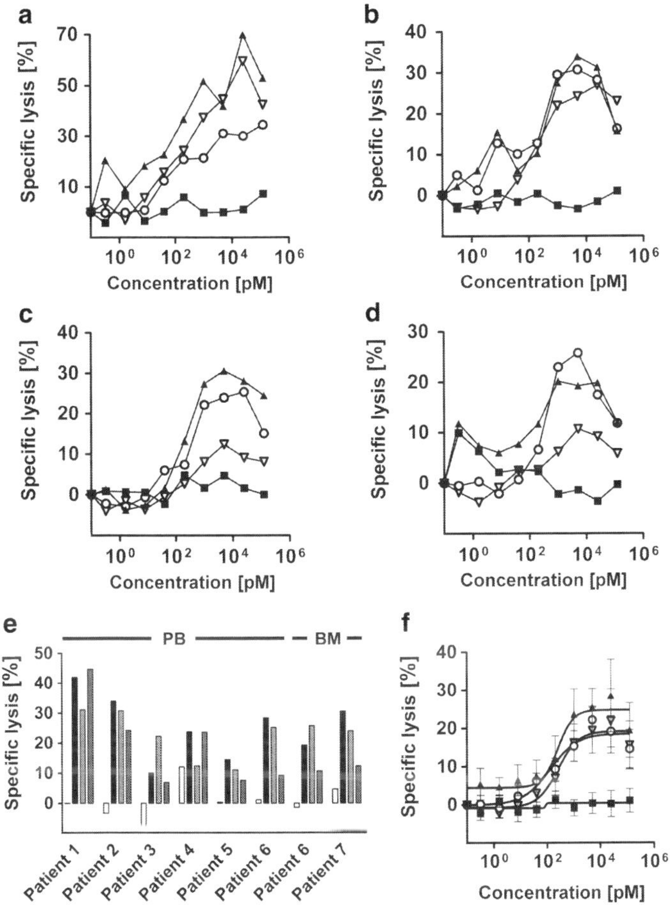

Fig. 4.6 Lysis of primary AML cells by single-chain triplebodies 123-16-33, 123-16-123, and 33-16-33. Sctbs 123-16-33 (*closed triangle*), 123-16-123 (*open circle*), and 33-16-33 (*open triangle*) mediated dose-dependent ADCC of primary AML cells, whereas a control sctb (*closed square*) failed to induce cellular lysis. (**a** and **b**) Induction of ADCC by sctbs of purified primary AML cells isolated from peripheral blood (patient 1 and 2). (**c** and **d**) Induction of ADCC by sctbs of purified primary AML cells isolated from bone marrow (patient 6 and 7). Data points represent percentage of specific lysis obtained with isolated MNCs from one healthy donor at an E:T ratio of 40:1. (**e**) Induction of ADCC by sctbs at a concentration of 5 nM for all eight samples. For patients 1–6, the cells analyzed were MNCs isolated from peripheral blood. For patients 6 and 7, bone marrow-derived cells were studied. *White bars*: control sctb; *black bars*: sctb 123-16-33; *light gray bars*: sctb 123-16-123; *dark gray bars*: sctb 33-16-33. (**f**) Induction of ADCC by sctbs of purified AML cells from peripheral blood of six different patients (patient 1–6), combined data. Data points represent mean percentage of specific lysis averaged over the six patients obtained with isolated MNCs from one healthy donor per patient sample at an E:T ratio of 40:1

combined (Fig. 4.6f). The EC_{50} values derived from this data set were ~250, ~130, and ~250 pM for the sctbs 123-16-33, 123-16-123 and 33-16-33, respectively. Maximum specific lysis was ~25%, ~19% and ~19% of input cells, respectively. Although the differences in maximum specific lysis between the dual-targeting sctb 123-16-33 and the mono-targeting sctbs 123-16-123 and 33-16-33 did not yet reach statistical significance, the dual-targeting molecule showed a remarkable tendency to mediate higher specific lysis of primary AML cells than its mono-targeting relatives. These results clearly demonstrate that the sctbs can induce potent ADCC of primary AML cells of different AML types in vitro, and point to small but distinct advantages for the dual-targeting agent over the corresponding mono-targeting agents.

It is not yet clear whether these agents will also be useful for applications against AML in vivo, because they have not yet been tested in experimental animals and humans. However, agents recruiting NK cells for the elimination of AML cells may be clinically attractive, because in a post-remission stage after a first round of chemotherapy, NK cells are present again in the periphery and the bone marrow, where most of the AML cells reside. It is considered unlikely that these agents would be useful for a front-line therapy, unless combined with other agents, because the numbers of AML blasts in an untreated patient by far exceed the number of available NK cells, and the NK cells will also be affected by the frontline chemotherapy. Therefore, if such agents ever reached clinical applications, they would most likely be used in a post-remission setting as adjuvant agents for consolidation therapy. In this setting they are attractive, because they may eliminate a fraction of MRD cells.

4.4.3 The Actual Benefits of Dual-Targeting Remain to Be Investigated

So far, our experiments have demonstrated that dual-targeting sctbs can be produced for a number of target antigens, that the three prototypes studied most extensively, 123-16-33, 33-16-19 and HLAII-16-19 (Schubert, unpublished data) all were able to bind to both antigens, and that both binding sites contributed to ADCC mediated by CD16-positive effector cells, mainly NK cells. This is a gratifying result, which was better than expected. We did not know before these experiments were performed, whether these antigens were in sufficient proximity on the surface of the cancer cells to be connected by a single sctb molecule. The distance between the two distal binding sites in our triplebodies is $\leq$20 nm (Singer et al. 2010), and it was not clear whether CD33 and CD19 on the surface of an MLL cell would come close to each other within 20 nm often enough, to permit simultaneous binding by a single sctb molecule and a measurable cytotoxic effect in an ADCC experiment with NK cells. The arrangement of these antigens on the surface of the leukemia cells is still unknown, but because an ADCC effect was observed, we must assume that the distribution was not random, but that these molecules must be clustered. There are currently three models for the arrangement of membrane proteins on the surface of human leukocytes. One is the fluid membrane model positing that proteins are arranged in a random fashion, but the membrane has sufficient fluidity to permit rapid lateral diffusion of all proteins. The second model is the lipid raft model positing that the membrane is a mosaic of rafts with distinct lipid composition, and that proteins are clustered in rafts, and that there are distinct marker proteins for different types of raft. This model is largely based on indirect evidence (density gradient centrifugation of membrane fractions) but not on direct visual analysis of intact membranes, and therefore it is not clear whether it provides a good description of an actual membrane. The leading model for human leukocytes is the "protein island model" (Lillemeier et al. 2010). It states that proteins are clustered in islands of 30–300 nm diameter, which have a distinct lipid composition, and that the islands are interspersed by large protein-free regions of different lipid composition. Within an island, the proteins can freely move by lateral diffusion, and entire islands can move in the membrane. Certain proteins home to distinct islands, and the raft-markers described above are found in distinct islands from the rest. Occasionally islands can come close to each other, and proteins can hop from one island to the next. Upon activation of lymphocytes, for example human T cells, proteins are redistributed and proteins located in separate islands in the nonactivated state can co-localize in the same island in the active state (Lillemeier et al. 2010). Therefore, it was not clear whether the target antigens listed above would co-localize in the same islands or segregate to different islands. The fact that all three dual-targeting sctbs were functional was therefore surprising, based on the protein island model. It suggests that these proteins are either located in the same islands, or that they may be located in different islands which come close to each other sufficiently often to permit their connection by one sctb, or that they hop from one island to the other. Alternatively, the protein island model may not be applicable at all to the membranes of leukemic cells, because these are likely to differ significantly from the membranes of normal cells. Viewed against this background, it was surprising and gratifying that the first three dual-targeting triplebodies constructed all were functionally active.

We have not yet achieved the objective to demonstrate the anticipated benefit of dual targeting, namely a preferential elimination of antigen double-positive cancer cells over single-positive normal cells or normal cells carrying the same pair of antigens in a lower combined density, but current efforts in our group are directed toward a rigorous test of this important claim.

4.5 Conclusion

Dual-targeting sctbs generated with the design shown in Fig. 4.1 are functional in ADCC tests with CD16-positive effector cells. They are capable of connecting two different target antigens on the same cell to a variable degree. The degree depends on the surface densities of the respective molecules, and probably also on their distribution in islands on the target cell, on their ability to co-localize in the same island or to hop from one island to the next. Adjustment of the linker length may be able to compensate for effects of surface density, although this has not yet been demonstrated directly. In the future, it may be possible to apply this design to a number of pairwise combinations of target antigens, not only on hematologic cancer cells, but possibly also on solid tumor cells. The format can accommodate binding sites for trigger molecules on other effector cells, and it will be interesting in the future to try recruiting macrophages, granulocytes, and T cells by molecules in this format. For different types of cancers, different effector cells may be suited best. Finally, sctbs may become useful not only for the treatment of malignant disorders, but possibly also for the treatment of chronic inflammatory and auto-immune disorders, as it has been the case for CD20 antibodies. The concept of dual-targeting is a major deviation from the mono-specific targeting of antibodies in current use for cancer therapy, and it may provide a qualitatively new means to selectively or preferentially target cancer stem cells, an important goal in tumor therapy.

References

Bargou R, Leo E, Zugmaier G, Klinger M, Goebeler M, Knop S, Noppeney R, Viardot A, Hess G, Schuler M, Einsele H, Brandl C, Wolf A, Kirchinger P, Klappers P, Schmidt M, Riethmuller G, Reinhardt C, Baeuerle PA, Kufer P (2008) Tumor regression in cancer patients by very low doses of a T cell-engaging antibody. Science 321(5891):974–977

Bruenke J, Barbin K, Kunert S, Lang P, Pfeiffer M, Stieglmaier K, Niethammer D, Stockmeyer B, Peipp M, Repp R, Valerius T, Fey GH (2005) Effective lysis of lymphoma cells with a stabilised bispecific single-chain Fv antibody against CD19 and FcgammaRIII (CD16). Br J Haematol 130(2):218–228

Chames P, Baty D (2009) Bispecific antibodies for cancer therapy: the light at the end of the tunnel? MAbs 1(6):539–547

Dick JE (2008) Stem cell concepts renew cancer research. Blood 112(13):4793–4807

Feuring-Buske M, Frankel AE, Alexander RL, Gerhard B, Hogge DE (2002) A diphtheria toxin-interleukin 3 fusion protein is cytotoxic to primitive acute myeloid leukemia progenitors but spares normal progenitors. Cancer Res 62(6):1730–1736

Fleit HB (1991) Monoclonal antibodies to human neutrophil Fc gamma RIII (CD16) identify polypeptide epitopes. Clin Immunol Immunopathol 59(2):222–235

Frankel A, Liu JS, Rizzieri D, Hogge D (2008) Phase I clinical study of diphtheria toxin-interleukin 3 fusion protein in patients with acute myeloid leukemia and myelodysplasia. Leuk Lymphoma 49(3):543–553

Guettinger Y, Barbin K, Peipp M, Bruenke J, Dechant M, Horner H, Thierschmidt D, Valerius T, Repp R, Fey GH, Stockmeyer B (2010) A recombinant bispecific single-chain fragment variable specific for HLA class II and Fc alpha RI (CD89) recruits polymorphonuclear neutrophils for efficient lysis of malignant B lymphoid cells. J Immunol 184(3):1210–1217

Handgretinger R, Zugmaier G, Henze G, Kreyenberg H, Lang P, von Stackelberg A (2010) Complete remission after blinatumomab-induced donor T-cell activation in three pediatric patients with post-transplant relapsed acute lymphoblastic leukemia. Leukemia 25:181–184

Hartmann F, Renner C, Jung W, Deisting C, Juwana M, Eichentopf B, Kloft M, Pfreundschuh M (1997) Treatment of refractory Hodgkin's disease with an anti-CD16/CD30 bispecific antibody. Blood 89(6):2042–2047

Hosen N, Park CY, Tatsumi N, Oji Y, Sugiyama H, Gramatzki M, Krensky AM, Weissman IL (2007) CD96 is a leukemic stem cell-specific marker in human acute myeloid leukemia. Proc Natl Acad Sci USA 104(26):11008–11013

Jin L, Lee EM, Ramshaw HS, Busfield SJ, Peoppl AG, Wilkinson L, Guthridge MA, Thomas D, Barry EF, Boyd A, Gearing DP, Vairo G, Lopez AF, Dick JE, Lock RB (2009) Monoclonal antibody-mediated targeting of CD123, IL-3 receptor alpha chain, eliminates human acute myeloid leukemic stem cells. Cell Stem Cell 5(1):31–42

Jordan CT, Upchurch D, Szilvassy SJ, Guzman ML, Howard DS, Pettigrew AL, Meyerrose T, Rossi R, Grimes B, Rizzieri DA, Luger SM, Phillips GL (2000) The interleukin-3 receptor alpha chain is a unique marker for human acute myelogenous leukemia stem cells. Leukemia 14(10):1777–1784

Kellner C, Bruenke J, Stieglmaier J, Schwemmlein M, Schwenkert M, Singer H, Mentz K, Peipp M, Lang P, Oduncu F, Stockmeyer B, Fey GH (2008) A novel CD19-directed recombinant bispecific antibody derivative with enhanced immune effector functions for human leukemic cells. J Immunother 31(9):871–884

Kugler M, Stein C, Kellner C, Mentz K, Saul D, Schwenkert M, Schubert I, Singer H, Oduncu F, Stockmeyer B, Mackensen A, Fey GH (2010) A recombinant trispecific single-chain Fv derivative directed against CD123 and CD33 mediates effective elimination of acute myeloid leukaemia cells by dual targeting. Br J Haematol 150(5):574–586

le Viseur C, Hotfilder M, Bomken S, Wilson K, Rottgers S, Schrauder A, Rosemann A, Irving J, Stam RW, Shultz LD, Harbott J, Jurgens H, Schrappe M, Pieters R, Vormoor J (2008) In childhood acute lymphoblastic leukemia, blasts at different stages of immunophenotypic maturation have stem cell properties. Cancer Cell 14(1):47–58

Lillemeier BF, Mortelmaier MA, Forstner MB, Huppa JB, Groves JT, Davis MM (2010) TCR and Lat are expressed on separate protein islands on T cell membranes and concatenate during activation. Nat Immunol 11(1):90–96

Marschalek R, Greil J, Lochner K, Nilson I, Siegler G, Zweckbronner I, Beck JD, Fey GH (1995) Molecular analysis of the chromosomal breakpoint and fusion transcripts in the acute lymphoblastic SEM cell line with chromosomal translocation t(4;11). Br J Haematol 90(2):308–320

Muller D, Kontermann RE (2007) Recombinant bispecific antibodies for cellular cancer immunotherapy. Curr Opin Mol Ther 9(4):319–326

Nimmerjahn F, Ravetch JV (2005) Divergent immunoglobulin g subclass activity through selective Fc receptor binding. Science 310(5753):1510–1512

Nimmerjahn F, Ravetch JV (2006) Fcgamma receptors: old friends and new family members. Immunity 24(1):19–28

Roberts AW, He S, Bradstock KF, Hertzberg MS, Durrant STS, Ritchie D, Lewis ID, Marlton P, McLachlan AJ, Yeadon TM, Busfield SJ, Barnden MJ, Davis R, Hosback S, Mirosa D, Biondo M, Bamford S, DeWitte M, Basser R (2008) A phase 1 and correlative biological study of CSL360 (anti-CD123 mAb) in AML. ASH annual meeting abstracts, 112, pp 2956–2964

Schubert I, Kellner C, Stein C, Kugler M, Schwenkert M, Saul D, Mentz K, Singer H, Stockmeyer B, Hillen W, Mackensen A, Fey GH (2011) A single-chain triplebody with specificity for CD19 and CD33 mediates effective lysis of mixed lineage leukemia cells by dual targeting. MAbs 3 (1):21–30

Sievers EL, Appelbaum FR, Spielberger RT, Forman SJ, Flowers D, Smith FO, Shannon-Dorcy K, Berger MS, Bernstein ID (1999) Selective ablation of acute myeloid leukemia using antibody-targeted chemotherapy: a phase I study of an anti-CD33 calicheamicin immunoconjugate. Blood 93(11):3678–3684

Silla LM, Chen J, Zhong RK, Whiteside TL, Ball ED (1995) Potentiation of lysis of leukaemia cells by a bispecific antibody to CD33 and CD16 (Fc gamma RIII) expressed by human natural killer (NK) cells. Br J Haematol 89(4):712–718

Singer H, Kellner C, Lanig H, Aigner M, Stockmeyer B, Oduncu F, Schwemmlein M, Stein C, Mentz K, Mackensen A, Fey GH (2010) Effective elimination of acute myeloid leukemic cells by recombinant bispecific antibody derivatives directed against CD33 and CD16. J Immunother 33(6):599–608

Sutlu T, Alici E (2009) Natural killer cell-based immunotherapy in cancer: current insights and future prospects. J Intern Med 266(2):154–181

Taussig DC, Pearce DJ, Simpson C, Rohatiner AZ, Lister TA, Kelly G, Luongo JL, Danet-Desnoyers GA, Bonnet D (2005) Hematopoietic stem cells express multiple myeloid markers: implications for the origin and targeted therapy of acute myeloid leukemia. Blood 106 (13):4086–4092

Topp MS, Goekbuget N, Kufer P, Zugmaier G, Klinger M, Degenhard E, Neumann S, Horst HA, Viardot A, Schmid M, Ottmann OG, Schmidt M, Reinhardt C, Baeuerle PA, Nagorsen D, Hoelzer D, Bargou R (2009) Blinatumomab (anti-CD19 bite) for targeted therapy of minimal residual disease (MRD) in patients with B precursor acute lymphoblastic leukemia (ALL): update of an ongoing phase II study. Hematologica 94:54–59

van Rhenen A, van Dongen GA, Kelder A, Rombouts EJ, Feller N, Moshaver B, Stigter-van Walsum M, Zweegman S, Ossenkoppele GJ, Jan Schuurhuis G (2007) The novel AML stem cell associated antigen CLL-1 aids in discrimination between normal and leukemic stem cells. Blood 110(7):2659–2666

Zwaan CM, Reinhardt D, Zimmerman M, Hasle H, Stary J, Stark B, Dworzak M, Creutzig U, Kaspers GJ (2010) Salvage treatment for children with refractory first or second relapse of acute myeloid leukaemia with gemtuzumab ozogamicin: results of a phase II study. Br J Haematol 148(5):768–776

Chapter 5
Diabodies, Single-Chain Diabodies, and Their Derivatives

Dafne Müller and Roland E. Kontermann

5.1 Introduction

The fragment variable (Fv) represents the smallest part of an antibody containing the entire antigen binding site formed by the variable heavy and light chain domains (V_H and V_L) (Filpula and McGuire 1999; Chames et al. 2009; Kontermann 2010). Because of the noncovalent interaction between the V_H and V_L domain, Fv molecules suffer from instability, which can be improved by connecting the two domains with a flexible peptide linker. These single-chain Fv (scFv) fragments can be easily produced in prokaryotic and eukaryotic systems and represent the prototype recombinant antibody format. The linker within scFv molecules must be long enough to span the distance between the C-terminus of the first domain and the N-terminus of the second domain. In the V_H–V_L arrangement this distance is ~3.5 nm, while in the V_L–V_H arrangement the distance is slightly larger (~4 nm). Thus, the standard linker is composed of 15 amino acid residues comprising three repeats of the sequence Gly-Gly-Gly-Gly-Ser [$(G_4S)_3$], allowing the formation of monomeric molecules through pairing of the V_H and V_L domain formed by one polypeptide chain. Size exclusion chromatography analyses of purified scFv molecules revealed that scFv molecules, e.g., produced in *Escherichia coli*, not only form monomeric, i.e., monovalent molecules but also assemble into dimeric molecules composed of two identical polypeptide chains leading to bivalent molecules (Essig et al. 1993; Whitlow et al. 1994; Arndt et al. 1998). In the early nineties, it was found that the formation of such bivalent molecules can be forced by reducing the linker length to approximately five residues, thus, preventing the pairing of the V_H and V_L domain within one chain (Holliger et al. 1993). These so-called diabodies (Db) possess a size of ~50 kDa and a rather compact structure as

D. Müller (✉) • R.E. Kontermann
Institut für Zellbiologie und Immunologie, Universität Stuttgart, Allmandring 31, 70569 Stuttgart, Germany
e-mail: dafne.mueller@izi.uni-stuttgart.de

R.E. Kontermann (ed.), *Bispecific Antibodies*,
DOI 10.1007/978-3-642-20910-9_5, © Springer-Verlag Berlin Heidelberg 2011

revealed by crystallographic studies (Perisic et al. 1994; Carmichael et al. 2003). Importantly, these studies showed that the two binding sites of a diabody are located at opposite sites facing away from each other (see also Fig. 5.1). Additional studies revealed that further reduction or complete removal of the linker results in the formation of homotrimeric and homotetrameric molecules (triabodies and tetrabodies) possessing a further increased size and valency (Iliades et al. 1997; Hudson and Kortt 1999; Todorovska et al. 2001; Kortt et al. 2001) (Fig. 5.1).

The standard linker connecting the V_H and V_L domain of a diabody has the sequence -GGGGS-, i.e., has a length of five residues (Holliger et al. 1993). However, also longer linkers comprising up to nine residues have been used for the generation of diabodies (Alfthan et al. 1995; Arndt et al. 1999; Wu et al. 1999). A detailed study employing libraries of single-chain diabodies (see below) containing linkers varying in sequence and size revealed that linkers of two to six residues and a preference for hydrophilic amino acids are applicable to generate functional molecules (Völkel et al. 2001).

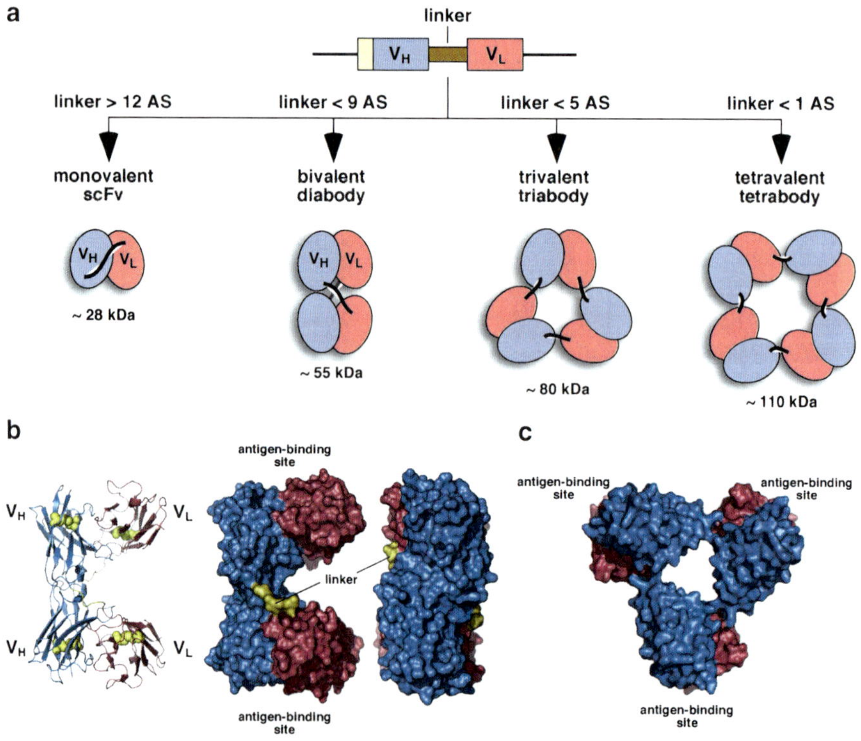

Fig. 5.1 (**a**) Generation of bivalent, monospecific diabodies as well as trivalent and tetravalent derivatives by the variation of the linker length. The signal sequence is shown in *light yellow*. (**b**) Structure of a bivalent diabody (Perisic et al. 1994; PDB entry 1LMK), (**c**) Structure of a trivalent triabody (Pei et al. 1997; PDB entry 1NQB)

5.2 Bispecific Diabodies and Single-Chain Diabodies

Besides the generation of bivalent, monospecific molecules, the diabody format can also be applied for the generation of bispecific molecules (Holliger et al. 1993). Because diabodies form by cross-over pairing of two chains in a head-to-tail orientation, the production of bispecific diabodies requires the expression of two polypeptide chains with either the arrangement $V_HA–V_LB$ and $V_HB–V_LA$ (HL configuration) or $V_LA–V_HB$ and $V_LB–V_HA$ (LH configuration), respectively, within the same cell (Fig. 5.2). However, although many bispecific diabodies have been successfully produced (see Table 5.1) (Kontermann 2005), the expression of two different chains within one cell results not only in heterodimeric and thus active molecules but also in the formation of inactive homodimeric molecules generated by the pairing of two identical chains, where the V_HA is paired with V_LB or the V_HB with V_LA. A strategy to facilitate correct pairing of the variable domains is provided by the genetic fusion of the two chains into a single-chain diabody (scDb) (Brüsselbach et al. 1999; Alt et al. 1999). The linker connecting the two chains has the same length as used for the generation of scFv molecules, i.e., ~15 residues. These scDb molecules have a similar size as tandem scFv molecules (taFv) generated by fusion of two scFv entities (see Chaps. 4 and 15), but differ by the arrangement of the variable domains and the length of the three linkers connecting the domains (Fig. 5.2). Thus, in a taFv, the two scFv represent separate building units linked by the middle linker, while in an scDb a compact structure is build by the four variable domains.

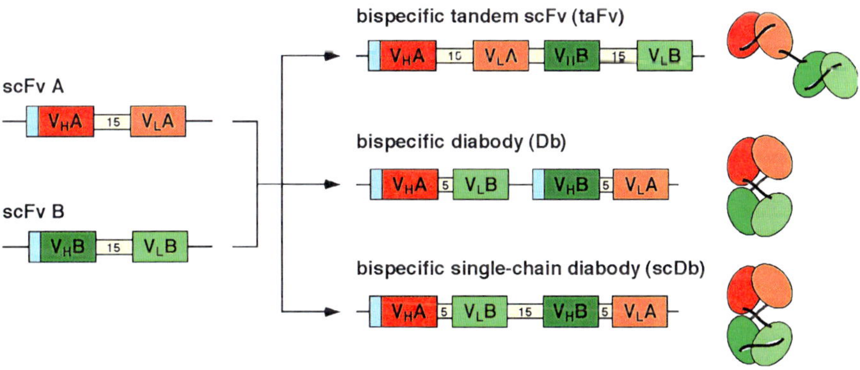

Fig. 5.2 Formation of bispecific antibodies from the variable domains of two antibodies, e.g., obtained from single-chain Fv molecules (scFv A and scFv B). Fusion of two different scFvs results in bispecific tandem scFv molecules (taFv). Bispecific diabodies (Db) are formed by the expression of two chains with the structure $V_HA–V_LB$ and $V_HB–V_LA$ (HL configuration) resulting in a heterodimeric molecule. Joining the two chains by an additional middle linker of 15 residues generates bispecific single-chain diabodies (scDb). The signal sequence is shown in *blue*

Table 5.1 Examples of diabodies, single-chain diabodies, and their derivatives for effector cell retargeting

Format	Specificity	Expression system	Application	References
Db	BCL-1 × CD3	*E. coli* (p)	In vitro	Holliger et al. (1996)
Db	Pgp × CD3	*E. coli* (p)	In vitro/in vivo	Gao et al. (2004) Guo et al. (2008)
dsDb	Pgp × CD3	*E. coli* (p)	In vitro/in vivo	Liu et al. (2009)
Db	EGP2 × CD3	*E. coli* (p)	In vitro	Helfrich et al. (1998)
Db	PSMA × CD3	*E. coli* (p)	In vitro/in vivo	Bühler et al. (2009) Bühler et al. (2008)
Db	CEA × CD3	*E. coli* (r)	In vitro	Asano et al. (2002)
Db	CEA × CD3	*E. coli* (p) 293T, PBL, MSC, HUVEC (s, in situ)	In vitro/in vivo	Holliger et al. (1999) Blanco et al. (2003) Compte et al. (2007, 2009, 2010)
scDb	CEA × CD3	*E. coli* (p) HEK293 (s)	In vitro	Müller et al. (2007)
scDb	EDG × CD3	*E. coli* (p)	In vitro	Korn et al. (2004b)
Db	erbB2 × CD3	*E. coli* (p)	In vitro	Zhu et al. (1996)
Db	EGFR × CD3	*E. coli* (r)	In vitro/in vivo	Hayashi et al. (2004)
Db-Fc scDb-Fc	EGFR × CD3	CHO (s)	In vitro	Asano et al. (2008)
Db	MUC1 × CD3	*E. coli* (r)	In vitro	Takemura et al. (2000)
SEA D227A-Db	MUC1 × CD3	*E. coli* (r)	In vitro/in vivo	Takemura et al. (2002)
Db	CD20 × CD3	*E. coli* (p)	In vitro/in vivo	Xiong et al. (2002) Liu et al. (2010)
Db	CD19 × CD3	*E. coli* (p)	In vitro/in vivo	Kipriyanov et al. (1998) Kipriyanov et al. (2002) Cochlovius et al. (2000b)
Tandem Db	CD19 × CD3	*E. coli* (p)	In vitro/in vivo	Kipriyanov et al. (1999) Cochlovius et al. (2000a), Reusch et al. (2004) Le Gall et al. (2004)
Db	CD30 × CD16	*E. coli* (p)	In vitro/in vivo	Arndt et al. (1999)
Db	CD19 × CD16	*E. coli* (p)	In vitro/in vivo	Kipriyanov et al. (2002) Schlenzka et al. (2004)
DART	CD32B × CD16	CHO (s)	In vitro/in vivo	Johnson et al. (2010)
Db	CD40 × CD28	*E. coli* (p)	In vitro	Houtenbos et al. (2008)

BCL-1 surface idiotype of the mouse lymphoma line BCL-1; *CEA* carcinoembryonic antigen; *CHO* Chinese hamster ovary cell line; *DART* dual-affinity retargeting; *Db* diabody; *dsDb* disulfide-stabilized diabody; *EDG* endoglin; *EGFR* epidermal growth factor receptor; *EGP2* epithelial glycoprotein 2; *erbB2* human epidermal growth factor receptor 2; *HUVEC* human umbilical venous cord endothelial cells; *MSC* mesenchymal stem cells; *MUC1* Mucin-1; *PBL* peripheral blood lymphocytes; *Pgp* P-glycoprotein; *PSMA* prostate-specific membrane antigen; *scDb* single-chain diabody; *SEA* D227A, staphylococcal enterotoxin A D227A mutant; *p* periplasmatic; *r* refolding; *s* secretion

5.3 Bispecific Diabody and Single-Chain Diabody Derivatives

The stability of bispecific diabodies was enhanced by introducing disulfide bonds covalently joining the two chains, which also helps to drive the correct pairing of the two different chains in bispecific diabodies (Fitzgerald et al. 1997) (Fig. 5.3).

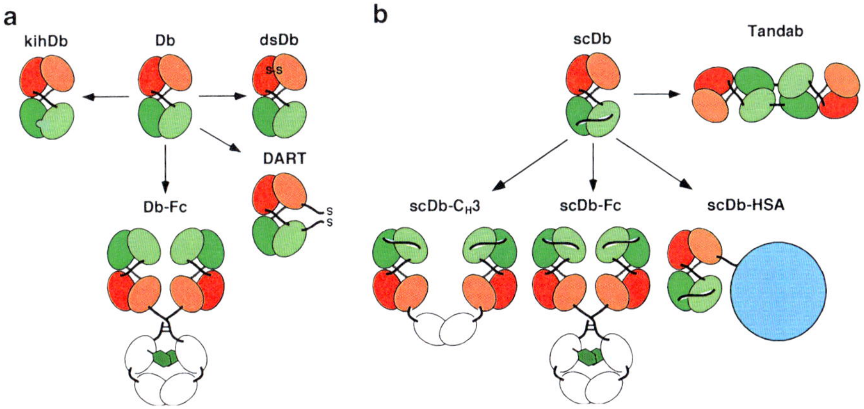

Fig. 5.3 Derivatives of bispecific diabodies and single-chain diabodies. (**a**) Derivatives of bispecific diabodies generated by introducing cysteine residues at the V_H–V_L interface (disulfide-stabilized diabodies, dsDb) or at the C-termini (dual-affinity retargeting, DART), or applying a knobs-into-holes strategy (kihDb). Bispecific diabodies have also been used to generate tetravalent molecules, e.g., by fusion to an Fc region. (**b**) Derivatives of bispecific single-chain diabodies. Tandab molecules are generated by reducing the length of the middle linker. Tetravalent scDb fusion proteins were generated by fusion of a C_H3 or Fc region to the C-terminus. Alternatively, other molecules, e.g., serum albumin, have been fused to scDb

The positions converted into cysteine residues are identical to those used to generate disulfide-stabilized Fv fragments (dsFv) (Reiter et al. 1996). For example, this strategy was applied to a bispecific diabody directed against p185^{HER2} and CD3 (Zhu et al. 1997). Cysteines were introduced at positions V_L L46C and V_H D101C and the diabody produced in *E. coli*. The purified protein consisted of >96% functional heterodimer compared with 83% of the parental diabody. Recently, Macrogenics presented a diabody derivative in which a short peptide tail containing a cysteine residue was added to the C-termini of the two different chains expressed in the LH configuration. The extensions had either the sequence LGGC or where derived from the IgG1 upper hinge (VEPKSC) and the kappa light chain (FNRGEC) (Johnson et al. 2010). These covalently linked structures were termed "dual-affinity re-targeting" (DART). Correct assembly into functional molecules was demonstrated, for example, for an anti-CD16 × anti-CD32 DART, which was stable under physiological conditions and highly potent in retargeting PBMCs to B-lymphoma cells. Diabodies were also stabilized applying the knobs-into-holes strategy targeting residues at the center of the V_L/V_H interface (Zhu et al. 1997) (Fig. 5.3). However, only some of the generated mutations exhibited an increased heterodimer assembly. In a similar approach, the V_L/V_H interface of a single-chain diabody format was engineered exchanging the hydrogen bonding between V_H39 and V_L38 with electrostatic interaction (Igawa et al. 2010).

Both the N- and C-termini of diabodies and single-chain diabodies are accessible for fusion of proteins and peptides. Monospecific diabodies have been fused, for example, to costimulatory molecules such as B7, resulting in homodimeric molecules possessing two identical binding sites and two fusion partners (Holliger et al. 1999; Blanco et al. 2003; Müller et al. 2007). Tetravalent and bispecific molecules were generated by fusion either bispecific Db or scDb to the IgG1 C_H3 domain or the Fc region. Db–C_H3 or Db–Fc fusion proteins are obtained by fusing the C_H3 domain or the Fc region (including the hinge region) to one of the two different chains (Lu et al. 2003; Lu et al. 2005) (Fig. 5.3). These fusion proteins are also termed "di-diabodies." However, as seen for diabodies, this approach also suffers from potential homodimerization of the two identical chains leading to inactive binding sites. This is avoided by employing single-chain diabodies fused at the C-terminus to the C_H3 domain or Fc region (Alt et al. 1999) (Fig. 5.3). In further studies, bispecific single-chain diabodies were also fused to other plasma proteins, e.g., serum albumin, resulting in molecules with strongly increased half-life (Müller et al. 2007).

Tetravalent bispecific molecules have been generated by varying the linker lengths of single-chain diabodies. Reducing the middle linker (see Fig. 5.2) connecting the V_HA–V_LB and V_HB–V_LA fragments from 15 residues to ~5 residues impedes assembly of the domains from one polypeptide chain and leads to the pairing of two chains into a dimeric single-chain diabody molecule in an antiparallel orientation (TandAbs) (Kipriyanov et al. 1999; Cochlovius et al. 2000a) (Fig. 5.3). Further experiments established that this dimerization takes place when the middle linker is reduced in length below nine residues and using flanking linkers composed of two to six residues (Völkel et al. 2001), although in another study tandAbs were also formed using a middle linker of 12 residues and flanking linkers of six or ten residues (Le Gall et al. 2004). Interestingly, it was found that dimers are also formed by chains possessing a middle linker of 13 or more residues but very short flanking linkers (zero or one residue), probably by assembly into a tetrabody-like circular structure (Völkel et al. 2001).

Diabodies have been mainly expressed in bacteria (*E. coli*). Production was conducted by secretion into the periplasmatic space, taking advantage of the oxidative environment that allows disulfide bound formation and therefore correct antibody folding. Alternatively, diabodies have been expressed cytoplasmatically as inclusion bodies, requiring a subsequent refolding step to generate functional molecules. In addition, diabodies have been produced in mammalian cells and secreted into the cell culture supernatant. To obtain purified recombinant protein, diabodies are usually equipped with tags (e.g., histidine-tag) and purified via respective affinity chromatography. Nevertheless, in the case of diabody transduced mammalian cells also in situ production has been reported. Originally, the periplasmatic bacterial production has been the method of choice for most of the diabody variants (ds-Db, scDb and tandem Db), but lately mammalian production is becoming more and more important as well (scDb, DARTs and Db/scDb–Fc) (see Tables 5.1 and 5.2).

5.4 Half-Life Extension Strategies

Several studies have shown that diabodies and single-chain diabodies are rapidly cleared from circulation owing to their small size and lack of FcRn-mediated recycling (Holliger et al. 1997; Müller et al. 2007). Therapeutic applications

Table 5.2 Examples of diabodies, single-chain diabodies, and their derivatives for effector molecule retargeting, dual targeting, gene therapy, and radioimmunotherapy

Format	Specificity	Expression system	Application	References
Retargeting of effector molecules				
Db	HEL × C1q	*E. coli* (p)	In vitro	Kontermann et al. (1997)
Db	HEL × LAM	*E. coli* (p)	In vitro	Holliger et al. (1997)
scDb	CEA × Gal	HEK293 (s)	In vitro	Brüsselbach et al. (1999)
scDb-C_H3 scDb-Fc	CEA × Gal	HEK293 (s)	In vitro	Alt et al. (1999)
Db	TAFI × PAI-1	*E. coli* (p)	In vitro	Develter et al. (2008)
Dual targeting				
Db	VEGFR2	*E. coli* (p)	In vitro	Lu et al. (1999)
Db	VEGFR2 × VEGFR3	*E. coli* (p)	In vitro	Jimenez et al. (2005)
Db	VEGFR1 × VEGFR2	*E. coli* (p)	In vitro	Lu et al. (2001)
Db–C_H3	VEGFR1 × VEGFR2	*E. coli* (p) COS (s)	In vitro	Lu et al. (2003)
Db	EGFR × IGFR	*E. coli* (p)	In vitro	Lu et al. (2004)
Db–Fc	EGFR × IGFR	NS0 (s)	In vitro/ in vivo	Lu et al. (2005)
DART	CD32B × CD79B	CHO (s)	In vitro/ in vivo	Veri et al. (2010)
Gene therapy (virus retargeting)				
scDb	EDG × Ad	*E. coli* (p)	In vitro	Nettelbeck et al. (2001)
scDb	CEA × Ad	*E. coli* (p)	In vitro	Korn et al. (2004a)
scDb	HMWMAA × Ad	*E. coli* (p)	In vitro	Nettelbeck et al. (2004)
scDb	erbB2 × Ad	HeLa (s)	In vitro	Kashentseva et al. (2009)
Radioimmunotherapy				
Db	HLA-DR × DOTA	*E. coli* (p)	In vitro	DeNardo et al. (2001)
Db	CEA × HSG	*E. coli* (p)	In vivo	Griffiths et al. (2004)

Ad Adenovirus fiber domain; *C1q* complement C1q; *CEA* carcinoembryonic antigen; *DART* dual-affinity retargeting; *Db* diabody; *scDb* single-chain diabody; *DOTA* 1,4,7,10-tetraazacyclo-dodecane-*N*,*N*′, *N*″,*N*‴-tetraacetic acid; *EDG* endoglin; *EGFR* epidermal growth factor receptor; *erbB2* human epidermal growth factor receptor 2; *Gal* β-galactosidase; *HEL* hen egg lysozyme; *HLA* human leukocyte antigen; *HMWMAA* high molecular weight melanoma-associated antigen; *HSG* histaminyl-succinyl-glycine; *IGFR* insulin-like growth factor receptor; *LAM* mouse IgMλ, IgG2aλ; *PAI-1* plasminogen activator inhibitor-1; *TAFI* thrombin activatable fibrinolysis inhibitor; *VEGFR* vascular endothelial growth factor receptor; *p* periplasmatic; *r* refolding; *s* secretion

require, therefore, infusions or repeated injections to maintain an effective dose over a prolonged period of time. Consequently, these molecules will benefit from strategies that extend the half-life and thus improve application and most likely also efficacy (Kontermann 2009). We have recently applied several half-life extension strategies to a bispecific single-chain diabody directed against carcinoembryonic antigen and CD3. These strategies included PEGylation, *N*-glycosylation, fusion to human serum albumin, and fusion to an albumin-binding domain (ABD) from streptococcal protein G (Müller et al. 2007; Stork et al. 2007; Stork et al. 2008). PEGylation was achieved by introducing an additional cysteine residue either in one of the flanking linkers or at the C-terminus of the scDb. This allowed for a side-directed and defined conjugation of a maleimide-functionalized 40 kDa branched PEG chain (Stork et al. 2008). Conjugation resulted in a strong increase of the hydrodynamic radius from 2.7 nm for the unmodified scDb to 7.9 nm for the PEGylated scDb. *N*-glycosylated derivatives were produced by introducing *N*-glycosylation sites (Asn-X-Thr) in the two flanking linkers as well as C-terminal tails containing 1, 4, or 7 *N*-glycosylation sites, respectively (Stork et al. 2008). Production in 293 cells resulted in *N*-glycosylated scDb molecules with an increased molecular mass, although SDS-PAGE and MALDI-MS analysis of the carbohydrates revealed a pronounced heterogeneity in respect to size and composition. Fusion to serum albumin increased the molecular mass from ~55 to 125 kDa resulting in an increase of the hydrodynamic radius to 3.9 nm. Fusion of an ABD possessing a molecular mass of 6 kDa enabled non-covalent binding of the scDb-ABD fusion protein to serum albumin, including human and mouse serum albumin (HSA, MSA), which increased the hydrodynamic radius to 4.8 nm. A comparative analysis of half-lives in mice showed that all modifications prolonged circulation time. Thus, the unmodified scDb exhibited a terminal half-life of 5–6 h, while that of PEGylated scDb, scDb-HSA, and scDb-ABD was increased two- to fourfold which resulted in an approximately 11-fold increased AUC_{0-24}. In contrast, the *N*-glycosylated derivatives showed only a moderate increase in half-life (6–9 h) and AUC (two- to threefold). The prolonged half-life also resulted in an increased accumulation in CEA-positive tumors, especially of the scDb-ABD fusion protein (Stork et al. 2009). Further studies with scDb-ABD derivatives exhibiting an increased or decreased affinity for albumin, or possessing two ABD fused to scDb, showed that affinity and valency have only a minor effect on half-life in mice (Hopp et al. 2010). In vitro studies with this particular scDb revealed for the modified molecules a reduced capacity to mediate target cell-dependent T-cell stimulation and cytotoxicity, indicating that the modifications and the interaction with albumin have negative effects on the interaction between target and effector cells.

5.5 Effector Cell Retargeting

Initially, the application of bispecific diabodies focused on effector cell retargeting in the context of cancer immunotherapy. According to this concept, simultaneous binding of the bispecific antibody to a tumor-associated antigen on the tumor cell and a trigger molecule on the immune effector cell leads to effector cell activation and consecutive tumor cell killing. The diabody format was designed to present the advantages of a small recombinant antibody molecule, lacking the Fc part, thus allowing better tumor penetration and more selective retargeting of immune cells, as well as avoiding side effects induced by the effector functions inherent of the Fc part (Segal et al. 1999). In addition, the recombinant expression offered a defined, more homogeneous production system than that provided by the hybridoma technology or chemically linkage strategy, facilitating purification and retrieval of higher yields of bispecific antibody molecules. Primarily, retargeting of T cells was attempted via the trigger molecule CD3 and tumor targeting was achieved by binding to diverse tumor-associated antigens, e.g., CEA, EGFR, PSMA, CD19, and CD20 (Table 5.1). For all of these constructs, diabody-mediated retargeting of activated T cells and tumor cell killing was demonstrated in vitro, confirming the feasibility of the strategy. Furthermore in several xenograft tumor mouse models, the application of the bispecific diabody together with preactivated human PBLs led to tumor growth inhibition and/or increased survival, thus indicating also in vivo efficacy. For example, a bispecific diabody DbCD3 × CD19 was reported to mediate in a concentration-dependent manner tumor cell lysis from a CLL patient by redirecting autologous T cell-mediated cytotoxicity in vitro. Thus at a E:T cell ratio of 10:1 more than 50% tumor cell lysis was achieved with a diabody concentration of 50 μg/ml. Furthermore, in a subcutaneous Raji Burkitt's lymphoma xenograft tumor mouse model, repeated administration of preactivated human PBLs followed by treatment with 50 μg bispecific diabody delayed tumor growth by approximately 20 days and increased the survival by around 5 weeks (Cochlovius et al. 2000a). In another case, bispecific diabody DbEGFR × CD3, retargeting preactivated T cells to EGFR-expressing TFK-1 cells, achieved 90% specific tumor cell killing at a concentration of 1 μg/ml and an E:T of 100:1. Also, in a subcutaneous TFK-1 xenograft tumor mouse model, repetitive treatment with 20 μg of this diabody, previously mixed with preactivated T cells, was started when tumor size reached 5 mm in diameter. Here, tumor growth could be suppressed up to 10 weeks at which in three of six mice complete tumor disappearance was observed (Hayashi et al. 2004).

Considering that under physiological conditions T-cell activation is dependent on appropriate costimulation, in several approaches the bispecific antibody in the diabody or single-chain diabody format was combined with a costimulatory signal. Therefore, a monoclonal antibody targeting the costimulatory receptor CD28 (Cochlovius et al. 2000b) or different antibody-ligand fusion proteins targeting costimulatory ligands such as B7 (Blanco et al. 2003; Müller et al. 2007) and 4-1BBL (Müller et al. 2008; Liu et al. 2010) to the tumor site were applied. Thus, enhanced T-cell activation was achieved as shown by increased cytokine release and cytotoxicity in vitro and delayed tumor growth in xenograft tumor mouse models in vivo. In another attempt to reinforce the antitumor response by T cells, a bispecific diabody (DbMUC1 × CD3) was fused to a mutated form of the superantigen staphylococcal enterotoxin A (SEA D227A). Also in this case, the antibody-mediated antitumor effects in a xenograft tumor mouse model were enhanced significantly in comparison to the bispecific diabody alone (Takemura et al. 2002).

Furthermore, retargeting immune effector cells other than T cells, e.g., NK cells, was achieved, generating bispecific diabodies with specificity for a leukemia-associated antigen (CD30, CD19) and the trigger molecule CD16 (Arndt et al. 1999;

Kipriyanov et al. 2002). Tumor growth inhibition and increased survival was reported in xenograft tumor mouse models. In addition, synergism in the antitumor response was observed for the combined application of DbCD19 × CD3 and DbCD19 × CD16 (Kipriyanov et al. 2002).

To improve the stability of the bispecific diabody, additional cysteine residues have been introduced either inside the variable regions themselves or at the end of the polypeptide chains to induce the covalent linkage of the dimer via disulphide bond formation. Thus, ds-Db (Pgp × CD3) was created introducing an additional cysteine residue in the variable regions of the CD3-specific antibody, respectively (Liu et al. 2009). Retaining full binding capacity and retargeting potential, highly improved serum stability was demonstrated in vitro. Furthermore, stabilization led to enhanced tumor targeting and twofold greater tumor growth inhibition in a xenograft tumor mouse model in vivo. According to the DART (dual-affinity re-targeting) design, dimerization takes place due to the cysteine residues localized at the C-terminus of the polypeptide chains. DART diabody CD16 × CD32B was generated, retargeting PBMCs to B-cell lymphoma cell lines (Johnson et al. 2010). High serum stability was demonstrated over several weeks. Also, in antibody-dependent cellular cytotoxicity (ADCC) assays, the cytotoxic potential of this diabody in comparison to an Fc-engineered CD32B-specific monoclonal antibody (increased CD16 binding) was shown to be far superior. In vivo, activity was further confirmed by B-cell depletion in transgenic mice expressing the respective human antigens CD16 and CD32B. Furthermore, antitumor response was analyzed in a Burkitt's lymphoma xenograft mouse model, monitoring survival. Here a single dose of 25 μg/mouse administrated weekly for 4 weeks led to complete protection by the Fc-enhanced monoclonal antibody and 87% protection by the DART antibody. Nevertheless, higher dose and treatment frequency of the DART antibody led also to complete tumor-free survival, indicating that the pharmacokinetic properties of these potent, small molecules have to be considered.

Other strategies aim to improve the potency of bispecific diabodies focusing on the generation of diabody forms with increased avidity. For example by modifying the expression conditions, non-covalent homodimerization of bispecific single-chain diabody chains can be induced, retrieving tetravalent tandem diabodies. Thus, tandem diabodies with CD3 × CD19 specificity were produced, which in compliance with the higher valency, showed stronger binding properties, leading in vitro to increased T cell retargeting in terms of proliferation and cytotoxicity of preactivated T cells. In vivo, due to the larger size longer blood retention was achieved (Kipriyanov et al. 1999). Nevertheless, in a Burkitt's lymphoma xenograft mouse model, the diabody and the tandem diabody (CD3 × CD19) achieved similar effects in terms of tumor regression and survival and only by coadministration of the costimulatory CD28-specific monoclonal antibody, a stronger antitumor response related to the tandem diabody could be observed (Cochlovius et al. 2000a). Ex vivo analysis with PBMCs from patients with B cell chronic lymphocytic leukemia, expressing high levels of costimulatory ligands (CD80 and CD86), showed tandem diabody (CD3 × CD19)-mediated T-cell proliferation and B-cell depletion, while the diabody and single-chain diabody (CD3 × CD19) showed no effect (Reusch et al. 2004). Interestingly, site-by-site comparison of the dimeric, tandem diabody (CD3 × CD19) and the monomeric BiTE, tandem scFv (CD3 × CD19), analyzing the antibody-mediated cellular cytotoxicity of unstimulated and preactivated PBMCs of healthy donors directed to several $CD19^+$ tumor cell lines, revealed an by several orders of magnitude stronger biological activity of the BiTE molecule (Mølhøj et al. 2007). Thus, highly efficient tumor cell killing by a monovalent bispecific antibody format is also feasible. Nevertheless, only the comparative evaluation of the antitumoral efficacy in vivo will, in the end, reveal which format might be more advantageous.

5.6 Effector Molecule Recruitment

Bispecific diabodies have also been applied in strategies for effector molecule recruitment. First approaches were inspired by the natural effector functions of monoclonal antibodies, seeking for alternative options to exploit these effector mechanisms by the upcoming recombinant antibody formats. Thus, for proof of principle, bispecific diabodies were generated with one binding side for hen egg lysozyme (HEL) as a model antigen and a second binding site for serum immunoglobulin (Ig) (Holliger et al. 1997) or complement C1 (Kontermann et al. 1997). In vitro, diabody-mediated retargeting of serum immunoglobulin to HEL-coated red blood cells induced complement recruitment, mononuclear phagocyte respiratory burst and phagocytosis. Similarly, diabody-mediated complement activation was achieved retargeting complement C1q to HEL-coated red blood cells.

Other strategy consisted in retargeting an enzyme to the tumor site to enable a localized enzyme-mediated prodrug conversion. Here, an scDb specific for the tumor associated carcinoembryonic antigen (CEA) and the enzyme β-galactosidase (Gal) were reported (Brüsselbach et al. 1999). In vitro, scDb-mediated recruitment of β-galactosidase to CEA-expressing tumor cells was shown, followed by the enzymatically activation of the prodrug into daunomycin and consecutive tumor cell killing.

Also implicating a two-step approach, but intended for radioimaging, a bispecific diabody binding to the CEA and a histaminyl-succinyl-glycine (HSG) hapten was generated (Griffiths et al. 2004). Here, first the bispecific antibody was injected into human colon tumor xenograft bearing athymic mice. Fifteen hours later, after unbound antibodies had been cleared, the peptide IMP 241 [DOTA-Phe-Lys(HSG)-D-Tyr-Lys(HSG)-NH_2] radiolabeled with ^{67}Ga was administered and tumor imaging performed by positron emission tomography (PET). Biodistribution of the radionuclide 3 h after the peptide IMP 241 injection retrieved a tumor-to-blood ratio of 137:1 and low background signal with a tumor-to-normal tissue ratio over 40:1 with the exception of the kidney (2,3:1). Thus, due to the rapid clearance from the blood, the bispecific diabody showed a considerably better performance than the corresponding $F(ab')_2$ molecule, in which case the maximal tumor-to-blood ratio determined was 12:1, while tumor-to-liver and tumor-to-lung ratios were 16:1 and 20:1, respectively.

5.7 Dual Targeting

The avidity of the bispecific diabody can be increased generating diabody– or single-chain diabody–Fc fusion proteins. Here, dimer formation occurs by covalent linkage of the Fc part in the hinge region, generating tetravalent, bispecific IgG-like antibody constructs (Fig. 5.3). Besides increased valency, improved pharmacokinetic properties and Fc-mediated effector functions come along with this format. Db–Fc and scDb–Fc with EGFR × CD3 specificity have been described in vitro (Asano et al. 2008). In comparison with the Db and scDb alone, they showed enhanced cancer cell growth inhibition retargeting preactivated T cells (10 fold) and PBMCs (100 fold). Thus, bispecific Db and scDb-mediated cytotoxicity can be enhanced significantly by increasing the avidity and introducing the additional antibody-dependent cellular cytotoxicity (ADCC) effector mechanism into this format. Nevertheless tumor target-dependent activation of the effector cells cannot be assured any more, as cross-linking of CD3 by the tetravalent antibody construct in solution might already be sufficient to activate T cells non-specifically and induce ADCC by Fc-receptor bearing cells in the absence of tumor target cells. Therefore, these formats are more likely to be applied to the dual targeting approach, where tumor growth inhibition is caused by enhanced antagonistic binding to two different tumor growth-related receptors. In vitro, bispecific diabodies targeting VEGFR1 and 2 have shown to be able to block the interaction of these receptors with the corresponding ligands VEGF and PlGF, inhibiting stronger than the parental antibodies the chemotactic and mitogenic activity of these ligands (Lu et al. 2001). Converting this diabody into a Db–C_H3 fusion protein resulted in a tetravalent bispecific molecule with stronger binding and blocking properties than the divalent bispecific diabody counterpart (Lu et al. 2003). Following the same principle, a Db–Fc fusion protein was reported, targeting EGFR and IGFR (Lu et al. 2005). Here, besides blocking receptor–ligand binding, ADCC and tumor growth inhibition in two xenograft mouse models was accomplished. Dual receptor targeting was clearly advantageous over single receptor targeting, as shown by the combination of the parental monoclonal antibodies. The capacity of the di-diabody to retain largely the functional properties of the individual parental antibodies and combine them into a single molecule confers it high potential for versatile applications, whereas the combinatorial effect is strongly influenced by the respective target constellation.

5.8 Viral Vector Retargeting and Generation of In Situ Producer Cells

Bispecific single-chain diabodies (scDb) have also been included in gene therapeutic approaches. By bridging adenoviral vector particles and target cells, scDb-mediated transduction showed improved selectivity and efficiency. This was observed for a bispecific scDbEDGxAd5 directed against the tumor vascular target endoglin (EDG) and the adenoviral (Ad5) capsid fiber knob domain that could selectively enhance (>sixfold) the adenovirus transduction in endoglin expressing HUVEC cells (Nettelbeck et al. 2001). The feasibility of this approach was also successfully confirmed in vitro, retargeting adenoviral vectors, e.g., to tumor cells expressing CEA, high molecular weight melanoma-associated antigen (HMWMAA) or erbB2 (Korn et al. 2004a; Nettelbeck et al. 2004; Kashentseva et al. 2009) (See also Chap. 18).

Another gene therapeutic approach proposes the generation and subsequent implantation of antibody producer cells into the patient, thus providing sustained concentrations of bispecific antibodies in the organism over an extended time period. Initially, 293T cells were genetically modified to secrete the bispecific diabody DbCEA × CD3, capable of retargeting T cells to CEA-expressing tumor cells. In vivo, irradiated producer cells were mixed with CEA-expressing human colon carcinoma cells and implanted in athymic nude mice. Once tumors became palpable, preactivated human T cells were injected intratumorally. In situ expression of the diabodies strongly inhibited the tumor growth (Blanco et al. 2003). In a similar mouse model, implantation of a mixture of tumor cells and activated human PBLs previously transduced to express DbCEA × CD3, led also to tumor growth inhibition (Compte et al. 2007). Other producer cells employed were human mesenchymal stem cells (MSCs) transduced with a lentiviral vector encoding DbCEA × CD3 that were seeded in a synthetic extracellular matrix scaffold and injected subcutaneously in immunodeficient mice at a distant location from the xenograft tumor implant. Next, preactivated human T cells were injected intravenously and tumor growth monitored. Plasma levels of the diabody were detectable for 6 weeks and therapeutic efficiency shown by tumor growth inhibition (Compte et al. 2009). Further development of this model used HUVECs as diabody producer cells embedded in matrigel together with MSCs, generating subcutaneous "vascular implants." Also here, significant plasma levels of the DbCEAxCD3 were measured over several weeks and tumor growth inhibition and enhanced survival were achieved (Compte et al. 2010). Nevertheless, inherent properties of the producer cells employed so far such as, e.g., short life span and low transduction efficiency (lymphocytes) or potential influence on tumor growth and metastasis at the tumor site (MSC) still constitute together with safety issues important restrictions for this approach in cancer therapy.

5.9 Conclusion

In vitro and in vivo studies have revealed the large and versatile application potential of diabodies, single-chain diabodies, and their derivatives, especially for cancer therapy. Although the proof of concept of these strategies are limited to preclinical studies so far, the advance of two BiTE (bispecific T cell engager) molecules of the tandem scFv format (taFv), into first clinical studies encourage further development in this direction. In fact, several companies, e.g., MacroGenics (DART), ImClone Systems (Db–Fc), and Affimed Therapeutics (tandAbs) focus on diabody derivatives in their product development. Thus, in the future, more bispecific diabody-related molecules are expected to find their way into the clinic.

References

Alfthan K, Takkinen K, Sizmann D, Söderlund H, Teeri TT (1995) Properties of a single-chain antibody containing different linker peptides. Protein Eng 8:725–731

Alt M, Müller R, Kontermann RE (1999) Novel tetravalent and bispecific IgG-like antibody molecules combining single-chain diabodies with the immunoglobulin g1 Fc or CH3 region. FEBS Lett 454:90–94

Arndt KM, Müller KM, Plückthun A (1998) Factors influencing the dimer to monomer transition of an antibody single-chain Fv fragment. Biochemistry 37:12918–12926

Arndt MA, Krauss J, Kipriyanov SM, Pfreundschuh M, Little M (1999) A bispecific diabody that mediates natural killer cell cytotoxicity against xenotransplantated human Hodgkin's tumors. Blood 94(8):2562–2568

Asano R, Kudo T, Nishimura Y, Makabe K, Hayashi H, Suzuki M, Tsumoto K, Kumagai I (2002) Efficient construction of a diabody using a refolding system: anti-carcinoembryonic antigen recombinant antibody fragment. J Biochem 132(6):903–909

Asano R, Kawaguchi H, Watanabe Y, Nakanishi T, Umetsu M, Hayashi H, Katayose Y, Unno M, Kudo T, Kumagai I (2008) Diabody-based recombinant formats of humanized IgG-like bispecific antibody with effective retargeting of lymphocytes to tumor cells. J Immunother 31(8):752–761

Blanco B, Holliger P, Vile RG, Alvarez-Vallina L (2003) Induction of human T lymphocyte cytotoxicity and inhibition of tumor growth by tumor-specific diabody-based molecules secreted from gene-modified bystander cells. J Immunol 171(2):1070–1077

Brüsselbach S, Korn T, Völkel T, Müller R, Kontermann RE (1999) Enzyme recruitment and tumor cell killing in vitro by a secreted bispecific single-chain diabody. Tumor Target 4:115–123

Bühler P, Wolf P, Gierschner D, Schaber I, Katzenwadel A, Schultze-Seemann W, Wetterauer U, Tacke M, Swamy M, Schamel WW, Elsässer-Beile U (2008) A bispecific diabody directed against prostate-specific membrane antigen and CD3 induces T-cell mediated lysis of prostate cancer cells. Cancer Immunol Immunother 57(1):43–52

Bühler P, Molnar E, Dopfer EP, Wolf P, Gierschner D, Wetterauer U, Schamel WW, Elsässer-Beile U (2009) Target-dependent T-cell activation by coligation with a PSMA x CD3 diabody induces lysis of prostate cancer cells. J Immunother 32(6):565–573

Carmichael JA, Power BE, Garrett TPJ, Yazaki PJ, Shively JE, Raubischek AA, Wu AM, Hudson PJ (2003) The crystal structure of an anti-CEA diabody assembled from T84.66 scFvs in VL-to-VH orientation: implications for diabody flexibility. J Mol Biol 326:341–361

Chames P, Van Regenmortel M, Weiss E, Baty D (2009) Therapeutic antibodies: successes, limitations and hopes for the future. Br J Pharmacol 157:220–233

Cochlovius B, Kipriyanov SM, Stassar MJ, Schuhmacher J, Benner A, Moldenhauer G, Little M (2000a) Cure of Burkitt's lymphoma in severe combined immunodeficiency mice by T cells, tetravalent CD3 x CD19 tandem diabody, and CD28 costimulation. Cancer Res 60(16): 4336–4341

Cochlovius B, Kipriyanov SM, Stassar MJ, Christ O, Schuhmacher J, Strauss G, Moldenhauer G, Little M (2000b) Treatment of human B cell lymphoma xenografts with a CD3 x CD19 diabody and T cells. J Immunol 165(2):888–895

Compte M, Blanco B, Serrano F, Cuesta AM, Sanz L, Bernad A, Holliger P, Alvarez-Vallina L (2007) Inhibition of tumor growth in vivo by in situ secretion of bispecific anti-CEA x anti-CD3 diabodies from lentivirally transduced human lymphocytes. Cancer Gene Ther 14(4):380–388

Compte M, Cuesta AM, Sánchez-Martín D, Alonso-Camino V, Vicario JL, Sanz L, Alvarez-Vallina L (2009) Tumor immunotherapy using gene-modified human mesenchymal stem cells loaded into synthetic extracellular matrix scaffolds. Stem Cells 27(3):753–760

Compte M, Alonso-Camino V, Santos-Valle P, Cuesta AM, Sánchez-Martín D, López MR, Vicario JL, Salas C, Sanz L, Alvarez-Vallina L (2010) Factory neovessels: engineered human blood vessels secreting therapeutic proteins as a new drug delivery system. Gene Ther 17(6):745–751

DeNardo DG, Xiong CY, Shi XB, DeNardo GL, DeNardo SJ (2001) Anti-HLA-DR/anti-DOTA diabody construction in a modular gene design platform: bispecific antibodies for pretargeted radioimmunotherapy. Cancer Biother Radiopharm 16(6):525–535

Develter J, Booth NA, Declerck PJ, Gils A (2008) Bispecific targeting of thrombin activatable fibrinolysis inhibitor and plasminogen activator inhibitor-1 by a heterodimer diabody. J Thromb Haemost 6(11):1884–1891

Essig NZ, Wood JF, Howard AJ, Raag R, Whitlow M (1993) Crystallization of single-chain Fv proteins. J Mol Biol 234:897–901

Filpula D, McGuire J (1999) Single-chain Fv designs for protein, cell and gene therapeutics. Exp Opin Ther Patents 9:231–245

Fitzgerald K, Holliger P, Winter G (1997) Improved tumour targeting by disulphide stabilized diabodies expressed in Pichia pastoris. Protein Eng 10:1221–1225

Gao Y, Xiong D, Yang M, Liu H, Peng H, Shao X, Xu Y, Xu C, Fan D, Qin L, Yang C, Zhu Z (2004) Efficient inhibition of multidrug-resistant human tumors with a recombinant bispecific anti-P-glycoprotein x anti-CD3 diabody. Leukemia 18(3):513–520

Griffiths GL, Chang CH, McBride WJ, Rossi EA, Sheerin A, Tejada GR, Karacay H, Sharkey RM, Horak ID, Hansen HJ, Goldenberg DM (2004) Reagents and methods for PET using bispecific antibody pretargeting and 68Ga-radiolabeled bivalent hapten-peptide-chelate conjugates. J Nucl Med 45(1):30–39

Guo H, Jiang W, Liu W, Gao Y, Yang M, Zhou Y, Wang J, Qi J, Cheng X, Zhu Z, Yang C, Xiong D (2008) Extracellular domain of 4-1BBL enhanced the antitumoral efficacy of peripheral blood lymphocytes mediated by anti-CD3 x anti-Pgp bispecific diabody against human multidrug-resistant leukemia. Cell Immunol 251(2):102–108

Hayashi H, Asano R, Tsumoto K, Katayose Y, Suzuki M, Unno M, Kodama H, Takemura S, Yoshida H, Makabe K, Imai K, Matsuno S, Kumagai I, Kudo T (2004) A highly effective and stable bispecific diabody for cancer immunotherapy: cure of xenografted tumors by bispecific diabody and T-LAK cells. Cancer Immunol Immunother 53(6):497–509

Helfrich W, Kroesen BJ, Roovers RC, Westers L, Molema G, Hoogenboom HR, de Leij L (1998) Construction and characterization of a bispecific diabody for retargeting T cells to human carcinomas. Int J Cancer 76(2):232–239

Holliger P, Prospero T, Winter G (1993) "Diabodies": small bivalent and bispecific antibody fragments. Proc Natl Acad Sci USA 90:6444–6448

Holliger P, Brissinck J, Williams RL, Thielemans K, Winter G (1996) Specific killing of lymphoma cells by cytotoxic T-cells mediated by a bispecific diabody. Protein Eng 9(3):299–305

Holliger P, Wing M, Pound JD, Bohlen H, Winter G (1997) Retargeting serum immunoglobulin with bispecific diabodies. Nat Biotechnol 15(7):632–636

Holliger P, Manzke O, Span M, Hawkins R, Fleischmann B, Qinghua L, Wolf J, Diehl V, Cochet O, Winter G, Bohlen H (1999) Carcinoembryonic antigen (CEA)-specific T-cell activation in colon carinoma induced by anti-CD3 x anti-CEA bispecific diabodies and B7 x anti-CEA bispecific fusion proteins. Cancer Res 59:2909–2916

Hopp J, Hornig N, Zettlitz KA, Schwarz A, Fuß N, Müller D, Kontermann RE (2010) The effects of affinity and valency of an albumin-binding domain (ABD) on the half-life of a single-chain diabody-ABD fusion protein. Protein Eng Des Sel 23:827–834

Houtenbos I, Santegoets S, Westers TM, Waisfisz Q, Kipriyanov S, Denkers F, Scheper RJ, de Gruijl TD, Ossenkoppele GJ, van de Loosdrecht AA (2008) The novel bispecific diabody alphaCD40/alphaCD28 strengthens leukaemic dendritic cell-induced T-cell reactivity. Br J Haematol 142(2):273–283

Hudson PJ, Kortt AA (1999) High avidity scFv multimers: diabodies and triabodies. J Immunol Meth 231:177–189

Igawa T, Tsunoda H, Kikuchi Y, Yoshida M, Tanaka M, Koga A, Sekimori Y, Orita T, Aso Y, Hattori K, Tsuchiya M (2010) VH/VL interface engineering to promote selective expression and inhibition of conformational isomerization of thrombopoeitin receptor agonist single-chain diabody. Protein Eng Des Sel 23:667–677

Iliades P, Kortt AA, Hudson PJ (1997) Triabodies: single chain Fv fragments wihtout a linker form trivalent trimers. FEBS Lett 409:437–441

Jimenez X, Lu D, Brennan L, Persaud K, Liu M, Miao H, Witte L, Zhu Z (2005) A recombinant, fully human, bispecific antibody neutralizes the biological activities mediated by both vascular endothelial growth factor receptors 2 and 3. Mol Cancer Ther 4(3):427–434

Johnson S, Burke S, Huang L, Gorlatov S, Li H, Wang W, Zhang W, Tuaillon N, Rainey J, Barat B, Yang Y, Jin L, Ciccarone V, Moore PA, Koenig S, Bonvini E (2010) Effector Cell recruitment with novel Fv-based dual-affinity re-targeting protein leads to potent tumor cytolysis and in vivo B-cell depletion. J Mol Biol 399(3):436–449

Kashentseva EA, Douglas JT, Zinn KR, Curiel DT, Dmitriev IP (2009) Targeting of adenovirus serotype 5 pseudotyped with short fiber from serotype 41 to c-erbB2-positive cells using bispecific single-chain diabody. J Mol Biol 388(3):443–461

Kipriyanov SM, Moldenhauer G, Strauss G, Little M (1998) Bispecific CD3 x CD19 diabody for T cell-mediated lysis of malignant human B cells. Int J Cancer 77(5):763–772

Kipriyanov SM, Moldenhauer G, Schuhmacher J, Cochlovius B, Von der Lieth CW, Matys ER, Little M (1999) Bispecific tandem diabody for tumor therapy with improved antigen binding and pharmacokinetics. J Mol Biol 293(1):41–56

Kipriyanov SM, Cochlovius B, Schäfer HJ, Moldenhauer G, Bähre A, Le Gall F, Knackmuss S, Little M (2002) Synergistic antitumor effect of bispecific CD19 x CD3 and CD19 x CD16 diabodies in a preclinical model of non-Hodgkin's lymphoma. J Immunol 169(1):137–144

Kontermann RE (2005) Recombinant bispecific antibodies for cancer therapy. Acta Pharmacol Sin 26(1):1–9

Kontermann RE (2009) Strategies to extend plasma half-lives of recombinant antibodies. BioDrugs 23(2):93–109

Kontermann RE (2010) Alternative antibody formats. Curr Opin Mol Ther 12(2):176–183

Kontermann RE, Wing MG, Winter G (1997) Complement recruitment using bispecific diabodies. Nat Biotechnol 15(7):629–631

Korn T, Nettelbeck DM, Völkel T, Müller R, Kontermann RE (2004a) Recombinant bispecific antibodies for the targeting of adenoviruses to CEA-expressing tumour cells: a comparative analysis of bacterially expressed single-chain diabody and tandem scFv. J Gene Med 6(6):642–651

Korn T, Müller R, Kontermann RE (2004b) Bispecific single-chain diabody-mediated killing of endoglin-positive endothelial cells by cytotoxic T lymphocytes. J Immunother 27(2):99–106

Kortt AA, Dolezal O, Power BE, Hudson PJ (2001) Dimeric and trimeric antibodies: high avidity scFvs for cancer targeting. Biomol Eng 18:95–108

Le Gall F, Reusch U, Little M, Kipriyanov SM (2004) Effect of linker sequences between the antibody variable domains on the formation, stability and biological activity of a bispecific tandem diabody. Protein Eng Des Sel 17:357–366

Liu J, Yang M, Wang J, Xu Y, Wang Y, Shao X, Yang C, Gao Y, Xiong D (2009) Improvement of tumor targeting and antitumor activity by a disulphide bond stabilized diabody expressed in *Escherichia coli*. Cancer Immunol Immunother 58(11):1761–1769

Liu R, Jiang W, Yang M, Guo H, Zhang Y, Wang J, Zhu H, Shi R, Fan D, Yang C, Zhu Z, Xie Y, Xiong D (2010) Efficient inhibition of human B-cell lymphoma in SCID mice by synergistic antitumor effect of human 4-1BB ligand/anti-CD20 fusion proteins and anti-CD3/anti-CD20 diabodies. J Immunother 33(5):500–509

Lu D, Kotanides H, Jimenez X, Zhou Q, Persaud K, Bohlen P, Witte L, Zhu Z (1999) Acquired antagonistic activity of a bispecific diabody directed against two different epitopes on vascular endothelial growth factor receptor 2. J Immunol Methods 230(1–2):159–171

Lu D, Jimenez X, Zhang H, Wu Y, Bohlen P, Witte L, Zhu Z (2001) Complete inhibition of vascular endothelial growth factor (VEGF) activities with a bifunctional diabody directed against both VEGF kinase receptors, fms-like tyrosine kinase receptor and kinase insert domain-containing receptor. Cancer Res 61(19):7002–7008

Lu D, Jimenez X, Zhang H, Atkins A, Brennan L, Balderes P, Bohlen P, Witte L, Zhu Z (2003) Di-diabody: a novel tetravalent bispecific antibody molecule by design. J Immunol Methods 279(1–2):219–232

Lu D, Jimenez X, Witte L, Zhu Z (2004) The effect of variable domain orientation and arrangement on the antigen-binding activity of a recombinant human bispecific diabody. Biochem Biophys Res Commun 318(2):507–513

Lu D, Zhang H, Koo H, Tonra J, Balderes P, Prewett M, Corcoran E, Mangalampalli V, Bassi R, Anselma D, Patel D, Kang X, Ludwig DL, Hicklin DJ, Bohlen P, Witte L, Zhu Z (2005) A fully human recombinant IgG-like bispecific antibody to both the epidermal growth factor receptor and the insulin-like growth factor receptor for enhanced antitumor activity. J Biol Chem 280 (20):19665–19672

Mølhøj M, Crommer S, Brischwein K, Rau D, Sriskandarajah M, Hoffmann P, Kufer P, Hofmeister R, Baeuerle PA (2007) CD19-/CD3-bispecific antibody of the BiTE class is far superior to tandem diabody with respect to redirected tumor cell lysis. Mol Immunol 44(8):1935–1943

Müller D, Karle A, Meissburger B, Höfig I, Stork R, Kontermann RE (2007) Improved pharmacokinetics of recombinant bispecific antibody molecules by fusion to human serum albumin. J Biol Chem 282(17):12650–12660

Müller D, Frey K, Kontermann RE (2008) A novel antibody-4-1BBL fusion protein for targeted costimulation in cancer immunotherapy. J Immunother 31(8):714–722

Nettelbeck DM, Miller DW, Jérôme V, Zuzarte M, Watkins SJ, Hawkins RE, Müller R, Kontermann RE (2001) Targeting of adenovirus to endothelial cells by a bispecific single-chain diabody directed against the adenovirus fiber knob domain and human endoglin (CD105). Mol Ther 3(6):882–891

Nettelbeck DM, Rivera AA, Kupsch J, Dieckmann D, Douglas JT, Kontermann RE, Alemany R, Curiel DT (2004) Retargeting of adenoviral infection to melanoma: combining genetic ablation of native tropism with a recombinant bispecific single-chain diabody (scDb) adapter that binds to fiber knob and HMWMAA. Int J Cancer 108(1):136–145

Pei XY, Holliger P, Murzin AG, Williams RL (1997) The 2.0-Å resolution crystal structure of a trimeric antibody fragment with noncognate VH-VL domain paris shows a rearrangement of VH CDR3. Proc Natl Acad Sci USA 94:9637–9642

Perisic O, Webb PA, Holliger P, Winter G, Williams RL (1994) Crystal structure of a diabody, a bivalent antibody fragment. Structure 2:1217–1226

Reiter Y, Brinkmann U, Lee B, Pastan I (1996) Engineering antibody Fv fragments for cancer detection and therapy: disulfide-stabilized Fv fragments. Nat Biotechnol 14:1239–1245

Reusch U, Le Gall F, Hensel M, Moldenhauer G, Ho AD, Little M, Kipriyanov SM (2004) Effect of tetravalent bispecific CD19xCD3 recombinant antibody construct and CD28 costimulation on lysis of malignant B cells from patients with chronic lymphocytic leukemia by autologous T cells. Int J Cancer 112(3):509–518

Schlenzka J, Moehler TM, Kipriyanov SM, Kornacker M, Benner A, Bähre A, Stassar MJ, Schäfer HJ, Little M, Goldschmidt H, Cochlovius B (2004) Combined effect of recombinant CD19 x CD16 diabody and thalidomide in a preclinical model of human B cell lymphoma. Anticancer Drugs 15(9):915–919

Segal DM, Weiner GJ, Weiner LM (1999) Bispecific antibodies in cancer therapy. Curr Opin Immunol 11:558–562

Stork R, Müller D, Kontermann RE (2007) A novel tri-functional antibody fusion protein with improved pharmacokinetic properties generated by fusing a bispecific single-chain diabody with an albumin-binding domain from streptococcal protein G. Protein Eng Des Sel 20(11): 569–576

Stork R, Zettlitz KA, Müller D, Rether M, Hanisch FG, Kontermann RE (2008) *N*-glycosylation as novel strategy to improve pharmacokinetic properties of bispecific single-chain diabodies. J Biol Chem 283(12):7804–7812

Stork R, Campigna E, Robert B, Müller D, Kontermann RE (2009) Biodistribution of a bispecific single-chain diabody and its half-life extended derivatives. J Biol Chem 284(38):25612–25619

Takemura S, Asano R, Tsumoto K, Ebara S, Sakurai N, Katayose Y, Kodama H, Yoshida H, Suzuki M, Imai K, Matsuno S, Kudo T, Kumagai I (2000) Construction of a diabody (small recombinant bispecific antibody) using a refolding system. Protein Eng 13(8):583–588

Takemura S, Kudo T, Asano R, Suzuki M, Tsumoto K, Sakurai N, Katayose Y, Kodama H, Yoshida H, Ebara S, Saeki H, Imai K, Matsuno S, Kumagai I (2002) A mutated superantigen SEA D227A fusion diabody specific to MUC1 and CD3 in targeted cancer immunotherapy for bile duct carcinoma. Cancer Immunol Immunother 51(1):33–44

Todorovska A, Roovers RC, Dolezal O, Kortt AA, Hoogenboom HR, Hudson RJ (2001) Design and application of diabodies, triabodies and tetrabodies for cancer targeting. J Immunol Meth 248:47–66

Veri MC, Burke S, Huang L, Li H, Gorlatov S, Tuaillon N, Rainey GJ, Ciccarone V, Zhang T, Shah K, Jin L, Ning L, Minor T, Moore PA, Koenig S, Johnson S, Bonvini E (2010) Therapeutic control of B cell activation via recruitment of Fcgamma receptor IIb (CD32B) inhibitory function with a novel bispecific antibody scaffold. Arthritis Rheum 62(7): 1933–1943

Völkel T, Korn T, Bach M, Müller R, Kontermann RE (2001) Optimized linker sequences for the expression of monomeric and dimeric bispecific single-chain diabodies. Protein Eng 14:815–823

Whitlow M, Filpula D, Rollence ML, Feng S-L, Wood JF (1994) Multivalent Fvs: characterization of single-chain Fv oligomers and preparation of a bispecific Fv. Protein Eng 7:1017–1026

Wu AM, Williams LE, Zieran L, Padma A, Sherman M, Bebb GG, Odom-MAryon T, Wong JYC, Shively JE, Raubitschek AA (1999) Anti-carcinoembryonic antigen (CEA) diabody for rapid tumor targeting and imaging. Tumor Target 4:47–58

Xiong D, Xu Y, Liu H, Peng H, Shao X, Lai Z, Fan D, Yang M, Han J, Xie Y, Yang C, Zhu Z (2002) Efficient inhibition of human B-cell lymphoma xenografts with an anti-CD20 x anti-CD3 bispecific diabody. Cancer Lett 177(1):29–39

Zhu Z, Zapata G, Shalaby R, Snedecor B, Chen H, Carter P (1996) High level secretion of a humanized bispecific diabody from Escherichia coli. Biotechnology 14(2):192–196

Zhu Z, Presta LG, Zapata G, Carter P (1997) Remodeling domain interfaces to enhance heterodimer formation. Protein Sci 6:781–788

Chapter 6
Bispecific Single Domain Antibodies

Patrick Chames and Daniel Baty

6.1 Introduction

Monoclonal antibodies are now widely recognized as therapeutic molecules and more than 25 molecules have been approved in the United States and other countries (Reichert 2010). These therapeutic achievements are the results of years of antibody engineering efforts aiming at decreasing the immunogenicity of these molecules and increasing their possibility to recruit the human effector cells of the immune systems. Despite very significant progresses, especially in cancer therapy, the clinical activity of these molecules is still far from optimal and new solutions have to be found. The creation of bispecific antibodies (bsAbs), capable of simultaneous binding to two different targets and thus capable of retargeting a large variety of payloads to cancer cells might be one the most promising solution to this issue (Chames and Baty 2009a, b). The potential of this approach has been demonstrated by several studies over the years but the difficulty to produce large amounts of homogenous preparation of bsAb using the available techniques (hybrid hybridomas, chemical cross-linking) has hindered a wider development of this approach. With the development of antibody engineering, several new recombinant formats have been designed and validated to a certain extent. These formats include tandem scFv, diabodies, tandem diabodies, dual variable domain antibodies, or heterodimerization using a motif such as the Dock and Lock motif. While most of these formats are characterized by various advantages, none of them has met so far all the requirements to become a standard format for clinical applications. Single domain antibodies, being the smallest fully functional antibody fragments, and characterized by many outstanding physical properties, might accelerate the way toward this goal.

P. Chames (✉) • D. Baty
INSERM U624, 163 avenue de Luminy – case 915, 13288 Marseille, Cedex 09, France
e-mail: patrick.chames@inserm.fr

R.E. Kontermann (ed.), *Bispecific Antibodies*,
DOI 10.1007/978-3-642-20910-9_6, © Springer-Verlag Berlin Heidelberg 2011

6.2 Single Domain Antibodies: Ideal Building Blocks

In a rather visionary work published in 1989, Sally Ward and collaborators demonstrated the possibility to reach affinities in the 20 nM range for an antigen using single variable domains isolated by phage display from small immune libraries of murine IgG heavy chain variable domains (VH) (Ward et al. 1989). The authors coined the name dAb for single domain antibodies for these molecules. In this paper, they already mention that such fragments could penetrate tissues more readily and could block "canyon" sites on some proteins. This discovery remained anecdotical until 1993, when Hamers and collaborators discovered the presence of antibodies devoid of light chain and first heavy constant domain in the blood of camels (Hamers-Casterman et al. 1993). These antibodies, named heavy chain antibodies (HcAb), also found in other camelids (dromedaries, llamas, and alpacas), bind their epitopes using a single variable domain from the heavy chain, called VHH (Fig. 6.1). Interestingly, these variable domains can be very efficiently produced in *E. coli* or in yeast. They are then named single domain antibodies (sdAb) or nanobodies. These sdAbs were subsequently shown to be endowed with outstanding physical properties making them very attractive as binding unit in antibody engineering for immunotherapy. First of all, despite their reduced paratope, these antibody fragments reach affinities typical for regular mAbs and are also capable of binding small molecules and haptens (Spinelli et al. 2000; Alvarez-Rueda et al. 2007; Behar et al. 2008). To do so, they often use a longer hypervariable loop (CDR3 for complementarity determining region 3) compared to regular mAbs, which also allow them to bind to buried cavities at the surface of their antigens (De Genst et al. 2006). Consequently, sdAbs are a good source of enzyme inhibitors and can yield antibodies against antigens which are invisible to conventional antibodies (Lauwereys et al. 1998). The single domain nature of these fragments permits the amplification and subsequent straightforward cloning of their genes, without requiring the use of artificial linker peptide (for single chain Fv fragments) or bi-cistronic constructs (for Fab fragments). This feature further allows direct cloning of immunized repertoire without being concerned by the usual disruption of VH/VL pairing faced by scFv and Fab fragments. It is also probably responsible for the high production yield reached by these fragments when expressed in *E. coli*. Moreover, sdAbs show extreme refolding capabilities and physical stability (Dumoulin et al. 2002; Dolk et al. 2005), which make them very attractive as detecting or targeting reagents. In fact, their high intrinsic stability allows them to fold properly in reducing environments that do not allow disulfide bond formation such as cell cytoplasm or nucleus. Consequently, and unlike conventional antibody fragments, most sdAbs can efficiently be used as intracellular antibodies (intrabodies) to probe or block a specific epitope of an intracellular protein (Muyldermans 2001; Gueorguieva et al. 2006; Rothbauer et al. 2006; Verheesen et al. 2006; Groot et al. 2008; Serruys et al. 2009).

Last but not least, the genes of these fragments show a large degree of homology with the VH3 subset of human VH genes (Ewert et al. 2002; Su et al. 2002), which

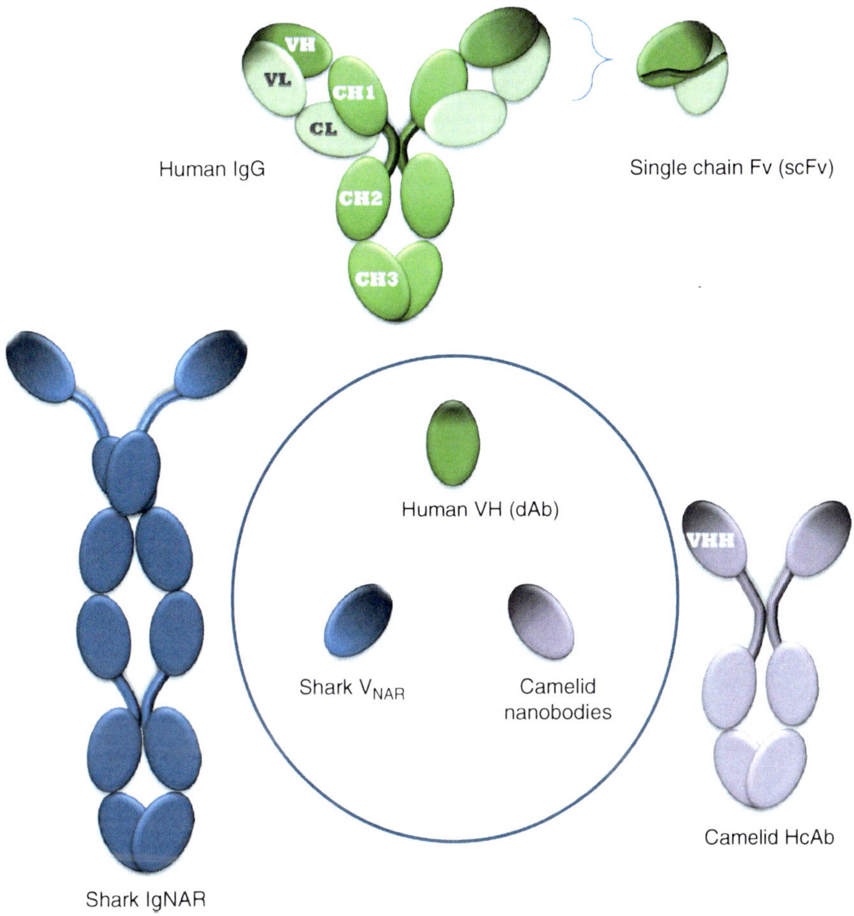

Fig. 6.1 *Sources of single domain antibodies.* Schematic structures of conventional human IgG, shark IgNAR, camelid heavy chain antibodies and their corresponding single domain antibodies [domain antibodies (dAb), V_{NAR} and nanobodies (Nb), respectively], *circled.* A single chain Fv fragment is depicted for comparison

confer them a low antigenicity in humans, a very attractive feature for immunotherapy approaches. If needed, they can also be humanized (Vincke et al. 2009). The immunization of llamas with recombinant proteins, intact cells, or cell lysates offers the possibility to take advantage of the very efficient in vivo affinity maturation to enrich the animal antibody repertoire with high affinity antibodies (in the nanomolar range) before the creation of the antibody gene library. The cloning of the resulting repertoires gives access to a panel of antigen-specific easy-to-produce and stable antibody fragments that can directly be used as targeting agents.

Camelids are not the only species producing antibodies devoid of light chain. Greenberg et al. (1995) described the presence of heavy chain antibodies in nurse shark, called IgNAR. Such antibodies were also found in wobbegong sharks

(Nuttall et al. 2001). The variable domain of IgNAR, called V_{NAR} (Fig. 6.1) shares many properties with camelid VHH, including their high stability, their long CDR3, and their tendency to bind cavities (Stanfield et al. 2004; Simmons et al. 2008).

The hope to confer these attractive features to fully human variable domains has been the goal of many antibody engineering team. Human VH domains (Fig. 6.1) have a propensity to aggregate when expressed in the absence of matching VL. To obtain human VH with favorable properties, Jespers et al. have selected a synthetic library of human VH based on a single scaffold with diversification of the binding loops. Once expressed at the surface of phage particles, these human VH were heat-denaturated, cooled, and non-aggregated VH domains were selected for their ability to bind protein A. This approach could select clones that were resistant to aggregation, soluble and that could be purified in good yields (Jespers et al. 2004). Because only 1 clone out of 1,000 survived this process, the authors restored a high diversity by combinatorial assembly of CDR building blocks from the repertoire of aggregation-resistant antibody domains, leading to a large library of VH containing around 80% of aggregation-resistant VH domains (Christ et al. 2007). The process of selection for resistant human VH was further refined by including selections at 37°C in acidic pH (3.2) for 2 h as the aggregation-promoting condition. This process was shown to select domains that were thermodynamically stable as well as aggregation-resistant, unlike the first method that was selecting aggregation-resistant clones but with melting temperatures similar to the parental clones (Famm et al. 2008).

Thus, single domain antibodies can be generated by immunization of animals naturally producing antibodies devoid of light chain, or from large human synthetic libraries. Immunization favors the direct selection of high affinity clones that have undergone in vivo affinity maturation, whereas the use of large synthetic libraries allows the selection of fully human domains with comparable properties. Because these domains are small, very efficiently produced, and highly stable, they represent ideal building tools to create more elaborate molecules such as bispecific antibodies (see Saerens et al. 2008 for a complete review on single domain antibodies).

6.3 Bispecific Antibodies

Single domain antibodies have been used to create various formats of bispecific antibodies. This chapter describes the main proposed formats.

6.3.1 Tandem and Linear Repeats of sdAbs

6.3.1.1 Proof of Concept

The most obvious way to create bispecific antibodies using single domain antibodies is to link two domains via a peptidic linker, thereby creating a tandem sdAb (Fig. 6.2). In 2001, a proof of concept was published by Muyldermans and

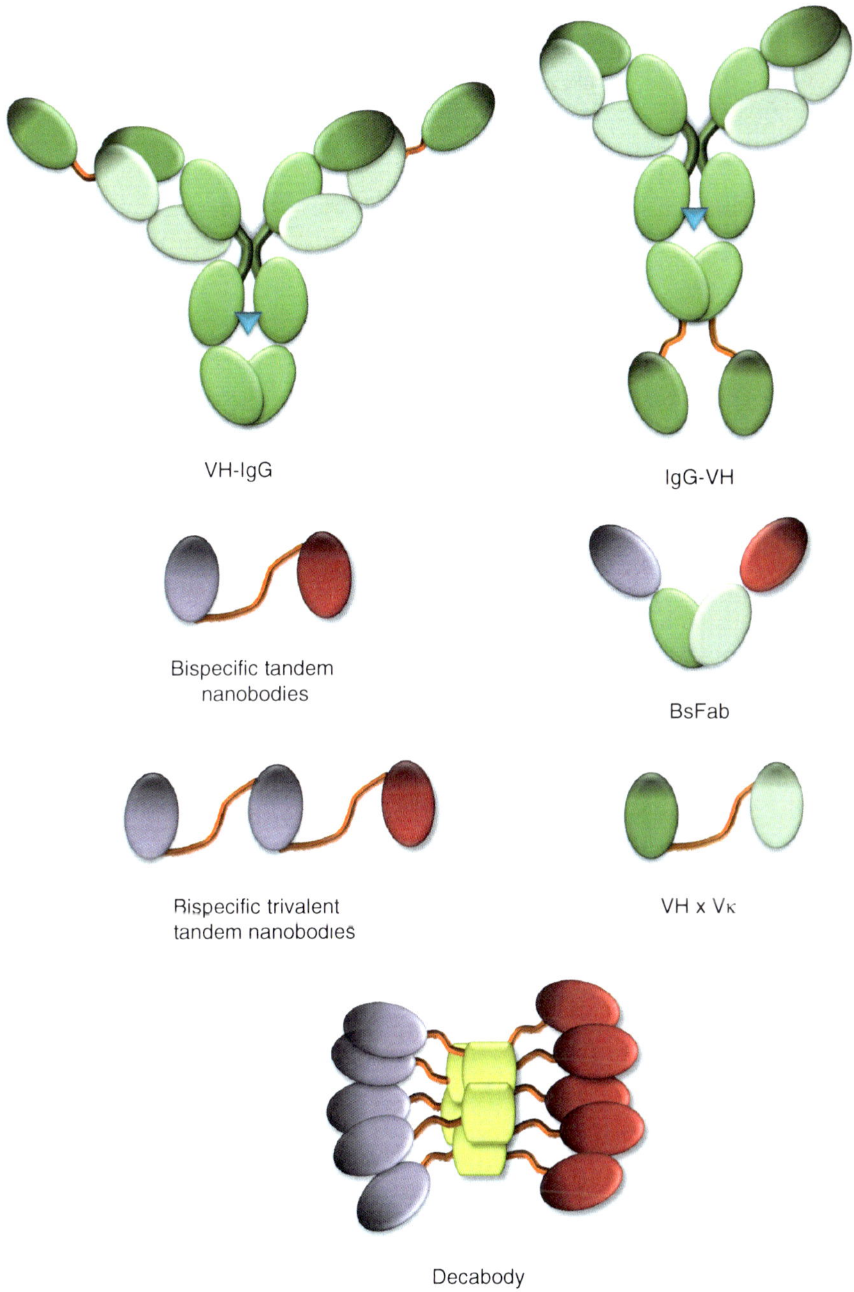

Fig. 6.2 *Formats of bispecific antibodies relying on the use of single domain antibodies.* Schematic structure of the various bispecific fragments discussed in this chapter. Human domains are colored in *dark* and *light* for heavy and light chain respectively. Camelid domains of different specificities are colored differently. The verotoxin B subunit part in the decabody construct is depicted as *cylinders*. The locations of hypervariable loops constituting the antigen binding sites near the N-terminus of Ig domains are shaded in *black*

coworkers who produced bispecific tandem camelid nanobodies by fusing them via a 29 amino acids peptide corresponding to the structural upper hinge of the llama IgG2a isotype (Els Conrath et al. 2001). BsAbs binding to hen egg lysozyme and porcine pancreatic α-amylase were produced in the periplasm of *E. coli* as efficiently as monomeric VHHs and remained fully active after 44 h of incubation in murine serum at 37°C. Simultaneous binding of these bsAbs to the two antigens was demonstrated by gel filtration and by biosensor experiments. This work clearly demonstrated the advantages of tandem nanobodies for the production of bsAbs.

6.3.1.2 αTNF × αTNF × αAlb

This concept was further developed a few years later in a more relevant setting for cancer therapy by Coppieters and co-workers (2006) who isolated an anti-tumor necrosis factor α (TNF) nanobody. This nanobody was expressed as a tandem to display avid binding to the trimeric antigen, yielding a 30 kDa bivalent monospecific molecule. The purpose of this molecule was to block TNF for the treatment of rheumatoid arthritis (RA). Unfortunately, small molecules with molecular weight under 60 kDa and devoid of Fc portion allowing binding to FcRn receptor are characterized by very short serum half-life. To solve this issues, the authors isolated a new nanobody binding to human and murine serum albumin and fused this domain via a flexible Gly-ser linker to the previous molecule yielding a trivalent bispecific tandem nanobody anti-TNF × Albumin (Fig. 6.2). The serum half-life of the tandem nanobody was substantially extended from 54 min to 2.2 days, most likely correlating with the murine albumin half-life. Interestingly, in vitro, the antagonistic potency of this molecule exceeded that of anti-TNF antibodies infliximab and adalimumab that are clinically used in RA. In a murine model of RA (murine collagen-induced arthritis), a murine surrogate of this construct capable of blocking murine TNF led to excellent therapeutic efficiency exceeding the one of etanercept, a Fc fusion with human TNF receptor cross-reacting with murine TNF.

6.3.1.3 αEGFR × αEGFR × αAlb

A very similar approach was developed by Roovers and collaborators. The authors isolated by phage display antagonistic anti-epidermal growth factor receptor (EGFR) nanobodies and dimerized them via the llama IgG3 hinge (12 amino acids). A bispecific trivalent molecule was also generated by fusion to a serum albumin binding domain to extend its serum half-life from less than 1 to 44 h (Roovers et al. 2007). In vitro, bsAbs could completely block the EGF-induced proliferation of A431 human tumor cells when added in excess. In an in vivo murine xenograft model, the bsAb injected twice a week for 4 weeks could effectively delay the growth of EGFR+ solid tumor until the end of the treatment. The tumor targeting properties of the same bsAb were further studied in a following work (Tijink et al. 2008). Here, the authors compared the biodistribution of the bivalent and trivalent nanobody with the well-described and FDA-approved cetuximab in a xenograft murine model. The bivalent anti-EGFR nanobody achieved an impressive tumor to blood ratio of 80 and a low tumor level of 5% ID/g, 6 h after injection. By contrast, and thanks to its binding activity for albumin, the bispecific trivalent construct yielded a very high tumor level (35% of ID/g), outperforming cetuximab (27%). Moreover, tumor sections demonstrated a uniform staining for the nanobody construct whereas only 60% of the tumor could be stained with cetuximab, reflecting a better tumor penetration of the 50 kDa nanobody construct vs. 150 kDa IgG.

6.3.1.4 αFMDV × αFMDV × αpIg

A similar approach was described by Harmsen and collaborators who designed a bispecific nanobody targeting the foot and mouth disease virus (FMDV) and porcine IgG to protect pigs by passive immunization. The authors showed that the addition of the anti-porcine IgG specificity yielded a 100-fold increase in serum half-life compared to the anti-FMDV monospecific nanobody (Harmsen et al. 2005, 2008). In a subsequent work, the team demonstrated that a mixture of two trivalent αFMDV × αFMDV × αpIg nanobody constructs injected 24 h before FMD challenge infection of pigs could reduce and delay the development of clinical disease, viremia, viral shedding, and FMD transmission (Harmsen et al. 2009). Such efficiency was clearly correlated with the long serum half-life conferred by the antiporcine Ig moiety of the construct.

6.3.1.5 αAahI′ × αAahII

Envenoming following scorpion sting is currently treated using a Fab′2 preparation of horse plasma antivenom serum (PAS). However, the beneficial effect of this serotherapy remains controversial. One of its main limitation might be the size of the antibody fragments (100 kDa) exceeding 14-fold the size of the main toxins named AahI′ and AahII. It is indeed demonstrated that, unlike the large Fab′2 fragments, these small toxic peptides distribute rapidly in blood and tissues. In an effort to propose a smaller and more active antivenom molecules, Hmila and collaborators (2010) have selected dromedary nanobodies against these two toxins and could isolate nanobodies able to fully protected mice against 100 LD_{50} of AahI′ administered intracerebroventricularly. To maximize the efficiency of nanobodies, the authors designed a bispecific tandem anti-AahI′ × AahII nanobody and compared its efficiency with PAS in a more relevant subcutaneous setting. Interestingly, the bispecific nanobody could successfully neutralize 5 LD_{50} where PAS failed at neutralizing 2 LD_{50}, and was still fully protective when mice with severe signs of envenoming were treated a few minutes before the untreated mice died

(Hmila et al. 2010). The dual targeting properties and the small molecular weight of this bispecific nanobody are probably to be accounted for its high efficiency.

6.3.1.6 αMP65 × αSAP2

Fully human variable domains have also been used as tandem domain antibodies. De Bernardis and collaborators (2007) have isolated by phage display human variable domains (VH and Vk) binding to the 62 kDa mannoprotein (MP65) or the secretory aspartyl proteinase (SAP2), two defined virulence traits of *Candida albicans*, an organism responsible of vaginal cadidiasis. Monovalent domain antibodies could inhibit fungus adherence to epithelial cells of rat vagina and accelerate the clearance of vaginal infection. When a tandem construct using a flexible 25-aa polypeptide linker (Fig. 6.2) and binding both antigens was tested in their in vivo model, they could achieve therapeutic efficiencies equivalent to treatment with the antifungal fluconazole, both in pre- and postchallenge treatment schedules and with fluconazole-sensitive and -resistant strains. Interestingly, the authors could demonstrate a synergy due to the simultaneous targeting of two different virulence factors afforded by the use of a bispecific construct.

6.3.2 IgG–sdAb Fusions

Because sdAbs are very small, they can also be easily fused to other molecules to add a new specificity. This concept was demonstrated with the description of full-length IgG whose heavy chains were N- or C-terminally fused to a single domain antibody of different specificity. The resulting IgG-like molecules are bispecific and still benefit from the presence of a Fc portion in terms of long half-life and effector functions.

6.3.2.1 αmPDGFRα × αmVEGFR2

This approach was first demonstrated by Shen and collaborators (2006) who produced a bispecific tetravalent molecule by fusing an antimouse platelet-derived growth factor receptor alpha to the N-terminus of the light chain of a conventional IgG binding to mouse vascular endothelial growth factor receptor 2 (Fig. 6.2). The authors first isolated a truncated clone from a large library of human Fab fragment panned against mPDGFRα. This clone was deleted from its VL domain but could still bind the antigen with high affinity (0.42 nM) and could block the interaction with the ligand PDGF-AA with an IC_{50} of 12 nM. The same library was used to select a regular high affinity (6.7 nM) blocking (IC_{50} of 3.5 nM) human Fab fragment against mVEGFR2 and was converted to the IgG format. The isolated VH of the truncated clone was subsequently fused to the N-terminus of the full-length

IgG light chains, yielding the bispecific IgG-like construct that could be efficiently produced in mammalian cells and retained the antigen binding specificity and the receptor neutralizing activity of the parent antibodies. Interestingly, on cells expressing both receptors, the bsAb displayed a better binding activity that the parent antibodies thereby demonstrating dual receptor-targeting.

6.3.2.2 αmPDGFRα × αmPDGFRβ

The same team constructed a similar bsAb by fusing the same anti-mPDGFRα VH to a human anti-mPDGFRβ IgG isolated by phage display (Shen et al. 2007). Two molecules were generated by fusing the isolated VH to the N terminus of the IgG light chain (VH-IgG) or the C terminus of the IgG heavy chain (IgG-VH) via short linkers (5 or 6 aa respectively) (Fig. 6.2). These constructs were purified from the supernatant of COS-7 cells with moderate yields of 0.5 and 3 mg/L for VH-IgG and IgG-VH, respectively. Both constructs could inhibit the ligand/receptors interactions (PDGF-AA with mPDGFRα, and PDGFR-BB with mPDGFRα and β) at nanomolar concentrations and importantly, both molecules could block the activation of PDGF receptors α and β on tumor cells stimulated by both platelet-derived factor AA and BB. On the down side, both constructs experienced a rather quick loss of mPDGFRα binding activity in 10% mouse serum in PBS at 37°C (49 and 56% of residual binding activity for PDGFRα for the N- and C-terminal fusion, respectively).

6.3.3 Multimerization Motives

6.3.3.1 Special Case of Variable Domains from Shark IgNAR

Other studies have relied on the use of dimerization motives to create bivalent or bispecific single domain antibody-based molecules. A work done using variable domains derived from shark IgNAR antibodies yielded surprising results (Simmons et al. 2006). The authors tried various ways to dimerize a V_{NAR} domain directed against the malarial vaccine candidate protein apical membrane antigen-1 (AMA1) from *Plasmodium falciparum*, including the use of C-terminal fusion with the first constant domain, the addition of a helix-turn-helix motif or of a hinge ending with a free cystein. They also produced a tandem V_{NAR} construct consisting of a first V_{NAR} C-terminally linked to another one via a 13-aa peptidic linker. Unfortunately, the authors demonstrated on several examples that in this configuration, the second V_{NAR} is not functional, possibly because of some steric hindrance between the linker and its paratope. Thus unlike the seemingly related camelid nanobodies, tandem constructs of shark V_{NAR} do not appear to lead to functional bsAbs.

6.3.3.2 Verotoxin B Subunit

Another multimerization motif proved more successful, this time using llama-derived nanobodies. Zhang et al. (2004) published a way to improve the apparent affinity of sdAbs of modest affinity selected from nonimmune library by means of avidity. To do so, they used the multimerization properties of verotoxin B subunit produced by *E. coli*. This toxin has a doughnut-shaped structure with N and C termini exposed at the periphery and on opposite sides of the molecules. The authors fused an antipeptide sdAb to the C-terminus via a 5-aa linker. The resulting 115 kDa pentameric molecules could be produced as soluble product in good yield in *E. coli* (5 mg/L). Interestingly, the so-called "pentabody" bound the immobilized antigen three to four orders of magnitude more strongly than the monomer sdAbs. The pentabody was also characterized by a good thermostability (Tm 59.7°C) and a high resistance to digestion by trypsin, chymotrypsin, and serum proteases. The logical sequel to this work was to use the same framework to create decavalent bispecific molecules by fusing two different sdAbs both at the N-terminus and the C-terminus of the toxin, thereby creating a 175 kDa molecule named decabody (Fig. 6.2) (Stone et al. 2007). By optimizing the nature of linkers between the sdAbs and the verotoxin B subunit, the authors could produce in *E. coli* with good yields (around 5 mg/L) a model bsAb showing no obvious aggregation or degradation during the production and purification process, despite the complexity of such a molecule. The performance of these constructs remains to be established in vivo. It should be stressed that the high avidity of decabodies might be an advantage for tumor targeting purposes when antigen densities on normal tissues are below the threshold level required for multivalent binding, leading to the exclusive targeting of tumor tissues expressing high densities of the antigen. However, a major concern with this type of bsAb might be their immunogenicity. Indeed the verotoxin part of this construct is predicted to be highly immunogenic, which might be a clear limitation of this construct for therapeutic approaches.

6.3.3.3 COMbodies

A similar approach has recently been published by Zhu et al. (2010), using this time a human motif to decrease the immunogenicity of the resulting molecules. The author chose the coiled-coil domain (Asp29-Gln76) of human cartilage oligomeric matrix protein (COMP) as pentameric scaffold. This unusually stable structure (Guo et al. 2000) is further stabilized by the formation of disulfide bonds between the five α-helices. A low-affinity (μM) single domain antibody targeting a MART-1/HLA-A2 peptide–MHC complex was pentamerized using this strategy. Strikingly, the resulting pentamer, named COMbody, yielded a strong and specific signal by flow cytometry whereas the parental low affinity monomer was negative. Surface plasmon resonance demonstrated that the functional affinity of the COMbody was increased by more than 10^5 fold compared to the monomer. More interestingly, the authors demonstrated the possibility to create bispecific molecules by fusing the scFv fragment of antihuman CD3 clone OKT3 at the C-terminus of the coiled coil domain. The resulting molecule was rather large (around 260 KDa) and was difficult to produce in eukaryotic cells but could be efficiently used to lyse target cells in vivo using preactivated human PBMCs. These results clearly demonstrate the possibility to use this 48 amino acid COMP fragment as pentamerization motif. Therefore, the fusion of different sdAbs at its N- and C-termini might be a straightforward approach to create high-avidity bispecific molecules of reasonable size (around 170 kDa).

6.3.3.4 CH1/Cκ Heterodimerization Motif

Most recombinant bispecific formats rely on the use of flexible peptidic linkers. Although these linkers have obvious advantages in terms of antibody engineering, their hydrophilic nature makes them prone to proteolytic cleavage, potentially leading to production issues, poor antibody stability, aggregation, and increased immunogenicity (Fischer and Leger 2007). Thus, an ideal domain antibody-based bispecific format for antibody therapy would use a heterodimerization motif of human nature devoid of linker and leading to relatively small molecules still able to diffuse efficiently into tumor tissues. The CH1 and Cκ domains are responsible for the interaction between the heavy and the light chains of human IgGs. This interaction is further stabilized by the formation of a covalent disulfide bond. Thus these two domains constitute an attractive heterodimerization motif. Our team is developing nanobody-based bispecific antibodies relying on this motif (Baty et al. 2006). One of the most exciting applications of bispecific antibodies is the retargeting and activation of effector cells of the patient's immune system against tumor cells. A large proportion of approved antitumor antibodies rely on the lysis of tumor cells by natural killer (NK) cells for their mode of action, i.e. antibody-dependent cell-mediated cytotoxicity (ADCC) (Chames et al. 2009). However, the efficiency of this process is limited by several factors, including the polymorphism of the main activating receptor FcγRIIIA expressed by human NK cells, variations in the glycosylation patterns of IgG Fc portions, the activation of inhibitory receptors such as FcγRIIB, and the competition between the therapeutic antibody and the high concentration of endogenous IgGs in patient sera. Bispecific antibodies capable of binding specific epitopes on FcγRIIIA can circumvent all these issues. Our aim is to develop bispecific antibodies retargeting NK cells to tumor cells via an interaction with a tumor marker and FcγRIIIA. We have recently isolated nanobodies binding to the well characterized tumor marker CEACAM5 (CEA) devoid of cross-reaction for the highly related CEACAM6 (NCA) (Behar et al. 2009) as well as anti-FcγRIII nanobodies capable of activating NK cells without being concerned by the above cited limitations (Behar et al. 2008). By directly fusing the C-terminus of these sdAbs to the N-terminus of human CH1 and Cκ domains, we have produced bispecific Fab-like antibody fragments (bsFabs) that can be efficiently produced in the cytoplasm of *E. coli* (yields in the 1–3 mg/L range) (Fig. 6.2). Interestingly, bsFabs remain fully active after 1 week of incubation at 37°C in 90% human serum and are capable of triggering the lysis of tumor cells by fresh NK cells in vitro at picomolar concentrations (Cornillon et al., in preparation). The in vivo activity of these very promising bispecific nanobodies is currently being evaluated in xenografted murine models.

6.4 Conclusion

Single domain antibodies are the smallest antibody fragment capable of recapitulating the binding activities of antibodies. Many of their characteristics including their small size, production yields, and very high stability qualifies them as ideal building blocks. These past few years have seen the emergence of innovative bispecific antibody formats integrating such domains. The next few years should confirm the rise of bispecific sdAb-based molecules. The first clinical studies involving such molecules are expected for soon and should tell if in vitro efficiencies demonstrated so far can be translated into therapeutic potential.

References

Alvarez-Rueda N, Behar G, Ferre V et al (2007) Generation of llama single-domain antibodies against methotrexate, a prototypical hapten. Mol Immunol 44:1680–1690

Baty D, Behar G, et al (2006) Production of antibody formats and immunological applications of said formats. WO/2006/064136

Behar G, Siberil S, Groulet A et al (2008) Isolation and characterization of anti-Fc{gamma}RIII (CD16) llama single-domain antibodies that activate natural killer cells. Protein Eng Des Sel 21:1–10

Behar G, Chames P, Teulon I et al (2009) Llama single-domain antibodies directed against nonconventional epitopes of tumor-associated carcinoembryonic antigen absent from nonspecific cross-reacting antigen. FEBS J 276:3881–3893

Chames P, Baty D (2009a) Bispecific antibodies for cancer therapy. Curr Opin Drug Discov Devel 12:276–283

Chames P, Baty D (2009b) Bispecific antibodies for cancer therapy: the light at the end of the tunnel? MAbs 1:539–547

Chames P, Van Regenmortel M, Weiss E et al (2009) Therapeutic antibodies: successes, limitations and hopes for the future. Br J Pharmacol 157:220–233

Christ D, Famm K, Winter G (2007) Repertoires of aggregation-resistant human antibody domains. Protein Eng Des Sel 20:413–416

Coppieters K, Dreier T, Silence K et al (2006) Formatted anti-tumor necrosis factor alpha VHH proteins derived from camelids show superior potency and targeting to inflamed joints in a murine model of collagen-induced arthritis. Arthritis Rheum 54:1856–1866

De Bernardis F, Liu H, O'Mahony R et al (2007) Human domain antibodies against virulence traits of Candida albicans inhibit fungus adherence to vaginal epithelium and protect against experimental vaginal candidiasis. J Infect Dis 195:149–157

De Genst E, Silence K, Decanniere K et al (2006) Molecular basis for the preferential cleft recognition by dromedary heavy-chain antibodies. Proc Natl Acad Sci USA 103:4586–4591

Dolk E, van Vliet C, Perez JM et al (2005) Induced refolding of a temperature denatured llama heavy-chain antibody fragment by its antigen. Proteins 59:555–564

Dumoulin M, Conrath K, Van Meirhaeghe A et al (2002) Single-domain antibody fragments with high conformational stability. Protein Sci 11:500–515

Els Conrath K, Lauwereys M, Wyns L et al (2001) Camel single-domain antibodies as modular building units in bispecific and bivalent antibody constructs. J Biol Chem 276:7346–7350

Ewert S, Cambillau C, Conrath K et al (2002) Biophysical properties of camelid V(HH) domains compared to those of human V(H)3 domains. Biochemistry 41:3628–3636

Famm K, Hansen L, Christ D et al (2008) Thermodynamically stable aggregation-resistant antibody domains through directed evolution. J Mol Biol 376:926–931

Fischer N, Leger O (2007) Bispecific antibodies: molecules that enable novel therapeutic strategies. Pathobiology 74:3–14

Greenberg AS, Avila D, Hughes M et al (1995) A new antigen receptor gene family that undergoes rearrangement and extensive somatic diversification in sharks. Nature 374:168–173

Groot AJ, Gort EH, van der Wall E et al (2008) Conditional inactivation of HIF-1 using intrabodies. Cell Oncol 30:397–409

Gueorguieva D, Li S, Walsh N et al (2006) Identification of single-domain, Bax-specific intrabodies that confer resistance to mammalian cells against oxidative-stress-induced apoptosis. FASEB J 20:2636–2638

Guo Y, Kammerer RA, Engel J (2000) The unusually stable coiled-coil domain of COMP exhibits cold and heat denaturation in 4-6 M guanidinium chloride. Biophys Chem 85:179–186

Hamers-Casterman C, Atarhouch T, Muyldermans S et al (1993) Naturally occurring antibodies devoid of light chains. Nature 363:446–448

Harmsen MM, Van Solt CB, Fijten HP et al (2005) Prolonged in vivo residence times of llama single-domain antibody fragments in pigs by binding to porcine immunoglobulins. Vaccine 23:4926–4934

Harmsen MM, Fijten HP, Dekker A et al (2008) Passive immunization of pigs with bispecific llama single-domain antibody fragments against foot-and-mouth disease and porcine immunoglobulin. Vet Microbiol 132:56–64

Harmsen MM, Fijten HP, Engel B et al (2009) Passive immunization with llama single-domain antibody fragments reduces foot-and-mouth disease transmission between pigs. Vaccine 27: 1904–1911

Hmila I, Saerens D, Ben Abderrazek R et al (2010) A bispecific nanobody to provide full protection against lethal scorpion envenoming. FASEB J 24:3479–3489

Jespers L, Schon O, Famm K et al (2004) Aggregation-resistant domain antibodies selected on phage by heat denaturation. Nat Biotechnol 22:1161–1165

Lauwereys M, Arbabi Ghahroudi M, Desmyter A et al (1998) Potent enzyme inhibitors derived from dromedary heavy-chain antibodies. EMBO J 17:3512–3520

Muyldermans S (2001) Single domain camel antibodies: current status. J Biotechnol 74: 277–302

Nuttall SD, Krishnan UV, Hattarki M et al (2001) Isolation of the new antigen receptor from wobbegong sharks, and use as a scaffold for the display of protein loop libraries. Mol Immunol 38:313–326

Reichert JM (2010) Antibodies to watch in 2010. MAbs 2:84–100

Roovers RC, Laeremans T, Huang L et al (2007) Efficient inhibition of EGFR signaling and of tumour growth by antagonistic anti-EFGR nanobodies. Cancer Immunol Immunother 56:303–317

Rothbauer U, Zolghadr K, Tillib S et al (2006) Targeting and tracing antigens in live cells with fluorescent nanobodies. Nat Meth 3:887–889

Saerens D, Ghassabeh GH, Muyldermans S (2008) Single-domain antibodies as building blocks for novel therapeutics. Curr Opin Pharmacol 8:600–608

Serruys B, Van Houtte F, Verbrugghe P et al (2009) Llama-derived single-domain intrabodies inhibit secretion of hepatitis B virions in mice. Hepatology 49:39–49

Shen J, Vil MD, Jimenez X et al (2006) Single variable domain-IgG fusion. A novel recombinant approach to Fc domain-containing bispecific antibodies. J Biol Chem 281:10706–10714
Shen J, Vil MD, Jimenez X et al (2007) Single variable domain antibody as a versatile building block for the construction of IgG-like bispecific antibodies. J Immunol Meth 318:65–74
Simmons DP, Abregu FA, Krishnan UV et al (2006) Dimerisation strategies for shark IgNAR single domain antibody fragments. J Immunol Meth 315:171–184
Simmons DP, Streltsov VA, Dolezal O et al (2008) Shark IgNAR antibody mimotopes target a murine immunoglobulin through extended CDR3 loop structures. Proteins 71:119–130
Spinelli S, Frenken LG, Hermans P et al (2000) Camelid heavy-chain variable domains provide efficient combining sites to haptens. Biochemistry 39:1217–1222
Stanfield RL, Dooley H, Flajnik MF et al (2004) Crystal structure of a shark single-domain antibody V region in complex with lysozyme. Science 305:1770–1773
Stone E, Hirama T, Tanha J et al (2007) The assembly of single domain antibodies into bispecific decavalent molecules. J Immunol Meth 318:88–94
Su C, Nguyen VK, Nei M (2002) Adaptive evolution of variable region genes encoding an unusual type of immunoglobulin in camelids. Mol Biol Evol 19:205–215
Tijink BM, Laeremans T, Budde M et al (2008) Improved tumor targeting of anti-epidermal growth factor receptor nanobodies through albumin binding: taking advantage of modular nanobody technology. Mol Cancer Ther 7:2288–2297
Verheesen P, de Kluijver A, van Koningsbruggen S et al (2006) Prevention of oculopharyngeal muscular dystrophy-associated aggregation of nuclear polyA-binding protein with a single-domain intracellular antibody. Hum Mol Genet 15:105–111
Vincke C, Loris R, Saerens D et al (2009) General strategy to humanize a camelid single-domain antibody and identification of a universal humanized nanobody scaffold. J Biol Chem 284: 3273–3284
Ward ES, Gussow D, Griffiths AD et al (1989) Binding activities of a repertoire of single immunoglobulin variable domains secreted from *Escherichia coli*. Nature 341:544–546
Zhang J, Tanha J, Hirama T et al (2004) Pentamerization of single-domain antibodies from phage libraries: a novel strategy for the rapid generation of high-avidity antibody reagents. J Mol Biol 335:49–56
Zhu X, Wang L, Liu R et al (2010) COMBODY: one-domain antibody multimer with improved avidity. Immunol Cell Biol 88:667–675

Chapter 7
Alternative Scaffolds as Bispecific Antibody Mimetics

John Löfblom and Fredrik Y. Frejd

7.1 Introduction

Alternative scaffold proteins are a class of antibody mimetics that share a number of essential properties with antibodies (Gebauer and Skerra 2009; Nygren and Skerra 2004), the most important being the ability to recognize molecular structures with high affinity and selectivity. In addition, alternative scaffolds often have improved properties in terms of stability, efficient recombinant production features, reduced size, few or no cysteines and an uncomplicated IP-situation. Due to their small size and ease of recombinant manipulation, they are well suited for generation of bispecific proteins. In fact, a number of alternative scaffolds are derived from natural precursors that are formatted as repeated domains, suggesting ability for function in a multimeric format, e.g., Affibody molecules (Löfblom et al. 2010; Nord et al. 1997), Adnectins (Koide et al. 1998), DARPins (Binz et al. 2004; Stumpp et al. 2008), and Avimers (Silverman et al. 2005).

7.2 Alternative Scaffolds as Antibody Mimetics

In combinatorial protein engineering, the term scaffold commonly defines a stable protein framework, harboring flexible loops or solvent accessible residues, which are randomized in order to construct a molecular library. Antibodies were the first,

J. Löfblom
Department of Molecular Biotechnology, School of Biotechnology, Royal Institute of Technology (KTH), AlbaNova University Center, 106 91 Stockholm, Sweden

F.Y. Frejd (✉)
Affibody AB, Lindhagensgatan 133, 112 51 Stockholm, Sweden

Unit of Biomedical Radiations Sciences, Rudbeck Laboratory, Uppsala University, 751 85 Uppsala, Sweden
e-mail: fredrik.frejd@affibody.se

R.E. Kontermann (ed.), *Bispecific Antibodies*,
DOI 10.1007/978-3-642-20910-9_7,

and are today still the most widely used, biomolecular scaffold for generation of affinity reagents. However, specific molecular recognition is not a unique property of immunoglobulins, and challenges associated with the antibody format have inspired researchers to investigate the possibilities of engineering specific affinity into other protein structures. In addition, in vitro display technologies (e.g., phage display) used for recombinant antibody generation, are readily modified for isolation of affinity proteins from non-Ig-based libraries, making isolation and identification of alternative affinity proteins relatively easy.

Alternative scaffolds can be designed to overcome several of the limitations associated with antibodies. The scaffold should preferably be small and composed of a single polypeptide chain in a fold that has an intrinsic biochemical and biophysical stability independent on post-translational modifications such as glycosylations. The small size and low complexity facilitates production using efficient prokaryotic systems and even chemical synthesis for the smallest scaffolds. To enable use in intracellular milieus and improve recombinant expression flexibility, the stability of the fold should if possible not be dependent on disulfide bonds. In addition, absence of natural cysteines allows for engineering of a cysteine into a suitable position for directed chemical modifications. It is also important that the protein scaffold demonstrates a high tolerance to randomizations in order to maintain high functionality of the resulting combinatorial library.

Non-antibody scaffolds can be classified in several different ways, e.g., based on origin, based on secondary structure composition, based on size, etc. One commonly used classification system is based on the site of diversification. The different types of scaffolds can be roughly divided into classes with a stable framework with one or several randomized loops (the immunoglobulin fold belongs to this class) and scaffolds in which randomizations are directed against surface-exposed residues on secondary structure elements. The antibody fold, with diversified loops displayed on a rigid framework, has been successfully employed for evolution of molecular recognition both in nature and in vitro. Several different alternative scaffolds are logically hence mimicking this molecular architecture. Diversifying loops generally require less information about the overall structure and the structure integrity and functionality might be less affected on average by the mutations. In addition, randomizing loops might result in additional diversity through the flexibility in the loops, i.e., one clone can adopt several structural forms and recognize several distinct target molecules. This may impact the specificity, something that has been demonstrated for antibodies where unmatured IgGs with relatively more flexible loops generally are more unspecific (James et al. 2003; Mouquet et al. 2010). In addition to scaffolds mimicking the antibody architecture, approaches in which randomizations are focused directly on secondary structure elements have also been described. Compared to flexible loops in antibodies and similar scaffolds, secondary structure surfaces are typically much more constrained, which might lead to less entropy losses upon binding and potentially a higher specificity.

A few examples of scaffolds from each group are listed in Table 7.1 and a more detailed description of the scaffolds that have been reported as bispecific agents is provided in the following section and in Table 7.2. Alternative scaffolds in general

Table 7.1 Examples of non-Ig-based alternative protein scaffolds

Name	Scaffold	Length (aa)	Disulfide stabilized	Bispecifics reported in literature	References	Commercialization
Randomizations on secondary structure elements						
Affibody molecules	Protein A domain	58	No	Yes	Löfblom et al. (2010)	http://www.affibody.se
Affilins	γB-crystallin/ubiquitin	198	No	No	Ebersbach et al. (2007)	http://www.scilproteins.com
DARPins	Ankyrin repeat protein	~133–199	No	Yes	Stumpp et al. (2008)	http://www.molecularpartners.com
Randomizations on loops						
Adnectins/monobodies	^{10}Fn3	94	No	Yes	Koide et al. (1998)	http://www.adnexustx.com
Anticalins	Lipocalin	170	Yes	Yes	Skerra (2008)	http://www.pieris-ag.com
Avimers	A domain	~105	Yes	Yes	Silverman et al. (2005)	http://www.amgen.com
Fynomers	SH3 domain	63	No	No	Bertschinger et al. (2007)	http://www.covagen.com
Knottins/microproteins	Cysteine knot peptides	30	Yes	No	Kolmar (2009)	http://www.amunix.com
Kunitz domains	Tissue factor pathway inhibitor	58	Yes	No	Lehmann (2008)	http://www.dyax.com

Table 7.2 Overview of bispecific alternative scaffolds and selected results at different stages of development

Scaffold	Target 1	Target 2	Biological mechanism	Biochemical data	Cell results	In vivo results	References
ABD	Albumin	TNF, Z-domain	Increased plasma half-life of binding function	Affinity capture on target column. High affinity binding against dual targets	–	–	Alm et al. (2010) Hober (personal communication)
Adnectins	EGFR	IGFR	Dual receptor blocking	K_D EGFR: 7.7 nM K_D IGFR: 0.4 nM Tm 57.5°C	Inhibition of pAKT. Degradation of EGFR and IGFR in cells.	Tumor xenograft growth retardation	Emanuel et al. (2011)[a]
Affibody molecules	EGFR	HER2	Increased targeting selectivity	Simultaneous binding shown in SPR	Stronger binding to cells co-expressing targets as well as cell-cell ligation shown	–	Friedman et al. (2009) Wållberg et al. (2010)[a]
Affibody molecules-ABD	HER2	Serum albumin	Increased plasma half life of targeting molecule	K_D HER2 = 0.02 nM K_D Albumin = 8 nM	Binding and retention on cells verified	50-fold increased $t_{1/2}$ in mice (36 h). Targeted tumor radiotherapy	Tolmachev et al. (2007a, b)
Anticalins	Digoxin	Fluorescein	Dual hapten capture	Simultaneous binding shown in sandwich ELISA	–	–	Schlehuber and Skerra (2001)
Avimer	IL-6, BAFF, CD40L	IgG	Increased plasma half life of ligand blocking molecule	K_D IL-6 < 0.2 nM	Inhibits IL6-stimulated TF-1 cell proliferation	Extended half life in monkeys (89 ± 15 h)	Silverman et al. (2005)
DARPins	FcεRI epitope "85"	FcεRI epitope "30"	More efficient receptor blocking	K_D FcεRI = 0.07 nM Do not aggregate the receptor	Inhibition of degranulation in rat basophilic cells	–	Eggel et al. (2009)
VEGF	VEGFR-2	avβ3	Dual receptor blocking	Simultaneous binding of both receptors in SPR experiment	Inhibition of VEGF mediated phosphorylation and cell proliferation	–	Cochran (Personal communication)

[a]Poster abstract

have been extensively reviewed elsewhere, where more comprehensive listings are found (Gebauer and Skerra 2009; Nygren and Skerra 2004). Below, we will focus on the relatively few cases described where alternative scaffolds have been used in bispecific formats to enhance the property of the resulting molecule.

7.3 Bispecific Antibody Mimetics

The concept of bispecific antibodies was postulated already in the beginning of the 1980s (Milstein and Cuello 1983), but the multi-chain structure of antibodies resulted in sample heterogeneity that hampered development of clinical products. Even though elaborate strategies for constructing defined homogenous antibody-based bispecific agents now exist, the task is still challenging.

Alternative scaffolds, however, are generally comprised of a short and single peptide chain, and can therefore be readily adapted into bispecific and multispecific formats. The possibility of facile construction of multispecific alternative scaffolds might potentially become a differentiating factor compared to traditional antibodies in the future. Applications of bispecific alternative scaffold proteins include increased binding selectivity, enhanced receptor blocking capacity, provision of one biomolecule that can block two different receptors and development toward recruitment of effector functions. Another general application for bispecific affinity molecules is to have one binding partner for prolonging serum half-life of the molecule, while the other partner is having a biological activity.

7.3.1 Bispecific Antibody Mimetics with Enhanced Biological Function

Biological function can be enhanced in many different ways. One example is to increase the selectivity of a targeting moiety by the simultaneous targeting of two receptors using fine-tuned affinities. The idea is that targeted cells, expressing both receptors, will be recognized with higher functional affinity due to the avidity effect compared to normal cells expressing only one of the receptors, hence improving the selectivity. Another strategy is to block two receptors involved in the same signaling pathway. Here, simultaneous binding is not a requirement, as long as both receptors are blocked by either part of the bispecific molecule. The main advantage is instead two functions in one molecule, facilitating preclinical development, clinical practice, and potentially lowering treatment costs if blocking of two pathways is needed for full treatment effect. Yet another approach is to use bispecific molecules to increase the functional affinity and ability to block one single receptor. Here it is important to have molecules binding to epitopes adjacent to each other for a favorable effect. Another example is bispecific molecules where one moiety is used to attract an effector function. Below, case studies using Adnectins, Affibody molecules, Anticalins, and DARPins are highlighted.

7.3.1.1 Adnectins

An example of an immunoglobulin-like alternative scaffold is the tenth human fibronectin type III domain (^{10}Fn3, also called monobody or Adnectin) (Koide et al. 1998). ^{10}Fn3 is a monomeric, 94 residue protein, based on a stable β-sandwich fold with seven strands connected by six loops (Fig. 7.1a). Similar to antibodies and antibody derivatives, Adnectins interact with the target molecule through the loops on the tip of the molecule. However, in contrast to antibodies and most antibody derivatives, ^{10}Fn3 is not dependent on disulfide bonds for stable structure. Using phage display (Koide et al. 1998), mRNA display (Xu et al. 2002), and yeast display (Lipovsek et al. 2007), specific ^{10}Fn3 molecules have been selected against a number of different targets including, ubiquitin, TNF-alpha, lysozyme, and Abelson (Abl) kinase SH2 domain. One of the most advanced Adnectins binds with high affinity and specificity to VEGFR2. It is currently in phase II clinical trials in patients with solid tumors or non-Hodgkin's lymphoma (by former Adnexus, now part of Bristol-Myers

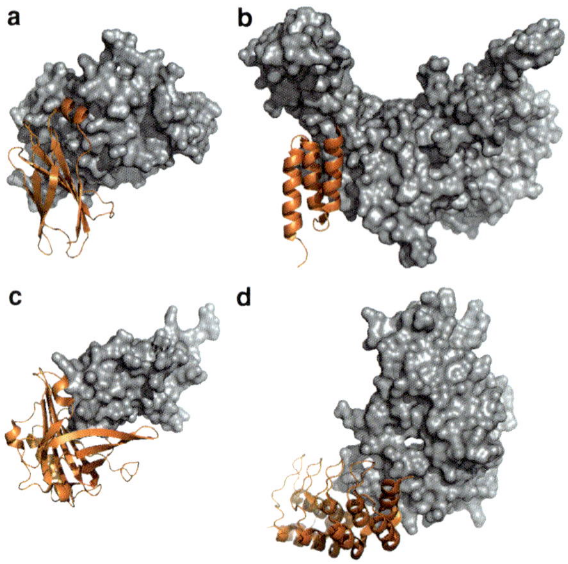

Fig. 7.1 Structures of alternative scaffolds that have been reported as bispecific constructs with enhanced biological function. Scaffold proteins are visualized as *orange cartoon* representations and respective target as *gray surface* representation. (**a**) Engineered ^{10}Fn3 in complex with human estrogen receptor alpha ligand-binding domain (PDB: 2OCF). (**b**) Affibody molecule in complex with HER2 extracelluar region (PDB: 3MZW). (**c**) Anticalin in complex with extracellular domain of human CTLA-4 (PDB: 3BX7). (**d**) DARPin in complex with aminoglycoside phosphotransferase (PDB: 2BKK)

Squibb). In addition to the commercial interest for this scaffold, the ^{10}Fn3 domain has been studied extensively in academia. An example is the work by Koide and coworkers, in which ^{10}Fn3 libraries have been constructed using a restricted diversity, in extreme cases using only serine and tyrosine in randomized positions (Koide et al. 2007), yielding high-affinity binders.

Recently, a bispecific Adnectin format targeting epidermal growth factor receptor (EGFR) and insulin-like growth factor receptor 1 (IGFR) has been presented (Emanuel et al. 2011). The rational is that both EGFR and IGFR play important roles in tumorigenesis of several human cancers. Blocking of either receptor reduces tumor growth in mouse models and human clinical studies. However, blocking of one receptor can induce compensatory activation of the other to drive tumor growth as seen in preclinical models. It would therefore be attractive to develop one therapeutic molecule with the capacity to block both receptors. After several rounds of selection and optimization using mRNA-display and a cell-based receptor activation assay, Adnectins that could bind and block either the EGF- or IGF1-receptor were isolated. Next, several different EGFR-binders were fused to one IGFR-binder and all constructs were PEGylated, resulting in EI-tandem Adnectins. The best construct inhibited phosphorylation of either receptor equally well as the reference blocking antibody. There was also a synergistic inhibition on phosphorylation of AKT, which is a shared downstream signaling node. The bispecific Adnectin was furthermore shown to induce degradation of both the EGF- and IGF-receptor in cell cultures within 7 h after administration of the EI-tandem molecule. When tested in mice, the plasma half-life of the PEGylated EI-tandem was 19.5 h, suggesting suitable exposure of the drug in vivo. Indeed, plasma levels of the receptor ligands IGF1 and TGFα increased due to blockade of their receptors and the kinetics of the increase correlated with the pharmacokinetics of EI-tandem in plasma. When tested in EGFR-driven tumor-bearing mice, the EI-tandem Adnectin showed equal efficacy in tumor growth retardation as the registered anti-EGFR drug cetuximab, and in addition decreased the tumor growth rate of an IGFR-driven cetuximab-insensitive tumor. Although the molar ratios of EI-tandem were 10- to 20-fold higher compared to the antibody, this clearly shows that alternative scaffolds potently can inhibit a receptor–ligand interaction and the benefit of providing two receptor-blocking ligands in one molecule. It still remains to be seen if the receptor blocking is simultaneous or if the bispecific molecule uses either the EGFR blocking or the IGFR blocking capacity. Since the EGFR binding part of the EI-tandem does not compete with cetuximab or several other EGFR-specific antibodies, this drug could in principle be administered as an add-on therapy to existing α-EGFR-mab therapies.

7.3.1.2 Affibody Molecules

Affibody molecules are based on small engineered three-helical proteins (Fig. 7.1b) (Löfblom et al. 2010). The basic scaffold was originally derived from the domain Z, a stabilized version of the B-domain of staphylococcal protein A, but has been subjected to protein engineering to further improve its scaffold properties (Ahlgren et al. 2010; Feldwisch et al. 2010). The small size of the proteins allows for inexpensive production in prokaryotic hosts. However, due to the small size and rapid folding kinetics, most Affibody molecules are also efficiently produced by peptide synthesis, providing means for addition of non-biological groups in order to modify and enhance functionality (Engfeldt et al. 2005; Nord et al. 2001).

Diversity is created by randomization of 13 solvent accessible residues [nine of which also participate in the native interaction with Fc (Deisenhofer 1981)] in helix one and two of the Z domain in order to build a combinatorial Affibody library (Nord et al. 1995). In a pioneering article from 1997, Nygren and coworkers describe the isolation of several specific Affibody molecules from such a library displayed on phage (Nord et al. 1997). Additional Affibody molecules have later been selected against various targets, such as EGFR (Friedman et al. 2007), HER2 (Wikman et al. 2004), amyloid beta (Grönwall et al. 2007), HIV gp-120 (Wikman et al. 2006), and CD28 (Sandström et al. 2003) and detailed biophysical characterization of several isolated clones have demonstrated intact three-helical structure and typically maintained stability and folding properties compared to the parental Z domain (Dincbas-Renqvist et al. 2004; Dogan et al. 2006; Eigenbrot et al. 2010; Hogbom et al. 2003; Lendel et al. 2004, 2006; Wahlberg et al. 2003).

A high-affinity (20 pM) Affibody molecule (Eigenbrot et al. 2010; Orlova et al. 2006) directed against HER2, has been investigated thoroughly in several preclinical and pilot clinical studies and is an interesting candidate for development of diagnostic and perhaps even therapeutic agents (Engfeldt et al. 2007; Orlova et al. 2007a, b; Steffen et al. 2006; Tolmachev et al. 2006, 2007a, b). In one therapy study by Nilsson and coworkers (Tolmachev et al. 2007a, b), the HER2-specific Affibody binder was fused to another small and robust protein, an albumin binding domain (ABD), to increase the half-life and avoid uptake of the radiolabeled bispecific construct in the kidneys. The kidney had indeed a 25-fold decreased uptake, and in addition the total tumor uptake was increased by a factor 3. Another example of bispecific Affibody molecules is the engineering of HER2–EGFR-binding molecules to increase the selectivity in tumor targeting. Expression of both HER2 and EGFR has been reported in several studies of cancers of the bladder, colon, and prostate, and in several cases the prognosis is worse for patients with cancer that co-express the receptors, as opposed to tumors expressing either HER2 or EGFR (Osaki et al. 1992; Tsutsui et al. 2003). Therefore, Ståhl and colleagues set out to create a bispecific molecule that would bind better, or preferably only, at co-expression of the receptors on the cell surface. Such a binder could present a new tool for increased tumor selectivity in diagnostic and therapeutic tumor targeting. Binders for HER2 and EGFR were isolated and affinity maturated (Friedman et al. 2008; Orlova et al. 2006). In a subsequent study, they were dimerized and fused to each other via a $(G_4S)_3$ linker (Friedman et al. 2009). The binding ability of the bispecific construct was tested against the parental binders. Whereas the HER2-binding moiety showed no difference in binding, the C-terminal EGFR-binding moiety had a lower association rate than the parental binders. The dissociation rate was however comparable. In cell studies, fluorescently labeled bispecific construct was shown to bind to both HER2-expressing SKOV-3 cells and to EGFR-expressing A431 cells. By using a device for real-time kinetics measurement of binding to live cells, it could be demonstrated that the bispecific HER2–EGFR-binder could mediate association of ^{14}C-labeled A431 cells in solution to immobilized SKOV-3 cells, thereby suggesting ability to bind simultaneously with both moieties. Recently, a new set of bispecific constructs was presented, where Affibody molecules with different affinities to EGFR or HER2 have been combined to further investigate the possibility for synergistic and therefore more selective binding (Wållberg et al. 2010). In vitro, on a mixed Biacore surface containing both receptors, the apparent K_D for the high-affinity bispecific construct was at least 30-fold higher compared to binding to surfaces containing only one of the two receptors. In real-time cell-binding experiments, using a lower affinity variant of the HER2-binding moiety, binding to HER2 high, EGFR low expressing SKBR-3 cells was blocked by addition of either monovalent HER2- or EGFR-binder. In addition, on A431 cells, the K_D of the bispecific construct was increased 30-fold when the high-affinity HER2-binding variant (K_D 22 pM) was used instead of the low affinity one (K_D 50 nM), likely reflecting the very small but still present expression of HER2 molecules on the EGFR high expressing A431 cells (Kong et al. 2008). Although the results are promising in the in vitro setting, it still remains to be seen if an increased selectivity will indeed be possible to achieve also in vivo in suitable animal models.

7.3.1.3 Anticalins

Anticalins are based on the protein architecture of natural lipocalins, demonstrating a rigid β-barrel fold displaying four loops targeted for diversification (Schlehuber and Skerra 2005; Skerra 2008) and specific binders have been isolated against both small molecules and proteins (Schlehuber and Skerra 2005; Skerra 2008) (Fig. 7.1c). Although structurally similar to antibodies in terms of displaying loops stabilized by a beta-sheet structure, Anticalins are smaller (around 180 amino acids), composed of only one polypeptide chain, are generally non-glycosylated and can be efficiently expressed in prokaryotic production systems. Natural lipocalins are generally secreted and often contain disulfide bridges. However, there are natural lipocalins lacking disulfide bonds and even examples where the disulfide bonds have been removed with retained functionality (Skerra 2008). Generally, the four loops of the lipocalin fold form a relatively deep pocket and specific Anticalins against small haptens and peptides have been successfully selected. However, as for antibodies, the loops demonstrate a relatively high flexibility, resulting in the ability to also interact with larger targets, such as proteins. An example is an anti-VEGF Anticalin with picomolar affinity to both the human and murine form of the growth factor. A pegylated variant (PRS-050) is currently in clinical trials for inhibition of angiogenesis in solid tumors by neutralization of VEGF (http://www.pieris-ag.com). The high stability of the lipocalin fold has motivated evaluation of Anticalins for local administration through alternative routes, such as pulmonary delivery. An Anticalin, specific for IL-4 receptor alpha (IL-4Ra), is in preclinical development for severe asthma. IL-4Ra plays a central role in chronic inflammation through interaction with both IL-4 and IL-13 in the lungs and pulmonary delivery of the IL-4Ra-specific Anticalin offers an appealing alternative to invasive routes with potentially improved patient compliance as well as directed local administration to the inflamed tissue (http://www.pieris-ag.com). Another Anticalin has been generated against mesenchymal-epithelial transition factor (cMET), a tumor-associated growth factor receptor. An advantage of the Anticalin protein in this setting, compared to many antibodies targeting this receptor is the monovalent nature of Anticalins. Bivalent antibodies tend to crosslink the receptor, causing an undesired agonistic response, which is avoided with the monovalent Anticalin.

In a relatively early publication, a bispecific Anticalin was described and denoted "duocalin" (Schlehuber and Skerra 2001). Here, an anti-digoxin and an anti-fluorescein Anticalin were fused by a peptidic linker and shown to be functionally expressed in bacteria, although with a somewhat lower yield. The binders were shown to retain the binding activity of each molecule in a sandwich ELISA and in a fluorescence titration experiment, demonstrating that Anticalins in principle can be formatted for bispecific binding. This concept is currently exploited for several therapeutic applications (www.pieris-ag.com). In particular, it appears attractive for a bispecific pre-targeting approach using a recently described Me•DTPA-binding Anticalin (Kim et al. 2009) together with tumor targeting molecules, for example, by fusion to the cMet-binding Anticalin (Skerra personal communication). Here, the affinity matured Anticalin binding DTPA with 400 pM affinity could be targeted to cMet expressing tumors to yield high tumor uptake. After the non-bound fusion protein is cleared from the blood-stream, a DTPA-complexed radiometal such as ^{90}Y or ^{177}Lu could be administered, and would be trapped by the DTPA-binding Anticalin at the tumor site, thus resulting in high tumor to non-tumor ratio of therapeutic activity.

7.3.1.4 DARPins

The DARPin scaffold is based on a class of proteins called ankyrin repeat proteins (Binz et al. 2004; Stumpp et al. 2008). Ankyrin repeat proteins are natural human proteins with a modular architecture, composed of repeats of a 33-residue domain. In protein engineering applications, the repeat approach has been kept and libraries are based on a protein composed of several randomized domains, hence denoted "designed ankyrin repeat proteins" (DARPins). Each domain is composed of a beta turn and two anti-parallel alpha helices. Two or three such domains generally constitute the binding surface with one capping domain on each side of the interacting domains, thus ranging in size from around 14 to 21 kDa (Fig. 7.1d). Randomizations are, as for Affibody molecules, to a large extent focused on secondary structure elements. Due to poor display level of DARPins on phages, specific binders were originally selected using ribosome display (Binz et al. 2004). However, in a recent study (Steiner et al. 2006), a modified phage display system was developed, in which recombinant proteins were exported via the signal recognition particle (SRP) translocation pathway instead of the typical Sec pathway, increasing the display level of fast-folding proteins such as DARPins considerably. Specific DARPins have been selected to numerous different targets and the engineered variants typically display favorable properties, such as high stability, efficient production in microbial hosts as well as high affinity. The most advanced DARPin is currently in phase I clinical trials: An anti-Vascular Endothelial Growth Factor-A (VEGF-A) DARPin (MP01112) that is being developed as an anti-angiogenic compound for treatment of wet age-related macular degeneration and diabetic macular edema (www.molecularpartners.com).

An interesting bispecific application of DARPins was recently described by Eggel et al. (2009), addressing a shortcoming of antibodies, which is the difficulty to block a receptor without crosslinking it with another one, often activating the resulting receptor pair. Several attempts using antibodies to block the interaction of immunoglobulin E (IgE) with its high affinity receptor (FcεRI) have been made. Upon allergen-induced crosslinking of IgE, FcεRI is aggregated on sensitized mast cells or basophils, and proinflammatory mediators that induce allergy symptoms are released. Antibodies binding to FcεRI and blocking IgE-binding have been isolated, but due to their bivalent nature, they cross-link the receptor and activate cell degranulation, leading to anaphylactic symptoms. Thus, the antibody format is not suitable for blocking IgE-mediated allergic symptoms. Instead, the DARPin technology was explored for this purpose. Since two binding sites have been mapped to the extracellular part of FcεRI, it has been suggested that it is necessary to block both IgE-binding sites to obtain an efficient antagonist. Hence, Eggel and co-workers set out to explore if a bispecific combination of two DARPins binding to different sites of the FcεRI could constitute a potent inhibitor of allergic syndromes. First, a set of binders was generated against recombinant FcεRIα. After initial ELISA screening, two DARPin binders, B-A4-85 (85) and C-A3-30 (30), were selected and produced for further characterization. They were shown not to compete for the same binding surface on the IgE-receptor, and were therefore subsequently tested for binding in monovalent, bivalent or bispecific formats, including both N- or C-terminal position of the respective binder in the bispecific format. The highest affinity was obtained for the bispecific construct with the DARPin 30 in N-terminal position (30/85), suggesting that this orientation was favorable, in contrast to DARPin 85 at the N-terminus (85/30), where the affinity was no better than for either monomeric binder alone. In a competition study using surface plasmon resonance, the bispecific constructs or a mixture of the bivalent (monospecific) molecules inhibited binding of IgE to its receptor much better than either of the bivalent monospecific molecules, suggesting that blocking of both epitopes is necessary for efficient inhibition of the receptor. In an HPLC aggregation analysis, it was shown that bispecific binder 30/85 was the only one that did not induce aggregation, suggesting that this binder could bind the two adjacent epitopes on the same receptor domain simultaneously. Indeed, in a rat basophilic cell granulation assay measuring inhibition of mediator release, the DARPin 30/85 showed best effect, with an inhibitory effect similar to the registered IgE-blocking drug

Omalizumab. Finally, it was confirmed that the 30/85 DARPin did not trigger cell degranulation, in contrast to two receptor-binding antibodies or crosslinked IgE. This study clearly shows that alternative scaffolds can provide a viable and sometimes superior bispecific alternative compared to antibodies.

7.3.1.5 Single-Domain Bispecific Protein Scaffolds

Most bispecific alternative scaffold molecules described to date are based on the use of several fused domains, each adding a binding specificity to the resulting fusion protein. A different approach is to engineer an additional binding specificity directly into a binding molecule. One example is the bispecific ABD molecule described below, taking advantage of the long half-life of ABD and adding onto it new specificities.

Another approach has been described by Cochran and co-workers at Stanford University, with their development of a bispecific agent antagonizing both the VEGFR-2 and αvβ3 integrin (Cochran, personal communication). Interaction between the VEGFR-2 and the αvβ3 integrin has been reported to increase VEGFR-2 phosphorylation and promote angiogenesis. In order to generate a molecule that would be able to target both of these receptors, a mutated single-chain format of VEGF (scVEGFmut), previously demonstrated to block VEGFR-2 signaling, was engineered for binding to αvβ3 integrin. A library of the mutated scVEGF protein was constructed by incorporating into one of the loops a peptide library containing the natural integrin binding RGD motif, flanked by randomized sequences. The library was displayed on yeast and integrin-binding variants were isolated using FACS. Several selected binders demonstrated retained affinity for VEGFR-2 as well as binding to αvβ3 integrin on cells, demonstrating a bispecific molecule. Using SPR technology and a sequential injection of the bispecific candidate followed by αvβ3 integrin over VEGFR-2 immobilized on the sensor chip surface, simultaneous binding to both receptors was verified. The bispecific variants were also able to inhibit VEGF-mediated VEGFR phosphorylation, cell proliferation as well as adhesion on vibronectin-coated surfaces.

In a relatively early study, Hufton et al. showed the suitability of the CTLA-4 protein for use as protein scaffold for engineering of new specificities (Hufton et al. 2000). As in the work described above, the authors first used the RGD motif flanked by a library of randomized amino acid residues to isolate the first binding specificity. They inserted the library in a CDR3 like loop on the CTLA-4 molecule, and then selected binders specific for the αvβ3 integrin. Using flow cytometry, it was showed that phage particles carrying the selected clone RGD-7B could bind to human umbilical vein endothelial cells. To investigate if the CTLA-4 protein scaffold could be developed toward bispecific applications, a short peptidic sequence was inserted in the loop formed by the A'B β-strands in the CTLA-4 framework opposite to the natural CDR-loops. Indeed, structural integrity was retained, and the CTLA-4 molecule could still bind its natural ligands B7-1 and B7-2, suggesting suitability for further development of bispecific ligands.

7.3.2 *Bispecific Antibody Mimetics with Enhanced Half-Life*

The small size of most alternative scaffolds results in good tissue penetration and rapid blood clearance, which in combination with strong and specific interaction enables high-contrast imaging of tumors for e.g., cancer diagnostics (Nilsson and Tolmachev 2007; Weissleder 2006) or therapeutic protocols where short exposure is key. In most therapeutic applications however, the short in vivo half-life may limit the therapeutic efficacy due to too low exposure, and if payloads are used, the high kidney uptake may result in toxic side effects. While a number of biologicals have been modified with PEG, or fused to a Fc-portion of Ig-molecules or to serum albumin for increased half-life, alternative scaffolds proteins can in addition be raised against proteins such as albumin or Fc to take advantage of the long circulation half-life of those serum proteins (Bertschinger et al. 2007; Jonsson et al. 2008; Tolmachev et al. 2007a, b). By fusing such a binder to another binder having a therapeutic function, a biotherapeutic with extended plasma half-life is created.

7.3.2.1 Avimers

The Avimer scaffold is based on a family of small protein domains called A-domains (Silverman et al. 2005). A-domains are typically expressed as multimers with a short linker region between each domain and are reported to be involved in interactions with several different types of targets, such as haptens, proteins, and viruses (Silverman et al. 2005). The A-domain is approximately 35 residues long and is rapidly folded into a stable structure stabilized by disulfide bridges and a calcium ion. In a study by Silverman and colleagues, the natural multimeric architecture of A-domains was exploited using a sequential selection strategy. The selection process was started using a randomized monomeric library from which binders were isolated by phage display. In following steps, additional randomized domains were fused to the first binder and the panning was repeated, thereby constructing multimeric Avimers with potentially growing interaction surfaces. Using this approach, a multidomain Avimer protein with subnanomolar affinity was selected against interleukin-6 (IL-6). To render the IL-6 binding Avimer a kinetic profile suitable for development into a pharmaceutical drug, the half-life needed to be increased. Therefore, a monomeric Avimer with low affinity binding for human IgG was developed. After fusion to the trimeric IL-6 binding Avimer, the resulting heterotetrameric construct had a terminal phase of the profile of 89 $\pm$ 15 h in cynomolgus monkeys and an expected human serum half-life of 178 $\pm$ 30 h, based on allometric scaling. The IgG-binding domain was also fused to Avimers binding to BAFF and to CD40L to enhance their half-life, showing that the approach is general.

7.3.2.2 Albumin-Binding Domain

Another approach to increase serum half-life is to use binding to serum albumin to confer the long half-life of albumin to the associated fusion protein of interest. This was recently used to increase the half-life of a radiolabeled HER2-binding Affibody molecule for therapeutic use in a murine xenograft model (Tolmachev et al. 2007a, b). It has also been used in conjunction with tandem scFvs, rendering a αCD3-αCEA-ABD bispecific antibody fragments a long half-life (Hopp et al. 2010; Stork et al. 2009). Although the ABD scaffold used for albumin binding originally bound to albumin, and as such is not a general alternative scaffold protein, it has been subject to extensive engineering, with an improved affinity to mid femtomolar K_D (Jonsson et al. 2008), as well as a complete removal of T-cell epitopes. As described below, it has also been used as a single domain scaffold into which another target binding property has been engineered while preserving the original specificity.

7.3.2.3 Bispecific Single-Domain ABD Molecules

As mentioned above, in vivo half-life extension by adding serum protein-binding capacity holds promise to offer an efficient and viable approach for non-Ig-based scaffolds to boost potency. A new strategy is to directly engineer the target-binding specificity into a serum protein-binding domain, hence reducing the overall size of the affinity protein even further. Hober and colleagues analyzed the albumin-binding properties of a natural albumin-binding domain (ABD) in order to identify positions suitable for such randomization (Alm et al. 2010). The ABD scaffold is merely 46 amino acids long and the task of engineering two high-affinity interactions, preferably onto two distinct surfaces on this domain, was indeed challenging. In a first proof-of-principle study, a phage-displayed library of randomized ABD molecules was constructed. Diversification was focused on residues on the opposite side of the albumin-binding site in the three-helical bundle in order to increase the probability of finding dual binding molecules. As a first step, the molecule was engineered into a small bispecific affinity handle for protein purification purposes. An engineered B-domain (the Z-domain) from staphylococcal protein A was used as target in the selections and ABD molecules demonstrating affinity for the Z-domain as well as having retained affinity for human serum albumin were isolated. By fusing the new bispecific ABD handle to different target proteins, excellent purification capacity could be demonstrated by affinity capture on a serially coupled protein A-based matrix column followed by an albumin-coupled column (Alm et al. 2010). The successful results inspired the group to investigate the approach for potential serum half-life extension of affinity proteins intended for therapy (JL, Sophia Hober personal communications). Using the same library, binders were selected against human tumor necrosis factor-alpha

(TNF-alpha). First-generation binders were isolated and identified, but although several were bispecific, a relatively high affinity for TNF-alpha seemed to limit the affinity for albumin and vice versa. To increase the affinity and to engineer binders with high affinity for both targets, a new library was designed based on the sequences from the first selection and the display format was changed from phage display technology to a bacterial-based cell display system. Since cell display enables the use of fluorescence-activated cell sorting (FACS), selections could be performed against both targets simultaneously by labeling TNF-alpha and HSA with different fluorophores followed by multiparameter sorting. The selection was successful and bispecific binders with high-affinity against both targets were isolated. Although challenging, the approach should be generally applicable, e.g., in selections against tumor-associated targets where a small overall size of the affinity protein is beneficial.

7.4 Conclusion

In this book chapter, the use of non-immunoglobulin-based protein scaffolds for engineering of bispecific recognition has been described. The use of alternative protein scaffolds as therapeutic antibody mimetics with improved properties over antibodies is now entering clinical validation. The promise of bispecific binding has initially been explored using the bivalent antibody format, but is increasingly expanded by adding new protein formats, both from immunoglobulin derivatives and completely unrelated scaffolds. Different problems can be addressed using bispecific alternative scaffolds, including half-life extension strategies, more efficient blocking of a receptor, better specificity for tumor cell targeting or blocking of two different receptors using only one biomolecule. A number of the alternative scaffolds in development holds promise for bispecific applications based on their ability to bind in the context of a multidomain format, the possibility to express them recombinantly as a single-chain polypeptide and their small size, allowing for multiepitope binding of surfaces too close to each other to be eligible for bispecific antibodies to access. An extension of engineering a scaffold protein for certain specificity and then to fuse it with another domain for e.g., half-life extension is to move the engineering of new specificity directly into an existing natural ligand, thus adding bispecificity to an already functional binding property. Alternative scaffolds are clearly useful and holds promise to further increase engineering possibilities to create new and optimized bispecific molecules. Many more examples of bispecific alternative scaffolds will likely be available in a near future, exploiting advantages scaffold proteins may hold over antibodies, and hopefully presenting new candidates for biotherapeutic drugs.

References

Ahlgren S, Orlova A, Wållberg H, Hansson M, Sandström M, Lewsley R, Wennborg A, Abrahmsén L, Tolmachev V, Feldwisch J (2010) Targeting of HER2-expressing tumors using ^{111}In-ABY-025, a second-generation affibody molecule with a fundamentally reengineered scaffold. J Nucl Med 51:1131–1138

Alm T, Yderland L, Nilvebrant J, Halldin A, Hober S (2010) A small bispecific protein selected for orthogonal affinity purification. Biotechnol J 5:605–617

Bertschinger J, Grabulovski D, Neri D (2007) Selection of single domain binding proteins by covalent DNA display. Protein Eng Des Sel 20:57–68

Binz HK, Amstutz P, Kohl A, Stumpp MT, Briand C, Forrer P, Grutter MG, Plückthun A (2004) High-affinity binders selected from designed ankyrin repeat protein libraries. Nat Biotechnol 22:575–582

Deisenhofer J (1981) Crystallographic refinement and atomic models of a human Fc fragment and its complex with fragment B of protein A from *Staphylococcus aureus* at 2.9- and 2.8-A resolution. Biochemistry 20:2361–2370

Dincbas-Renqvist V, Lendel C, Dogan J, Wahlberg E, Härd T (2004) Thermodynamics of folding, stabilization, and binding in an engineered protein–protein complex. J Am Chem Soc 126:11220–11230

Dogan J, Lendel C, Hard T (2006) Thermodynamics of folding and binding in an affibody:affibody complex. J Mol Biol 359:1305–1315

Ebersbach H, Fiedler E, Scheuermann T, Fiedler M, Stubbs MT, Reimann C, Proetzel G, Rudolph R, Fiedler U (2007) Affilin-novel binding molecules based on human gamma-B-crystallin, an all beta-sheet protein. J Mol Biol 372:172–185

Eggel A, Baumann MJ, Amstutz P, Stadler BM, Vogel M (2009) DARPins as bispecific receptor antagonists analyzed for immunoglobulin E receptor blockage. J Mol Biol 393:598–607

Eigenbrot C, Ultsch M, Dubnovitsky A, Abrahmsén L, Härd T (2010) Structural basis for high-affinity HER2 receptor binding by an engineered protein. Proc Natl Acad Sci USA 107: 15039–15044

Emanuel SL et al. (2011) A fibronectin scaffold approach to bispecific inhibitors of epidermal growth factor receptor and insulin-like growth factor-I receptor. MAbs 3:38–48

Engfeldt T, Renberg B, Brumer H, Nygren PÅ, Karlström AE (2005) Chemical synthesis of triple-labelled three-helix bundle binding proteins for specific fluorescent detection of unlabelled protein. Chembiochem 6:1043–1050

Engfeldt T, Orlova A, Tran T, Bruskin A, Widstrom C, Karlström AE, Tolmachev V (2007) Imaging of HER2-expressing tumours using a synthetic affibody molecule containing the ^{99m}Tc-chelating mercaptoacetyl-glycyl-glycyl-glycyl (MAG3) sequence. Eur J Nucl Med Mol Imaging 34:722–733

Feldwisch J, Tolmachev V, Lendel C, Herne N, Sjöberg A, Larsson B, Rosik D, Lindqvist E, Fant G, Höidén-Guthenberg I et al (2010) Design of an optimized scaffold for affibody molecules. J Mol Biol 398:232–247

Friedman M, Nordberg E, Höidén-Guthenberg I, Brismar H, Adams GP, Nilsson FY, Carlsson J, Ståhl S (2007) Phage display selection of affibody molecules with specific binding to the extracellular domain of the epidermal growth factor receptor. Protein Eng Des Sel 20:189–199

Friedman M, Orlova A, Johansson E, Eriksson TL, Höidén-Guthenberg I, Tolmachev V, Nilsson FY, Ståhl S (2008) Directed evolution to low nanomolar affinity of a tumor-targeting epidermal growth factor receptor-binding affibody molecule. J Mol Biol 376:1388–1402

Friedman M, Lindström S, Ekerljung L, Andersson-Svahn H, Carlsson J, Brismar H, Gedda L, Frejd FY, Ståhl S (2009) Engineering and characterization of a bispecific HER2 x EGFR-binding affibody molecule. Biotechnol Appl Biochem 54:121–131

Gebauer M, Skerra A (2009) Engineered protein scaffolds as next-generation antibody therapeutics. Curr Opin Chem Biol 13:245–255

Grönwall C, Jonsson A, Lindström S, Gunneriusson E, Ståhl S, Herne N (2007) Selection and characterization of affibody ligands binding to Alzheimer amyloid beta peptides. J Biotechnol 128:162–183

Hogbom M, Eklund M, Nygren PÅ, Nordlund P (2003) Structural basis for recognition by an in vitro evolved affibody. Proc Natl Acad Sci USA 100:3191–3196

Hopp J, Hornig N, Zettlitz KA, Schwarz A, Fuss N, Muller D, Kontermann RE (2010) The effects of affinity and valency of an albumin-binding domain (ABD) on the half-life of a single-chain diabody-ABD fusion protein. Protein Eng Des Sel. 23:827–834

Hufton SE, van Neer N, van den Beuken T, Desmet J, Sablon E, Hoogenboom HR (2000) Development and application of cytotoxic T lymphocyte-associated antigen 4 as a protein scaffold for the generation of novel binding ligands. FEBS Lett 475:225–231

James LC, Roversi P, Tawfik DS (2003) Antibody multispecificity mediated by conformational diversity. Science 299:1362–1367

Jonsson A, Dogan J, Herne N, Abrahmsén L, Nygren PÅ (2008) Engineering of a femtomolar affinity binding protein to human serum albumin. Protein Eng Des Sel 21:515–527

Kim HJ, Eichinger A, Skerra A (2009) High-affinity recognition of lanthanide(III) chelate complexes by a reprogrammed human lipocalin 2. J Am Chem Soc 131:3565–3576

Koide A, Bailey CW, Huang X, Koide S (1998) The fibronectin type III domain as a scaffold for novel binding proteins. J Mol Biol 284:1141–1151

Koide A, Gilbreth RN, Esaki K, Tereshko V, Koide S (2007) High-affinity single-domain binding proteins with a binary-code interface. Proc Natl Acad Sci USA 104:6632–6637

Kolmar H (2009) Biological diversity and therapeutic potential of natural and engineered cystine knot miniproteins. Curr Opin Pharmacol 9:608–614

Kong A, Calleja V, Leboucher P, Harris A, Parker PJ, Larijani B (2008) HER2 oncogenic function escapes EGFR tyrosine kinase inhibitors via activation of alternative HER receptors in breast cancer cells. PLoS One 3. doi:10.1371/journal.pone.0002881

Lehmann A (2008) Ecallantide (DX-88), a plasma kallikrein inhibitor for the treatment of hereditary angioedema and the prevention of blood loss in on-pump cardiothoracic surgery. Expert Opin Biol Ther 8:1187–1199

Lendel C, Dincbas-Renqvist V, Flores A, Wahlberg E, Dogan J, Nygren PA, Härd T (2004) Biophysical characterization of Z(SPA-1)–a phage-display selected binder to protein A. Protein Sci 13:2078–2088

Lendel C, Dogan J, Härd T (2006) Structural basis for molecular recognition in an affibody: affibody complex. J Mol Biol 359:1293–1304

Lipovsek D, Lippow SM, Hackel BJ, Gregson MW, Cheng P, Kapila A, Wittrup KD (2007) Evolution of an interloop disulfide bond in high affinity antibody mimics based on fibronectin type III domain and selected by yeast surface display: molecular convergence with single-domain camelid and shark antibodies. J Mol Biol 368:1024–1041

Löfblom J, Feldwisch J, Tolmachev V, Carlsson J, Ståhl S, Frejd FY (2010) Affibody molecules: engineered proteins for therapeutic, diagnostic and biotechnological applications. FEBS Lett 584:2670–2680

Milstein C, Cuello AC (1983) Hybrid hybridomas and their use in immunohistochemistry. Nature 305:537–540

Mouquet H, Scheid JF, Zoller MJ, Krogsgaard M, Ott RG, Shukair S, Artyomov MN, Pietzsch J, Connors M, Pereyra F et al (2010) Polyreactivity increases the apparent affinity of anti-HIV antibodies by heteroligation. Nature 467:591–595

Nilsson FY, Tolmachev V (2007) Affibody molecules: new protein domains for molecular imaging and targeted tumor therapy. Curr Opin Drug Discov Devel 10:167–175

Nord K, Nilsson J, Nilsson B, Uhlén M, Nygren PÅ (1995) A combinatorial library of an alpha-helical bacterial receptor domain. Protein Eng 8:601–608

Nord K, Gunneriusson E, Ringdahl J, Ståhl S, Uhlén M, Nygren PÅ (1997) Binding proteins selected from combinatorial libraries of an alpha-helical bacterial receptor domain. Nat Biotechnol 15:772–777

Nord K, Nord O, Uhlén M, Kelley B, Ljungqvist C, Nygren PÅ (2001) Recombinant human factor VIII-specific affinity ligands selected from phage-displayed combinatorial libraries of protein A. Eur J Biochem 268:4269–4277

Nygren PÅ, Skerra A (2004) Binding proteins from alternative scaffolds. J Immunol Methods 290:3–28

Orlova A, Magnusson M, Eriksson TL, Nilsson M, Larsson B, Höidén-Guthenberg I, Widström C, Carlsson J, Tolmachev V, Ståhl S et al (2006) Tumor imaging using a picomolar affinity HER2 binding affibody molecule. Cancer Res 66:4339–4348

Orlova A, Tran T, Widström C, Engfeldt T, Eriksson Karlström A, Tolmachev V (2007a) Pre-clinical evaluation of [^{111}In]-benzyl-DOTA-Z(HER2:342), a potential agent for imaging of HER2 expression in malignant tumors. Int J Mol Med 20:397–404

Orlova A, Tolmachev V, Pehrson R, Lindborg M, Tran T, Sandstrom M, Nilsson FY, Wennborg A, Abrahmsen L, Feldwisch J (2007b) Synthetic affibody molecules: a novel class of affinity ligands for molecular imaging of HER2-expressing malignant tumors. Cancer Res 67:2178–2186

Osaki A, Toi M, Yamada H, Kawami H, Kuroi K, Toge T (1992) Prognostic significance of co-expression of c-erbB-2 oncoprotein and epidermal growth factor receptor in breast cancer patients. Am J Surg 164:323–326

Sandström K, Xu Z, Forsberg G, Nygren PÅ (2003) Inhibition of the CD28-CD80 co-stimulation signal by a CD28-binding affibody ligand developed by combinatorial protein engineering. Protein Eng 16:691–697

Schlehuber S, Skerra A (2001) Duocalins: engineered ligand-binding proteins with dual specificity derived from the lipocalin fold. Biol Chem 382:1335–1342

Schlehuber S, Skerra A (2005) Lipocalins in drug discovery: from natural ligand-binding proteins to “anticalins”. Drug Discov Today 10:23–33

Silverman J, Liu Q, Bakker A, To W, Duguay A, Alba BM, Smith R, Rivas A, Li P, Le H et al (2005) Multivalent avimer proteins evolved by exon shuffling of a family of human receptor domains. Nat Biotechnol 23:1556–1561

Skerra A (2008) Alternative binding proteins: anticalins – harnessing the structural plasticity of the lipocalin ligand pocket to engineer novel binding activities. FEBS J 275:2677–2683

Steffen AC, Orlova A, Wikman M, Nilsson FY, Ståhl S, Adams GP, Tolmachev V, Carlsson J (2006) Affibody-mediated tumour targeting of HER-2 expressing xenografts in mice. Eur J Nucl Med Mol Imaging 33:631–638

Steiner D, Forrer P, Stumpp MT, Plückthun A (2006) Signal sequences directing cotranslational translocation expand the range of proteins amenable to phage display. Nat Biotechnol 24:823–831

Stork R, Campigna E, Robert B, Muller D, Kontermann RE (2009) Biodistribution of a bispecific single-chain diabody and its half-life extended derivatives. J Biol Chem 284:25612–25619

Stumpp MT, Binz HK, Amstutz P (2008) DARPins: a new generation of protein therapeutics. Drug Discov Today 13:695–701

Tolmachev V, Nilsson FY, Widström C, Andersson K, Rosik D, Gedda L, Wennborg A, Orlova A (2006) ^{111}In-benzyl-DTPA-ZHER2:342, an affibody-based conjugate for in vivo imaging of HER2 expression in malignant tumors. J Nucl Med 47:846–853

Tolmachev V, Orlova A, Nilsson FY, Feldwisch J, Wennborg A, Abrahmsén L (2007a) Affibody molecules: potential for in vivo imaging of molecular targets for cancer therapy. Expert Opin Biol Ther 7:555–568

Tolmachev V, Orlova A, Pehrson R, Galli J, Baastrup B, Andersson K, Sandström M, Rosik D, Carlsson J, Lundqvist H et al (2007b) Radionuclide therapy of HER2-positive microxenografts using a ^{177}Lu-labeled HER2-specific affibody molecule. Cancer Res 67:2773–2782

Tsutsui S, Ohno S, Murakami S, Kataoka A, Kinoshita J, Hachitanda Y (2003) Prognostic value of the combination of epidermal growth factor receptor and c-erbB-2 in breast cancer. Surgery 133:219–221

Wahlberg E, Lendel C, Helgstrand M, Allard P, Dincbas-Renqvist V, Hedqvist A, Berglund H, Nygren PÅ, Härd T (2003) An affibody in complex with a target protein: structure and coupled folding. Proc Natl Acad Sci USA 100:3185–3190

Wållberg H et al. (2010) Bi-specific affibody molecules for targeting of EGFR and HER2. Sixth annual protein engineering summit PEGS, May 17–21, Boston. Poster Abstract 78

Weissleder R (2006) Molecular imaging in cancer. Science 312:1168–1171

Wikman M, Steffen AC, Gunneriusson E, Tolmachev V, Adams GP, Carlsson J, Ståhl S (2004) Selection and characterization of HER2/neu-binding affibody ligands. Protein Eng Des Sel 17:455–462

Wikman M, Rowcliffe E, Friedman M, Henning P, Lindholm L, Olofsson S, Ståhl S (2006) Selection and characterization of an HIV-1 gp120-binding affibody ligand. Biotechnol Appl Biochem 45:93–105

Xu L, Aha P, Gu K, Kuimelis RG, Kurz M, Lam T, Lim AC, Liu H, Lohse PA, Sun L et al (2002) Directed evolution of high-affinity antibody mimics using mRNA display. Chem Biol 9:933–942

Chapter 8
Tribodies: Fab–scFv Fusion Proteins as a Platform to Create Multifunctional Pharmaceuticals

Nico Mertens

8.1 Introduction

Our proper immune system is a powerful but highly complex regulated system. Powerful since it is capable of specifically repel bulky cell masses (think of transplantation) and defend us quiet successfully against a pleitropy of invaders, and this with minimal damage to healthy tissue. This is due to a high specificity and self-regulation build into a complex system, i.e., a system with a multitude of redundancies and feed-back loops. Antibodies are one way in which the system incorporates specificity. The main function of antibodies is to opsonize bacteria and deliver them to neutrophils and macrophages. They also have a natural function of clearing toxic compounds or circulating viruses out of the blood and target them for clearance. Cellular malignancies such as viral infected cells are more effectively halted by the cellular arm of the system, mainly by the action of T-cells and NK-cells. Only the latter carry a receptor for opsonizing antibodies.

Due to the high specificity and high affinity, we can find in antibodies and the ease of developing recombinant proteins and adapt them to our needs antibodies became a first means for specific immunotherapy, with a special interest in trying to specifically target cancer cells. Through their Fc part, antibodies are able to induce antibody-induced cell death (ADCC) and complement-depended cytotoxicity (CDC). Target-cell bound antibodies are recognized through Fc-receptors on effector cells, such as the high-affinity Fc-receptor FcγRIII on macrophages and activated neutrophils, and also to inhibiting receptors such as FcγRIIb and on FcγRIIa complexes on non-cytotoxic cells such as platelets and B-cells. However, experience with antitumor antibodies has shown that effective antibodies also have an influence on the target cell by the nature of the receptor (or even epitope) they target. Ways this can be done is by growth inhibition (e.g., by downregulation of

N. Mertens (✉)
Biotecnol SA, Lagoas Park, Edificio 7, 2741–901 Oeiras (Lisbon), Portugal
e-mail: nm@biotecnol.com

R.E. Kontermann (ed.), *Bispecific Antibodies*,
DOI 10.1007/978-3-642-20910-9_8, © Springer-Verlag Berlin Heidelberg 2011

growth factor receptors), rendering the cancer cells more susceptible to apoptosis, or even induce cell lysis.

Receptors without such a function have been targeted with antibodies with much less clinical antitumor effectiveness. This created a possibility to develop molecules with a comparable affinity and specificity for these receptors as antibodies can develop, but with a different effector mechanism. Effective routes under investigation are to couple toxic payloads or functions to the targeting molecules (such as the antibodies) that either need to be internalized or stay at the cell surface. The toxins can be of variable nature, such as protein toxins with enzymatic activity, chemical drugs or radionucleotides. Although such fusions are effective one problem is to target enough into the tumor. The percentage of injected dose (%ID) that actually reaches the tumor is often very low in human patients (can be in the order of less than 1%) (Steffens et al. 1997). As long as these "armed" antibodies or antibody derivatives are not bound to the tumor or excreted they remain circulating and can cause toxic effects elsewhere. Since monoclonal antibodies have a special rescue mechanism by the FcRn binding in liver cell endosomes, they can have half lives of up to 3 weeks (Borvak et al. 1998). This eventually allows good tumor targeting but results in a bad ratio of targeted versus nontargeted (circulating or deposited) antibody.

So there is a clear need to develop targeting reagents which have a good tumor binding but are cleared more rapidly from the rest of the body. Smaller fragments still containing the antigen binding properties of the antibody have been used as carrying vehicles of toxic functions. However, when these proteins are below 60–80 kDa in size they will be cleared rapidly by the kidneys. Kidneys work as an ultrafiltration device with a 6–10 nm cut-off value. As a comparison: a Fab molecules is about 8×5 nm, so it can pass readily through the pores. With a glomerular filtration rate in humans of about 7 L/h, a short half life can be predicted for these small fragments. This is indeed experimentally noticed using a variety of small antibody fragments. In most cases, half lives were determined to be less than 1 h, with disappointing tumor accumulation as a result (Smith et al. 2001).

An intermediate-sized molecule should avoid kidney clearance and ideally provide a half life sufficient for improved tumor accumulation while also avoiding a body detainment which would hamper targeting toxic function to the tumor.

One way of combating tumor cells different from the normal monoclonal antibody functions is by using antibody functions to cross-link cytotoxic T-cells with tumor cells. This is done using bispecific antibodies (BsAb) targeting both the tumor and the T-cell receptor (TCR). Such a strategy involves T-cells in the induced immunotherapy. T-cells are naturally involved in combating cellular targets but are left inactive when invoicing a cure with monoclonal antibodies (moAb). T-cell engaging bispecific antibodies are thus a complementary approach to moAb therapy. In this strategy, it is important to have a monovalent binding toward the T-cell since bivalent binding will lead to initial T-cell activation but then induces T-cell death or anergy. This is why Orthoclone (OKT3 moAb anti-TCR) or its $(Fab)_2$ fragments are used with the aim to tolerize tissue transplants (Todd and Brogden 1989). On the other hand, a bivalent binding toward the tumor cell can be more advantageous since the avidity will increase the functional binding and targeting toward the tumor.

We have created a manifold that combines two scFv fragments with a Fab fragment. The Fab fragment serves as a specific heterodimerization signal, and the two scFv fragments are each fused to a different Fab chain. In this way, we obtain a molecule of intermediate molecular weight (100 kDa) which allows incorporating three different antibody fragments (Schoonjans et al. 2000a, b). This manifold, which we baptized "tribody," can be used to create trivalent molecules as well as bispecific molecules with bivalent binding to only one target, as trispecific molecules.

8.2 Tribody Structure, Production, and Stability

8.2.1 What is a Tribody?

Tribodies are multifunctional recombinant antibody derivatives, which utilize the natural in vivo heterodimerization of the heavy chain (Fd fragment) and light chain (L) of a Fab fragment, to form a scaffold, upon which additional functions can be incorporated, such as additional binders – e.g., scFv binding domains.

Each chain can be extended preferably at the C terminus with an additional scFv binder. This leads to a very specific heterodimerization, using molecules abundantly present in serum (and hence non-immunogenic and non-antigenic). The only nonnatural sequences in the molecule are the peptide linkers connecting the VH and VL domains in the scFv moiety and the linkers connecting the Fab chains and the scFv (Schoonjans et al. 2000a, b, 2001). These sequences can, however, be chosen to resemble natural human linker sequences or be composed of glycin-polymers which are considered to be neither antigenic (recognized by antibodies) nor immunogenic (recognized by T-cells when presented by antigen-presenting cells), probably due to the lack of amino acid side chains.

This design allows easy engineering of multispecificity in a single molecule. The Fab chains (Fd and L chain) can be extended at both the N- and C-terminal side. However, N-terminal extension at the Fab chains often leads to a hampered binding function of the Fab fragment (unpublished observations), so our preferred model will have the extensions made at the C-terminal side of the Fd and L-chain (Fig. 8.1).

The genetic structure of these fusion protein encoding genes thus starts with the coding sequence of either the Fd or the L chain, extended with a linker sequence separating the Fab from the scFv and the scFv molecule. Of course, the scFv molecule can be substituted by other binding structures. New developments are producing single domain binders (human, camel, or shark derived) that can substitute for the two-domain scFv molecule. Furthermore, non-antibody domains often selected from more stable proteins allow the insertion of randomized loops and thus can be used for selecting an antibody-like type of binding to any desired target

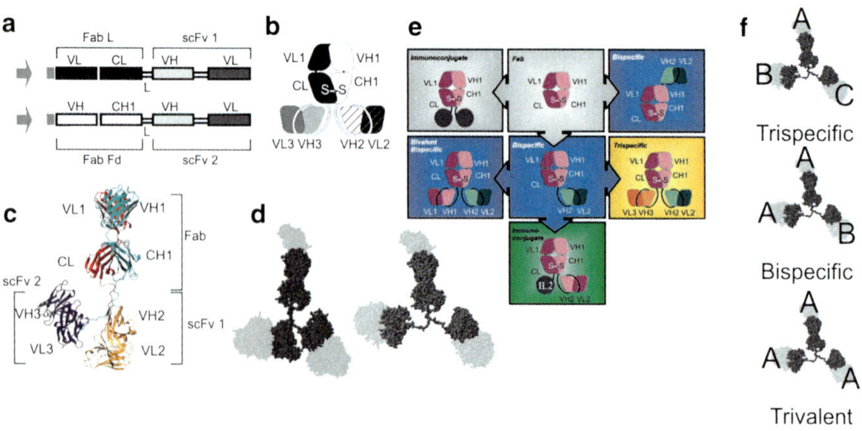

Fig. 8.1 (**a**) Tribodies are genetically constructed by fusing entities (i.e., to the Fd and L chain of a Fab molecule. As a Fab is a natural disulphide stabilized heterodimer, the position of each function is exactly determined and a single homogeneous product is produced. (**b**) Schematic representation of a Tribody Structure. (**c**) Models of the molecule predict no sterical hindrance between the subunits. (**d**) Even as the scFv are N-terminally linked, their binding site is facing outside. (**e**) Tribodies can lead to 3 × 1 (trispecific), 2 + 1 (bispecific) or trivalent molecules, as well as immunotoxins, immunocytokines, enzyme fusions, etc. (**f**) The model easily allows trispecific, bispecific, and trivalent antibody derivatives to be produced

Table 8.1 Examples of alternative binding domains in development

Domain for targeting	Derived from	Platform trade name	Company
Antibody derived			
scFv	Human antibody VH + VL domains	SCA	Micromet/Enzon
Human V_H or V_L	Stable human V domains	dAbs	Domantis GSK
Camel/Llama VHH	Camel or Llama heavy chain mAbs	Nanobodies	Ablynx
Non-antibody derived			
Ankyrin repeats	Synthetic human ankyrin repeats	DARPins	Molecular Partners
Fibronectin	Trimeric human plasminogen binding protein	Tetranectins	Borean Pharma
Fibronectin	10th fibronectin type III domain (human)	AdNectins	Adnexus Therapeutics
Transferrin	Non-glycosylated form of human transferrin	Trans-bodies	BioRexis (Pfizer)
Lipocalins	Derived from *P. brassicae* butterfly	Anticallins	Pieris Proteolab
γ-Crystallin	Stable human protein from eye lens	Affilins	Scil Proteins
Cystein knots	*E. elaterium* trypsin inhibitor II – cyclotide EETI-II	Microbodies	NascaCell Technologies
Synthetic coiled coil	De novo design	Alphabodies	Complix
Protein A domains	*S. aureus* protein A domain	Affibodies	Affibody

(Table 8.1). Also other domains or proteins can be included into the scaffold, depending whether the total fusion product can be produced. Examples here are natural protein ligands, cytokines, receptor domains, tags, or enzymes.

A molecular model of a tribody molecule is shown in Fig. 8.1. Noteworthy is the orientation of the binding planes as present in the scFv molecules, which are oriented toward the outside (as opposed to orientation toward the Fab fragment), predicting the possibility to cross-link cells and large molecules. Through the choice of flexible linkers of a reasonable size (e.g., 15 amino acids), a large span of the molecule can be predicted, comparable to the span range of a monoclonal antibody. We observed, however, no difference in expression or functionality when using shorter five amino acid linkers (Schoonjans et al. 2000a, b).

8.2.2 Tribody Expression

The two fusion genes Fd-scFv and L-scFv can be produced in a variety of host cells. Production in *Escherichia coli*, however, resulted in a large fraction of precipitated protein as inclusion bodies. Also, a significant fraction of the produced material was degraded. In one attempt, a final purified batch of *E. coli* produced tribody yielded only 0.1 mg/L and had a lower specific activity when compared with the same protein produced in mammalian cell culture.

Mammalian cell culture is becoming more and more advanced for high-level production of antibodies (especially IgG formats). We found that tribodies are easily expressed in a range of mammalian cells, including HEK293, NS0, SP2/0, and CHO cells. Mammalian cells all posses the BiP protein, an HSP70 class of chaperone present in the endoplasmic reticulum. This chaperone is involved in guiding the correct assembly of the antibody Fab fragment. It mainly binds to the CH1 domain of the Fd chain and is only displaced after displacement by correct Fd: L chain pairing (Lee et al. 1999). As a consequence, no free Fd chains are found in the culture medium. This quality control mechanism does not appear to be inhibited by the C-terminal extensions to the Fab chains. L:L dimers can be secreted by normal and malignant B-cells and also L-scFv:L:scFv dimers are found in the supernatant of transfected cells. We never found them to interfere with overall production since they are easily removed by simple ion exchange chromatography. It is, however, a good strategy to select a final cell line which produces less fee light chain and base a purification strategy on the Fd-chain properties. For ease of evaluation, we often incorporate a hexa-histidine tag at the C terminus of Fd-containing fusion protein (e.g., Fd-scFv-H6).

Typical transient expression levels in either adherent or suspension HEK239 cells growing in T-flasks or shake flasks are in the tens of mg/L range. These values are obtained using a vector with an actin promoter, a Kozak-optimized mRNA, and a standard IgG kappa chain signal sequence. Expression yields using stable transformed cells are of course highly dependent on the cell selection procedure and thereafter from the bioreactor conditions. Without putting effort in the selection of high-level producing cells and using laboratory shake flask conditions, we have examples where yields higher than 100 mg/L were obtained. This leads us to assume that high-level yielding bioreactor runs can be obtained with a >1 g/L yield as is the case for most IgG productions.

Tribodies can also be expressed in yeast cells (Schoonooghe et al. 2009). We have demonstrated more or less similar overall yields using the yeast *Pichia pastoris* with an AOX promoter and the α-mating type prepro sequence. We constructed bicistronic expression plasmids to be used for integration since double integration events occur with very low yield. Also in *Pichia*, no free Fd chains were observed in the medium. This indicates that the yeast chaperone system can substitute for the quality control mechanisms as seen in mammalian cells. Even when using defined medium for growing both host cells, the resulting cleared medium from a *Pichia* fermentation contained considerably more contaminants than cleared medium from serum-free growing mammalian cells. This resulted in a lower overall recovery yield after purification (20% in *Pichia pastoris* when compared with 70% from mammalian culture supernatants). The tribodies produced in *Pichia* showed the same functionality as those derived from mammalian cells. The α-mating prepro sequence is highly efficient in guiding recombinant proteins to the medium (as opposed to a major route to the vacuoles, which is the natural deposit for most yeast-secreted proteins). This prepro-sequence is removed by KEX-2 protease. This protease prefers a Gln-Ala-Gln-Ala repeat after the cleavage site. These four amino acids can be removed by the di-aminopeptidase STE-13. We found that the N terminus of antibody fragments produced with this strategy to be heterogeneous as a result of partial or no removal of the extra amino acids. Also, when producing glycosylated tribodies in yeast cells, it must be taken into account that yeast glycosylation differs from mammalian type of glycosylation which can lead to severe effects on half live in vivo and antigenicity.

8.2.3 Purification and Stability

Since most tribodies have a p*I* value of 7–8, we found a cation exchange fast flow column at pH 5.5 resolved with discrete step elutions an ideal and practical capture step. Further purification was dependent on the antibody. As mentioned, the inclusion of a hexahistidine tag simplifies this by using a common IMAC column. In this way, a simple purification step based on immobilized metal chromatography can already lead to a highly enriched protein and produces only Fd-scFv:L-scFv heterodimers. Other purification strategies, however, could also be designed, e.g., based on protein-L, MEP-hypercel, or a combination of ion exchange steps (Willems et al. 2003). As a polishing step, gel filtration has the advantage to be able to separate dimers and possible aggregates that might have been copurifying. Typical, a purified and concentrated sample of tribody shows a small fraction of dimer. The amount of dimerized or even multimerized protein is highly dependent on the nature of the scFv proteins used. This is a behavior known to be associated with scFv molecules (Arndt et al. 1998). It is possibly a result of the inherent domain swapping property of the scFv molecules, which leads to a diabody type of dimerization. We have noticed that this behavior to be even more pronounced when making $sc(Fv)_2$ constructs. Using the same scFv sequences in a tribody structure (i.e., fusing them to the Fab) reduces the dimers and multimers already considerably. The dependency on the nature of the scFv used thus stresses the importance of proper selection of lead candidate scFv molecules, which should not only be characterized for binding and epitope selection but hence also for its lack of tendency to multimerize when incorporated in a more complex manifold [such as even a $sc(Fv)_2$]. Remarkably, for many tribodies, we could purify the monomeric fraction and this fraction remained monomeric upon concentration and storage.

Tribody preparations were stored at 4°C in PBS at 1 mg/ml concentration and as such stable for at least 3 months. However, an even higher stability can be achieved in a storage buffer with a pH of 5–6, well away of the p*I* of 7–8 of most molecules. Freeze–thawing stress testing again revealed large differences in behavior depending on the antibody V-domain sequences used and even on the position of these domains (Fig. 8.2), again indicating the need for extensive characterization and monitoring starting with building block selection, but even extending to selecting the optimal configuration within the manifold. Some tribodies could be repeatedly frozen and thawed in PBS buffer lacking any cryoprotectant or stabilizing additive without any sign of aggregation, while others were very prone to aggregate upon such treatment.

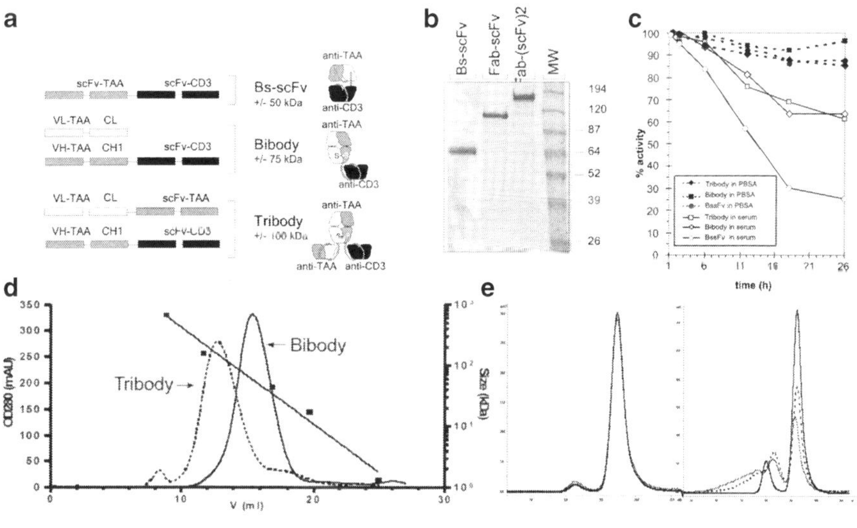

Fig. 8.2 Production and stability. (**a**) Gene and protein structure of an $sc(Fv)_2$ (BiTE), Fab–scFv (bibody), and Fab–$(scFv)_2$ (tribody) format. (**b**) Coomassie brilliant blue stained gel of 10 μg of purified fractions. (**c**) Tribody (*squares*) and bibody (*diamonds*) residual activity in T-cell activation assay after 24 h incubation at 37°C in PBS (*dotted lines*) and freshly prepared mouse serum (*solid lines*) (**d**) Preparative gel filtration of bibody and tribody structures and column calibration. (**e**) Two different tribody structure analyzed on analytical size exclusion before (*solid line*) and after one (*dashed line*) or two (*dotted line*) freeze–thaw cycles (−80 to 37°C)

8.3 Tribody Activity

8.3.1 In Vitro Binding and T-Cell Activation

To study the use of these tribodies in a natural immune environment, we constructed BsAbs to target BCL1 (an IgM/λ idiotypic determinant expressed on the murine myeloma cell line BCL1) and murine TCR determinant CD3ε (Fig. 8.3). The BCL1 system has been well characterized and is a model for NHL in a syngeneic immunocompetent Balb/c mouse model (Brissinck et al. 1991). As a comparison, we used the "gold standard" for small bispecific antibody formats: the bispecific scFv (BsscFv or BiTE) format. This is the smallest BsAb (50 kDa) used

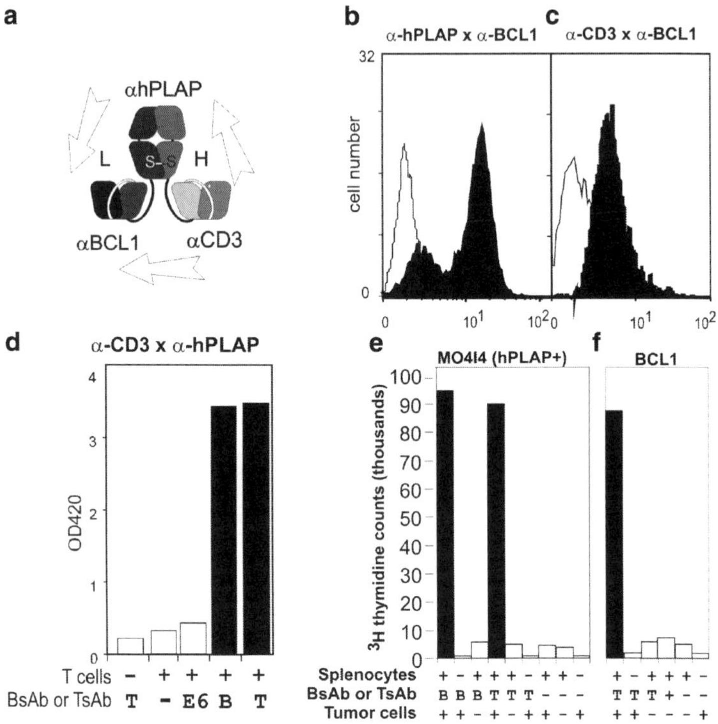

Fig. 8.3 Binding and T-cell activation. (**a**) A tribody was constructed having specificity for two different TAA: mouse BCL1 and human placental alkaline phosphatase (hPLAP). The functionality of each of the indicated axes is demonstrated by binding to one antigen and detecting with the second. (**b**) binding of the hPLAPxCD3xBCL1 Tribody to mouse T cells and detecting with hPLAP (which has alkaline phosphatase activity). (**c**) Binding of the tribody to MO4I4 hPLAP-transfected cells and detected with fluorescent BCL1 antigen, and the same detection after binding the tribody to mouse T-cells. (**d**) Tribodies (T) and Bibodies (B) bound to T-cells via the anti-CD3 can be detected with the hPLAP antigen. (**e**) The hPLAPxCD3xBCL1-trispecific antibody acts as a bispecific antibody in a mouse T-cell proliferation assay primed both with hPLAP-positive cells and with BCL1 cells (**f**)

and was composed of two scFv linked together by a peptide linker. This BsscFv (BCL1xCD3) was previously reported to be successful in treating BCL1 lymphoma bearing mice. We engineered BsAbs of intermediate size (75–100 kDa) in the tribody format by fusing single-chain variable fragments (scFv) to the C terminus of one or both of the Fd and L chains of a Fab fragment. Starting from the anti-BCL1 scFv, a chimeric anti-BCL1 Fab-fragment was constructed by grafting the variable domains onto murine CL and CH1 constant domains of a Fab fragment, We then fused an anti-mouse CD3 2C11 scFv to the C terminus of the heavy chain of the chimeric Fab-fragment and an anti-BCL1 scFv to the C terminus of the light chain of the chimeric Fab-fragment. Coexpression of the heavy chain Fab–scFv fusion gene with, respectively, the light-chain Fab gene or the light-chain Fab–scFv fusion gene, lead to the production of an (BCL1 × CD3) Fab–scFv bibody, or a (BCL1 × BCL1 × CD3) Fab–$(scFv)_2$ tribody (Fig. 8.4). The tumor binding mode (anti-BCL1) is different in all three BsAbs: In the BsscFv the tumor is recognized via a scFv, in the bibody via a Fab-fragment, and in the tribody via both a Fab-fragment and a scFv. An ELISA experiment determined whether these differences resulted in a different functional affinity for the tumor antigen BCL1. To this end, the BCL1 tumor antigen was coated and subsequently incubated with a serial dilution of the respective BsAbs, using equimolar amounts. The bound Bs-Abs were then detected via the C-terminal His-tag, with a His-tag-specific mAb. The B_{50} value of the bivalent binding Fab–$(scFv)_2$ tribody was 3 nM, while the monovalent binding Fab-fragment in the bibody had a ten times higher B_{50} (30 nM). Binding through the scFv alone in the BsscFv lead to a further decrease in apparent binding affinity.

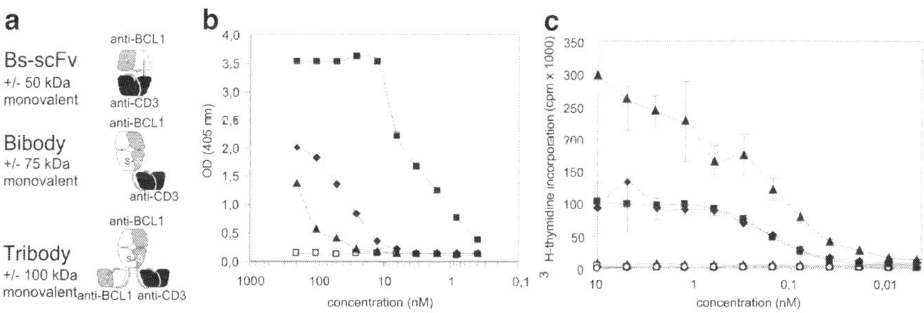

Fig. 8.4 (**a**) In vitro comparison of bispecific antibodies crosslinking BCL1-cells to T-cells in an $sc(Fv)_2$ (BiTE) format, a Fab–scFv (bibody), and Fab–$(scFv)_2$ (tribody) format. (**b**) Apparent affinity of the BsscFv (BCL1xCD3), bibody (BCL1xCD3), and tribody (BCL1xBCL1xCD3). An ELISA plate was coated with the BCL1 IgM tumor antigen and subsequently incubated with, respectively, the BsscFv (*filled dashed triangle*), the bibody (*dashed bullet*), the tribody (*filled dashed square*), and an irrelevant bibody (*open dashed square*). The bound BsAbs were detected via their His-tag, with an anti-Histag Ab and a secondary alkaline phosphatase-conjugated anti-mouse-IgG1. Data are representative for three independent experiments. (**c**) In vitro T-cell activating potential of the BsscFv, bibody, and tribody. Mitomycin C-treated BCL1 cells were coincubated with syngeneic Balb/c spleen cells in the presence of decreasing concentrations of the indicated BsAb. Each condition was tested in triplicate

In a T-cell proliferation assay, it was determined whether these differences in tumor binding affinity had an effect on the T-cell activating potential of the different BsAbs in vitro. A serial dilution of the different BsAbs, starting from equimolar amounts, was incubated with mitomycine-inactivated BCL1 tumor cells and syngeneic spleen cells. All three BsAbs were found to be capable of activating T-cells in the presence of tumor cells. Despite the difference in affinity seen with ELISA, no difference in T-cell activating potential could be observed using both the monovalent binding bibody and the BsscFv. However, the bivalent binding tribody exerted a clearly improved T-cell activating potential. The maximum T-cell proliferation reached with the bivalent tribody was twofold higher than with the monovalent bibody or BsscFv (Fig. 8.4). Moreover, the bivalent tribody remained capable of inducing T-cell proliferation at fourfold lower concentration compared with the monovalent bibody and the BsscFv. Also in targeting other tumor markers, the increase in valency lead to better tumor binding (Schoonooghe et al. 2010).

8.3.2 *In Vivo Model: Mouse Bcl1 Lymphoma*

In a next step, it was investigated whether the differences in molecular weight and tumor avidity would also lead to a different therapeutic potential of the BsAbs. Groups of mice ($n = 13$) were inoculated i.p. with 5000 BCL1 cells on day 0 and i.v. (in the tail vein) treated with four daily injections on days 2–5. Treatment was performed with different equimolar amounts (200, 100, or 50 pmol per injection) of the BsscFv, the bibody, or the tribody. In addition, a group of six mice were injected with PBS or 200 pmol per injection of anti-Id mAb that targets bivalently to the BCL1 tumor surface. All the animals in the control group receiving PBS or anti-Id mAb treatment developed terminal illness and were euthanized by day 60 or 140, respectively (Fig. 8.5). After dissection the spleen was found to be enormously

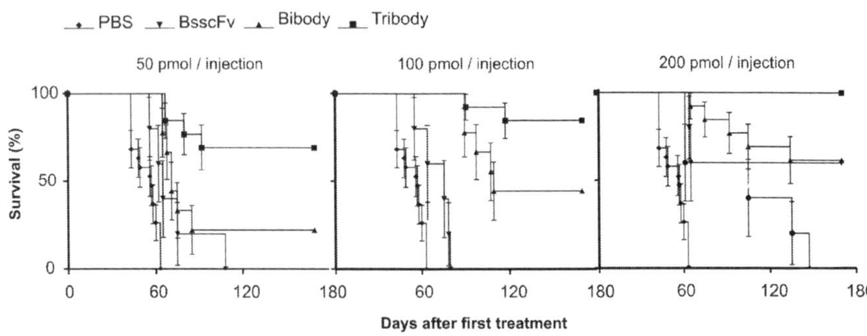

Fig. 8.5 Balb/C mice were inoculated with 5000 BCL1 lymphoma cells i.p. in 100 μl sterile and endotoxin-free PBS on day 1 of each experiment. During treatment, the BsAbs (aggregate and endotoxin free) were injected i.v. in the tail vein on days 2–5 in a volume of 200 μl sterile PBS. Animals were followed until a swollen abdomen could be observed, and then euthanized by cervical dislocation. Results are presented as a Kaplan–Meyer plot

enlarged by the massive presence of tumor cells. In contrast, the mice treated with 200 pmol per injection of each of the BsAbs demonstrated significant ($P < 0.0001$) protection against tumor formation in the spleen: after treatment with the BsscFv or the Fab–scFv bibody, 60% survival was observed, while treatment with the Fab–(scFv)$_2$ tribody even resulted in 100% protection of the mice. Lowering the dose of BsscFv treatment to 100 pmol per injection and 50 pmol per injection resulted in the loss of protection. The Fab–scFv bibody treatment, however, remained active at 100 pmol per injection resulting in 45% survival, which was found to be statistically significant ($P < 0.0001$). At 50 pmol per injection, the Fab–scFv bibody protected only 20% of the mice. However, the bivalent Fab–(scFv)$_2$ tribody manifold was able to protect mice at concentrations were other manifolds were ineffective (70%, 85%, and 100% survival at 50, 100, or 200 pmol injected four times, respectively) ($P = 0.0190$, $P = 0.0371$, and $P = 0.0059$). Mice surviving after tribody treatment at the 200 pmol per dose treatment were 100% cured and did not develop tumors during a follow-up of 200 days. This argues for a effective tumor elimination and against the antibody induced dormancy of lymphoma cells.

8.3.3 PK Measurements

We also determined whether the size of the different BsAbs significantly influences the in vivo blood clearance rate. Mice were i.v. injected in the tail vein with 800 pmol of either the BsscFv, the bibody, or the tribody. At different time points after injection, mice were bled and the remaining biological activity in the serum was determined by a T-cell proliferation assay. A typical biphasic blood clearance was observed for all three antibody derivatives (Fig. 8.5). The initial blood distribution phase half-life ($T_{1/2}\alpha$) was found to be comparable for both the bibody and tribody, i.e., 56 and 72 min, respectively, but was considerably lower for the BsscFv (13 min). The terminal blood elimination phase half-life ($T_{1/2}\beta$) was found to correlate directly with the molecular weight of the antibody derivatives. The BsscFv was cleared fast ($T_{1/2}\beta = 1.5$ h), while the bibody and tribody remained longer in circulation ($T_{1/2}\beta = 2.9$ and 5.7 h, respectively).

As a test for biodistribution, we radiolabeled the anti-BCL1 IgG1, BsscFv, and tribody with ^{123}I and analyzed six mice 1 h after injection. The distribution in selected tissues is shown in Fig. 8.6b. It can be concluded that the BsscFv format is rapidly accumulating in the kidneys while the IgG1 as well as the tribody is not.

We also compared serum clearance and tumor accumulation of these radiolabeled proteins (Fig. 8.6c). As expected the IgG1 clears very slowly. The BsscFv, however, is quickly eliminated while the tribody has an intermediate clearing time. The accumulation in a limb-injected tumor nodule was followed over a period of 24 h and shows low accumulation of the BsscFv in contrast to an increased accumulation of the IgG1 and the tribody.

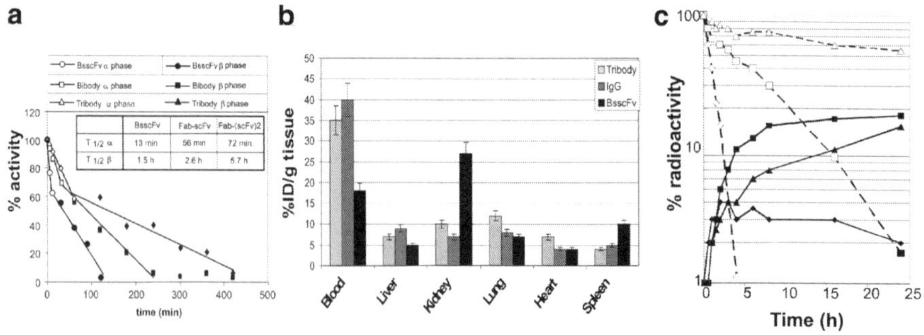

Fig. 8.6 (**a**) Healthy Balb/c mice were i.v. injected with 800 pmol of the Ab-fragments into the tail vein. At various time points, mice were bled and serum was frozen. The remaining biological activity of the Bs-Abs in the serum was analyzed by a T-cell proliferation assay. The activity present 5 min after injection (i.e., time required for anesthetic to take effect) was used as zero time point activity. (**b**) ^{123}I labeled tribody, IgG1, and BsscFv were injected in the tail vein and a group of six mice was sacrificed after 1 h. Selected tissues were counted and compared. (**c**) Comparison of ^{123}I labeled IgG1 (*triangles*), tribody (*squares*), and BsscFv (*diamonds*) serum clearance (*dotted lines*) and accumulation into a limb tumor nodule (*solid lines*) over a 24 h period, plotted on a logarithmic scale

These data indicate that introducing bivalency and increasing half life can increase the potency of a reagent in vivo. The tribody manifold already does this by incorporating a third binding function.

8.4 Tribody Use and Potential

As mentioned, there are a lot of molecules that can be recombinantly fused to be heterodimerized by the Fab chains. I will discuss one example where we engineered a cross-interacting pretargeting system using this scaffold. Bispecific antibodies binding when crosslinking a T-cell via CD3 to the tumor cell do activate the T-cell both for proliferation and cytolytic response. This has been shown as well in vitro as in clinical trails (Baeuerle et al. 2009). However, T-cells are more readily activated when a costimulatory signal is present (Gimmi et al. 1993). The CD28 coreceptor is one of the first costimulatory signals identified and is probably the most potent one. Most antibodies targeting CD28 alone do not activate T-cells and are as such not toxic, although superagonistic CD28 binders have been described and found to be dangerous. In general, the activation trigger for a T-cell comes from CD3 engagement. Without a proper environment (co-stimulus), T-cells can go into anergy or cell death. These costimulatory signals are given by activated antigen presenting cells, another control loop to avoid accidental activation of the immune system. A single bispecific or trispecific antibody crosslinking both the CD3 TCR and the CD28 costimulatory receptor does activate T-cells even without crosslinking to a tumor cell. This might induce systemic T-cell activation with associated cytokines storms and be potentially life-threatening (Stebbings et al. 2007).

We have created a couple of tribodies that had a build-in cross-reactivity (cross-reactive bispecific antibodies, CriBs). This was done by including a small peptide in the first and an anti-peptide scFv in the second. One of both targeted tumor and CD28 and was completely harmless, unable to activate T-cells on its own.

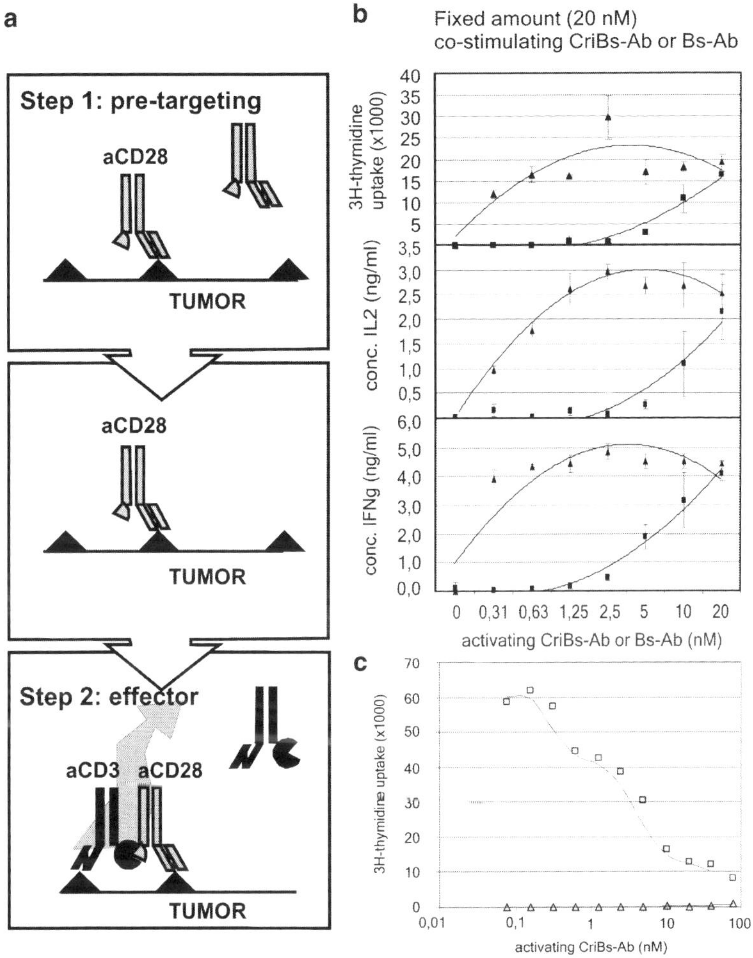

Fig. 8.7 (**a**) A couple of cross-interacting bispecific antibodies (CRIBs) was constructed where a nonactivating tribody is pretargeted to crosslink the tumor cell with the T-cell CD28 co-stimulus. This tribody contains a unique peptide P. A second tribody bind tumor, CD3 on the T-cell and the P-peptide and can crosslink with the first upon binding by the P–αP interaction. (**b**) Mitomycine-treated tumor cells were preincubated for 1 h with a fixed amount (20 nM) of anti-(TAAxCD28) BsAb (*open dashed square*) or anti-(TAAxCD28)P CriBs-Ab (*open dashed triangle*) followed by a 1 h incubation with a serial dilution (0–20 nM) of anti-(TAAxCD3) BsAb (*open dashed square*), or anti-(TAAxCD3xP) CriBs-Ab (*open dashed triangle*). Preincubated tumor cells were then coincubated with T cells. T-cell activation was measured by proliferation, IL-2 production, and IFNγ production. Each condition was tested in triplicate. (**c**) Dose-dependent activation of T-cells in the presence (*squares*) and absence (*triangles*) of tumor cells in a range from 0.1 to 100 nM

The second could bind the tumor and CD3 and also the P-peptide includes in the first. This allowed us to pretarget the anti-CD28 activity on the tumor cell, and then give the TAAxCD3xP tribody to activate the T-cell. We found that using this CriBs system, efficient T-cell activation could be achieved at 30-fold lower concentrations of the anti-CD3-containing bispecific when compared with a pair of non-cross interacting bispecific antibodies (Willems et al. 2005). Also, an optimized stoichiometry of co-stimulus versus TCR engagement could be achieved with a large window between effective concentrations and start of marginal non-tumor cell-induced T-cell activation (Fig. 8.7). This might illustrate the potential and flexibility of tribody molecules.

8.5 Conclusion

Clearly, above mentioned data illustrate that the tribody manifold has the potential to generate more active and potent molecules, by possibilities to create better binding (through multivalency) as well as through more binding (by targeting more antigens). The molecule also has more favorable properties for toxic payload delivery. It can be argued that delivering a T-cell activating activity can be regarded as a toxic function. Antibodies with such a function remaining in the body for a longer time have the potency to accumulate aspecifically or form deposits in healthy tissue.

To be eligible for drug development, the tribody needs to be able to be produced to high titers (g/L) and remain stable under standard manufacturing conditions. The tendency to form dimers or multimers can be a problem both in manufacturing and hamper several applications, tumor cell-dependent T-cell activation being one example.

Surprising differences can be seen with distinct building blocks (scFv as well as Fab moieties). This can be situated on the expression level, binding performance as well as on the level of tendency to form multimers. Moreover, permutations within the scaffold of the same moieties can induce the same differences. In our experience, a well integrated and professionally run technology platform which closely monitors these factors can produce drug development ready tribody lead candidates in relative short-time frames.

References

Arndt KM, Muller KM, Pluckthun A (1998) Factors influencing the dimer to monomer transition of an antibody single-chain Fv fragment. Biochemistry 37(37):12918–12926

Baeuerle PA, Kufer P, Bargou R (2009) BiTE: teaching antibodies to engage T-cells for cancer therapy. Curr Opin Mol Ther 11(1):22–30

Borvak J, Richardson J, Medesan C, Antohe F, Radu C, Simionescu M, Ghetie V, Ward ES (1998) Functional expression of the MHC class I-related receptor, FcRn, in endothelial cells of mice. Int Immunol 10(9):1289–1298

Brissinck J, Demanet C, Moser M, Leo O, Thielemans K (1991) Treatment of mice bearing BCL1 lymphoma with bispecific antibodies. J Immunol 147(11):4019–4026

Gimmi CD, Freeman GJ, Gribben JG, Gray G, Nadler LM (1993) Human T-cell clonal anergy is induced by antigen presentation in the absence of B7 costimulation. Proc Natl Acad Sci USA 90(14):6586–6590

Lee YK, Brewer JW, Hellman R, Hendershot LM (1999) BiP and immunoglobulin light chain cooperate to control the folding of heavy chain and ensure the fidelity of immunoglobulin assembly. Mol Biol Cell 10(7):2209–2219

Schoonjans R, Willems A, Grooten J, Mertens N (2000a) Efficient heterodimerization of recombinant bi- and trispecific antibodies. Bioseparation 9(3):179–183

Schoonjans R, Willems A, Schoonooghe S, Fiers W, Grooten J, Mertens N (2000b) Fab chains as an efficient heterodimerization scaffold for the production of recombinant bispecific and trispecific antibody derivatives. J Immunol 165(12):7050–7057

Schoonjans R, Willems A, Schoonooghe S, Leoen J, Grooten J, Mertens N (2001) A new model for intermediate molecular weight recombinant bispecific and trispecific antibodies by efficient heterodimerization of single chain variable domains through fusion to a Fab-chain. Biomol Eng 17(6):193–202

Schoonooghe S, Kaigorodov V, Zawisza M, Dumolyn C, Haustrate J, Grooten J, Mertens N (2009) Efficient production of human bivalent and trivalent anti-MUC1 Fab-scFv antibodies in *Pichia pastoris*. BMC Biotechnol 9(1):70

Schoonooghe S, Burvenich I, Vervoort L, De Vos F, Mertens N, Grooten J (2010) PH1-derived bivalent bibodies and trivalent tribodies bind differentially to shed and tumour cell-associated MUC1. Protein Eng Des Sel 23(9):721–728

Smith BJ, Popplewell A, Athwal D, Chapman AP, Heywood S, West SM, Carrington B, Nesbitt A, Lawson AD, Antoniw P, Eddelston A, Suitters A (2001) Prolonged in vivo residence times of antibody fragments associated with albumin. Bioconjug Chem 12(5):750–756

Stebbings R, Findlay L, Edwards C, Eastwood D, Bird C, North D, Mistry Y, Dilger P, Liefooghe E, Cludts I, Fox B, Tarrant G, Robinson J, Meager T, Dolman C, Thorpe SJ, Bristow A, Wadhwa M, Thorpe R, Poole S (2007) "Cytokine storm" in the phase I trial of monoclonal antibody TGN1412: better understanding the causes to improve preclinical testing of immunotherapeutics. J Immunol 179(5):3325–3331

Steffens MG, Boerman OC, Oosterwijk-Wakka JC, Oosterhof GO, Witjes JA, Koenders EB, Oyen WJ, Buijs WC, Debruyne FM, Corstens FH, Oosterwijk E (1997) Targeting of renal cell carcinoma with iodine-131-labeled chimeric monoclonal antibody G250. J Clin Oncol 15(4):1529–1537

Todd PA, Brogden RN (1989) Muromonab CD3. A review of its pharmacology and therapeutic potential. Drugs 37(6):871–899

Willems A, Leoen J, Schoonooghe S, Grooten J, Mertens N (2003) Optimizing expression and purification from cell culture medium of trispecific recombinant antibody derivatives. J Chromatogr B 786(1–2):161–176

Willems A, Schoonooghe S, Eeckhout D, De Jaegher G, Grooten J, Mertens N (2005) CD3 × CD28 cross-interacting bispecific antibodies improve tumor cell dependent T-cell activation. Cancer Immunol Immunother 54:1059–1071

Chapter 9
The Design and Engineering of IgG-Like Bispecific Antibodies

Pei Jin and Zhenping Zhu

9.1 Introduction

Therapeutic monoclonal antibodies (mAb) have become an increasingly important asset for clinicians to fight cancer, inflammation, and infectious diseases (Chan and Carter 2010; Beck et al. 2010; Weiner et al. 2010). mAb as therapeutics exploit one of several mechanisms of action for their biologic effects: (1) Sequester antigens including growth factors and cytokines; (2) Act as receptor antagonists or agonists; (3) Act as activators of effector functions such as antibody-dependent cellular cytotoxicity (ADCC) and complement-dependent cytotoxicity (CDC); (4) Act as carriers for targeted delivery of cytotoxic agents, in the forms of immunoconjugates or immunotoxins to diseased tissues; and (5) Act as chemotherapy or radiotherapy adjuncts (Chan and Carter 2010; Beck et al. 2010; Weiner et al. 2010; Muller and Kontermann 2010). Despite recent success in commercial development of antibody therapeutics, current single-targeted mAb have demonstrated limited efficacy in clinic. This is because cells are evolved with multiple and redundant signaling pathways and are capable of escape from monotherapy. Although small molecule drugs can be developed to impact multiple signaling pathways, they often cause serious off-target toxicities. Therefore, therapeutics are needed that are capable of modifying multiple signaling pathways without augmented side effects. Solutions to circumvent limitations of current mAb therapies include antibody combination and the creation of dual-targeting bispecific antibodies (BsAb). In this chapter, we review the current status and future perspectives of IgG-like BsAb and their

P. Jin
Protein Sciences and Design, Novartis Biologics, Novartis Pharma AG, 200 Technology Square, Cambridge, MA 02139, USA

Z. Zhu (✉)
Kadmon Pharmaceuticals LLC, 400 Madison Avenue, 11th Floor, New York, NY 10017, USA
e-mail: zzhu.mab@gmail.com

R.E. Kontermann (ed.), *Bispecific Antibodies*,
DOI 10.1007/978-3-642-20910-9_9,

potential therapeutic applications. Advantages and limitations of the various approaches will be reviewed in the context of each BsAb format.

9.2 Antibody Combination Strategies and Drawbacks

Evidences are accumulating that the combination of mAb to impact multiple targets in disease pathways provides improved efficacy while limiting systemic toxicity. A clinical example is the combination of the anti-epidermal growth factor receptor (anti-EGFR, Erbitux®) and the anti-vascular endothelial growth factor (anti-VEGF, Avastin®) antibodies, both of which have been approved for colorectal cancer (CRC). A Phase II study among the irinotecan-refractory CRC patients receiving Erbitux®, Avastin®, and irinotecan showed a 37% response rate (Saltz et al. 2007) as compared to a 23% response among patients receiving Erbitux® and irinotecan only. Currently, multiple clinical trials of antibody combinations, e.g., the combination of anti-HER2 antibody Herceptin® and Avastin®, and the combination of Erbitux® and Avastin® are being conducted in breast cancer and CRC patients, respectively.

Most available data supporting the antibody combination therapies are from preclinical studies, such as the combination of anti-EGFR and anti-vascular endothelial growth factor receptor-2 (anti-VEGFR2) (Ahmad et al. 2001; Jung et al. 2002; Tonra et al. 2009; Lamszus et al. 2005; Penar et al. 1997; Kunkel et al. 2001), the combination of anti-EGFR with anti-insulin-like growth factor receptor (IGFR) (Chakravarti et al. 2002; Krane et al. 1991; Kobari et al. 1998; Burtrum et al. 2003; Lu et al. 2005), and the combination of anti-EGFR with anti-HER2 (Ye et al. 1999), to treat a variety of xenograft tumors and other disease models. In each of the reported cases, the combination treatment achieved improved efficacy with additive or synergistic inhibitory activities. Antibody combination therapy has also been applied to target two distinct non-overlapping epitopes on the same molecule such as EGFR (Dechant et al. 2008; Pedersen et al. 2010) and HER2 (Nahta et al. 2004) to achieve synergistic therapeutic efficacy. However, therapeutic development of antibody combination encounters limitations, such as high cost, development delay, and regulatory hurdles. An alternative to antibody combination is the creation of dual-targeting BsAb capable of delivering biological activities of two mAb in the form of one molecule (see below).

If combination of mAb is therapeutically more effective than monotherapy, then the question is if more is better. Bulk of evidence in favor of oligoclonal antibodies (defined mixtures of mAb) comes from clinical indications such as neurotoxin poisoning in which escape from therapy can be instantly life threatening. In one case, the oligoclonal antibody cocktails were 200-fold more potent at neutralizing tetanus toxin in vivo than the monoclonal treatment (Volk et al. 1984). In another case, a cocktail of three antibodies was about 20,000-fold more potent against botulinum toxin than the monotherapy (Nowakowski et al. 2002). An alternative to defined mixtures of mAb is the use of polyclonal antibodies. Often regarded as primitive reagents best suited for molecular biology applications, polyclonal antibodies are regaining favor as potential therapeutic agents (Williams and Sharon 2002; Williams et al. 2002). While the concept of oligoclonal or polyclonal antibodies is appealing, therapeutic development of such mixtures might encounter huge regulatory challenges. Characterization of individual components in the mixtures are prohibitively expensive (oligoclonal cocktails) or nearly impossible (polyclonal antibodies). An undefined or under-defined mixture of recombinant IgGs is both a safety and an efficacy concern. Also, batch consistency in the manufacturing process would be difficult to control from the regulatory perspective.

9.3 Engineering IgG-Like Bispecific Antibodies

BsAb are antibodies or antibody-like molecules specific for two distinctive epitopes or antigens. Initially, they were generated by fusing two antibody-producing hybridoma cell lines, each with a different specificity (Cotton and Milstein 1973). However, due to random pairing between the heavy chains (HC) and light chains (LC), the resulting hybrid hybridoma can produce only a minor population of heterodimeric BsAb among a pool of randomly formed active or inactive monomers and dimers. Purification of the BsAb out of such a pool is difficult and expensive. Chemical crosslinking can also be used to covalently link two mAb of different specificities to form BsAb (Nisonoff and Rivers 1961). However, chemical conjugation tends to inactivate the molecules. Further, the resulting antibody conjugates are heterogenous, and extensive purification is required to obtain limited amounts of BsAb.

Approaches to generating BsAb by recombinant technology have been more fruitful (Marvin and Zhu 2006; Coloma and Morrison 1997; Muller and Kontermann 2010). Recombinant BsAb are generated by protein design and genetic engineering. The resulting cDNA constructs are expressed in bacteria or mammalian cells. The expressed active BsAb are easily purified to homogeneity. While BsAb have been studied for over 2 decades, their development into clinically useful therapeutics has been hampered by limitations such as low expression, poor stability, heavy aggregation, and insufficient safety and efficacy profiles. However, recent progress in the field of antibody engineering and expression has resulted in new approaches to address these issues (Miller et al. 2009, 2010; Ewert et al. 2004; He et al. 2010; Michaelson et al. 2009). This has led to a renewed enthusiasm in engineering and therapeutic application of novel BsAb.

Recombinant BsAb can be broadly categorized into two classes: (1) the small bispecific formats derived from Fv or Fab that do not have an IgG Fc domain; and (2) the Fc-containing IgG-like bispecific formats. While small bispecific formats have the advantages of improved penetration into disease tissues or cells, they have significantly shorter half-life in vivo and thus require repeated dosing, or further genetic or chemical modification to extend their circulating half-life in vivo. This review focuses on the Fc-containing IgG-like bispecific formats and their potential therapeutic applications.

9.3.1 Heterodimeric IgG-Like Bispecific Antibody

Since one of the major barriers in the production of BsAb in a single cell system, e.g., hybrid hydridoma, is the undesired formation of HC homodimers, therefore a first and obvious solution is to re-engineer the CH3 domain of the Fc, so as to favor the formation of HC heterodimers versus homodimers.

9.3.1.1 "Knobs-Into-Holes" BsAb IgG

An elegant experiment was designed by Carter and colleagues 15 years ago (Ridgway et al. 1996; Merchant et al. 1998) to facilitate the heterodimer pairing of two different IgG HC. In the experiment, single amino acid substitutions were introduced into the CH3 domains of the opposite Fc chain, so that the steric complementarity required for CH3/CH3 association obligates the CH3 domain with mutations to pair with a CH3 domain that has different accommodating mutations (Fig. 9.1b). Co-expression of the pair of HC mutants resulted in the formation of a mostly (92%) heterodimeric product, but with a stability that is significantly lower than that of the wild-type (knob:T366W + hole:Y407A mutant, Tm = 65.4°C; wild-type, Tm = 80.4°C). Further optimization using a phage display library resulted in a variant (knob: T366W; hole: T366S, L368A, Y407V), which had an improved Tm of 69.4°C (Atwell et al. 1997). These results demonstrate the utility of the knobs-into-holes approach. It is possible that other residue sets in the CH3 domain could also be optimized, to further improve stability and the heterodimer to homodimer ratio.

While this approach solves the HC homodimerization problem, it did not address the equally problematic mispairing between the LC and HC from two different antibodies (e.g., LC_A-HC_B and LC_B-HC_A mispairing, instead of the correct LC_A-HC_A and LC_B-HC_B pairing). This can be resolved by similarly redesigning the LC-HC interaction to incorporate knobs and holes at the VL-VH interface (Zhu et al. 1997). Based on inspection of the VL-VH interface of the anti-HER2 Fab fragment, 4D5 Fab (Eigenbrot et al. 1993), knob mutations, and the complementary hole mutations were designed for a diabody fragment containing the Fvs of 4D5 and an anti-CD3 antibody (Zhu et al. 1996). For one of the variants (termed "v5"; knob: VH-V37F, L45W; hole: VL-Y87A, F98M), 92% of the products were heterodimers, and near wild-type affinity for antigens was retained (Zhu et al. 1997). Although these mutations were used to promote heterodimerization in a bispecific diabody, the principle is applicable to a full length IgG format.

9.3.1.2 Common Light Chain "Knobs-Into-Holes" BsAb IgG

An alternative approach to solve the problem of LC-HC mispairing is to construct BsAb using two antibodies of different specificities that can share the same LC (Merchant et al. 1998) (Fig. 9.1c). In a proof-of-concept study, a large single chain Fv (scFv) phage library was used to screen for a variety of antigen specificities. The VL regions of the selected scFv were compared. Those with identical amino acid sequences, but paired with the VH regions of different antigen-binding specificities, were selected to create the BsAb. Functional IgG-like BsAb were produced at

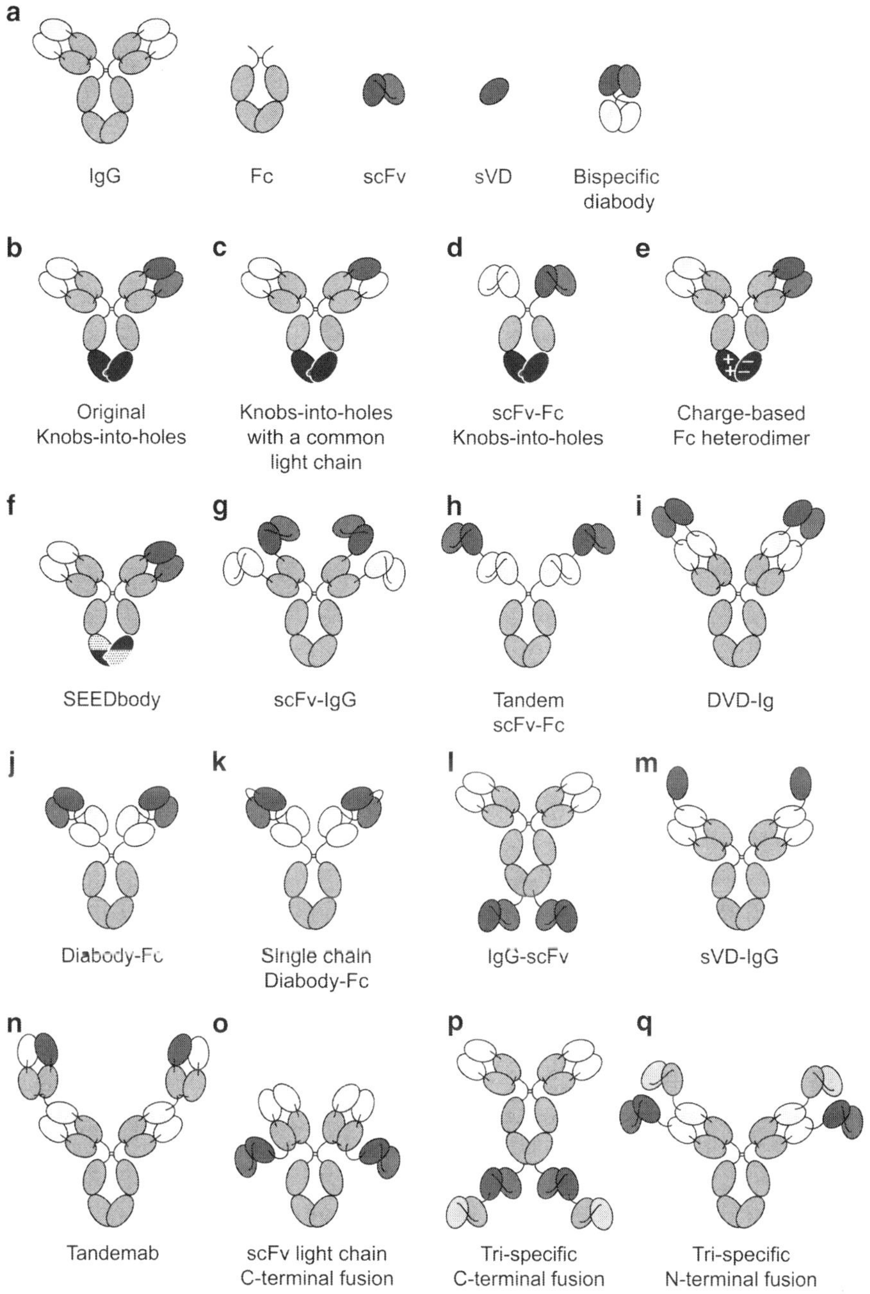

Fig. 9.1 Schematic drawings of the IgG-like bi- and tri-specific antibodies. (**a**), Components as building blocks to construct bi- and tri-specific antibodies; (**b–q**), Engineered IgG-like bi- and tri-specific antibodies that are discussed in this chapter

a high percentage (>95% of total IgGs produced), after a single LC was co-transfected into a host with two HC of different specificity that were engineered to incorporate the original knobs and holes mutations (knob: T366W, hole: T366S, L368A, Y407V), together with a heterodimer stabilizing disulfide bond (S354C, Y349'C). However, it is quite a challenge in practice to identify antibodies that can share common LC. This is particularly true for high affinity antibodies.

9.3.1.3 scFv-Fc "Knobs-Into-Holes" BsAb

The LC-HC mispairing problem may also be circumvented by fusing the VL and VH in a scFv format (Xie et al. 2005). In this construct, an anti-HER2 scFv is fused to an Fc with the "knob" mutation T366Y, and an anti-CD16 scFv is fused to an Fc with the "hole" mutation Y407T (Fig. 9.1d). Efficient pairing of the two chains, via the knobs-into-holes mechanism, was demonstrated, and specificity for both antigens was confirmed by cell surface binding analysis. The scFv-Fc "knobs-into-holes" molecule is also able to crosslink the two target antigens, as shown by its ability to induce higher NK cell-mediated cytotoxicity than a monospecific anti-HER2 antibody alone. This format, which is smaller (120 kDa) than a normal IgG (150 kDa), might have slight gains in tumor penetration, while maintaining the longer half-life and effector functions provided by the intact Fc region.

9.3.1.4 Other Novel Fc-Heterodimer Formats

In addition to the knobs-into-holes technology, other approaches have been developed to achieve Fc heterodimerization. For example, heterodimeric IgG-like pairing can be made by charge polarity engineering of the opposite Fc chains within the CH3 domain (Fig. 9.1e) (Gunasekaran et al. 2010). It was demonstrated that mutations of a few charged residues along the interface of the opposite chains of the CH3 domains can promote Fc heterodimer formation, while effectively suppressing formation of homodimers (see Chap. 11 for a detailed description). If the knobs-into-holes technology is combined with the charge polarity engineering, a higher ratio of heterodimer formation may be achieved. However, a drawback of these methods is the inclusion of multiple mutations in the CH3 domains, which may pose an immunogenic risk in therapeutic settings.

In another approach, a SEED technology (the Strand-Exchange Engineered Domain) was developed, in which a pair of asymmetric fusion proteins were created based on structure modeling between the CH3 domains of the IgG and the IgA (Davis et al. 2010) (Fig. 9.1f). This approach resulted in the complementary AG and GA SEED CH3 domain-containing Fc chains, which preferentially form heterodimers with few or no homodimers when the opposite SEED chains were co-expressed in mammalian cells. Since the natural IgA Fc region does not bind FcRn and Protein A, the relevant surface residues of the AG SEED CH3 domain were mutated back to the IgG sequences to restore these binding activities.

The SEED CH3 domain-containing heterodimer scaffold can be genetically linked to one or more antigen-binding domains, resulting in the bi- or multi-specific antibodies. This novel scaffold was expressed at high levels, purified on recombinant Protein A resin, and had a long serum half-life comparable to that of the IgG1 controls.

9.3.2 *N-Terminal IgG-Like Fusion*

The fact that scFv can often retain the antigen binding affinity and specificity of their parent antibodies, coupled with their small size, makes them the ideal building blocks for construction of BsAb. The scFv can be covalently linked to the N-termini of IgGs resulting in the tetravalent BsAb.

9.3.2.1 scFv-IgG Fusion

In one such format, two scFv, each with a different specificity, were fused, respectively, to the N-termini of the constant light (CL) and the first constant heavy chain (CH1) domains of an IgG: ($scFv_A$)-CL and ($scFv_B$)-CH1-CH2-CH3. Co-expression of the two recombinant chains in a single host resulted in a tetravalent BsAb, Bs$(scFv)_4$-IgG (Zuo et al. 2000) (Fig. 9.1g). Only homogenous BsAb were generated due to the natural heterodimerization between the CL and the CH1 domains.

Two similar Bs$(scFv)_4$-IgG molecules were constructed using two scFv fragments specific for EGFR and IGFR, respectively (Lu et al. 2004b). The two constructs were in different fusion orientations: i.e., $scFv_{anti\text{-}EGFR}$-CL plus $scFv_{anti\text{-}IGFR}$-CH1-CH2-CH3, or $scFv_{anti\text{-}IGFR}$ CL plus $scFv_{anti\text{-}EGFR}$-CH1-CH2-CH3. Both constructs were expressed in mammalian cells and purified via one-step Protein-A chromatography. The BsAb prevented the binding of EGF and IGF to their respective receptors and blocked the ligand-induced receptor activation, whereas a parent mAb only inhibited receptor activation induced by an individual ligand, but not both. Further, both Bs $(scFv)_4$-IgGs inhibited proliferation of cultured tumor cell at levels similar to that of the combination of the two parental IgG antibodies. Both molecules demonstrated excellent stability when incubated in mouse serum in vitro at 37°C for up to 7 days. This Bs$(scFv)_4$-IgG format can be widely applied for construction of other BsAb.

In a more recent report (Dimasi et al. 2009), a scFv of different specificity was linked to the N terminus of either a full-length IgG LC (scFv-VL-CL), or a full-length IgG HC (scFv-VH-CH1-CH2-CH3), or both the LC and HC (scFv-VL-CL plus scFv-VH-CH1-CH2-CH3). Co-expression of the scFv-VL-CL fusion and the HC, or the scFv-VH-CH1-CH2-CH3 fusion with the LC, or both scFv-VL-CL and scFv-VH-CH1-CH2-CH3 fusions, led to the production of homogenous molecules that are either bispecific or trispecific (see Sect. 9.3.5 for more discussion).

9.3.2.2 Tandem scFv-Fc Fusion

As an alternative to the N-terminal scFv-IgG fusion, tandem scFv [scFv(A)-scFv (B)] was fused to the Fc domain of an IgG to form a tetravalent BsAb (Fig. 9.1h) (Natsume et al. 2006). For example, a scFv specific for TAG-72 was fused to a second scFv specific for MUC1 and then to an Fc. The resulting scFv(A)-scFv(B)-Fc or tandem scFv-Fc was expressed efficiently in CHO cells, recognized both target antigens, and was able to mediate ADCC to tumor cells expressing either TAG-72 or MUC-1.

9.3.2.3 Dual-Variable-Domain Immunoglobulin

In this format, the VL and VH domains of a mAb specific for one antigen are covalently linked, via short linkers, to the N terminus of the respective VL and VH of a second mAb, which is specific for a different antigen, resulting in the dual-variable-domain immunoglobulin (DVD-Ig). Several different DVD-Ig constructs were produced and showed activities similar to that of the parental antibodies (Wu et al. 2007) (Fig. 9.1i). The DVD-Igs can be efficiently produced and exhibits good in vivo stability, suggesting that the approach may be widely applicable (see Chap. 12 for more details).

9.3.2.4 Diabody-Fc Fusions

Another method for homogenous production of an IgG-like BsAb is to replace the Fab fragment with a bispecific diabody. Diabodies are a derivative of the scFv fragment (Holliger et al. 1993). A scFv is composed of a VH and a VL domain, connected by a flexible linker of approximately 15 amino acids, e.g., [(Glycine) 4-Serine] $\times$ 3, that permits self assembly into an antigen-binding competent form. If the linker is shortened to 5 amino acids, e.g., (Glycine)4-Serine, self assembly between VL and VH becomes impossible. This allows the two scFv to interact with each other to form a bivalent molecule of two interlinked polypeptides, the VL of one chain associating with the VH of the other. If VL and VH of different specificities comprise the diabody, i.e., VH_A-VL_B and VH_B-VL_A (the two so-called "cross-over" scFv), bispecific bivalent diabodies are formed with one binding site for each antigen. In addition to assembling into functional heterodimers, the cross-over scFv can also assemble into non-functional homodimers. Fortunately, purification of properly heterodimerized molecules can easily be achieved by a single round affinity chromatography. Further, as discussed earlier, the "knobs-into-holes" technique can be used to re-engineer the Fv interface to promote the correct heterodimeric VH-VL pairing (Zhu et al. 1997). Diabodies have shown to be useful for antigen cross-linking (Zhu et al. 1996; Holliger et al. 1993) and their small size is ideal for tumor penetration (Colcher et al. 1990). Diabodies can be fused to the Fc domain of an IgG to create tetravalent IgG-like BsAb, or so-called di-diabody (Lu et al. 2005) (Fig. 9.1j). A similar but smaller di-diabody construct can also be created by using the CH3 domain for dimerization, i.e., VL_B-VH_A plus VL_A-VH_B-CH3 (Lu et al. 2003). As a precautionary measure to avoid potential immunogenicity in human therapy, the unstructured first five amino acids of the human IgG CL(Kappa) domain, instead of a "standard" (Glycine)4-Serine linker, can be used as the linker between the variable domains to create the cross-over scFv (Zhu et al. 1996). A di-diabody that binds both EGFR and IGFR was constructed using the variable domains of an anti-EGFR and an anti-IGFR antibody as the building blocks. The di-diabody bound both EGFR and IGFR with affinity similar to that of their parent antibodies. It blocked binding of both EGF and IGF to their respective receptors and prevented the ligand-induced receptor activation. In addition, the di-diabody could efficiently induce ADCC activity against tumor cells that express EGFR and/or IGFR, indicating that the di-diabody possesses an intact and unhindered Fc domain. It also had an in vivo half-life that is equivalent to that of an intact human IgG. Lastly, the di-diabody effectively inhibited the growth of two different human tumor xenografts in nude mice. The di-diabody construct was transiently expressed in mammalian cells with excellent yield (>400 mg/L in non-optimized conditions), which could greatly facilitate the transition from "proof-of-concept" to therapeutic application. Unfortunately, the di-diabody construct has a tendency to form inactive molecules in vivo that lack the smaller cross-over scFv chain, a likely result of dissociation between the two cross-over chains. This shortcoming may be surmounted by introduction of the disulfide bonds (Zhu et al. 1997; Jung et al. 1994; FitzGerald et al. 1997) or packing improvement (Worn and Pluckthun 2001) to stabilize the VL-VH interfaces in the diabody.

9.3.2.5 Single Chain Diabody-Fc Fusion

In this approach, a single chain diabody (scDb) is first constructed by fusing both "cross-over" scFv of a bispecific diabody with a flexible linker (Kipriyanov et al. 1999; Alt et al. 1999, Kontermann and Muller 1999). The resulting fragment is fused to an Fc fragment (or just a CH3 domain) to create a tetravalent bispecific IgG-like molecule (Fig. 9.1k). In this format, one polypeptide with six domains is produced: VH_A-VL_B-$VH_B$$VL_A$-CH2-CH3, which assembled into IgG-like dimers via the Fc domains. The scDb-Fc is bispecific and bivalent for both antigens, and has a full Fc. However, the stability of the scDb-Fc was not reported and the expression level was also low (~5 mg/L).

9.3.3 C-Terminal IgG-scFv Fusion

scFv can be fused to the C terminus of IgGs to create tetravalent BsAb with one antigen-binding specificity at the N-terminal and the other at the C-terminal scFv of the antibodies (Fig. 9.1l). In one such a BsAb, the specificity at the N terminus was for dextran and the specificity at the C-terminal scFv was for dansyl (Coloma and Morrison 1997). The C-terminal scFv in this BsAb showed a tenfold lower affinity for dansyl than that of the parent IgG owing to a slower on-rate. This may result from steric hindrance of the scFv upon fusing to the IgG C-terminal end or from the conversion of the anti-dansyl IgG to a scFv. Regardless, the IgG-scFv construct is bispecific. Furthermore, it is capable of binding C1q, apparently through its intact Fc fragment, and triggering a partial complement cascade. This format has the potential to allow the creation of tetravalent molecules with effector function, but the full potential of the latter remains to be investigated.

A similar IgG-scFv antibody was constructed (Mabry et al. 2010) by fusing a scFv specific for IL-23 to the C terminus of a full-length anti-17A IgG. Before the construction of the bispecific molecule, a stability assessment was performed using a small group of phage library-derived scFv to select the most thermostable binder as the building blocks. The resulting anti-IL-17A/IL-23 BsAb showed an identical N-terminal affinity for IL-17A, and an eightfold lower C-terminal affinity of the scFv for IL-23, compared to their parental IgGs. The decrease in the C-terminal affinity of scFv for IL-23, however, did not affect its bioactivity.

While a BsAb may be functionally equivalent to the simple combination of their parental mAbs, in certain cases, BsAb can achieve significantly better efficacy. For example, an IgG specific for the TNF-Related Apoptosis Inducing Ligand Receptor-2 (TRAIL-R2) was linked at C terminus to a scFv specific for the Lymphotoxin-beta Receptor (LTβR). The BsAb was further engineered for improved thermal stability and reduced aggregation. The resulting anti-TRAIL-R2/LTβR BsAb demonstrated a much potent effect (IC50 = 46 pM) in inhibiting the growth of tumor cells e.g., MDA-MB-231, Me180, and WiDr, while the individual parental mAb or the combination of them achieved only negligible or lower inhibitory activity (Michaelson et al. 2009). In a mouse MDA-MB-231-derived tumor xenograft model, the bispecific IgG-scFv demonstrated significant anti-tumor activity achieving up to 50% tumor inhibition, whereas the individual or combination of the mAb showed no efficacy. Interestingly, in the same xenograft model no tumor inhibition was observed when a differently formatted bispecific anti-LTβR/TRAIL-R2 antibody was administrated. In the atter format, the scFv specific for LTβR was linked to the N terminus of the anti- TRAIL-R2 IgG. The lack of efficacy in the xenograft model correlated well to the in vitro cell-based results. Both bispecific formats are capable of dual binding to the soluble TRAIL-R2 and LTβR. The difference in inhibitory activity between the IgG-scFv and the scFv-IgG formats in this case may be related to the proximity of the cell surface TRAIL-R2 and LTβR, which favor the IgG-scFv configuration for effective antibody-antigen interaction (Michaelson et al. 2009).

9.3.4 Other Novel IgG-Like BsAb Constructs

9.3.4.1 Single Domain Antibodies as Building Blocks

Some camel or shark antibodies are composed of only HCs (Muyldermans et al. 1994; Els et al. 2001). This observation has led to the development of human single variable domain (sVD) antibody fragments, in which a VL or a VH alone comprises the binding unit (Ward et al. 1989; Riechmann and Muyldermans 1999; Jespers et al. 2004). These fragments can then be used to construct the IgG-like BsAb by fusing a sVD to the CL, and a second sVD of a different specificity to the CH1 of an IgG. Preferably, one sVD would be derived from a VL and the other from a VH to provide increased stability via a VL-VH interface, in addition to that of the CL-CH1 interface, i.e., VL_A-CL plus VH_B-CH1-CH2-CH3). sVD binders can also be directly fused to the N or C terminus of an IgG of a different specificity (Fig. 9.1m). In one such report, a single VH domain binder to the platelet-derived growth factor receptor-alpha (PDGFR-α) was isolated from a phage library, and fused to the N terminus of VL of an IgG antibody specific for PDGFR-β (Shen et al. 2006), or to either the N terminus of VH or the C terminus of the CH3 of an IgG antibody specific for mouse VEGFR-2 (Shen et al. 2007). All three BsAb constructs were expressed well in mammalian cells, and demonstrated the binding ability to both targets with affinity comparable with that of their parent antibodies. Further, the BsAb constructs were able to cross-link both target antigens in solution and also inhibited the function of both targets in cell-based assays.

In theory, sVD antibodies have the potential to satisfy the production criteria essential for drug development. In practice, however, while sVD binders are effective binding modules in isolation, their reconstitution into full IgG-like bispecific molecules often requires further engineering for improved product expression, stability, and retention of biological activity. This may be due to global instability resulting from the incompatible pairing of variable domains identified in isolation or local steric conflicts among the antigen-binding loops. In addition to VL and VH single domain scaffolds, other novel scaffold binders, such as fibronectin, ankyrin repeats, and lipocalin can be developed as building blocks for the construction of IgG-like BsAb molecules. For reviews, see Holt et al. (2003), Holliger and Hudson (2005), Saerens et al. (2008), and Wesolowski et al. (2009).

9.3.4.2 Two-in-One Antibodies

The first two-in-one antibody was developed by phage display of the anti-HER2 antibody Herceptin® with random mutations introduced into the CDRs of the VL domain. The phage library was screening for binding to both HER2 and VEGF (Bostrom et al. 2009). A variant with high affinity for both VEGF (Kd = 3 nM) and HER2 (Kd = 0.2 nM) was generated. The binding surface on the antibody for each antigen overlapped, with VEGF binding primarily mediated by the VL residues and HER2 binding by the VH residuals (see Chap. 13 for details).

9.3.4.3 Tandemabs

Another interesting bispecific format is the tandem Fab-Fc antibodies (tandemab; Fig. 9.1n). In such a format, a tetravalent IgG-like antibody was constructed (Zapata et al. 1995; Miller et al. 2003), where two VH-CH1 units with distinct specificities were linked in tandem, which was then fused to the Fc. Each of the VH-CH1 units in this construct retained the binding affinity of the parent antibody when a common light chain (VL-CK) was co-expressed to produce the tetravalent tandemab. While tetravalent BsAb can be constructed into this format using Fab fragments of different specificities, the challenge remains to identify a common light chain that can be shared by both of the VH-CH1 unites used, so that the high affinity and dual specificities of the tandemabs for the targets can be retained.

9.3.4.4 IgG Light Chain C-Terminal-scFv Fusion

In this format, a scFv is fused to the C terminus of the LC of an IgG via a [(Glycine) 4-Serine] $\times$ 2 linker resulting in a tetravalent BsAb (Orcutt et al. 2010) (Fig. 9.1o). The fusion of a scFv to the LC C terminus partially blocks formation of the disulfide bond between the LC and the HC, and thus requires engineering an additional disulfide bond between the IgG VL and VH domain to stabilize the LC-HC association. Two BsAb constructed in this format demonstrated their dual binding ability with the affinity of their parent IgGs and scFv. However, their expression levels was low (5–7 mg/L) in a transient 293 expression system.

9.3.5 Triple Specific IgG-Like Antibodies

Two trispecific IgG-like antibodies were recently constructed (Dimasi et al. 2009). In the first trispecific antibody (Ts1Ab), two scFv of different specificities were linked each other via a (Glycine)4-Serine linker. The tandem scFv was then fused to the CH3 domain of an IgG (Fig. 9.1p). In the second trispecific antibody (Ts2Ab), one scFv was linked via (Glycine)4-Serine linker to the N terminus of the VL domain of an IgG mAb, and the second scFv to the N terminus of the VH domain of the same IgG (Fig. 9.1q). The Ts1Ab and Ts2Ab were expressed at 120–150 mg/L, compared to 130–200 mg/L of the three parent mAb in a transient HEK293T expression system. Both Ts1Ab and Ts2Ab bound to their respective antigens simultaneously with reduced affinity (5- to 97-fold lower for Ts1Ab antigens and 2- to 4-fold lower for Ts2Ab antigens) as compared to that of their respective parent mAb. Further study showed that Ts2Ab bound to the FcRn and Fcγ receptors with the affinity of a classic IgG. Pharmacokinetic analysis in tumor-bearing nude mice confirmed an in vivo half-life that is comparable to that of their parent mAb. It is reasonable to expect that such a trispecific antibody, impacting multiple signal pathways, can provide additive or synergistic effect in therapy.

9.3.6 *Novel Application of BsAb*

9.3.6.1 Binding Avidity Enhancement

One of the valuable benefits of BsAb is the enhanced avidity for their antigens (Lu et al. 1999). High binding affinity (or avidity) is usually beneficial and, in some cases, required for the biological activity of a therapeutic antibody. In addition to the intrinsic high affinity for their antigens, normal IgG antibodies also exploit the avidity effect to increase their association with antigens because of their bivalent binding towards the targets. Thus, a BsAb directed against two separate (non-overlapping) epitopes on the same target molecule may most likely possess an increased avidity due to its bivalent paratopic (bi-paratopic) binding. Similarly, a tetravalent IgG-like BsAb, with bivalent binding for each antigen, are expected to have a very slow dissociation rate because two dissociation events must occur simultaneously for the BsAb to be free from the cell. In addition, bivalent binding to a target, particularly a cell-surface receptor, is under many circumstances a prerequisite for antibody function, e.g., cross-linking the receptors in order to stimulate activation, to induce apoptosis, or to promote receptor internalization.

9.3.6.2 Epitope Cross-linking for Acquired Antagonistic Activity

In addition to avidity enhancement, a BsAb directed against two non-overlapping epitopes on the same antigen molecule may also acquire novel properties that are not associated with either of the parent antibodies. For example, a bispecific diabody that binds two distinct epitopes on VEGFR-2 cross-linked the epitopes and effectively blocks the binding of VEGF to the receptor, whereas the parent scFv from which the bispecific diabody was derived did not, on their own or in combination, block VEGF binding (Lu et al. 1999). Cross-linking two separate epitopes on the same receptor molecule may introduce new steric hindrance for ligand binding, and/or induce conformation changes in the receptor, preventing it from binding ligand. It will be interesting to see if this principle is applicable to other molecules for which ligand-blocking antibodies are difficult to identify.

9.3.6.3 Fine-Tuning Antibody Specificity Towards Tumor Cells

An emerging and intriguing concept is the use of BsAb to further fine-tune the specificity of anti-tumor antibodies to target cells. In this context, a BsAb is constructed from two antibodies specific for two different tumor antigens, each with low to moderate binding affinity. Only simultaneous bivalent binding (or cross-linking) of the two target antigens on the same tumor cell surface by the BsAb would result in strong association that is required for initiating biological processes, such as down-regulation of signaling pathways, internalization of BsAb-conjugated toxins, ADCC, and CMC. In contrast, monovalent binding of the BsAb to cells that only express one of the target antigens would only result in weak association, which may not be sufficient for induction of any meaningful cellular activity, thus would preserve the cells from the side effects of antibody mediated activities. This is significant since most of the targets currently being used as tumor cell identifiers are, in fact, not truly "tumor-specific," but rather "tumor-associated," i.e., they are also expressed in certain normal tissues/cells, albeit at lower density than in tumor cells. By identifying and constructing BsAb to pairs of targets simultaneously expressed on a given type of tumor, one could expect the enhanced antibody specificity towards the targeted cells, while sparing normal cells of unwanted side effects. This concept was elegantly demonstrated in a recent report using an ant-HER2 × anti-HER3 BsAb (Robinson et al. 2008). In this report, the authors constructed an anti-ErbB3 × anti-ErbB2 bispecific scFv dimer, and demonstrated that the BsAb selectively targeted tumor cells in vitro and in vivo that co-expressed the two target antigens over tumor cells that express only one target antigen or normal cells that express low levels of both antigens. Further, the bispecific scFv also exhibited significantly greater in vivo targeting of ErbB2/ErbB3 double positive tumors than the derivative molecules that contain one functional arm only targeting either ErbB2 or ErbB3, but not both. Together, these results clearly suggest that BsAb selected to co-target critical functional pairs of tumor-associated antigens could enhance the targeting specificity of antibody-based cancer therapeutics.

9.3.7 Stability Optimization of Bispecific Antibodies

scFv and it derivatives are commonly used as the building blocks for construction of BsAb. However, in the absence of the stabilizing CH1 and CL domains and the disulfide bond that covalently joins these two constant domains, scFv are usually less stable than the Fab fragment of an IgG. The insufficient stability of scFv in the BsAb constructs often leads to their poor expression, multimeric aggregation, and short serum half-life, limiting their manufacturability and therapeutic applications. Attempts have been made to improve the stability of scFv molecules, for example, by interface engineering to improve packing (Worn and Pluckthun 2001), by introducing new disulfide bonds between the VL-VH (Zhu et al. 1997; Jung et al. 1994; FitzGerald et al. 1997; Michaelson et al. 2009), by CDR-grafting to the proven stable frameworks, and/or further structure-based framework engineering (Ewert et al. 2004; Michaelson et al. 2009; Jordan et al. 2009). Further, it has been shown that the orientation of the VH and VL could also affect the expression and antigen-binding activity of the final scFv constructs (Lu et al. 2004a). A new technology was recently reported to improve the stability of IgG-like BsAb using a combination of the sequence-based analyses of residue co-variations in IgG variable domains, as well as other structure-, and knowledge-based methods (Miller et al. 2010). Thermally stable scFv were identified using this technology that not only retained full antigen-binding affinity, but also were much less prone to aggregation. For example, a mutant with the substitutions of S16E in the VH and S46L in the VL were identified that increased the thermal stability as much as 26°C, and decrease the soluble aggregation from ~40% of the parent scFv to ~7%. These results demonstrate that by applying appropriate strategies, scFv-based BsAb can be developed with satisfying stability and manufacturability.

9.4 Conclusions

BsAb equip two disease-targeting/modifying capabilities within one molecule, and thus may be able to circumvent certain limitations encountered by current mAb therapeutics, such as insufficient efficacy, and challenges associated with antibody combinations. BsAb development may open up new target space for the development of mAb therapeutics. For example, there are individual targets (pathways) that may not provide sufficient therapeutic benefit when being modified alone. However, significant therapeutic benefits of these targets may be achieved, due to additive and/or synergy effects, if they are combined with other targets/pathways. These targets thus represent excellent candidates for the development of dual-targeting BsAb therapeutics. Recent progress in the field of antibody engineering has led to novel approaches to optimizing the formatting, the expression, the stability, and the manufacturability of IgG-like BsAb. Many issues, however, remain to be carefully addressed, e.g., manufacturing and downstream process development (purification and formulation), analytical and quality control, preclinical toxicology and pharmacology studies, regulatory and clinical development pathways, etc., before we see that BsAb become bonafide mainstream therapeutics.

References

Ahmad SA, Liu W, Jung YD, Fan F, Wilson M, Reinmuth N, Shaheen RM, Bucana CD, Ellis LM (2001) The effects of angiopoietin-1 and -2 on tumor growth and angiogenesis in human colon cancer. Cancer Res 61:1255–1259

Alt M, Muller R, Kontermann RE (1999) Novel tetravalent and bispecific IgG-like antibody molecules combining single-chain diabodies with the immunoglobulin gamma1 Fc or CH3 region. FEBS Lett 454:90–94

Atwell S, Ridgway JB, Wells JA, Carter P (1997) Stable heterodimers from remodeling the domain interface of a homodimer using a phage display library. J Mol Biol 270:26–35

Beck A, Wurch T, Bailly C, Corvaia N (2010) Strategies and challenges for the next generation of therapeutic antibodies. Nat Rev Immunol 10:345–352

Bostrom J, Yu SF, Kan D, Appleton BA, Lee CV, Billeci K, Man W, Peale F, Ross S, Wiesmann C, Fuh G (2009) Variants of the antibody herceptin that interact with HER2 and VEGF at the antigen binding site. Science 323:1610–1614

Burtrum D, Zhu Z, Lu D, Anderson DM, Prewett M, Pereira DS, Bassi R, Abdullah R, Hooper AT, Koo H, Jimenez X, Johnson D, Apblett R, Kussie P, Bohlen P, Witte L, Hicklin DJ, Ludwig DL

(2003) A fully human monoclonal antibody to the insulin-like growth factor I receptor blocks ligand-dependent signaling and inhibits human tumor growth *in vivo*. Cancer Res 63:8912–8921

Chakravarti A, Loeffler JS, Dyson NJ (2002) Insulin-like growth factor receptor I mediates resistance to anti-epidermal growth factor receptor therapy in primary human glioblastoma cells through continued activation of phosphoinositide 3-kinase signaling. Cancer Res 62:200–207

Chan AC, Carter PJ (2010) Therapeutic antibodies for autoimmunity and inflammation. Nat Rev Immunol 10:301–316

Colcher D, Bird R, Roselli M, Hardman KD, Johnson S, Pope S, Dodd SW, Pantoliano MW, Milenic DE, Schlom J (1990) *In vivo* tumor targeting of a recombinant single-chain antigen-binding protein. J Natl Cancer Inst 82:1191–1197

Coloma MJ, Morrison SL (1997) Design and production of novel tetravalent bispecific antibodies. Nat Biotechnol 15:159–163

Cotton RG, Milstein C (1973) Letter: fusion of two immunoglobulin-producing myeloma cells. Nature 244:42–43

Davis JH, Aperlo C, Li Y, Kurosawa E, Lan Y, Lo KM, Huston JS (2010) SEEDbodies: fusion proteins based on strand-exchange engineered domain (SEED) CH3 heterodimers in an Fc analogue platform for asymmetric binders or immunofusions and bispecific antibodies. Protein Eng Des Sel 23:195–202

Dechant M, Weisner W, Berger S, Peipp M, Beyer T, Schneider-Merck T, Lammerts van Bueren JJ, Bleeker WK, Parren PW, van de Winkel JG, Valerius T (2008) Complement-dependent tumor cell lysis triggered by combinations of epidermal growth factor receptor antibodies. Cancer Res 68:4998–5003

Dimasi N, Gao C, Fleming R, Woods RM, Yao XT, Shirinian L, Kiener PA, Wu H (2009) The design and characterization of oligospecific antibodies for simultaneous targeting of multiple disease mediators. J Mol Biol 393:672–692

Eigenbrot C, Randal M, Presta L, Carter P, Kossiakoff AA (1993) X-ray structures of the antigen-binding domains from three variants of humanized anti-p185HER2 antibody 4D5 and comparison with molecular modeling. J Mol Biol 229:969–995

Els CK, Lauwereys M, Wyns L, Muyldermans S (2001) Camel single-domain antibodies as modular building units in bispecific and bivalent antibody constructs. J Biol Chem 276:7346–7350

Ewert S, Honegger A, Pluckthun A (2004) Stability improvement of antibodies for extracellular and intracellular applications: CDR grafting to stable frameworks and structure-based framework engineering. Methods 34:184–199

FitzGerald K, Holliger P, Winter G (1997) Improved tumour targeting by disulphide stabilized diabodies expressed in *Pichia pastoris*. Protein Eng 10:1221–1225

Gunasekaran K, Pentony M, Shen M, Garrett L, Forte C, Woodward A, Ng SB, Born T, Retter M, Manchulenko K, Sweet H, Foltz IN, Wittekind M (2010) and Yan, W. Applications to bispecific molecules and monovalent IgG. J. Biol. Chem, Enhancing antibody Fc heterodimer formation through electrostatic steering effects

He F, Hogan S, Latypov RF, Narhi LO, Razinkov VI (2010) High throughput thermostability screening of monoclonal antibody formulations. J Pharm Sci 99:1707–1720

Holliger P, Hudson PJ (2005) Engineered antibody fragments and the rise of single domains. Nat Biotechnol 23:1126–1136

Holliger P, Prospero T, Winter G (1993) "Diabodies": small bivalent and bispecific antibody fragments. Proc Natl Acad Sci USA 90:6444–6448

Holt LJ, Herring C, Jespers LS, Woolven BP, Tomlinson IM (2003) Domain antibodies: proteins for therapy. Trends Biotechnol 21:484–490

Jespers L, Schon O, James LC, Veprintsev D, Winter G (2004) Crystal structure of HEL4, a soluble, refoldable human V(H) single domain with a germ-line scaffold. J Mol Biol 337: 893–903

Jordan JL, Arndt JW, Hanf K, Li G, Hall J, Demarest S, Huang F, Wu X, Miller B, Glaser S, Fernandez EJ, Wang D, Lugovskoy A (2009) Structural understanding of stabilization patterns in engineered bispecific Ig-like antibody molecules. Proteins 77:832–841

Jung SH, Pastan I, Lee B (1994) Design of interchain disulfide bonds in the framework region of the Fv fragment of the monoclonal antibody B3. Proteins 19:35–47

Jung YD, Mansfield PF, Akagi M, Takeda A, Liu W, Bucana CD, Hicklin DJ, Ellis LM (2002) Effects of combination anti-vascular endothelial growth factor receptor and anti-epidermal growth factor receptor therapies on the growth of gastric cancer in a nude mouse model. Eur J Cancer 38:1133–1140

Kipriyanov SM, Moldenhauer G, Schuhmacher J, Cochlovius B, Von der Lieth CW, Matys ER, Little M (1999) Bispecific tandem diabody for tumor therapy with improved antigen binding and pharmacokinetics. J Mol Biol 293:41–56

Kobari M, Kullenberg B, Bjorkman A, Matsuno S, Ihse I, Axelson J (1998) The inhibitory effect of an EGF receptor-specific tyrosine kinase inhibitor on pancreatic cancer cell lines was more potent than inhibitory antibodies against the receptors for EGF and IGF I. Int J Pancreatol 24:85–95

Kontermann RE, Muller R (1999) Intracellular and cell surface displayed single-chain diabodies. J Immunol Methods 226:179–188

Krane JF, Murphy DP, Carter DM, Krueger JG (1991) Synergistic effects of epidermal growth factor (EGF) and insulin-like growth factor I/somatomedin C (IGF-I) on keratinocyte proliferation may be mediated by IGF-I transmodulation of the EGF receptor. J Invest Dermatol 96:419–424

Kunkel P, Ulbricht U, Bohlen P, Brockmann MA, Fillbrandt R, Stavrou D, Westphal M, Lamszus K (2001) Inhibition of glioma angiogenesis and growth *in vivo* by systemic treatment with a monoclonal antibody against vascular endothelial growth factor receptor-2. Cancer Res 61:6624–6628

Lamszus K, Brockmann MA, Eckerich C, Bohlen P, May C, Mangold U, Fillbrandt R, Westphal M (2005) Inhibition of glioblastoma angiogenesis and invasion by combined treatments directed against vascular endothelial growth factor receptor-2, epidermal growth factor receptor, and vascular endothelial-cadherin. Clin Cancer Res 11:4934–4940

Lu D, Kotanides H, Jimenez X, Zhou Q, Persaud K, Bohlen P, Witte L, Zhu Z (1999) Acquired antagonistic activity of a bispecific diabody directed against two different epitopes on vascular endothelial growth factor receptor 2. J Immunol Methods 230:159–171

Lu D, Jimenez X, Zhang H, Atkins A, Brennan L, Balderes P, Bohlen P, Witte L, Zhu Z (2003) Di-diabody: a novel tetravalent bispecific antibody molecule by design. J Immunol Methods 279:219–232

Lu D, Jimenez X, Witte L, Zhu Z (2004a) The effect of variable domain orientation and arrangement on the antigen-binding activity of a recombinant human bispecific diabody. Biochem Biophys Res Commun 318:507–513

Lu D, Zhang H, Ludwig D, Persaud A, Jimenez X, Burtrum D, Balderes P, Liu M, Bohlen P, Witte L, Zhu Z (2004b) Simultaneous blockade of both the epidermal growth factor receptor and the insulin-like growth factor receptor signaling pathways in cancer cells with a fully human recombinant bispecific antibody. J Biol Chem 279:2856–2865

Lu D, Zhang H, Koo H, Tonra J, Balderes P, Prewett M, Corcoran E, Mangalampalli V, Bassi R, Anselma D, Patel D, Kang X, Ludwig DL, Hicklin DJ, Bohlen P, Witte L, Zhu Z (2005) A fully human recombinant IgG-like bispecific antibody to both the epidermal growth factor receptor and the insulin-like growth factor receptor for enhanced antitumor activity. J Biol Chem 280:19665–19672

Mabry R, Lewis KE, Moore M, McKernan PA, Bukowski TR, Bontadelli K, Brender T, Okada S, Lum K, West J, Kuijper JL, Ardourel D, Franke S, Lockwood L, Vu T, Frank A, Appleby MW, Wolf A, Reardon B, Hamacher NB, Stevens B, Lewis P, Lewis KB, Gilbertson DG, Lantry M, Julien SH, Ostrander C, Chan C, Byrnes-Blake K, Brody J, Presnell S, Meengs B, Levin SD, Snavely M (2010) Engineering of stable bispecific antibodies targeting IL-17A and IL-23. Protein Eng Des Sel 23:115–127

Marvin JS, Zhu Z (2006) Bispecific antibodies for dual-modality cancer therapy: killing two signaling cascades with one stone. Curr Opin Drug Discov Devel 9:184–193

Merchant AM, Zhu Z, Yuan JQ, Goddard A, Adams CW, Presta LG, Carter P (1998) An efficient route to human bispecific IgG. Nat Biotechnol 16:677–681

Michaelson JS, Demarest SJ, Miller B, Amatucci A, Snyder WB, Wu X, Huang F, Phan S, Gao S, Doern A, Farrington GK, Lugovskoy A, Joseph I, Bailly V, Wang X, Garber E, Browning J, Glaser SM (2009) Anti-tumor activity of stability-engineered IgG-like bispecific antibodies targeting TRAIL-R2 and LTbetaR. MAbs 1:128–141

Miller K, Meng G, Liu J, Hurst A, Hsei V, Wong WL, Ekert R, Lawrence D, Sherwood S, DeForge L, Gaudreault J, Keller G, Sliwkowski M, Ashkenazi A, Presta L (2003) Design, construction, and *in vitro* analyses of multivalent antibodies. J Immunol 170:4854–4861

Miller BR, Glaser SM, Demarest SJ (2009) Rapid screening platform for stabilization of scFvs in *Escherichia coli*. Methods Mol Biol 525:279–289, xiv

Miller BR, Demarest SJ, Lugovskoy A, Huang F, Wu X, Snyder WB, Croner LJ, Wang N, Amatucci A, Michaelson JS, Glaser SM (2010) Stability engineering of scFvs for the development of bispecific and multivalent antibodies. Protein Eng Des Sel 23(7):549–557

Muller D, Kontermann RE (2010) Bispecific antibodies for cancer immunotherapy: current perspectives. BioDrugs 24:89–98

Muyldermans S, Atarhouch T, Saldanha J, Barbosa JA, Hamers R (1994) Sequence and structure of VH domain from naturally occurring camel heavy chain immunoglobulins lacking light chains. Protein Eng 7:1129–1135

Nahta R, Hung MC, Esteva FJ (2004) The HER-2-targeting antibodies trastuzumab and pertuzumab synergistically inhibit the survival of breast cancer cells. Cancer Res 64:2343–2346

Natsume A, Wakitani M, Yamane-Ohnuki N, Shoji-Hosaka E, Niwa R, Uchida K, Satoh M, Shitara K (2006) Fucose removal from complex-type oligosaccharide enhances the antibody-dependent cellular cytotoxicity of single-gene-encoded bispecific antibody comprising of two single-chain antibodies linked to the antibody constant region. J Biochem 140:359–368

Nisonoff A, Rivers MM (1961) Recombination of a mixture of univalent antibody fragments of different specificity. Arch Biochem Biophys 93:460–462

Nowakowski A, Wang C, Powers DB, Amersdorfer P, Smith TJ, Montgomery VA, Sheridan R, Blake R, Smith LA, Marks JD (2002) Potent neutralization of botulinum neurotoxin by recombinant oligoclonal antibody. Proc Natl Acad Sci USA 99:11346–11350

Orcutt KD, Ackerman ME, Cieslewicz M, Quiroz E, Slusarczyk AL, Frangioni JV, Wittrup KD (2010) A modular IgG-scFv bispecific antibody topology. Protein Eng Des Sel 23:221–228

Pedersen MW, Jacobsen HJ, Koefoed K, Hey A, Pyke C, Haurum JS, Kragh M (2010) Sym004: a novel synergistic anti-epidermal growth factor receptor antibody mixture with superior anticancer efficacy. Cancer Res 70:588–597

Penar PL, Khoshyomn S, Bhushan A, Tritton TR (1997) Inhibition of epidermal growth factor receptor-associated tyrosine kinase blocks glioblastoma invasion of the brain. Neurosurgery 40:141–151

Ridgway JB, Presta LG, Carter P (1996) 'Knobs-into-holes' engineering of antibody CH3 domains for heavy chain heterodimerization. Protein Eng 9:617–621

Riechmann L, Muyldermans S (1999) Single domain antibodies: comparison of camel VH and camelised human VH domains. J Immunol Methods 231:25–38

Robinson MK, Hodge KM, Horak E, Sundberg AL, Russeva M, Shaller CC, von MM, Shchaveleva I, Simmons HH, Marks JD, Adams GP (2008) Targeting ErbB2 and ErbB3 with a bispecific single-chain Fv enhances targeting selectivity and induces a therapeutic effect *in vitro*. Br J Cancer 99:1415–1425

Saerens D, Ghassabeh GH, Muyldermans S (2008) Single-domain antibodies as building blocks for novel therapeutics. Curr Opin Pharmacol 8:600–608

Saltz LB, Lenz HJ, Kindler HL, Hochster HS, Wadler S, Hoff PM, Kemeny NE, Hollywood EM, Gonen M, Quinones M, Morse M, Chen HX (2007) Randomized phase II trial of cetuximab, bevacizumab, and irinotecan compared with cetuximab and bevacizumab alone in irinotecan-refractory colorectal cancer: the BOND-2 study. J Clin Oncol 25:4557–4561

Shen J, Vil MD, Jimenez X, Iacolina M, Zhang H, Zhu Z (2006) Single variable domain-IgG fusion. A novel recombinant approach to Fc domain-containing bispecific antibodies. J Biol Chem 281:10706–10714

Shen J, Vil MD, Jimenez X, Zhang H, Iacolina M, Mangalampalli V, Balderes P, Ludwig DL, Zhu Z (2007) Single variable domain antibody as a versatile building block for the construction of IgG-like bispecific antibodies. J Immunol Methods 318:65–74

Tonra JR, Corcoran E, Deevi DS, Steiner P, Kearney J, Li H, Ludwig DL, Zhu Z, Witte L, Surguladze D, Hicklin DJ (2009) Prioritization of EGFR/IGF-IR/VEGFR2 combination targeted therapies utilizing cancer models. Anticancer Res 29:1999–2007

Volk WA, Bizzini B, Snyder RM, Bernhard E, Wagner RR (1984) Neutralization of tetanus toxin by distinct monoclonal antibodies binding to multiple epitopes on the toxin molecule. Infect Immun 45:604–609

Ward ES, Gussow D, Griffiths AD, Jones PT, Winter G (1989) Binding activities of a repertoire of single immunoglobulin variable domains secreted from *Escherichia coli*. Nature 341:544–546

Weiner LM, Surana R, Wang S (2010) Monoclonal antibodies: versatile platforms for cancer immunotherapy. Nat Rev Immunol 10:317–327

Wesolowski J, Alzogaray V, Reyelt J, Unger M, Juarez K, Urrutia M, Cauerhff A, Danquah W, Rissiek B, Scheuplein F, Schwarz N, Adriouch S, Boyer O, Seman M, Licea A, Serreze DV, Goldbaum FA, Haag F, Koch-Nolte F (2009) Single domain antibodies: promising experimental and therapeutic tools in infection and immunity. Med Microbiol Immunol 198:157–174

Williams BR, Sharon J (2002) Polyclonal anti-colorectal cancer Fab phage display library selected in one round using density gradient centrifugation to separate antigen-bound and free phage. Immunol Lett 81:141–148

Williams BR, Sompuram SR, Sharon J (2002) Generation of anti-colorectal cancer fab phage display libraries with a high percentage of diverse antigen-reactive clones. Comb Chem High Throughput Screen 5:489–499

Worn A, Pluckthun A (2001) Stability engineering of antibody single-chain Fv fragments. J Mol Biol 305:989–1010

Wu C, Ying H, Grinnell C, Bryant S, Miller R, Clabbers A, Bose S, McCarthy D, Zhu RR, Santora L, Davis-Taber R, Kunes Y, Fung E, Schwartz A, Sakorafas P, Gu J, Tarcsa E, Murtaza A, Ghayur T (2007) Simultaneous targeting of multiple disease mediators by a dual-variable-domain immunoglobulin. Nat Biotechnol 25:1290–1297

Xie Z, Guo N, Yu M, Hu M, Shen B (2005) A new format of bispecific antibody: highly efficient heterodimerization, expression and tumor cell lysis. J Immunol Methods 296:95–101

Ye D, Mendelsohn J, Fan Z (1999) Augmentation of a humanized anti-HER2 mAb 4D5 induced growth inhibition by a human-mouse chimeric anti-EGF receptor mAb C225. Oncogene 18:731–738

Zapata G, Ridgway JB, Mordenti J, Osaka G, Wong WL, Bennett GL, Carter P (1995) Engineering linear F(ab')2 fragments for efficient production in Escherichia coli and enhanced antiproliferative activity. Protein Eng 8:1057–1062

Zhu Z, Zapata G, Shalaby R, Snedecor B, Chen H, Carter P (1996) High level secretion of a humanized bispecific diabody from *Escherichia coli*. Biotechnology (N Y) 14:192–196

Zhu Z, Presta LG, Zapata G, Carter P (1997) Remodeling domain interfaces to enhance heterodimer formation. Protein Sci 6:781–788

Zuo Z, Jimenez X, Witte L, Zhu Z (2000) An efficient route to the production of an IgG-like bispecific antibody. Protein Eng 13:361–367

Chapter 10
Dual-Variable Domain Immunoglobulin (DVD-Ig™) Technology: A Versatile, Novel Format for the Next Generation of Dual-Targeting Biologics

Edit Tarcsa, Wolfgang Fraunhofer, Tariq Ghayur, Jochen Salfeld, and Jijie Gu

Abbreviations

bsAb	Bispecific antibody
CHO	Chinese hamster ovary cell line
CL	Clearance
C_{max}	Observed maximum plasma concentration
DLL4	Delta-like protein 4
DSC	Differential scanning calorimetry
DVD-Ig™	Dual-variable domain immunoglobulin
EGFR	Epidermal growth factor receptor
F	Bioavailability
HEK293	Human embryonic kidney 293
hIL	Human interleukin
IC_{50}	Half maximal inhibitory concentration
IEC	Ion exchange chromatography
IL	Interleukin
IV	Intravenous
K_D	Equilibrium dissociation constant

E. Tarcsa
Department of Drug Metabolism and Pharmacokinetics, Abbott Bioresearch Center, 100 Research Drive, Worcester, MA 01605, USA

W. Fraunhofer
Global Pharmaceutical Sciences, Abbott Laboratories, Abbott Park, IL 60064, USA

T. Ghayur • J. Salfeld • J. Gu (✉)
Department of Biologics, Abbott Bioresearch Center, 100 Research Drive, Worcester, MA 01605, USA
e-mail: jijie.gu@abbott.com

R.E. Kontermann (ed.), *Bispecific Antibodies*,
DOI 10.1007/978-3-642-20910-9_10,

mAb	Monoclonal antibody
MS	Mass spectrometry
MTX	Methotrexate
PEG	Polyethylene glycol
PGE_2	Prostaglandin E_2
SC	Subcutaneous
scFv	Single-chain variable fragment
SDS-PAGE	Sodium dodecyl sulfate polyacrylamide gel electrophoresis
SEC	Size-exclusion chromatography
$T_{½}$	Half-life
T_m	Thermal unfolding
T_{max}	Time to reach C_{max}
TNF	Tumor necrosis factor
VEGF	Vascular endothelial growth factor
VH/VL	Heavy/light chain variable domain
V_{ss}	Volume of distribution at steady state

10.1 Introduction

Monospecific monoclonal antibodies (mAbs) have been remarkably successful in providing clinical benefits across a vast array of diseases, including immunologic diseases (e.g., autoimmunity, transplantation), cancer, neurologic disorders, infectious diseases, cardiovascular diseases, and bone metabolic abnormalities (Reichert 2008; Chan and Carter 2010; Weiner et al. 2010; Reichert 2011). In addition, mAbs represent one of the fastest growing segments of the pharmaceutical business, with hundreds of mAbs in various stages of development for many other indications. However, with emerging clinical data, it is becoming increasingly apparent that mAbs show either partial efficacy or benefit only a subset of patients with a given disease. These data likely reflect the limitations of targeting a single disease mechanism (i.e., monospecificity of mAbs) and the mechanistic complexity/hetero geneity of human diseases where redundant and distinct mechanisms may determine the disease outcome.

The bispecific antibody (bsAb) field has evolved over the past two decades, mainly for applications that cannot be achieved by monospecific mAbs, including redirected toxicity and the ability to target two or more disease mechanisms with a single agent to enhance efficacy and provide benefits to a larger number of patients. Several formats for generating bsAbs have been reported recently (Morrison 2007; Presta 2008; Chan and Carter 2010). The first bsAb, catumaxomab, received approval in Europe for the treatment of malignant ascites in 2009 (Sebastian et al. 2009), and several other bsAbs are in various stages of clinical development. The success of the bsAb field will depend upon identification of the right target or target pair for a given disease indication. The bsAb field will continue to grow as we learn more about the advantages and limitations of the various bsAb formats and as clinical data from mAbs identify the contribution(s) of individual targets to the given disease pathology and the correct epitope(s) on a given target. Collectively, these developments will enable us to make better bi- and multispecific biologic therapeutics.

Several of the bsAb formats reported to date have encountered technical challenges. Recently, Chan and Carter (2010) outlined the requirements for an ideal bsAb platform. These requirements include (a) broad applicability, without a need for extensive customization for individual antibody pairs or for different combinations of target antigens; (b) high level of expression (~1 g/L); (c) retention of biologic activities of the parental mAbs; (d) simultaneous binding of both antigens (essential for certain applications such as redirected toxicity); (e) ease of purification to homogeneity and recovery in high yield; (f) favorable physico-chemical properties (e.g., high solubility, in vitro and in vivo stability, low propensity to aggregate); and (g) appropriate pharmacokinetic properties (e.g., long serum half-life). In this chapter we will describe the properties of DVD-Ig™ molecules, with particular emphasis on how the DVD-Ig™ format compares with the requirement for an ideal bsAb platform.

10.2 bsAb Formats

In general, three different approaches have been used to generate bsAbs: cell fusion, chemical conjugation, and recombinant DNA techniques (Gu and Ghayur 2010).

10.2.1 Cell Fusion

An initial approach for generating bsAbs was based on the somatic fusion of two different hybridoma cell lines expressing murine mAbs (Milstein and Cuello 1983). This approach generates up to ten different antibodies because of the random pairing of two different Ig heavy and light chains within the resulting hybrid hybridoma (or quadroma) cell line. Extensive, sophisticated purification procedures are required to isolate the bsAb of interest from the mispaired by-products of the cell fusion approach.

10.2.2 Chemical Conjugation

Chemical conjugation of two different mAbs (Staerz et al. 1985) or smaller antibody fragments (Brennan et al. 1985) has also been used as an approach for generating bsAbs. Another method is the coupling of two parent antibodies with a heterobifunctional crosslinker. However, molecular heterogeneity is a problem with these approaches. To obtain more homogeneous preparations of bsAbs, two different Fab fragments may be chemically crosslinked at their hinge cysteine residues in a site-directed manner (Glennie et al. 1987), but this method results in Fab′2 fragments rather than a full IgG molecule.

10.2.3 Recombinant DNA Techniques

More recently, a wide variety of recombinant bsAb formats have been developed. Tandem single-chain variable fragment (scFv) bsAbs and diabodies are the most common recombinant bsAb constructs (Kriangkum et al. 2001).

10.2.3.1 scFv bsAbs

Generation of scFv bsAbs have been described using a peptide linker to connect two scFv fragments that recognize different antigens (Kriangkum et al. 2001). Various linkers can be optimized to connect the two scFv fragments (Völkel et al. 2001). Although the parental scFv fragments can normally be expressed in soluble form in bacteria, the two tandem scFv molecules form separate folding entities and thus are often insoluble in bacteria. Hence, refolding protocols or the use of mammalian expression systems are routinely applied to produce soluble tandem scFv molecules. For example, one study reported in vivo expression (serum concentrations of up to 100 mg/L) by transgenic rabbits and cattle of a tandem scFv directed against CD28 and a melanoma-associated proteoglycan connected by a CH1 linker (Gracie et al. 1999). Variations of the domain order and the use of linkers with varying length or flexibility allowed soluble expression in bacteria (Völkel et al. 2001). Other studies reporting expression of soluble tandem scFv molecules in bacteria used either a very short Ala3 linker or long glycine/serine-rich linkers (Korn et al. 2004). Phage display of tandem scFv molecules in combination with directed mutagenesis has been used to enrich for molecules that are produced in soluble and active form in bacteria (Arndt and Krauss 2003).

10.2.3.2 Bispecific Diabodies

Bispecific diabodies are generated from scFv fragments by reducing the length of the linker connecting the heavy chain variable (VH) and light chain variable (VL) domains to approximately five residues (Holliger et al. 1993). Thus, the linker is too short to allow pairing between domains on the same polypeptide chain and instead facilitates dimerization of two polypeptide chains by crossover pairing of the VH and VL domains, resulting in bispecific diabodies with either a VH–VL or VL–VH configuration. A large variety of different bispecific diabodies have been generated, most of which can be expressed in soluble form in bacteria.

However, a comparative study demonstrated that the orientation of the variable domains influenced expression and formation of active binding sites (Mack et al. 1995). Nevertheless, soluble expression in bacteria represents an important advantage over tandem scFv molecules.

Knob-into-hole diabodies limit the production of inactive homodimeric by-products and maximize the generation of active heterodimers (Zhu et al. 1997). For example, they were used to generate a bispecific diabody directed against HER2 and CD3. A large knob was introduced in the VH domain by exchanging Val37 with Phe and Leu45 with Trp and a complementary hole was produced in the VL domain by mutating Phe98 to Met and Tyr87 to Ala, either in the anti-HER2 or anti-CD3 variable domains. This approach increased the production of bispecific diabodies from 72% by the parental diabody to more than 90% by the knob-into-hole diabody. However, concerns with the knob-into-hole approach include the possibility of reduced antigen-binding activity, increased immunogenicity, and poor in vivo stability as a result of mutational modification of the Ig sequence at the constant region.

Single-chain diabodies represent an alternative strategy to improve the generation of bispecific diabody-like molecules (Holliger and Winter 1997; Wu et al. 1996). Bispecific single-chain diabodies are produced by connecting the two diabody-forming polypeptide chains with an additional middle linker with a length of approximately 15 amino acid residues. Thus, all molecules with a molecular weight corresponding to monomeric single-chain diabodies (50–60 kDa) are bispecific. Several studies have demonstrated that bispecific single-chain diabodies are expressed in bacteria in soluble and active form, with the majority of purified molecules present as monomers (Holliger and Winter 1997; Wu et al. 1996; Plückthun and Pack 1997; Ridgway et al. 1996). Thus, single-chain diabodies combine the advantages both of tandem scFv bsAbs (all monomers are bispecific) and diabodies (soluble expression in bacteria).

More recently, diabodies have been fused to Fc to generate more Ig-like molecules, referred to as di-diabodies (Lu et al. 2004). In addition, multivalent antibody constructs comprising two Fab repeats in the heavy chain of an IgG and capable of binding four antigen molecules have been described (Miller et al. 2003).

10.3 DVD-Ig™ Format

We have recently reported a novel bsAb format: the DVD-Ig™ design (Wu et al. 2007) (Fig. 10.1). The DVD-Ig™ design differs from other bispecific Ig formats in that each DVD-Ig™ Fab binds two targets. A DVD-Ig™ molecule can be generated from any two mAbs and is designed such that variable domains from the two parent mAbs are linked in tandem via a peptide linker by recombinant DNA techniques, followed by the constant domain of the heavy and light chains, respectively. The resulting DVD-Ig™ molecule retains the features (affinity and potency) of both parental mAbs. In addition, and unlike several previously described bsAb formats,

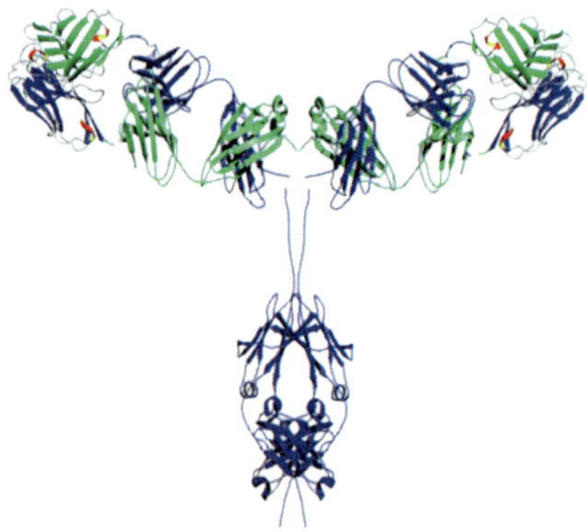

Fig. 10.1 DVD-Ig™ design. A DVD-Ig™ molecule is generated from two parent mAbs, with the two variable domains from the parent antibodies linked in tandem in each heavy and light chain. The DVD-Ig™ molecule retains the activities of both parent antibodies

a well-designed DVD-Ig™ construct is highly stable in vivo, displays good, IgG-like physicochemical and pharmacokinetic properties, and is amenable to large-scale manufacturing.

The key challenges for any bi- or multispecific antibody format are general applicability, manufacturability, and drug-like physicochemical and pharmacokinetic properties. We have now made several hundred DVD-Ig™ molecules and have subjected >20 molecules to thorough characterization to understand these features of the DVD-Ig™ format. We describe the properties of the DVD-Ig™ format in the context of the various challenges for the bsAb field. As examples, we provide details from our previously published study of the DVD-Ig™ molecule directed against interleukin-12 (IL-12) and interleukin-18 (IL-18) (Wu et al. 2007) and a brief summary of our experience with more than 100 DVD-Ig™ molecules (Ghayur et al. 2009a, b; 2010; Wu et al. 2009b).

10.4 General Applicability of the DVD-Ig™ Format

10.4.1 Source of Parent mAbs

We examined the construction of DVD-Ig™ molecules using the variable domains of antibodies derived from various sources and murine and human Fc (Table 10.1) (Ghayur et al. 2009a, b; 2010; Wu et al. 2009b). DVD-Ig™ molecules were generated by placing two variable domains linked in tandem in each heavy and light chain, such that each variable domain was placed in both the outer (VD1) and inner (VD2) positions. The fully human mAbs can be derived from either phage/yeast display libraries or transgenic mice with human Ig genes. In this instance, the linkers between the two variable domains were the short peptides derived from the N-termini of the constant domains CH1 (for heavy chain) or CL (for light chain) otherwise naturally connected to the VH1 or VL1 of the variable domain at the outer position.

Table 10.1 Source of antibodies for selected variable domain pairing combinations tested in the DVD-Ig™ format

Source of antibodies for selected variable domain pairs	
VD1	VD2
Mouse	Mouse
Rat	Mouse
Mouse	Human
Humanized	Humanized
Humanized	Fully human
Fully human (κ light chain)	Fully human (κ light chain)
Fully human (κ light chain)	Fully human (λ light chain)
Fully human (λ light chain)	Fully human (κ light chain)

Table 10.2 Examples of DVD-Ig™ constructs with various target pair combinations

Variable domain pairing		
VD1	VD2	Selected examples
Soluble	Soluble	IL-12/IL-18, IL-1α/IL-1β, TNFα/PGE_2
Membrane	Membrane	CD3/CD19, CD3/EGFR
Membrane	Soluble	DLL4/VEGF
Soluble	Membrane	VEGF/DLL4
Smaller size	Larger size	IL-18/IL-12
Larger size	Smaller size	IL-12/IL-18

DLL4 delta-like protein 4; *EGFR* epidermal growth factor receptor; *IL* interleukin; *PGE_2* prostaglandin E_2; *TNF* tumor necrosis factor; *VEGF* vascular endothelial growth factor

Other DVD-Ig™ molecules with different IgG light chain combinations (i.e., κ/λ or λ/κ hybrid light chains) were also generated (Wu et al. 2007). In this instance, the linker chosen was consistent with the isotype of the variable domain at the outer position and the constant region was consistent with the isotype of the variable domain at the inner position (i.e., Vκ-TVAAP [derived from the first several amino acid residues of C_κ]-V_λ-C_λ and V_λ-QPKAAP [derived from first several amino acid residues of C_λ]-V_κ-C_κ) (Wu et al. 2007). In general, DVD-Ig™ proteins expressed well in mammalian cells (COS and human embryonic kidney 293 [HEK293], described below).

10.4.2 Nature of Target Molecules

We also examined the applicability of the DVD-Ig™ format to a variety of target pair combinations (e.g., soluble/soluble targets, cell surface/cell surface targets, soluble/cell surface targets) (Table 10.2). In addition, DVD-Ig™ constructs against target molecules of different sizes were also generated (Table 10.2). In general, the DVD-Ig™ format allows the two targeting binding domains to bind their respective targets. (For further details about functional properties of DVD-Ig™ molecules, see the discussion and references.)

10.5 Expression and Purification of DVD-Ig™ Molecules

Poor expression, inadequate manufacturability (scalable expression), and protein homogeneity after simple purification steps are major hurdles for some of the bsAb formats. We have now transiently expressed hundreds of DVD-Ig™ constructs in COS cells and/or HEK293 cells and have also investigated the manufacturing feasibility of >20 DVD-Ig™ constructs in Chinese hamster ovary (CHO) cells. The current understanding of DVD-Ig™ expression in mammalian cells is briefly outlined later.

10.5.1 Expression, Purification, and Manufacturing Feasibility of DVD-Ig™ Molecules

The heavy and light chains of hundreds of DVD-Ig™ constructs have been subcloned into mammalian expression vectors and cotransfected into COS or HEK293 cells. In general, expression levels of well-designed DVD-Ig™ molecules are comparable to those of the parent mAbs. For example, expression profiles of a DVD-Ig™ molecule against IL-12 and IL-18 from transfected COS and HEK293 cells were similar to that of the parent IL-18 mAb (Wu et al. 2007). Certain aspects of DVD-Ig™ architecture appear to be important in determining the level of expression, including (a) variable domain combinations (i.e., sequences and canonical structures); (b) variable domain orientation (i.e., which domain is placed on the inside or outside position); and (c) the length of linkers connecting the VH1–VH2 and VL1–VL2. Based on these considerations, unoptimized DVD-Ig™ molecules can be broadly classified into three groups: (1) those that express well regardless of variable domain combination/orientation; (2) those that express well in only one specific variable domain orientation; and (3) those that have low expression regardless of variable domain combination/orientation.

Although transient expression in HEK293 cells is a good initial indicator of the expression level of individual DVD-Ig™ molecules in mammalian cells, it may not always translate to the expression level in manufacturing CHO cell lines. Therefore, we selected 25 DVD-Ig™ constructs and further evaluated their initial expression and amplification potential with increasing concentrations of methotrexate (MTX) in CHO cells. Four DVD-Ig™ molecules have now gone through initial MTX amplification (at 100 nM of MTX) and achieved expression levels similar to those of the parent mAbs. The IL-12/IL-18 DVD-Ig™ molecule, for example, was similar to its parent IL-18 mAb with regard to the number of positive clones generated from one single transfection at various stages of CHO cell line generation and the distribution pattern of protein expression level (Wu et al. 2007). Yield of this DVD-Ig™ molecule achieved >1 g/L in a 25-L wave-bag cell culture system with limited downstream CHO line development.

Because DVD-Ig™ molecules have intact Fc domains, they can easily be purified to homogeneity by protein A chromatography, with high post-purification yields (70–90%) from HEK293 and CHO cell supernatants. Most DVD-Ig™ molecules behave just as mAbs during purification, although some have a slightly greater aggregation propensity. After the standard three-step purification process (protein A, Q column, and ceramic hydroxyapatite chromatography) for generating good laboratory practice-grade materials, the recovery of DVD-Ig™ molecules in general is approximately 40–70%. Details regarding the IL-12/IL-18 DVD-Ig™ molecule (Wu et al. 2007) and the IL-1α/IL-1β DVD-Ig™ molecule (Wu et al. 2009a) have been reported previously.

10.6 Physicochemical Properties of DVD-Ig™ Molecules

Good drug-like physiochemical properties are essential for drug substance and drug product formulation, storage and shipment, applicability of delivery devices, and ultimately clinical use of any biologic therapeutic. An extensive array of methodologies has been applied for assessing physicochemical and drug-like properties of therapeutic mAbs (Table 10.3). To date, we have extensively characterized more than 30 DVD-Ig™ molecules using a battery of approaches to determine DVD-Ig™ protein identity, purity (levels of monomer, aggregate and fragments, and charge heterogeneity), stability (intrinsic unfolding temperature, propensity for aggregation/fragmentation over broad ranges of pH, temperature, protein concentration, and ice–water and air–water interfaces), and solubility. This investigation demonstrated that DVD-Ig™ molecules with desired physicochemical and drug-like properties can be generated. The IL-12/IL-18 DVD-Ig™ molecule is an example of a DVD-Ig™ molecule with physicochemical properties similar to a stable therapeutic mAb (Table 10.3) (Wu et al. 2007). The IL-12/IL-18 DVD-Ig™ molecule demonstrated an appropriate profile in terms of molecular weight, primary sequence, disulfide bonds, glycosylation profile, and charge heterogeneity. In addition, the purity, solubility, and stability of the IL-12/IL-18 DVD-Ig™ molecule are similar to those of stable mAbs. These physicochemical and drug-like properties suggest that well-designed DVD-Ig™ molecules are suitable for therapeutic applications.

10.7 Functional Activities of the Two Variable Domains of the DVD-Ig™ Molecule

The antigen-binding properties of ~50 DVD-Ig™ molecules have been determined using enzyme-linked immunosorbent assay (direct-binding or sandwich), surface plasmon resonance-based measurements with a Biacore 3000 instrument (Biacore AB, Uppsala, Sweden), flow cytometry (for cell surface molecules), and cell-based

Table 10.3 Protein analytics and physicochemical characterization of the IL-12/IL-18 DVD-Ig™ molecule

Parameters	Methodology	Findings
Molecular weight	MS	Matches with theoretical molecule weight
Primary sequence	Sequencing–MS	All matched
Disulfide bonds	Peptide mapping	All 20 disulfide bonds are matched
Glycosylation profile		NGA2F and NGA1F glycans observed as the major forms (similar to human antibodies)
Charge heterogeneity	Cation exchange (Dionex ProPac® WCX-10)	Homogeneity profile similar to mAbs)
Purity	SDS-PAGE	Homogeneity on both reducing (~64 kDa heavy chain and ~36 kDa light chain bands) and nonreducing (~200 kDa) gels
	SEC	>99% single peak monomer after protein A purification
Solubility	PEG precipitation assay	Predicted good solubility (at least 100 mg/mL in pH6, 10-mM phosphate, 10-mM citrate buffer)
Stability		
Intrinsic (T_m)	DSC	DSC thermogram of DVD-Ig™ protein at pH4, 6, and 8 is similar to that of mAbs
Freeze/thaw	SEC	Monomer, aggregate, and fragmentation formed at 2 mg/mL of DVD-Ig™ protein in pH6, 10-mM phosphate, 10-mM citrate buffer after up to 4 cycles of freeze (–80°C) and thaw (room temperature) resemble that of mAbs
pH	SEC and IEC	Monomer, aggregate, and fragmentation formed at 2 mg/mL of DVD-Ig™ protein in 10-mM phosphate, 10-mM citrate buffer at pH4, 6, or 8 resemble that of mAbs
Storage	SEC and IEC	Monomer, aggregate, and fragmentation formed at 2 mg/mL in 10-mM phosphate, 10-mM citrate buffer at 5°C, 40°C, and 50°C for up to 3 weeks resemble those of mAbs

DSC differential scanning calorimetry; *IEC* ion exchange chromatography; *MS* mass spectrometry; *PEG* polyethylene glycol; *SDS-PAGE* sodium dodecyl sulfate polyacrylamide gel electrophoresis; *SEC* size-exclusion chromatography; T_m thermal unfolding

bioassays for both targets. Some of DVD-Ig™ molecules were further tested in acute in vivo pharmacodynamic models and relevant disease models.

In general, the two target-binding domains (or variable domains) are functional in a DVD-Ig™ molecule. In some instances, the inner variable domain may lose some target binding relative to the parent mAb. However, through appropriate selection of antibody variable domains and adjusting variable domain orientation and linker length, a DVD-Ig™ molecule can be generated that preserves the

Table 10.4 Affinity and potency of optimized DVD-Ig™ molecules

DVD-Ig™ molecule	Linker	Affinity, K_D (pM): DVD-Ig™ Molecule/Parent mAb	Potency, IC_{50} (pM): DVD-Ig™ Molecule/Parent mAb
hIL-12/hIL-18 DVD-Ig™ molecule 1[a]	VH: short	hIL-12: 130/120	hIL-12: 220/420
	VL: short	hIL-18: 160/140	hIL-18: 120/220
hIL-12/hIL-18 DVD-Ig™ molecule 2[b]	VH: short	hIL-12: 38/65	hIL-12: 7/5
	VL: short	hIL-18: 622/137	hIL-18: 180/200
hIL-1α/hIL-1β DVD-Ig™ molecule[c]	VH: long	hIL-1α: 560/600	hIL-1α: 350/330
	VL: long	hIL-1β: 130/260	hIL-1β: 500/600

K_D equilibrium dissociation constant; *hIL* human interleukin; IC_{50} half maximal inhibitory concentration[a]Humanized anti-IL-12 antibody and fully human anti-IL-18 antibody[b]Fully human anti-IL-12 antibody and fully human anti-IL-18 antibody[c]DVD4b-Ig construct derived from 18F4.2C8 (anti-IL-1α) and 1B12.4H4 (anti-IL-1β)

antigen-binding specificity, affinity, and biologic activities of both parent antibodies.

10.7.1 In Vitro Characterization of DVD-Ig™ Molecules

The binding affinities and in vitro potencies of three optimized DVD-Ig™ molecules reported in previous publications (Wu et al. 2007, 2009a) are summarized in Table 10.4. A list of affinity and in vitro potency results for hundreds of newly constructed DVD-Ig™ molecules can be found in recent US patent publications (Ghayur et al. 2009a, b; 2010; Wu et al. 2009a).

10.7.2 In Vivo Characterization of DVD-Ig™ Molecules

We have also assessed the in vivo therapeutic efficacy of several DVD-Ig™ proteins in preclinical animal models. As previously reported (Wu et al. 2007), both IL-12/IL-18 DVD-Ig™ constructs were as potent as anti-IL-12 and anti-IL-18 mAbs in combination and almost completely abrogated *Staphylococcus aureus* dried cell-induced interferon-γ production in a severe combined immunodeficient mouse model engrafted with human peripheral blood mononuclear cells. In addition, a mouse IL-1α/IL-1β DVD-Ig™ molecule potently prevented disease progression in collagen-induced arthritis in mice (Wu et al. 2007). Several DVD-Ig™ molecules have been tested in relevant preclinical animal models, and each showed efficacy similar to the parental mAb combination; moreover, the efficacy/potency profiles were in alignment with the molecules' in vitro antigen-binding affinities and potencies (data not shown).

10.8 Pharmacokinetic Properties of DVD-Ig™ Molecules

To gain an understanding of the pharmacokinetic properties of DVD-Ig™ molecules, we have assessed the pharmacokinetic profiles of more than 20 DVD-Ig™ molecules in rats. Following intravenous administration, the DVD-Ig™ proteins exhibited biphasic elimination profiles similar to mAbs. Clearance and volumes of distribution were low, with half-lives ranging from 3 to 15 days. The pharmacokinetic parameters of most DVD-Ig™ molecules examined were within the range of those observed for normal human mAbs (unpublished data). For example, as reported previously (Wu et al. 2007), the hIL-12/hIL-18 DVD-Ig™ molecule – upon either intravenous or subcutaneous dosing at 4 mg/kg in rats – exhibited low clearance (0.26 mL/h/kg) with an average terminal half-life of 11 days, and good bioavailability ranging from 85 to 92%.

Recently, we have also evaluated the pharmacokinetic properties of hIL-12/hIL-18 DVD-Ig™ molecule in male and female cynomolgus monkeys upon either intravenous or subcutaneous administration at 5 mg/kg. The serum concentration-time profiles of the DVD-Ig™ protein was very similar when determined by either hIL-12 or hIL-18 (data not shown) capture-based enzyme-linked immunosorbent assay methods, indicating that the DVD-Ig™ molecule was intact and capable of binding both antigens. The pharmacokinetic profile shown in Fig. 10.2 and the pharmacokinetic parameters summarized in Table 10.5 are similar to those observed for human mAbs in cynomolgus monkeys. Overall, the pharmacokinetic parameters observed for DVD-Ig™ proteins in rats and now in monkeys are sufficient to support toxicology studies in monkeys and potential clinical studies.

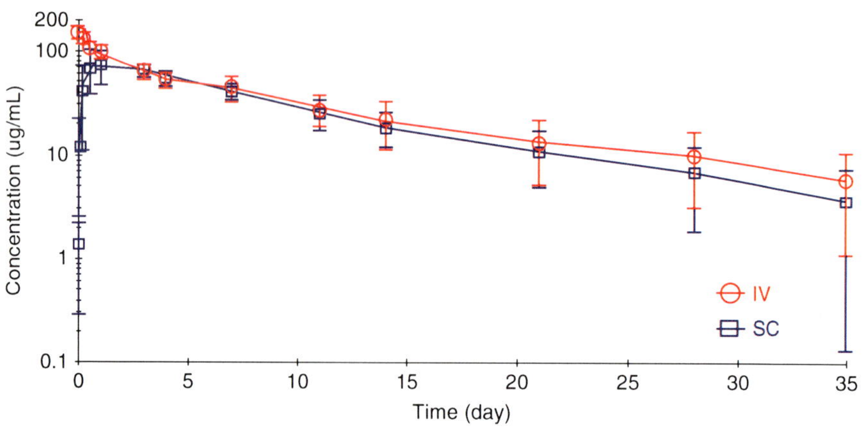

Fig. 10.2 Pharmacokinetic profile of a hIL-12/hIL-18 DVD-Ig™ protein in cynomolgus monkeys. Serum concentration (±standard deviation) time profile of hIL-12/hIL-18 DVD-Ig™ protein after a 5-mg/kg intravenous or subcutaneous dose in cynomolgus monkeys ($n = 4$; time points with likely anti-drug antibody response [$t \geq 21$ days; $n = 2$ monkeys] were not included in the mean calculations). *IV* intravenous; *SC* subcutaneous

Table 10.5 Pharmacokinetic parameters of a hIL-12/hIL-18 DVD-Ig™ protein in cynomolgus monkeys

Route of administration	Intravenous			Subcutaneous			
Parameter	$T_{½}$, d	V_{ss}, mL/kg	CL, mL/h/kg	$T_{½}$, d	T_{max}, d	C_{max}, μg/mL	F, %
IL-12 capture ELISA ($n = 4^a$)	6.7	47.2	0.25	6.9	1.9	82.0	85.1

CL clearance; C_{max} observed maximum plasma concentration; *F* bioavailability; $T_{½}$ half-life; T_{max} time to reach C_{max}; V_{ss} volumes of distribution at steady state[a]Four monkeys at start of study; data from two animals were omitted for time points at which anti-drug antibody response was likely ($t \geq 21$ days)

10.9 Discussion

bsAbs and bsAb fragments offer therapeutic opportunities and advancements not possible with monospecific mAbs. Some unique applications described for bsAbs include (a) enhancing the efficacy of clinically validated mAbs; (b) recruiting immune effector cells to eliminate cancer; (c) providing functionalities unattainable by a combination of two mAbs (e.g., enhancing affinity/specificity through avidity); (iv) site-specific targeting of therapeutic agents (e.g., small molecules, growth factors, stem cells); and advances in imaging and diagnostics (Ford et al. 2001; Goldenberg et al. 2006; Morrison, 2007; Goldenberg et al. 2008; Chan and Carter, 2010). Over the past two decades, tremendous effort has been expended in developing bsAbs and their fragments. Several promising formats have been developed, and some are moving into clinical development. With the recent approval of catumaxomab in Europe for the treatment of malignant ascites (Sebastian et al. 2009) and the very exciting results in Phase I/II clinical trials of the bispecific T-cell engager format (Brischwein et al. 2007; Baeuerle et al. 2009), the bsAb field is witnessing a new revival.

In general, bsAb formats are complex owing to the architectural requirements of putting together two target-binding domains with linkers of various lengths. The ideal bispecific platform will successfully address issues related to (a) manufacturability; (b) drug-like properties and physical stability; (c) pharmacokinetics; and (d) high throughput potential (Chan and Carter 2010). The DVD-Ig™ format appears to be robust and versatile. We have now converted the DVD-Ig™ format into a higher throughput platform wherein we can make and characterize hundreds of these molecules in a short period of time. Our current experience suggests that a DVD-Ig™ molecule can be generated against a variety of target pairs (e.g., two soluble targets, two cell surface targets, or one soluble and one cell surface target). The DVD-Ig™ architecture can also accommodate two targets of various sizes. DVD-Ig™ molecules have been shown to display good functional, physicochemical, drug-like, and pharmacokinetic properties (in rats, mice, and cynomolgus monkeys) and good manufacturing feasibility. The key to the future success of bsAbs will depend upon identifying appropriate target pairs that will enhance efficacy without affecting safety (Morrison 2007; Chan and Carter 2010). In this regard, clinical data from >100 mAbs now in various stages of clinical (and preclinical) development will be crucial. Such information will enhance our understanding of not only the target biology but also of the role of specific targets in particular diseases. The enhanced understanding of disease biology should allow for selection of preferred pairs to be incorporated in multispecific biologics, including DVD-Ig™ proteins.

Acknowledgements We acknowledge the significant contributions by Chengbin Wu, Rong Rong Zhu, and Jerry Carson. Editorial support was provided by Robin Stromberg, of Arbor Communications, Inc. (Ann Arbor, MI, USA) and funded by Abbott Laboratories.

References

Arndt M, Krauss J (2003) Bispecific diabodies for cancer therapy. Methods Mol Biol 207:305–321

Baeuerle PA, Kufer P, Bargou R (2009) BiTE: teaching antibodies to engage T-cells for cancer therapy. Curr Opin Mol Ther 11:22–30

Brennan M, Davison PF, Paulus H (1985) Preparation of bispecific antibodies by chemical recombination of monoclonal immunoglobulin G1 fragments. Science 229:81–83

Brischwein K, Parr L, Pflanz S, Volkland J, Lumsden J, Klinger M, Locher M, Hammond SA, Kiener P, Kufer P, Schlereth B, Baeuerle PA (2007) Strictly target cell-dependent activation of T cells by bispecific single-chain antibody constructs of the BiTE class. J Immunother 30:798–807

Chan AC, Carter PJ (2010) Therapeutic antibodies for autoimmunity and inflammation. Nat Rev Immunol 10:301–316

Ford CH, Osborne PA, Rego BG, Mathew A (2001) Bispecific antibody targeting of doxorubicin to carcinoembryonic antigen-expressing colon cancer cell lines in vitro and in vivo. Int J Cancer 92:851–855

Ghayur T, Hahn A, Mueller B (2009a) inventors; Abbott Laboratories, assignee. Dual variable domain immunoglobulin and uses thereof. US patent application 20090304693, 3 June 2009

Ghayur T, Isakson PC, Olson LM, Wu C, Hugunin M, Kamath RV, Cuff CA, Allen HJ (2009b) inventors; Abbott Laboratories, assignee. Dual variable domain immunoglobulin and uses thereof. US patent application 20090311253, 3 June 2009

Ghayur T, Reilly EB, Bell RL, Wang J, Kingsbury GA, Morgan-Lappe SE, Phillips A, Norvell SM, Li Y, Liu J, Ying H (2010) inventors; Abbott Laboratories, assignee. Dual variable domain immunoglobulin and uses thereof. US patent application 20100076178, 25 March 2010

Glennie MJ, McBride HM, Worth AT, Stevenson GT (1987) Preparation and performance of bispecific F(ab' gamma)2 antibody containing thioether-linked Fab' gamma fragments. J Immunol 139:2367–2375

Goldenberg DM, Sharkey RM, Paganelli G, Barbet J, Chatal JF (2006) Antibody pretargeting advances cancer radioimmunodetection and radioimmunotherapy. J Clin Oncol 24:823–834

Goldenberg DM, Rossi EA, Sharkey RM, McBride WJ, Chang CH (2008) Multifunctional antibodies by the Dock-and-Lock method for improved cancer imaging and therapy by pretargeting. J Nucl Med 49:158–163

Gracie JA, Forsey RJ, Chan WL, Gilmour A, Leung BP, Greer MR, Kennedy K, Carter R, Wei XQ, Xu D, Field M, Foulis A, Liew FY, McInnes IB (1999) A proinflammatory role for IL-18 in rheumatoid arthritis. J Clin Invest 104:1393–401

Gu J, Ghayur T (2010) Rationale and development of multispecific antibody drugs. Expert Rev Clin Pharmacol 3:491–508

Holliger P, Winter G (1997) Diabodies: small bispecific antibody fragments. Cancer Immunol Immunother 45:128–130

Holliger P, Prospero T, Winter G (1993) "Diabodies": small bivalent and bispecific antibody fragments. Proc Natl Acad Sci USA 90:6444–6448

Korn T, Nettelbeck DM, Völkel T, Müller R, Kontermann RE (2004) Recombinant bispecific antibodies for the targeting of adenoviruses to CEA-expressing tumour cells: a comparative analysis of bacterially expressed single-chain diabody and tandem scFv. J Gene Med 6:642–651

Kriangkum J, Xu B, Nagata LP, Fulton RE, Suresh MR (2001) Bispecific and bifunctional single chain recombinant antibodies. Biomol Eng 18:31–40

Lu D, Zhang H, Ludwig D, Persaud A, Jimenez X, Burtrum D, Balderes P, Liu M, Bohlen P, Witte L, Zhu Z (2004) Simultaneous blockade of both the epidermal growth factor receptor and the insulin-like growth factor receptor signaling pathways in cancer cells with a fully human recombinant bispecific antibody. J Biol Chem 279:2856–2865

Mack M, Riethmüller G, Kufer P (1995) A small bispecific antibody construct expressed as a functional single-chain molecule with high tumor cell cytotoxicity. Proc Natl Acad Sci USA 92:7021–7025

Miller K, Meng G, Liu J, Hurst A, Hsei V, Wong WL, Ekert R, Lawrence D, Sherwood S, DeForge L, Gaudreault J, Keller G, Sliwkowski M, Ashkenazi A, Presta L (2003) Design, construction, and in vitro analyses of multivalent antibodies. J Immunol 170:4854–4861

Milstein C, Cuello AC (1983) Hybrid hybridomas and their use in immunohistochemistry. Nature 305:537–540

Morrison SL (2007) Two heads are better than one. Nat Biotechnol 25:1233–1234

Plückthun A, Pack P (1997) New protein engineering approaches to multivalent and bispecific antibody fragments. Immunotechnology 3:83–105

Presta LG (2008) Molecular engineering and design of therapeutic antibodies. Curr Opin Immunol 20:460–470

Reichert JM (2008) Monoclonal antibodies as innovative therapeutics. Curr Pharm Biotechnol 9:423–430

Reichert JM (2011) Antibody-based therapeutics to watch in 2011. MAbs 3(1):76–99

Ridgway JB, Presta LG, Carter P (1996) 'Knobs-into-holes' engineering of antibody CH3 domains for heavy chain heterodimerization. Protein Eng 9:617–621

Sebastian M, Kuemmel A, Schmidt M, Schmittel A (2009) Catumaxomab: a bispecific trifunctional antibody. Drugs Today (Barc) 45:589–597

Staerz UD, Kanagawa O, Bevan MJ (1985) Hybrid antibodies can target sites for attack by T cells. Nature 314:628–631

Völkel T, Korn T, Bach M, Müller R, Kontermann RE (2001) Optimized linker sequences for the expression of monomeric and dimeric bispecific single-chain diabodies. Protein Eng 14:815–823

Weiner LM, Surana R, Wang S (2010) Monoclonal antibodies: versatile platforms for cancer immunotherapy. Nat Rev Immunol 10:317–327

Wu AM, Chen W, Raubitschek A, Williams LE, Neumaier M, Fischer R, Hu SZ, Odom-Maryon T, Wong JY, Shively JE (1996) Tumor localization of anti-CEA single-chain Fvs: improved targeting by non-covalent dimers. Immunotechnology 2:21–36

Wu C, Ying H, Grinnell C, Bryant S, Miller R, Clabbers A, Bose S, McCarthy D, Zhu RR, Santora L, Davis-Taber R, Kunes Y, Fung E, Schwartz A, Sakorafas P, Gu J, Tarcsa E, Murtaza A, Ghayur T (2007) Simultaneous targeting of multiple disease mediators by a dual-variable-domain immunoglobulin. Nat Biotechnol 25:1290–1297

Wu C, Ying H, Bose S, Miller R, Medina L, Santora L, Ghayur T (2009a) Molecular construction and optimization of anti-human IL-1alpha/beta dual variable domain immunoglobulin (DVD-Ig) molecules. MAbs 1:339–347

Wu C, Ghayur T, Dixon RW, Salfeld J (2009b) inventors; Abbott Laboratories, assignee. Dual variable domain immunoglobulin and uses thereof. US patent 7,612,181, 3 Nov 2009

Zhu Z, Presta LG, Zapata G, Carter P (1997) Remodeling domain interfaces to enhance heterodimer formation. Protein Sci 6:781–788

Chapter 11
Two-in-One Antibodies

Patrick Koenig and Germaine Fuh

11.1 Background

Natural monoclonal antibodies as IgGs have two equivalent antigen-binding fragments or Fab arms (bivalency) and typically recognize a single target antigen (monospecificity). Engineered antibodies by assembling two antigen-binding domains of different binding specificity into one molecule creates bi-specificity. Bi-specific therapeutic antibodies are thought to offer several advantages over monospecific antibody drugs (Beck et al. 2010; Chames and Baty 2009a; Cochran 2010). These advantages include greater efficacy as two instead of one molecules are targeted and novel modes of action as the two targeted antigens are brought together close in space (Johnson et al. 2010; Lindhofer et al. 1995; Mack et al. 1995). The standard approach to generate bi-specific antibodies is by linking up two or more antigen-binding domains, each of which contains a monospecific high affinity binding site. Monospecific binding sites are those that exhibit high affinity towards one distinct epitope, while the affinity for other proteins is so low that it is beyond the detection limit and/or physiologically inconsequential. Many bi-specific antibody formats that assemble antigen-binding domains in various configurations have been generated. Several of these formats are discussed in detail in different chapters of this book. However, one disadvantage of all these approaches is that they require either modification of or deviation from the standard IgG format, which often results in difficulties in the production of these antibodies and/or undesired pharmacokinetic properties (Chames and Baty 2009b). This hurdle has contributed to the long interval between the first description of bi-specific antibodies (Staerz et al. 1985) and the first regulatory approval of a bi-specific antibody for use in humans (Seimetz et al. 2010). To overcome these limitations

P. Koenig • G. Fuh (✉)
Department of Antibody Engineering, Genentech, Inc, 1 DNA Way, South San Francisco, CA 94080, USA
e-mail: gml@gene.com

R.E. Kontermann (ed.), *Bispecific Antibodies*,
DOI 10.1007/978-3-642-20910-9_11, © Springer-Verlag Berlin Heidelberg 2011

a new approach was developed to generate antibodies, which exhibit specific high affinity binding towards two different epitopes – termed dual-specificity – in each Fab arm (Bostrom et al. 2009). These antibodies are called two-in-one antibodies. In contrast to bi-specific antibodies, the two-in-one (dual-specific) antibodies are natural IgG antibodies in format, wherein the two antigen-binding arms are identical in amino acid sequence. But, different from a typical monospecific antibody, each Fab arm is capable of recognizing two antigens. Dual-specificity allows the antibodies to interact with high affinity with two different antigens as a single Fab or IgG molecule and may provide a good option to satisfy the increasing demand for dual targeting therapeutics.

The dual-specificity of the two-in-one antibody can be seen as multispecificity limited to two distinct antigens. Multispecificity can result from the recognition of chemically and structurally conserved epitopes on different proteins; examples are species cross-reactive antibodies. This type of multispecificity is essentially monospecificity towards one epitope present on different proteins (Fuh et al. 2006; Kramer et al. 1997; Lee et al. 2006; Mohan et al. 2009; Trinh et al. 1997). More relevant for the concept of multispecificity is a single binding site being able to recognize multiple epitopes that are chemically and structurally distinct. An example of such a well-characterized multispecific binding site is the complement-binding fragment (Fc) of IgG (DeLano et al. 2000). This site has been reported to interact with protein A, protein G, rheumatoid factor, and the FcRn (neonatal Fc receptor). Another interesting example is one region of human growth hormone receptor being able to bind two completely different areas of its cognate ligand, thereby assembling the two-receptor-to-one-hormone complex required for the activation of signaling cascade (de Vos et al. 1992). In contrast, the role of antibodies exhibiting multispecific antigen binding in immune response has been appreciated only recently. Multispecific antigen binding has been originally thought to be limited to germline antibodies as many immature B-cells express multispecific (or polyreactive) antibodies, which are removed from the B-cell repertoire during B-cell maturation (Wardemann et al. 2003). However, some multispecific antibodies are preferably selected in the development of neutralizing multispecific antibodies as demonstrated for antibodies against the human immune deficiency virus (HIV) protein (Mouquet et al. 2010). In addition, IgG+ memory B-cells have been shown to acquire multispecificity through somatic hypermutations (Tiller et al. 2007).

Structural studies revealed some of the mechanisms that may lead to antibody multispecificity. One study of antibody multispecificity showed that an antigen-binding site may contain two separate paratopes. A human antibody against the CD4-gp120 complex binds the gp120 portion of the complex using it's heavy chain (HC) CDRs, while binding the CD4 portion using the light chain (LC) CDRs (Diskin et al. 2010). Interestingly this antibody binds CD4 even in the absence of gp120, confirming that the CD4 paratope on the light chain can function independently from the gp120 paratope on the heavy chain. Another way to allow multispecific antigen binding is structural plasticity of the antigen-binding site. A study on an IgE antibody against the hapten, 2,4-dinitrophenyl, demonstrated that distinct conformations of the antigen-binding site exist in equilibrium before encountering its antigen (James et al. 2003). These different conformations are then selectively used to bind two structurally unrelated antigens: the hapten 2,4-dinitrophenyl and a repertoire selected protein. However, overt structural plasticity is not a strict requirement for multispecificity. Rearrangement of paratope side chain residues allows a given paratope conformation to bind a number of unreleated antigens as demonstrated for a germline antibody (Sethi et al. 2006).

These structural studies suggest that antibodies can carry a well-defined dual-specific binding site and bind tightly to two distinct antigens which lack recognizable structural or chemical homology. We therefore developed a strategy to generate these dual-specific antibodies based on the observation that the antigen-binding site of antibodies is often asymmetric: many antibodies use on one hand mainly the heavy chain to interact with the antigen, likely due to its greater combinatorial natural diversity (Janeway 2001). One the other hand there are examples where the light chain locus can contribute to specificity of the antigen binding (Nemazee 2006; Senn et al. 2003). Thus the chosen approach to produce a dual-specific antibody was to first obtain an antibody with a paratope that primarily uses heavy chain CDRs for binding the first antigen. The next step is to evolve the antibody by limited mutation and repertoire selection for clones that also bind another antigen. The two paratopes might overlap in binding function to some extent. However, studies on protein-protein interfaces showed that they exhibit a degree of robustness in tolerating mutation (Pal et al. 2006). It should be thus possible to select for a paratope that consists of residues compatible with the binding of either of the two antigens.

11.2 Her2/VEGF Dual Specific Antibody Derived from Herceptin

A proof-of-concept study was initiated to select a dual specific antibody from a library based on Herceptin (trastuzumab), an antibody against the human epidermal growth factor receptor 2 (Her2). Herceptin was chosen because the antibody is an approved therapeutic against Her2 positive breast cancer (Baselga et al. 1998), and an excellent example of an antibody that binds its antigen mainly using the heavy chain residues (Cho et al. 2003). The library was subjected to selection and screening, and the antibody bH1 dual-specific toward vascular endothelial growth factor (VEGF) and Her2 was isolated. In the rest of the chapter we will focus on details of bH1 generation and will discuss the structural and biophysical basis for the antibody's dual-specificity focusing mainly on data published by the original proof-of-concept study (Bostrom et al. 2009). We will also give an outlook on the future role of dual-specific antibodies as therapeutic molecules.

The first step in generating bH1 was the diversification of the light chain of Herceptin and the display of the resulting library in a bivalent single chain F_v format on phage. Note that bivalent Fab phage display format has been used in all subsequent studies. Randomized light chain positions include CDR-L1 positions

28, 29, 30, 31 and 32, CDR-L2 positions 50, 51, 53, and CDR-L3 91, 92, 93 and 94. While most residues were randomized based on natural amino acid diversity, the L3 positions 91 and 94–96 were biased to retain Herceptin sequence, because these are important for Her2 binding; L1 positions 30, 31, and L3 positions 92 and 93 were randomized to allow all twenty amino acids. Length variations (inserting 0–5 amino acids between positions 30 and 31 as well as 0–2 amino acids between positions 93 and 94) were introduced to mimic natural CDR length variation. Screening the resulting phage library showed that 25% of the functionally displayed clones kept the ability to bind Her2. The library was panned against three proteins clearly different from Her2 in structure: the VEGF, the death receptor 5 (DR5), and Fc of immunoglobulin G (human IgG). After four rounds of panning on the three secondary antigens individually, randomly picked clones were screened for their ability to bind the respective secondary antigen and Her2-ECD (extracellular domain). We observed that most of the clones bound to the secondary antigen indicating that the libraries were functionally productive, yielding a high hit rate of binding clones against the new antigen. From approximately 100 clones, 29 dual-specific Her2/VEGF, two dual-specific Her2/DR5 and one Her2/Fc specific binders were identified. The dual-specific clones were enriched upon an additional panning on Her2. Most clones retained a high binding affinity for Her2 (low nano-molar affinity) and showed somewhat lower affinity for the second antigen (affinity in the high nanomolar to the low micromolar range) by phage ELISA. The dual-specific binder bH1 exhibited the highest affinity against its two antigens. A K_d of 26 nM for Her2 binding and a K_d of 300 nM for VEGF binding has been measured using surface plasmon resonance. bH1 was subsequently selected for affinity maturation, which lead to the dual-specific high affinity bH1-44 (Fab $K_d = 0.2/3$ nM for Her2/VEGF).

11.3 Molecular Details of HER2/VEGF Dual Interaction

Protein crystal structures of bH1 in complex with the extracellular domain of Her2 and with VEGF reveal the molecular basis for the dual-specificity of bH1. They give insight how the paratope of bH1 adapts to bind VEGF on the existing Her2 binding site. Here it is important to distinguish between the structural paratope, which consists of the amino acids, which are in direct contact with the antigen as shown by the X-ray crystal structures and the functional paratope, which consists of the amino acids, which are important for antigen binding *in vitro* as determined by mutagenesis screening (Cunningham et al. 1989).

The structure of bH1 in complex Her2-ECD at a resolution of 2.9 Å shows that the mode of Her2 binding is structurally conserved between bH1 and Herceptin (Cho et al. 2003) (Fig. 11.1). The main chain Cα's root mean square deviation between Her2-ECD:Herceptin and Her2-ECD:bH1 is only 0.8 Å^2. Both antibodies, bH1 and Herceptin, recognize the same epitope on domain IV of Her2, suggesting that the biological activity of Herceptin on Her2-expressing cells should remain intact in bH1. The structural paratopes for Her2 binding on Herceptin as well as on

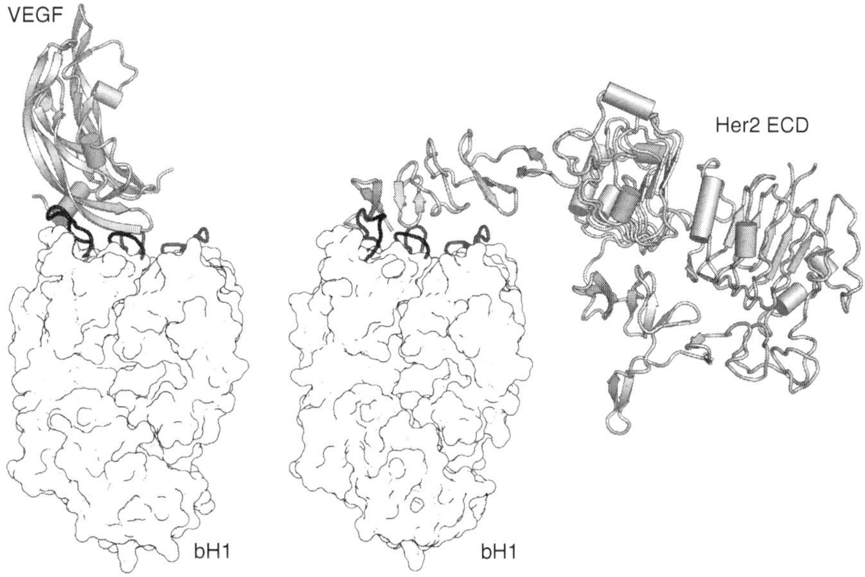

Fig. 11.1 Overview of bH1 bound to VEGF or the ECD of Her2. The figure shows the dual specific bH1 Fab (Bostrom et al. 2009) (PDB codes: 3BDY, 3BE1) in surface representation apart from the CDRs, which are shown in *black* in cartoon representation. Both complexes are shown from the same perspective. The two antigens VEGF and Her2 are shown in cartoon representation in *gray*. Note the different L1 loop conformation in the two different antigen–antibody complexes. The figure has been generated using Pymol http://www.pymol.org)

bH1 include all six CDRs. CDR H1-H3 residues in bH1 and Herceptin are identical, as these loops remained unchanged in the library used to generate bH1 (Fig. 11.2). Although the amino acids of CDR loops L1 and L2 differ in both antibodies, several residues are similarly part of the structural paratopes for Her2. Moreover, the CDR-loops H1, H2, H3, L2, and L3 in Herceptin and bH1 adopt the same conformation. The CDR-L1 loop conformation is clearly different in the two bH1 complex structures. However, consistent with this, CDR-L1 carries a four amino acid insertion in bH1, when compared to Herceptin. In a direct comparison of the functional paratope comprised primarily of residues that result in significantly reduced antigen binding upon mutation to alanine, i.e., hot spot residues, it is evident that there is some modification in the use of paratope residues for interaction with Her2 (Fig. 11.3). While hotspot residues important for Her2 binding of Herceptin include the heavy chain residues Arg50, Trp95, and Tyr100a, but also covers the light chain residue His91 (Kelley and O'Connell 1993), in bH1 the functional important residues are located solely on the heavy chain: Tyr33, Tyr56, Arg50, Arg58, Trp95, Gly99, Phe100, and Tyr100a. Thus while the structural paratope for Her2 binding of bH1 and Herceptin are very similar and include both heavy and light chain residues, the distribution of functional important residues differ between both antibodies: the functional epitope of Herceptin involves both heavy and light chain residues, while bH1 functional eptiope is mainly located on the heavy chain.

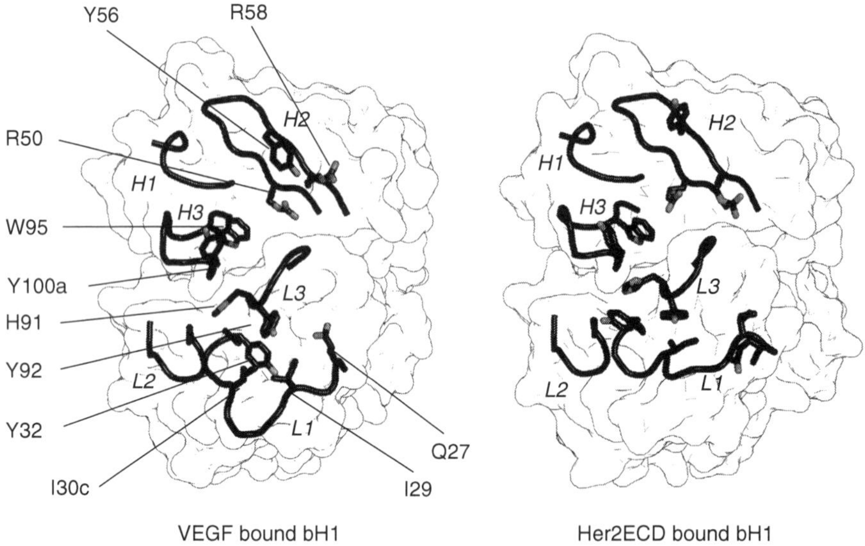

Fig. 11.2 Distinct conformations of the CDR loops in the bH1:VEGF and the bH1:Her2ECD complex. The figure shows the antigen binding region of the bH1 Fab in the VEGF-bound state (*left*) and the Her2-bound state (*right*) (Bostrom et al. 2009) (PDB codes: 3BDY, 3BE1). The CDRs are shown in *black* in cartoon representation, while the rest of the Fab is in surface representation. Some selected residues are shown as *sticks*. While the CDR-L1 adopts a different conformation in the two complexes, conformational change in the other CDR loops is limited to side chains. The figure has been generated using Pymol (http://www.pymol.org)

The structure of bH1 in complex with VEGF (at 2.6 Å resolution) reveals that bH1 binds an epitope on VEGF similar to that recognized by the VEGF receptor (Wiesmann et al. 1997) and many VEGF neutralizing antibodies (e.g., Avastin Fab, G6 or B20-4 Fab) (Fuh et al. 2006) (Fig. 11.1). VEGF blocking activity of bH1 was confirmed in ELISA assays using VEGFR1 and VEGFR2 Fc fusion proteins. Alike the bH1 structural paratope for Her2, the structural paratope for VEGF includes all six CDR loops. However, while in bH1-Her2, binding the contribution of the heavy and light chain to the structural paratope area is nearly equal (HC: 47%, LC: 53%), two-thirds of the buried surface area of the VEGF binding paratope consists of light chain residues (HC: 32% LC: 68%). The contribution of the light chain becomes even more evident in the functional paratope determined by alanine scanning mutagenesis. Only 3 out of 13 hotspot residues are located on the heavy chain. All three residues (Try95, Phe100, Tyr100a) are located on the CDR-H3 and thus lie directly adjacent to the light chain. The rest of the residues comprising the functional paratope site on all three light chain CDR loops (CDR-L1: Ser30b, Ile30c, Gly31, Ile29 and Tyr32, CDR-L2: Try50, and Tyr53, CDR-L3L: His91 and Tyr92). Hence, in contrast to the bH1-Her2 paratope, the structural and especially the functional paratope for bH1-VEGF binding is mainly made up by light chain residues, consistent with the fact that VEGF binding is recruited from light chain diversity library.

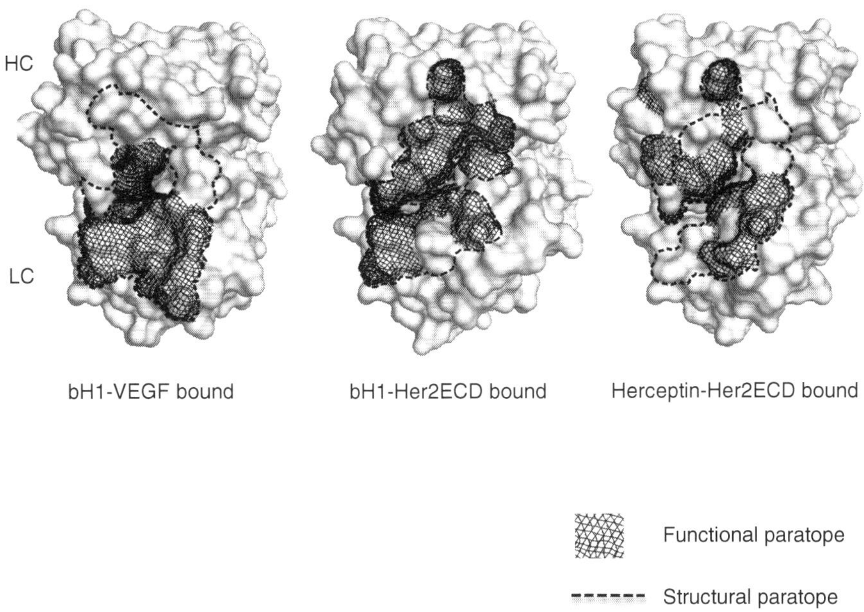

Fig. 11.3 Comparison of the structural paratope with the functional paratope. The figure shows the antigen binding region of bH1 Fab in the VEGF-bound state (*left*) and the Her2-bound state (*middle*) (Bostrom et al. 2009) (PDB codes: 3BDY, 3BE1) as wells as the antigen binding region of Herceptin in the Her2-bound state (*right*) (Cho et al. 2003) (PDB code 1N8Z). Both Fabs are shown in surface representation. The area surrounded by the *dotted line* represents the structural paratope – the area that is in contact with the antigen as determent by the crystal structures, while the *shaded area* is the functional paratope consisting of the amino acids essential for binding as determined by alanine scanning mutagensis (Bostrom et al. 2009). The figure has been generated using Pymol (http://www.pymol.org)

A comparison of the epitopes recognized by bH1 on Her2ECD and VEGF shows that they share no structural similarity beyond an elongated rod like shape, and a similar size: 800 Å^2 for the Her2 epitope and 760 Å^2 for the VEGF eptiope. The Her2 epitope is made up largely by three loops of domain IV: a loop of mostly hydrophobic residues is found between two polar loops, which give the epitope an overall negatively charged electrostatic character (Cho et al. 2003). The bH1 epitope on the VEGF dimer is made up by residues located on a loop and a helix, and is positively charged. Interestingly, the ability of bH1 to bind the negative Her2 epitope and the positive VEGF epitope is apparent in the surface charge distribution of the bH1 antigen-binding site and the way it is different from Herceptin. The Herceptin binding site exhibits a high charge complementarity to the Her2 epitope; it contains two positively charged patches, which bind the negatively charged Her2-ECD loops. One patch is located on the heavy chain whereas the second patch is on the light chain. bH1 maintained the heavy chain patch, while the positively charged patch on the light chain is converted to a slightly negatively charged patch. This patch binds the positively charged VEGF epitope.

The shape complementarity (sc) (Lawrence and Colman 1993) between the respective bH1 paratope conformation and the epitope on Her2 and VEGF (bH1: Her2 sc = 0.72 and bH1:VEGF sc = 0.68) is only slightly lower than between Herceptin and Her2 (sc = 0.75), or the anti-VEGF antibody G6, which binds to a similar epitope on VEGF as bH1 (sc = 0.72) (Fuh et al. 2006). The extent to which the paratope of bH1 adapts to each antigen is demonstrated by the low shape complementary between bH1 in the Her2 bound conformation and the bH1 VEGF epitope, and between bH1 in the VEGF bound conformation and the Her2 epitope (sc = 0.40 and sc = 0.44, respectively). The great structural plasticity of the bH1 antigen-binding site is exhibited in two ways. The first is the CDR-L1 loop conformation (Fig. 11.2). L1 is the only CDR-loop, which requires large main chain rearrangements between the two different bound states of bH1; LC-Ile30c and LC-Tyr32 appear to be involved. In the Her2 bound conformation, LC-Tyr32 is flipped "out" and involved in the interaction with the main chain of Her2-Cys601, while LC-Ile30c is folded back into a hydrophobic pocket. The flipped out conformation of Tyr32 is not compatible with VEGF binding, as it would cause a severe clash with main chain of VEGF residue His90. In the VEGF bound conformation LC-Tyr32 is flipped back in to the very same hydrophobic pocket that is occupied by LC-Ile30c in the Her2 bound L1-conformation. Ile30c on the other hand is involved in VEGF binding, being part of the short beta-sheet like hydrogen-bond network between LC residues 28–30 and VEGF residues 91–93. The importance of the LC-Ile30c/LC-Tyr32-flip for VEGF binding is demonstrated by the alanine mutagenesis scanning. Mutating Ile30c and Tyr32 abolished VEGF binding in ELISA, presumably because the L1-loop conformation for VEGF binding is less likely to form.

The second facet of the great adaptability of the bH1 antigen-binding site is the adaptation of side chains. Side chain movement allows the formation of different interactions with the two different antigens and avoids steric clashes. One example of side chain rearrangement involves the CDR-H2 residues HC-Arg50 and HC-Arg58. While the H2-loop back bone conformation is the same in the bH1: Her2 complex and the bH1:VEGF complex, the position of HC-Arg50 and HC-Arg58 side chains differ significantly in both complexes (Fig. 11.2). The two arginine residues are part of the positively charged patch of the bH1:Her2 paratope and are important for bH1 Her2 binding. Mutation of both residues in the alanine and the homolog scan drastically reduced Her2 binding. The bH1 HC-Arg50Ala/ HC-Arg58Ala mutant for example shows no Her2 binding in ELISA. In the bH1: Her2 complex the arginine side chains are positioned in the vicinity of negatively charged Her2 side chains (Asp560 and Glu558). HC-Arg50 seems to form a salt bridge with Asp560. However, the side chain conformation of both residues as observed in the bH1:Her2 complex would cause clashes in the bH1:VEGF complex. Therefore, both side chains adapt to a different conformation in the bH1: VEGF complex. Another example is the CDR-H2 Tyr56 residue, which is important for both Her2 and VEGF binding and its side chain is significantly different in conformation in the two bH1 complex structures. The adaptation of this residue is important, since the unique side chain conformation in the VEGF-bound bH1 is not compatible for both bH1 and Herceptin to bind HER2.

An extensive investigation on the biophysical properties of bH1/bH1-44 revealed further differences in antigen binding between bH1-44 and Herceptin. While the kinetics of antigen binding are similar for both antibodies, the thermodynamic properties are different: the bH1-44 interaction with Her2 and VEGF is largely entropy driven, while the Herceptin Her2 interaction is based on large enthalpy changes. Further, the data suggest that both paratope conformations of bH1 are of equivalent energetic states as no significant entropic penalty restricting conformational freedom could be measured (Bostrom et al. 2011).

In summery the structural and biophysical investigation of the dual-specific bH1 shows that the binding sites for the two antigens overlap structurally. However, the large L1-loop rearrangements as well as smaller conformational changes of various CDR-residue side chains, i.e. structural plasticity, allows and tolerates the paratope overlap. These factors together with the small overlap of the functional paratope and equivalent conformational energies result in a high affinity antibody-binding site specific against two structurally and chemically unrelated antigens.

11.4 Validation of HER2/VEGF Dual Binding with the Antibody's Dual Action *In Vitro* and *In Vivo*

The characterization of the dual-specific Her2-VEGF antibody bH1 was complemented by demonstrating the dual action of bH1 affinity improved variant bH1-44 to inhibit Her2- and VEGF-mediated cell proliferation *in vitro* and tumor growth *in vivo* (Bostrom et al. 2009). The human umbilical vein endothelial cells (HUVEC) and the human breast cancer cell line BT474 were used to examine the VEGF- and the Her2-mediated proliferation, respectively. The affinity improved variant of bH1-44 (human IgG form) inhibited cell growth in a concentration depended manner to the same extent as the respective positive controls Avastin (bevacizumab) (a FDA approved anti-VEGF antibody) and Herceptin (trastuzumab). To examine the two biological activities of bH1-44 *in vivo*, two xenograph mouse models were used. The human colorectal cancer cell line Colo205 that responds to anti-VEGF, but not to Herceptin treatment was used to examine the anti-VEGF activity of bH1-44, whereas the human breast cancer cell line BT474M1 that responds to Herceptin, but not to anti-VEGF treatment was utilized to test the anti-Her2 Herceptin-like action of bH1-44. In both models, bH1-44 exhibited equivalent efficacy as Avastin, Herceptin or the combination of both. The results are consistent to the fact that the affinity-improved variant of bH1 has a similar binding affinity and *in vitro* potency as Avastin and Herceptin. Current clinical trials of the combination of Avastin and Herceptin have to answer the question whether targeting Her2 and VEGF simultaneously brings an improvement in anti-cancer treatment (see identification numbers NCT00625898, NCT00364611 and NCT00670982 on http://clinicaltrials.gov).

11.5 Conclusion

The dual-specific two-in-one antibody offers several advantages over other bi-specific antibody formats. The dual-specific antibody is in a naturally occurring IgG format and therefore does not require new expression and purification/manufacturing/quality control protocols. Furthermore, the *in vivo* properties, such as pharmacokinetics, stability, and effector function, the dual- specific antibodies are expected to more closely resemble those of the standard monoclonal antibodies, when compared to other bi-specific formats. In addition, the dual-specific antibody could be combined with a knob-in-whole bi-specific technology (Ridgway et al. 1996) to create tri- and tetra-specific antibodies, in an IgG-like format. A single two-in-one antibody as IgG can engage either antigen bi-valently using its two Fab arms, which may counter the desired mode of action. In fact, the actual complex, the two-in-one antibody forms *in vivo* will not only depend on the affinity towards the two antigens, but also on the local antigen concentration, which is hard to predict. One should certainly look out for unexpected effects due to the various possible binding configurations, but this is normal for antibody development. For cell surface antigens or when the antigens are dimeric molecules, two-in-one antibody may present advantages over two monospecific antibodies in combination as it can harness avidity to enhance binding activity in many different settings with varying presence and density of its two antigens.

The ease of production and characterization of two-in-one antibodies as conventional IgGs is indeed a significant advantage to speed up development of therapeutics. Since, the report of the proof-of-concept two-in-one antibody bH1 in 2009, the first such dual action antibody aiming for therapeutic development generated by the strategy described in this chapter co-targeting human epidermal growth factor receptor 3 and 1 has entered clinical trial in 2010 (see identification number NCT01207323 on http://clinicaltrials.gov).

Acknowledgement We thank Charles Eigenbrot for helpful input and discussion.

References

Baselga J, Norton L, Albanell J, Kim YM, Mendelsohn J (1998) Recombinant humanized anti-HER2 antibody (Herceptin) enhances the antitumor activity of paclitaxel and doxorubicin against HER2/neu overexpressing human breast cancer xenografts. Cancer Res 58:2825–2831

Beck A, Wurch T, Bailly C, Corvaia N (2010) Strategies and challenges for the next generation of therapeutic antibodies. Nat Rev Immunol 10:345–352

Bostrom J, Yu SF, Kan D, Appleton BA, Lee CV, Billeci K, Man W, Peale F, Ross S, Wiesmann C et al (2009) Variants of the antibody herceptin that interact with HER2 and VEGF at the antigen binding site. Science 323:1610–1614

Bostrom J, Haber L, Koenig P, Kelley RF, Fuh G (2011) High affinity antigen recognition of the dual specific variants of herceptin is entropy-driven in spite of structural plasticity. PLoS One 6(4):e17887

Chames P, Baty D (2009a) Bispecific antibodies for cancer therapy. Curr Opin Drug Discov Devel 12:276–283

Chames P, Baty D (2009b) Bispecific antibodies for cancer therapy: the light at the end of the tunnel? MAbs 1:539–547

Cho HS, Mason K, Ramyar KX, Stanley AM, Gabelli SB, Denney DW Jr, Leahy DJ (2003) Structure of the extracellular region of HER2 alone and in complex with the Herceptin Fab. Nature 421:756–760

Cochran JR (2010) Engineered proteins pull double duty. Sci Transl Med 2:17ps15

Cunningham BC, Jhurani P, Ng P, Wells JA (1989) Receptor and antibody epitopes in human growth hormone identified by homolog-scanning mutagenesis. Science 243:1330–1336

de Vos AM, Ultsch M, Kossiakoff AA (1992) Human growth hormone and extracellular domain of its receptor: crystal structure of the complex. Science 255:306–312

DeLano WL, Ultsch MH, de Vos AM, Wells JA (2000) Convergent solutions to binding at a protein-protein interface. Science 287:1279–1283

Diskin R, Marcovecchio PM, Bjorkman PJ (2010) Structure of a clade C HIV-1 gp120 bound to CD4 and CD4-induced antibody reveals anti-CD4 polyreactivity. Nat Struct Mol Biol 17:608–613

Fuh G, Wu P, Liang WC, Ultsch M, Lee CV, Moffat B, Wiesmann C (2006) Structure-function studies of two synthetic anti-vascular endothelial growth factor Fabs and comparison with the Avastin Fab. J Biol Chem 281:6625–6631

James LC, Roversi P, Tawfik DS (2003) Antibody multispecificity mediated by conformational diversity. Science 299:1362–1367

Janeway C (2001) Immunobiology: the immune system in health and disease, 5th edn. Garland Publishing, New York

Johnson S, Burke S, Huang L, Gorlatov S, Li H, Wang W, Zhang W, Tuaillon N, Rainey J, Barat B et al (2010) Effector cell recruitment with novel Fv-based dual-affinity re-targeting protein leads to potent tumor cytolysis and *in vivo* B-cell depletion. J Mol Biol 399:436–449

Kelley RF, O'Connell MP (1993) Thermodynamic analysis of an antibody functional epitope. Biochemistry 32:6828–6835

Kramer A, Keitel T, Winkler K, Stocklein W, Hohne W, Schneider-Mergener J (1997) Molecular basis for the binding promiscuity of an anti-p24 (HIV-1) monoclonal antibody. Cell 91:799–809

Lawrence MC, Colman PM (1993) Shape complementarity at protein/protein interfaces. J Mol Biol 234:946–950

Lee CV, Hymowitz SG, Wallweber HJ, Gordon NC, Billeci KL, Tsai SP, Compaan DM, Yin J, Gong Q, Kelley RF et al (2006) Synthetic anti-BR3 antibodies that mimic BAFF binding and target both human and murine B cells. Blood 108:3103–3111

Lindhofer H, Mocikat R, Steipe B, Thierfelder S (1995) Preferential species-restricted heavy/light chain pairing in rat/mouse quadromas. Implications for a single-step purification of bispecific antibodies. J Immunol 155:219–225

Mack M, Riethmuller G, Kufer P (1995) A small bispecific antibody construct expressed as a functional single-chain molecule with high tumor cell cytotoxicity. Proc Natl Acad Sci USA 92:7021–7025

Mohan S, Kourentzi K, Schick KA, Uehara C, Lipschultz CA, Acchione M, Desantis ME, Smith-Gill SJ, Willson RC (2009) Association energetics of cross-reactive and specific antibodies. Biochemistry 48:1390–1398

Mouquet H, Scheid JF, Zoller MJ, Krogsgaard M, Ott RG, Shukair S, Artyomov MN, Pietzsch J, Connors M, Pereyra F et al (2010) Polyreactivity increases the apparent affinity of anti-HIV antibodies by heteroligation. Nature 467:591–595

Nemazee D (2006) Receptor editing in lymphocyte development and central tolerance. Nat Rev Immunol 6:728–740

Pal G, Kouadio JL, Artis DR, Kossiakoff AA, Sidhu SS (2006) Comprehensive and quantitative mapping of energy landscapes for protein–protein interactions by rapid combinatorial scanning. J Biol Chem 281:22378–22385

Ridgway JB, Presta LG, Carter P (1996) 'Knobs-into-holes' engineering of antibody CH3 domains for heavy chain heterodimerization. Protein Eng. 9(7):617–621

Seimetz D, Lindhofer H, Bokemeyer C (2010) Development and approval of the trifunctional antibody catumaxomab (anti-EpCAM x anti-CD3) as a targeted cancer immunotherapy. Cancer Treat Rev 36:458–467

Senn BM, Lopez-Macias C, Kalinke U, Lamarre A, Isibasi A, Zinkernagel RM, Hengartner H (2003) Combinatorial immunoglobulin light chain variability creates sufficient B cell diversity to mount protective antibody responses against pathogen infections. Eur J Immunol 33:950–961

Sethi DK, Agarwal A, Manivel V, Rao KV, Salunke DM (2006) Differential epitope positioning within the germline antibody paratope enhances promiscuity in the primary immune response. Immunity 24:429–438

Staerz UD, Kanagawa O, Bevan MJ (1985) Hybrid antibodies can target sites for attack by T cells. Nature 314:628–631

Tiller T, Tsuiji M, Yurasov S, Velinzon K, Nussenzweig MC, Wardemann H (2007) Autoreactivity in human IgG+ memory B cells. Immunity 26:205–213

Trinh CH, Hemmington SD, Verhoeyen ME, Phillips SE (1997) Antibody fragment Fv4155 bound to two closely related steroid hormones: the structural basis of fine specificity. Structure 5:937–948

Wardemann H, Yurasov S, Schaefer A, Young JW, Meffre E, Nussenzweig MC (2003) Predominant autoantibody production by early human B cell precursors. Science 301:1374–1377

Wiesmann C, Fuh G, Christinger HW, Eigenbrot C, Wells JA, de Vos AM (1997) Crystal structure at 1.7 A resolution of VEGF in complex with domain 2 of the Flt-1 receptor. Cell 91:695–704

Chapter 12
The Dock-and-Lock (DNL) Approach to Novel Bispecific Antibodies

Chien-Hsing Chang, Edmund A. Rossi, Robert M. Sharkey, and David M. Goldenberg

12.1 Introduction

The basic format of a divalent, bispecific antibody (bsAb) comprises two half molecules of IgG, each with a different antigen-binding specificity. Although such bsAbs may form in nature as a result of dynamic Fab-arm exchange involving two different IgG4 molecules (van der Neut Kolfschoten et al. 2007; Burton and Wilson 2007; Labrijn et al. 2009), they were at first generated for potential applications either from quadromas (Milstein and Cuello 1983; Staerz and Bevan 1986) via fusing two hybridomas, or through chemical crosslinking of IgG (Perez et al. 1985) or Fab‘ (Brennan et al. 1985; Glennie et al. 1987). Subsequent efforts were directed primarily toward recombinant engineering of Fc-lacking (Kriangkum et al. 2001; Muller and Kontermann 2007), as well as Fc-containing, bsAbs (Hollander 2009), with a more recent interest in the construction of tetravalent, IgG-like bsAbs (Marvin and Zhu 2005) that vary in design, structure, and antigen-binding constituents (Coloma and Morrison 1997; Lu et al. 2004, 2005; Shen et al. 2006, 2007; Asano et al. 2007; Wu et al. 2007).

We have advanced an alternative approach of constructing bsAbs using the Dock-and-Lock (DNL) method (Rossi et al. 2006; Chang et al. 2007), which enables site-specific self-assembly of two modular components only with each other, resulting, after combining under mild redox conditions, in a covalent structure of defined composition with retained bioactivity. The initial validation of DNL was provided by linking a stabilized dimer of Fab specific for one antigen to a monomeric Fab with specificity for a different antigen, resulting in a trivalent bsAb

C.-H. Chang (✉) • E.A. Rossi
Immunomedics, Inc., 300 The American Road, Morris Plains, NJ 07950, USA
e-mail: kchang@immunomedics.com

R.M. Sharkey • D.M. Goldenberg
Garden State Cancer Center, Center for Molecular Medicine and Immunology, 300 The American Road, 07950, Morris Plains, NJ, USA

R.E. Kontermann (ed.), *Bispecific Antibodies*,
DOI 10.1007/978-3-642-20910-9_12,

composed of three stably tethered Fab-arms (Rossi et al. 2006). Since then, we have applied the DNL method successfully to generate hexavalent bsAbs by derivatizing a divalent IgG to contain two stabilized dimers of Fab with a different specificity from the IgG (Chang et al. 2007; Rossi et al. 2009). In this chapter, we present select examples of these new classes of bsAbs made by DNL, and discuss the advantages of DNL as a tool to build multivalent, multifunctional, and multi-specific complexes that are bioactive.

12.2 DDD/AD Modules Based on PKA and A-kinase Anchoring Protein

The essence of the DNL method relies on the specific protein/protein interactions occurring in nature between the regulatory (R) subunits of cAMP-dependent protein kinase A (PKA) and the anchoring domain (AD) of an interactive A-kinase anchoring protein (AKAP) (Baillie et al. 2005; Wong and Scott 2004). Two types of R subunits (RI and RII) are found in PKA and each has α and β isoforms. The R subunits have been isolated only as stable dimers with the dimerization domain shown to consist of the first 44 amino-terminal residues (Newlon et al. 1999). The AD of AKAPs for PKA is an amphipathic helix of 14–18 residues (Carr et al. 1991), which binds only to dimeric R subunits. For human RIIα, the AD binds to a hydrophobic surface formed by the 23 amino-terminal residues (Colledge and Scott 1999). Thus, the dimerization domain and AKAP-binding domain of human RIIα are both located within the same N-terminal 44 amino acid sequence (Newlon et al. 1999, 2001), and are herein referred to as the dimerization and docking domain (DDD).

We recognized the prospect of exploring a DDD and its cognate AD as an attractive pair of linkers and envisioned the feasibility of specifically docking a module containing the DDD of human RIIα, referred to as DDD1, with a module containing *AKAP-IS* (Alto et al. 2003), a synthetic peptide optimized for RII-selective binding with a reported K_D of 4×10^{-10} M, referred to as AD1, to form a noncovalent complex. This could be locked into a covalently tethered structure to improve in vivo stability by introducing cysteine residues into DDD1 and AD1, resulting in DDD2 and AD2, respectively, to facilitate the formation of disulfide bonds. The amino acid sequences of DDD1, DDD2, AD1, and AD2, as well as a schematic of a basic DNL complex, are shown in Fig. 12.1.

12.3 Trivalent BsAbs Composed of Three Fab-Arms

For proof-of-concept (Rossi et al. 2006), we engineered three DDD-Fab modules of hMN-14 (Sharkey et al. 1995), a humanized antibody targeting CEACAM5, and two AD-Fab modules of h679 (Rossi et al. 2003), a humanized antibody specific for

AD1: QIEYLAKQIVDNAIQQA

AD2: CGQIEYLAKQIVDNAIQQAGC

DDD1: HIQIPPGLTELLQGYTVEVLRQQPPDLVEFAVEYFTRLREARA

DDD2: CGHIQIPPGLTELLQGYTVEVLRQQPPDLVEFAVEYFTRLREARA

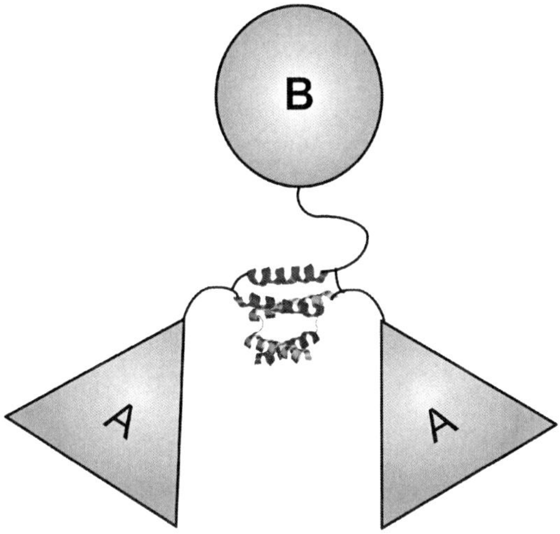

Fig. 12.1 Amino acid sequences (AD1, AD2, DDD1, and DDD2) and a drawing of a basic DNL conjugate comprising a dimer of A-DDD2 (*triangle*) linked to a monomer of B-AD2 (*sphere*)

histamine-succinyl-glycine (HSG). Schematics of these five modules, C_H1-DDD1-Fab-hMN-14, C_H1-DDD2-Fab-hMN-14, DDD2-V_H-Fab-hMN-14, C_H1-AD1-Fab-h679, and C_H1-AD2-Fab-h679, are shown in Fig. 12.2a. As expected, C_H1-DDD1-Fab-hMN-14 and C_H1-AD1-Fab-h679 were purified from culture media exclusively as a homodimer of Fab and a monomer of Fab, respectively. When C_H1-DDD1-Fab-hMN-14 was combined with C_H1-AD1-Fab-h679, the formation of a trimeric complex was demonstrated readily by size-exclusion high performance liquid chromatography (SE-HPLC) with the apparent dissociation constant determined by equilibrium gel filtration analysis (Gegner and Dahlquist 1991) to be about 8 nM, which is presumably too weak of an affinity to keep the noncovalent complex intact at concentrations typical (<1 μg/ml) for in vivo applications.

Our subsequent experiments with the Fab-DDD2 and Fab-AD2 modules produced, in nearly quantitative yields, two stably tethered trivalent bsAbs (Fig. 12.2b), referred to as TF1 for the conjugate of DDD2-V_H-Fab-hMN-14 and CH_1-AD2-Fab-h679, and TF2 for the conjugate of CH_1-DDD2-Fab-hMN-14 and CH_1-AD2-Fab-h679. TF1 and TF2 were each shown by SE-HPLC to be a single peak of the expected molecular size (~150 kDa), by BIAcore to be bispecific, and by competition enzyme-linked immunosorbent assay (ELISA) to be equivalent to hMN-14 IgG and h679 Fab, reflecting the full retention of valency and binding affinity. Furthermore, TF1 and

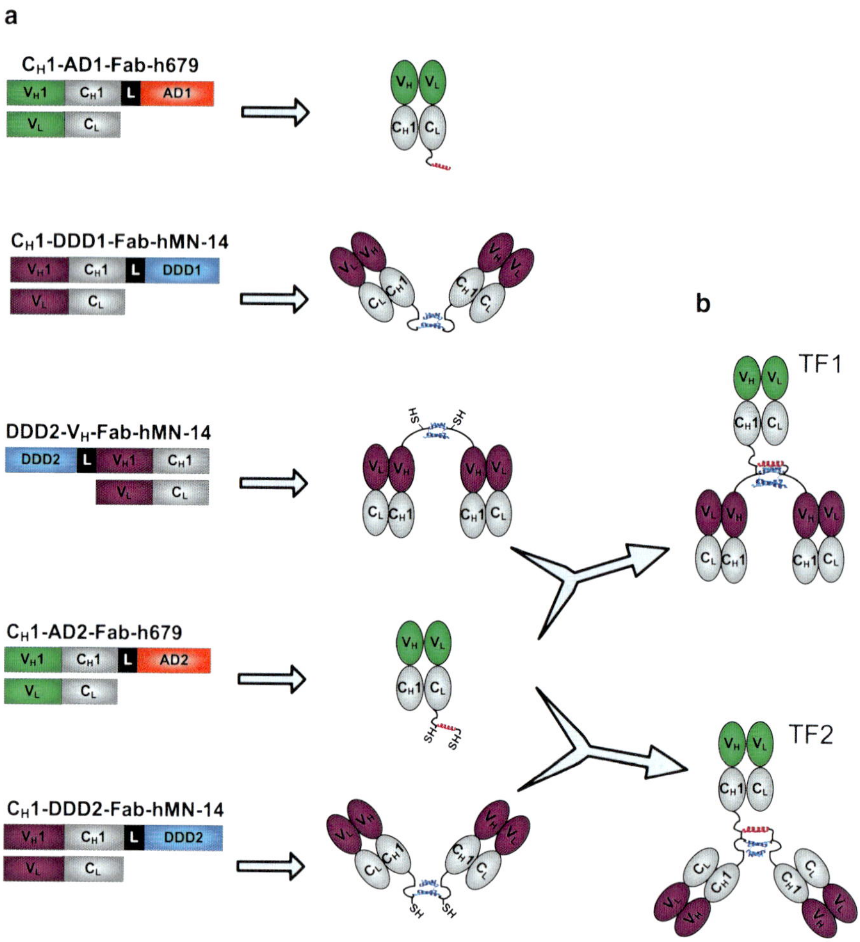

Fig. 12.2 (**a**) Drawings of expression cassettes and DNL modules used to make bispecific T-Fabs that bind divalently to CEACAM5 via hMN-14 Fab (*purple* variable domains) and monovalently to HSG via h679 Fab (*green* variable domains). The GS-linker (L) comprises 14–16 amino acid residues. (**b**) Drawings of TF1 and TF2 made by DNL conjugation of C_H1-AD2-Fab-h679 with DDD2-V_H-Fab-hMN-14 and C_H1-DDD2-Fab-hMN-14, respectively

TF2 were stable for at least 7 days when incubated at 37°C in human or mouse serum, and the superiority of TF2 as a pretargeting agent was demonstrated in several imaging studies with nude mice bearing CEACAM5-expressing, human colonic cancer xenografts, using di-HSG peptides labeled with ^{99m}Tc (Rossi et al. 2006; Sharkey et al. 2008), ^{124}I (Sharkey et al. 2008), ^{68}Ga (Schoffelen et al. 2010), and ^{18}F (McBride et al. 2009).

Since the generation of TF1 and TF2, we have applied the DNL method to produce additional trivalent antibodies that are either monospecific or bispecific (Table 12.1) with similar results and expected properties. Among these T-Fabs, TF4, TF10, and TF12 also show promise as pretargeting agents in preclinical

Table 12.1 Modules, targets, and codes of the T-Fabs made by DNL

	AD-module		DDD-module	
Code	Format	Target	Format	Target
TF18	C_H1-AD2-Fab-h679	HSG	C_H1-DDD2-Fab-hAFP	AFP
TF16	C_H1-AD2-Fab-h679	HSG	C_H1-DDD2-Fab-hA19	CD19
TF14	C_H1-AD2-Fab-h679	HSG	C_H1-DDD2-Fab-hMN-15	CEACAM6
TF12	C_H1-AD2-Fab-h679	HSG	C_H1-DDD2-Fab-hRS7	TROP-2
TF10	C_H1-AD2-Fab-h679	HSG	C_H1-DDD2-Fab-hPAM4	MUC1
TF8	C_H1-AD2-Fab-h679	HSG	C_H1-DDD2-Fab-hMN-3	CEACAM6
TF6	C_H1-AD2-Fab-h679	HSG	C_H1-DDD2-Fab-hLL2	CD22
TF4	C_H1-AD2-Fab-h679	HSG	C_H1-DDD2-Fab-hA20	CD20
TF2	C_H1-AD2-Fab-h679	HSG	C_H1-DDD2-Fab-hMN-14	CEACAM5
TF1	C_H1-AD2-Fab-h679	HSG	DDD2-V_H-Fab-hMN-14	CEACAM5
TF3	C_H1-AD2-Fab-hLL2	CD22	C_H1-DDD2-Fab-hA20	CD20
TF5	C_H1-AD2-Fab-hA20	CD20	C_H1-DDD2-Fab-hLL2	CD22

human xenograft models of B-cell lymphoma (Sharkey et al. 2009), pancreatic cancer (Gold et al. 2008), and prostate cancer (van Rij et al. 2010), respectively. We also found that TF3, the trivalent bsAb that simultaneously targets both CD20 and CD22, displayed direct toxicity in vitro against several B-cell non-Hodgkin lymphoma (NHL) lines without the need for further crosslinking with a secondary antibody (Qu et al. 2008).

TF2 is currently in clinical trials (Sharkey et al. 2010) at the University of Radboud Medical Center, Nijmegen, The Netherlands, using ^{111}In-labeled di-HSG peptide for imaging and the same di-HSG peptide labeled with either ^{90}Y or ^{177}Lu for therapy. The first patient, who had a history of metastatic colorectal cancer, and presented with several metastases in the lungs and a large liver lesion, was given 75 mg of TF2 (~35 mg/m^2) by intravenous infusion over 2 h, and 5 days later received ^{111}In-labeled di-HSG peptide (185 MBg, 100 μg) by a 2-min IV push. All sites of known disease were visualized and there was no adverse effect observed. Importantly, no anti-TF2 responses have been detected, but more studies are required to assess the immunogenicity of the T-Fabs. The pretargeting protocol is being optimized.

12.4 IgG-Like BsAbs Composed of Six Fab-Arms

Another application of the DNL platform is the generation of hexavalent antibodies (HexAbs) that can be either monospecific (Rossi et al. 2008) or bispecific (Rossi et al. 2009). All such HexAbs comprise a pair of Fab-DDD2 dimers linked to a full IgG at the carboxyl termini of the two heavy chains, thus having six Fab-arms and a common Fc domain, as illustrated in Fig. 12.3. To identify these HexAbs, we presently assign each of them a code of X-(Y)-(Y), where X and Y are specific numbers given to differentiate the antibodies, and a designated number enclosed in parentheses represents that antibody is in the Fab format. For example, the code of 20-(22)-(22) designates the bispecific HexAb comprising a divalent anti-CD20

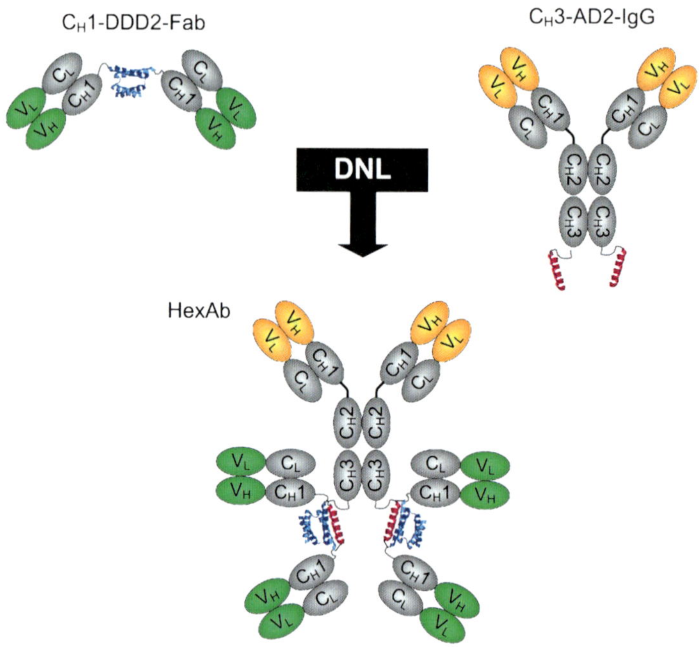

Fig. 12.3 Schematic of a HexAb made by DNL conjugation of C_H3-AD2-IgG with C_H1-DDD2-Fab

Table 12.2 Various designations of select antibodies used in DNL

Antigen	Antibody		
	Trivial name	USAN	DNL code
CD20	hA20	Veltuzumab	20
CD22	hLL2	Epratuzumab	22
CEACAM5	hMN-14	Labetuzumab	14
Indium-DTPA	h734	–	734

humanized IgG (veltuzumab or hA20) and a pair of dimeric anti-CD22 humanized Fab (epratuzumab or hLL2); 22-(20)-(20) specifies the bispecific HexAb comprising a divalent hLL2 IgG and a pair of dimeric hA20 Fab; and 20-(20)-(20) denotes the monospecific HexAb comprising a divalent hA20 IgG and a pair of dimeric hA20 Fab. The various designations of the antibodies that are pertinent to the HexAbs described here are provided in Table 12.2.

12.4.1 Generation and Study Design

For the initial validation, we made seven different HexAbs, namely, 20-(22)-(22), 22-(20)-(20), 20-(20)-(20), 22-(22)-(22), 20-(14)-(14), 22-(14)-(14), and 734-(20)-(20), by combining a C_H3-AD2-IgG module with a C_H1-DDD2-Fab module as indicated in Table 12.3 under mild redox conditions followed by purification with Protein A

Table 12.3 Modules, targets, and codes of HexAbs made by DNL

Code	Previous designation[a]	AD-module		DDD-module	
		Format	Target	Format	Target
20-(22)-(22)	20-22	C_H3-AD2-IgG-hA20	CD20	C_H1-DDD2-Fab-hLL2	CD22
22-(20)-(20)	22-20	C_H3-AD2-IgG-hLL2	CD22	C_H1-DDD2-Fab-hA20	CD20
20-(20)-(20)	Hex-hA20	C_H3-AD2-IgG-hA20	CD20	C_H1-DDD2-Fab-hLL1	CD20
22-(22)-(22)	Hex-hLL2	C_H3-AD2-IgG-hLL2	CD22	C_H1-DDD2-Fab-hLL2	CD22
20-(14)-(14)	20-14	C_H3-AD2-IgG-hA20	CD20	C_H1-DDD2-Fab-hMN-14	CEACAM5
22-(14)-(14)	22-14	C_H3-AD2-IgG-hLL2	CD22	C_H1-DDD2-Fab-hMN-14	CEACAM5
734-(20)-(20)	734-20	C_H3-AD2-IgG-h734	Indium-DTPA	C_H1-DDD2-Fab-hA20	CD20

[a]Rossi et al. (2008, 2009)

affinity chromatography. The individual modules used to assemble these HexAbs were produced in mammalian cell cultures and the ensuing DNL reaction typically proceeded uneventfully, resulting in each final conjugate shown by SE-HPLC to consist of predominantly a single peak of the expected molecular size (~365 kDa) and by SDS-PAGE to be of high purity with the IgG and the two dimeric Fab covalently linked.

Apart from assessing the feasibility of using the DNL approach to create novel IgG-like bsAbs, this panel of HexAbs was also designed to evaluate the effect of increased valency of the anti-CD20 or anti-CD22 component on binding avidity. A further aim was to investigate the signal transduction pathways induced with the bispecific 20-(22)-(22) and 22-(20)-(20) upon simultaneous engagement of both CD20 and CD22 on target cells, which may or may not be the same as those induced by the monospecific 20-(20)-(20) or by crosslinking the bound parental hA20 IgG with a secondary antibody. In addition, various studies were performed to define the in vitro and in vivo properties of 20-(22)-(22) and 22-(20)-(20), which include direct toxicity, antibody-dependent cellular cytotoxicity (ADCC), complement-dependent cytotoxicity (CDC), serum stability, pharmacokinetics (PK), and antitumor efficacy in animals bearing human B-cell lymphoma xenografts. The results presented below for the HexAbs in Sects. 12.4.2 and 12.4.3 are based on the published reports of Rossi et al. (2008, 2009), and Gupta et al. (2010).

12.4.2 *In Vitro Characterizations*

12.4.2.1 Binding Avidity, Off-Rates, Bispecificity, and Internalization

Using an antiidiotype antibody to hA20 as the surrogate antigen, we determined by competition ELISA the binding avidity of the hexavalent 20-(20)-(20) to be 1.4 nM, compared to 2.2 nM and 4.8 nM of the tetravalent 22-(20)-(20) and the divalent 20-(22)-(22), respectively. Similarly, with an antiidiotype to hLL2 as the surrogate antigen, we found the hexavalent 22-(22)-(22) to display the highest binding avidity (0.30 nM), compared to 0.43 nM and 0.66 nM of the tetravalent 20-(22)-(22) and the divalent 22-(20)-(20), respectively. In addition, 20-(22)-(22) and 22-(20)-(20) showed nearly the same binding avidity as the divalent hA20 (4.8 vs. 4.9 nM) and hLL2 (0.66 vs. 0.81 nM), respectively. These results indicate that each Fab-arm in a HexAb retains its binding activity and the avidity of a divalent IgG can be conveniently enhanced by increasing the valency with the DNL method.

The enhanced binding avidity of a HexAb due to multiple valencies should slow its dissociation from the surface of bound cells, which can be measured by flow cytometry. Using Raji cells as the CD20/CD22 target, the off-rates of 20-(20)-(20), 22-(20)-(20), 20-(22)-(22), and hA20 IgG were determined to be 322, 268, 152, and 145 min, respectively. The difference between the hexavalent 20-(20)-(20) and the tetravalent 22-(20)-(20) was statistically significant ($P = 0.0042$), as was the difference between the tetravalent 22-(20)-(20) and the divalent hA20 IgG ($P < 0.0001$). However, the difference between 20-(22)-(22) and hA20 IgG was not significant ($P = 0.3810$). Thus, the relative off-rates appear to correlate well with the number of CD20-binding arms, but not the number of CD22-binding arms, which may be due to the tenfold higher expression of CD20 than CD22 in Raji cells.

The bispecificity of 20-(22)-(22) and 22-(20)-(20) was demonstrated by performing cell-binding analysis with flow cytometry. Raji cells were preincubated with excess C_H1-DDD2-Fab-hA20, C_H1-DDD2-Fab-hLL2, or both, to block CD20, CD22, and CD20/CD22 binding, respectively. Cells were then stained with a saturating amount of PE (phycoerythrin)-conjugated 22-(20)-(20), 20-(22)-(22), hA20 IgG, or hLL2 IgG, and the resulting fluorescence intensity was measured and compared to that of unblocked control. Without blocking, the median fluorescence intensity (MFI) of cells stained with hA20 IgG was approximately tenfold higher than that observed with hLL2 IgG, indicating that CD20 is expressed at a considerably higher level than CD22 on Raji cells. Whereas C_H1-DDD2-Fab-hA20, C_H1-DDD2-Fab-hLL2, and the combination completely inhibited hA20, hLL2, and all PE-labeled antibodies, respectively, either DDD2-Fab module alone could only partially block 20-(22)-(22) or 22-(20)-(20) from binding to Raji cells, providing evidence that 20-(22)-(22) and 22-(20)-(20) can bind to both CD20 and CD22 on the target.

We made an intriguing observation when investigating the extent of internalization of 20-(22)-(22) and 20-(22)-(22) into Raji cells by flow cytometry. Live cells were incubated with PE-conjugated antibodies at 37°C for 1 h before trypsin digestion to remove noninternalized antibodies. The MFI of cells stained with PE-hA20 and PE-22-(20)-(20) was reduced by 90% and 85%, respectively, indicating that 22-(20)-(20) behaves like hA20 with a slow internalization rate. On the other hand, we found approximately 50% of the 20-(22)-(22) internalized, similar to the results obtained for hLL2 IgG. We tentatively concluded that the internalization property of a bsAb composed of a rapid internalizing antibody, such as hLL2, and a slowly or not internalizing antibody, such as hA20, would depend on the relative valency of the two antibodies with different internalization characteristics.

12.4.2.2 Direct and Indirect Cytotoxicity

As shown by a cell counting assay, 22-(20)-(20) and 20-(22)-(22) effectively inhibited the growth of three Burkitt lymphoma cell lines, Ramos, Raji, and Daudi, at 15 nM or lower, whereas under the same conditions the individual parental antibodies alone, or together, were either ineffective (hLL2) or not as effective (hA20 or combined with hLL2). Based on the EC_{50} values determined from the dose-response curves, 22-(20)-(20) was more potent than 20-(22)-(22) in the three lymphoma cell lines examined, and the observed direct toxicity was not appreciably affected for either 22-(20)-(20) or 20-(22)-(22) upon the addition of a crosslinking antihuman Fc antibody which, however, markedly increased the inhibitory activity of hA20.

To assess only the effect of divalent, tetravalent, and no binding to CD20 on cell proliferation, we evaluated the potency of 20-(14)-(14), 734-(20)-(20), and 22-(22)-(22), which have the molecular structure of the HexAbs, to inhibit the growth of Ramos cells by the MTS assay. Our finding that the effect of 20-(14)-(14) was similar to that of hA20 IgG suggests that the formation of a heterocomplex of CD20 and CD22 on the cell surface is essential for the direct toxicity of 20-(22)-(22). We also confirmed the capability of a tetravalent CD20-binding molecule, such as 734-(20)-(20), to display direct toxicity and observed no statistical difference between the dose-response curves obtained for 734-(20)-(20) and 22-(20)-(20). Unlike 20-(20)-(20), the monospecific 22-(22)-(22) showed no direct toxicity against NHL cell lines in vitro. Direct toxicity of 20-(20)-(20), 20-(22)-(22), and 22-(20)-(20) was also evaluated on eight CLL patient specimens, which varied in their CD20 expression. The three specimens expressing moderate to high CD20 showed 30–60% inhibition by the three HexAbs, whereas no significant inhibition was observed in the other five specimens with low CD20 expression. Interestingly, neither rituximab nor hA20 IgG, with or without hypercrosslinking, produced measurable inhibition.

As an IgG1, hA20, but not hLL2, exhibited potent CDC. Similarly, C_H3-AD2-IgG-hA20, but not C_H3-AD2-IgG-hLL2, induced CDC. However, the addition of four Fab-arms to C_H3-AD2-IgG-hA20, as exemplified by 20-(20)-(20), 20-(22)-(22), and 20-(14)-(14), abolished CDC, which also occurs for 22-(20)-(20). Nevertheless, we could show that 20-(22)-(22), 20-(14)-(14), and 20-(20)-(20) preserved the potent ADCC of hA20 IgG; and 734-(20)-(20) and 22-(20)-(20) had similar ADCC, either of which was statistically higher ($P = 0.004$) than hLL2 IgG. These findings support the notion that ADCC is governed primarily by CD20, which has a higher antigen density than CD22, and thus HexAbs based on hA20 IgG mediate ADCC more efficiently.

12.4.2.3 Type I vs. Type II Anti-CD20 Antibodies

Based on their efficacy in certain in vitro assays (Cragg et al. 2003; Cragg and Glennie 2004), monoclonal antibodies (mAbs) targeting CD20 have been classified

(Cragg and Glennie 2004) as either Type I, represented by rituximab, or Type II, represented by tositumomab (Cardarelli et al 2002). Type I anti-CD20 mAbs are characterized by potent CDC, rapid calcium mobilization upon the addition of a second antibody (Hofmeister et al. 2000; Unruh et al. 2005), weak induction of homotypic adhesion (Deans and Polyak 2002), enhanced association of CD20 with lipid rafts (Li et al. 2004), and a general requirement of crosslinking with a second antibody for efficient antiproliferation or apoptosis (Deans and Polyak 2002). In contrast, Type II anti-CD20 mAbs lack CDC (Cragg and Glennie 2004), do not stimulate calcium mobilization upon further crosslinking (Unruh et al. 2005), trigger a high level of homotypic adhesion (Deans and Polyak 2002), fail to bring CD20 into lipid rafts (Teeling et al. 2004; Mössner et al. 2010), and induce significant apoptosis via direct toxicity (Cardarelli et al. 2002). A further differentiation is that apoptotic cell death mediated by Type I mAbs involves both caspase-dependent and -independent pathways, but only caspase-independent pathways appear to be associated with Type II mAbs (Mössner et al. 2010). However, both Type I and Type II anti-CD20 mAbs are effective in ADCC and those of mouse origin apparently bind to overlapping epitopes in the large extracellular loop of CD20 (Polyak and Deans 2002; Ernst et al. 2005; Teeling et al. 2006; Du et al. 2007). Interestingly, at saturation the number of a Type II mAb that binds to each CD20-positive B cell only approaches half of that observed with a Type I mAb (Teeling et al. 2004; Mössner et al. 2010), suggesting that a Type II mAb may act virtually as a tetravalent antibody by engaging two adjacent extracellular loops of CD20 with each antigen-binding arm.

The three hA20-containing HexAbs, namely, 20-(20)-(20), 22-(20)-(20), and 20-(22)-(22), share properties of both Type I and Type II anti-CD20 mAbs. Consistent with Type II, 20-(20)-(20), 22-(20)-(20), and 20-(22)-(22) are negative for CDC and calcium mobilization, do not require crosslinking for growth inhibition or apoptosis, and induce strong homotypic adhesion; yet they translocate CD20 to lipid rafts as Type I. Thus, one effective approach to convert a Type I anti-CD20 mAb to a Type II can be achieved by making the Type I mAb multivalent, such as 20-(20)-(20) or 22-(20)-(20), or part of a bispecific HexAb, such as 20-(22)-(22). It is noted that the cell-counting assay showed that 20-(20)-(20) was considerably more potent than the Type II tositumomab (EC_{50} = 0.17 nM vs. 4.65 nM) when evaluated in Ramos cells for 3 days.

12.4.2.4 Signal Transduction

We have explored the signaling pathways that are involved in evoking direct toxicity of 20-(20)-(20), 22-(20)-(20), and 20-(22)-(22) in Daudi cells. For comparison, selective studies included cells treated with anti-IgM antibodies to activate the B-cell receptor, or with hA20 or rituximab in the presence of a crosslinking antibody to enhance the apoptotic potency. Our key findings are summarized as follows. (a) The signaling events triggered by 20-(20)-(20), 22-(20)-(20), or 20-(22)-(22) are quantitatively and qualitatively similar in Daudi cells, but distinct from those induced by anti-IgM. (b) Although hA20 and rituximab modify the signaling events in Daudi cells similar to the three CD20-targeting HexAbs, as observed for the ERK and NF-κB pathways, both require a higher concentration to be effective and are less efficient in modulating the cell cycle regulators that promote growth arrest (e.g., upregulation of p21, p27, and Kip2 and downregulation of cyclin D1 and phosphorylated Rb). In addition, the divalent hA20 IgG and rituximab fail to alter the levels of phosphorylated p38 and PTEN from untreated control, whereas all three HexAbs increase phosphorylated p38 and PTEN levels significantly. Similar results were obtained in Raji cells for the decrease in phosphorylated ERKs and the increase in phosphorylated p38. No appreciable change in the basal expression of signaling molecules was observed in Daudi cells upon ligation of CD22 by hLL2. (c) The apoptosis and inhibition of cell proliferation resulting from crosslinking hA20 or rituximab with a secondary antibody involves signaling events that are distinguishable from those associated with the HexAbs, as manifested in phosphorylated ERK (increase vs. decrease), intracellular calcium (increase vs. no change), and mitochondrial membrane potential (loss vs. no change).

Collectively, our findings are consistent with the view that the potent direct toxicity of the three CD20-targeting HexAbs is due to their multivalent binding ability, which lowers the threshold for modifying multiple signaling pathways, resulting in a new distribution of pro- and antiapoptotic proteins that promotes growth arrest, apoptosis, and eventually cell death. Intriguingly, these effects translated to notable differences with regard to their relative potency for killing normal human B cells vs. human Burkitt lymphoma cells ex-vivo, as the bispecific 22-(20)-(20) and 20-(22)-(22) showed a higher toxicity to malignant than normal B cells, compared to hA20 and rituximab. Noting also the potential advantages of lacking CDC and the moderate but significant enhancement of ADCC observed for 22-(20)-(20) as compared to hLL2, we speculate that a bispecific anti-CD20/CD22 HexAb may be a more potent class of antilymphoma therapeutic antibodies for clinical use.

12.4.3 PK and In Vivo Antitumor Efficacy

We used a bispecific ELISA to quantify the amount of 20-(22)-(22) and 22-(20)-(20) in serum samples collected from PK studies in BALB/c mice and found that the two HexAbs display a shorter circulating half-life than their parental antibodies (24–37 h vs. 46–52 h). Because both HexAbs are stable in serum when assessed in vitro, their faster blood clearance is likely due to intracellular breakdown of the modular components, which presumably occurs after their uptake into the vascular endothelium of mice. Evidence for the slow dissociation of the HexAbs in vivo was provided by SE-HPLC analysis of the PK samples obtained 72 h after injecting radiolabeled 20-(22)-(22), which identified the presence of a new peak with a size of an IgG shown to be derived from hA20, not hLL2, as it failed to bind the antiidiotype antibody to hLL2. A parallel study using radiolabeled 22-(20)-(20) also revealed the presence in the 72-h PK samples of a new peak with a size of an IgG shown to be derived from hLL2, not hA20, using an antiidiotype antibody to hA20.

The in vivo antitumor efficacy of 22-(20)-(20) and 20-(22)-(22) was evaluated in SCID mice with a disseminated Daudi model in three studies. In the first study, 22-(20)-(20) and 20-(22)-(22) were each administered i.v. in a single dose of 10 pmol (~3.7 μg) and compared with various controls including 22-(14)-(14), 20-(14)-(14), 22-(22)-(22), and hA20 IgG, giving the equimolar dose. With therapy starting 1 day after inoculating Daudi cells i.v., the group treated with 22-(20)-(20) had a significantly extended median survival time (MST) compared to the saline control (36 vs. 29 days, $P = 0.005$), whereas the two groups treated with 22-(22)-(22) or 22-(14)-(14) did not improve survival over saline (29 days, $P = 1.0$). The MST of mice treated with 20-(22)-(22) was significantly longer than the saline control (50 vs. 29 days, $P = 0.005$), but there was no statistically significant difference in the MST of 20-(22)-(22), 20-(14)-(14), and hA20 IgG.

In the second study, we compared the efficacy of 22-(20)-(20) to hLL2 IgG, 22-(14)-(14), 734-(20)-(20), and a mixture of 22-(14)-(14) and 734-(20)-(20). Groups of mice (ten in each) were administered 10 μg doses of 22-(20)-(20), 734-(20)-(20), 22-(14)-(14), or 10 μg of both 734-(20)-(20) and 22-(14)-(14) on days 1, 4, and 7. Additional groups received an equimolar dose of hLL2 (4 μg) or saline. All mice in the saline-treated group died within 4 weeks (MST = 25 days). The MSTs for mice treated with 22-(20)-(20), 734-(20)-(20), 22-(14)-(14), the combination, and hLL2 IgG were 66.5 days, 42 days, 32 days, 68.5 days, and 32 days, respectively. A statistically significant difference ($P < 0.001$) in the MST was found between 22-(20)-(20) and each of 734-(20)-(20), 22-(14)-(14), and hA20. Notably, the MST of 22-(20)-(20) was about the same as the combination of 734-(20)-(20) and 22-(14)-(14); despite that, the latter provides the same number of CD20- and CD22-binding arms with twice the number of Fc groups.

In the third study, we examined the role of effector cells in the ability of 22-(20)-(20) or 20-(22)-(22) to inhibit tumor growth. Groups of mice (five in each) that had been depleted of NK cells and neutrophils were administered i.v. with 230 μg of 22-(20)-(20) or 20-(22)-(22) on days 1, 3, 5, and 9. As controls, four groups of mice without depletion of NK cells and neutrophils were each treated with saline, hLL2 (100 μg), or the two bispecific HexAbs at the same dose and schedule as the depleted groups. Treatments with 22-(20)-(20) or 20-(22)-(22) resulted in no survival benefit to animals in the depleted groups, as there was insignificant difference in the MST from the saline control (18 days vs. 21 days). In contrast, treatments of animals in the nondepleted groups with either 22-(20)-(20) or 20-(22)-(22) significantly ($P < 0.002$) increased their survival, with MST of 63 and 91 days, respectively, compared to 21 days of the saline control and 28 days of the hLL2 control. However, the observed difference in the MSTs between the 22-(20)-(20) and 20-(22)-(22) was not statistically significant, perhaps because of the relatively small number of animals included in these groups. These initial results underscore the importance of ADCC as the major mechanism of action in retarding tumor growth in animal models by antibodies that target CD20, CD22, or both.

12.5 The Advantages of DNL

Besides the unique feature that a module derivatized with the DDD is always presented in two copies, there are additional advantages of the DNL method as discussed below.

DNL is modular. Each DDD- or AD-containing entity serves as a module and any DDD module can be paired with any AD module. Such modules can be produced independently, stored separately "on shelf," and combined "on demand." There is essentially no limit on the types of precursors that can be converted into a DDD- or AD-module, so long as the resulting modules do not interfere with the dimerization of DDD or the binding of DDD to AD. In addition to the DDD sequence of human RIIα, other DDD sequences may be selected from human RIα, human RIβ, or human RIIβ. The DDD sequence of choice will be matched with a highly interactive AD sequence, which can be deduced from the literature (Burns-Hamuro et al. 2003) or determined experimentally.

DNL is versatile. Modules can be made recombinantly or chemically. Recombinant modules, which may be produced in mammalian or microbial systems, may include fusions of antibodies or antibody fragments, cytokines, enzymes, carrier proteins (such as human serum albumin and human transferrin), or a variety of natural or artificial nonantibody binding or scaffold proteins (Binz et al. 2005; Binz and Pluckthun 2005; Hey et al. 2005; Hosse et al. 2006). Although each recombinant module would usually be produced in a separate expression system, certain pairs of DDD- and AD-modules may be co-expressed in the same host cell without affecting the formation of the DNL conjugates. Furthermore, DDD or AD can be coupled to the amino terminal or carboxyl terminal end or even positioned internally within the fusion protein, preferably with a spacer containing an appropriate length and composition of amino acid residues, provided that the binding activity of the DDD or AD and the desired activity of the polypeptide fusion partners are not compromised.

Modules may also be made synthetically, as demonstrated with linking AD2 to either polyethylene glycol or peptides (Chang et al. 2009), and it should be feasible to develop chemistries for preparing modules that contain peptide mimetics, oligo- or polynucleotides, small interfering RNA, chelators with or without radioactive or nonradioactive metals, drugs, dyes, oligosaccharides, natural or synthetic polymeric substances, nanoparticles, dendrimers, fluorescent molecules, or quantum dots, depending on the intended applications.

DNL manufacture is easy. The DNL method is basically a one-pot preparation and requires three simple steps to recover the product from the starting materials: (a) combine DDD- and AD-modules in stoichiometric amounts; (b) add redox agents to facilitate the self-assembly of the DNL conjugate; and (c) purify by an appropriate affinity chromatography.

DNL results in quantitative yields of a homogeneous product with a defined composition and retained bioactivities. The facile binding between the DDD and AD modules effects nearly 100% conversion of each into the desired DNL product.

The site-specific conjugation also assures that the full activity of each module is preserved, the molecular size is homogeneous, the composition is defined, and in vivo integrity is largely sustained.

12.6 Outlook

The DNL method utilizes both recombinant engineering and site-specific conjugation chemistry to provide a facile, efficient, and modular platform technology for the production of multivalent antibodies that can be monospecific, bispecific, Fc-lacking, or Fc-containing, as demonstrated here for the T-Fabs and the HexAbs.

The imaging data obtained from tumor-bearing mice with select T-Fabs and from cancer patients with TF2 indicate these bsAbs of ~150 kDa are able to localize at their primary tumor and subsequently capture the HSG-containing haptens. The in vivo therapeutic efficacy observed for 20-(22)-(22), 22-(20)-(20), and 20-(20)-(20) in lymphoma models also led us to believe these HexAbs of ~365 kDa are capable of penetrating and accumulating at the target sites. We have more recently determined by dynamic light scattering that the diameter of a HexAb is between 15 and 16 nm. Noting that the average diameter of a Y-shaped monomeric IgG as measured by electrospray differential mobility analysis is between 9 and 10 nm (Pease et al. 2008) and that 10–100 nm is considered the optimal range of the diameter for nanoparticle drug carriers in cancer therapy (Davis et al. 2008), the concern that a HexAb made by DNL would be too large to penetrate into tumor may be overstated, but nevertheless needs to be addressed in further studies. Indeed, it is far more likely that the penetration of the DNL constructs will be impeded by the so-called "binding site barrier" (Fujimori et al. 1989; Weinstein and van Osdol 1992; Saga et al. 1995), since this natural phenomenon will have a larger impact on the movement of agents that bind to surface proteins, particularly agents with multivalent binding capabilities.

Whether a HexAb can be made by DNL to retain CDC and show more potent antitumor effect in vivo is worthy of pursuit. On the other hand, because CDC has been suggested to be a major factor in infusion-related toxicity for rituximab, it is tempting to speculate that a HexAb lacking CDC may be a viable alternative where reduced side-effects are desired.

Although all of the DNL complexes that have been generated to date are highly stable in vitro, the Fc-bearing HexAbs lose the appended Fab-arms in experimental animals with time, resulting in a less favorable PK than a monomeric IgG as well as a concurrent disappearance of the multivalent advantages. Thus, a key goal for future development is to enhance the integrity of the HexAbs in vivo.

In an effort to minimize immunogenicity, the DDD and AD used for the DNL method consist of the smallest functional peptides derived from human protein sequences. However, the DDD2 and AD2 peptides possess additional cysteine residues, which are not part of the natural domains, and both are fused to a modular component via flexible Gly-Ser linker peptides, which are foreign but generally considered to be largely nonimmunogenic. Whether the DNL complexes in general, and a particular T-Fab or HexAb would be immunogenic in humans can be best assessed in clinical testing.

References

Alto NM, Soderling SH, Hoshi N, Langeberg LK, Fayos R, Jennings PA, Scott JD (2003) Bioinformatic design of A-kinase anchoring protein-in silico: a potent and selective peptide antagonist of type II protein kinase A anchoring. Proc Natl Acad Sci USA 100(8):4445–4450

Asano R, Watanabe Y, Kawaguchi H, Fukuzawa H, Nakanishi T, Umetzu M, Hayashi H, Katayose Y, Unno M, Kudo T, Kumahai I (2007) Highly effective recombinant format of a humanized IgG-like bispecific antibody for cancer immunotherapy with retargeting of lymphocytes to tumor cells. J Biol Chem 282(38):27659–27665

Baillie GS, Scott JD, Houslay MD (2005) Compartmentalisation of phosphodiesterases and protein kinase A: opposites attract. FEBS Lett 579(15):3264–3270

Binz HK, Pluckthun A (2005) Engineered proteins as specific binding reagents. Curr Opin Biotechnol 16(4):459–469

Binz HK, Amstutz P, Pluckthun A (2005) Engineering novel binding proteins from nonimmunoglobulin domains. Nat Biotechnol 23(10):1257–1268

Brennan M, Davison PF, Paulus H (1985) Preparation of bispecific antibodies by chemical recombination of monoclonal immunoglobulin G1 fragments. Science 229(4708):81–83

Burns-Hamuro LL, Ma Y, Kammerer S, Reineke U, Self C, Cook C, Olson GL, Cantor CR, Braun A, Taylor SS (2003) Designing isoform-specific peptide disruptors of protein kinase A localization. Proc Natl Acad Sci USA 100(7):4072–4077

Burton DR, Wilson IA (2007) Immunology. Square-dancing antibodies. Science 317(5844): 1507–1508

Cardarelli PM, Quinn M, Buckman D, Fang Y, Colcher D, King DJ, Bebbington C, Yarronton G (2002) Binding of CD20 by anti-B1 antibody or $F(ab')_2$ is sufficient for induction of apoptosis in B-cell lines. Cancer Immunol Immunother 51:15–24

Carr DW, Stofko-Hahn RE, Fraser ID, Bishop SM, Acott TS, Brennan RG, Scott JD (1991) Interaction of the regulatory subunit (RII) of cAMP-dependent protein kinase with RII-anchoring proteins occurs through an amphipathic helix binding motif. J Biol Chem 266(22):14188–14192

Chang C-H, Rossi EA, Goldenberg DM (2007) The dock and lock method: a novel platform technology for building multivalent, multifunctional structures of defined composition with retained bioactivity. Clin Cancer Res 13(18 Pt2):5586s–5591s

Chang CH, Rossi EA, Cardillo TM, Nordstrom DL, McBride WJ, Goldenberg DM (2009) A new method to produce monoPEGylated dimeric cytokines shown with human interferon-α2b. Bioconjug Chem 20(10):1899–1907

Colledge M, Scott JD (1999) AKAPs: from structure to function. Trends Cell Biol 9(6):216–222

Coloma MJ, Morrison SL (1997) Design and production of novel tetravalent bispecific antibodies. Nat Biotechnol 15(2):159–163

Cragg MS, Glennie MJ (2004) Antibody specificity controls in vivo effector mechanisms of anti-CD20 reagents. Blood 103(7):2738–2743

Cragg MS, Morgan SM, Chan HT, Morgan BP, Filatov AV, Johnson PWM, French RR, Glennie MJ (2003) Complement-mediated lysis by anti-CD20 mAb correlates with segregation into lipid rafts. Blood 101(3):1045–1052

Davis ME, Chen Z, Shin DM (2008) Nanoparticle therapeutics: an emerging treatment modality for cancer. Nat Rev Drug Discov 7(9):771–782

Deans JP, Polyak MJ (2002) CD20-mediated apoptosis: signaling through lipid rafts. Immunology 107(2):176–182

Du J, Wang H, Zhong C, Peng B, Zhang M, Li B, Huo S, Guo Y, Ding J (2007) Structural basis for recognition of CD20 by therapeutic antibody Rituximab. J Biol Chem 282(20): 15073–15080

Ernst JA, Li H, Kim HS, Nakamura GR, Yansura DG, Vandlen RL (2005) Isolation and characterization of the B-cell marker CD20. Biochemistry 44(46):15150–15158

Fujimori K, Covell DG, Fletcher JE, Weinstein JN (1989) Modeling analysis of the global and microscopic distribution of immunoglobulin G, F(ab')2, and Fab in tumors. Cancer Res 49(20):5656–5663

Gegner JA, Dahlquist FW (1991) Signal transduction in bacteria: CheW forms a reversible complex with the protein kinase CheA. Proc Natl Acad Sci USA 88(3):750–754

Glennie MJ, McBride HM, Worth AT, Stevenson GT (1987) Preparation and performance of bispecific F(ab' gamma)2 antibody containing thioether-linked Fab' gamma fragments. J Immunol 139(7):2367–2375

Gold DV, Goldenberg DM, Karacay H, Rossi EA, Chang C-H, Cardillo TM, McBride WJ, Sharkey RM (2008) A novel bispecific, trivalent antibody construct for targeting pancreatic carcinoma. Cancer Res 68(12):4819–4826

Gupta P, Goldenberg DM, Rossi EA, Chang C-H (2010) Multiple signaling pathways induced by hexavalent, monospecific, anti-CD20 and hexavalent, bispecific, anti-CD20/CD22 humanized antibodies correlate with enhanced toxicity to B-cell lymphomas and leukemias. Blood 116(17):3258–3267

Hey T, Fiedler E, Rudolph R, Fiedler M (2005) Artificial, non-antibody binding proteins for pharmaceutical and industrial applications. Trends Biotechnol 23(10):514–522

Hofmeister JK, Cooney D, Coggeshall KM (2000) Clustered CD20 induces apoptosis: src-family kinase, the proximal regulator of tyrosine phosphorylation, calcium influx, and caspase-dependent apoptosis. Blood Cells Mol Dis 26(2):133–143

Hosse RJ, Rothe A, Power BE (2006) A new generation of protein display scaffold for molecular recognition. Protein Sci 15(1):14–27

Hollander N (2009) Bispecific antibodies for cancer therapy. Immunotherapy 1(2):211–222

Kriangkum J, Xu B, Nagata LP, Fulton RE, Suresh MR (2001) Bispecific and bifunctional single chain recombinant antibodies. Biomol Eng 18(2):31–40

Labrijn AF, Buijsse AO, van den Bremer ETJ, Verwilligen AYW, Bleeker WK, Thorpe SJ, Killestein J, Polman CH, Aalberse RC, Schuurman J, van de Winkel JGJ, Parren PWHI (2009) Therapeutic IgG4 antibodies engage in Fab-arm exchange with endogenous human IgG4 *in vivo*. Nat Biotechnol 27(8):767–771

Li H, Ayer LM, Polyak MJ, Mutch CM, Petrie RJ, Gauthier L, Shariat N, Hendzel MJ, Shaw AR, Patel KD, Deans JP (2004) The CD20 calcium channel is localized to microvilli and constitutively associated with membrane rafts: antibody binding increases the affinity of the association through an epitope-dependent cross-linking-independent mechanism. J Biol Chem 279(19):19893–19901

Lu D, Zhang H, Ludwig D, Persaud A, Jimenez X, Burtrum D, Balderes P, Liu M, Bohlen P, Witte L, Zhu Z (2004) Simultaneous blockade of both the epidermal growth factor receptor and the insulin-like growth factor receptor signaling pathways in cancer cells with a fully human recombinant bispecific antibody. J Biol Chem 279(4):2856–2865

Lu D, Zhang H, Koo H, Tonra J, Balderes P, Prewett M, Corcoran E, Mangalampalli V, Bassi R, Anselma D, Patel D, Kang X, Ludwig DL, Hicklin DJ, Bohlen P, Witte L, Zhu Z (2005) A fully human recombinant IgG-like bispecific antibody to both the epidermal growth factor receptor and the insulin-like growth factor receptor for enhanced antitumor activity. J Biol Chem 280(20):19665–19672

Marvin JS, Zhu Z (2005) Recombinant approaches to IgG-like bispecific antibodies. Acta Pharmacol Sin 26(6):649–658

McBride WJ, Sharkey RM, Karacay H, D'Souza CA, Rossi EA, Laverman P, Chang C-H, Boerman OC, Goldenberg DM (2009) A novel method of ^{18}F radiolabeling for PET. J Nucl Med 50(6):991–998

Milstein C, Cuello AC (1983) Hybrid hybridomas and their use in immunochemistry. Nature 305 (5934):537–540

Mössner E, Brünker P, Moser S, Püntener U, Schmidt C, Herter S, Grau R, Gerdes C, Nopora A, van Puijenbroek E, Ferrara C, Sondermann P, Jäger C, Strein P, Fertig G, Friess T, Schüll C, Bauer S, Dal Porto J, Del Nagro C, Dabbagh K, Dyer MJ, Poppema S, Klein C, Umaña P (2010) Increasing the efficacy of CD20 antibody therapy through the engineering of a new type II anti-CD20 antibody with enhanced direct and immune effector cell-mediated B-cell cytotoxicity. Blood 115(22):4393–4402

Muller D, Kontermann RE (2007) Recombinant bispecific antibodies for cellular cancer immunotherapy. Curr Opin Mol Ther 9(4):319–326

Newlon MG, Roy M, Morikis D, Hausken ZE, Coghlan V, Scott JD, Jennings PA (1999) The molecular basis for protein kinase A anchoring revealed by solution NMR. Nat Struct Biol 6(3):222–227

Newlon MG, Roy M, Morikis D, DW C, Westphal R, Scott JD, Jennings PA (2001) A novel mechanism of PKA anchoring revealed by solution structures of anchoring complexes. EMBO J 20(7):1651–1662

Pease LF III, Elliot JT, Tsai D-H, Zachariah MR, Tarlov MJ (2008) Determination of protein aggregation with different mobility analysis: application to IgG antibody. Biotechnol Bioeng 101(6):1214–1222

Perez P, Hoffman RW, Shaw S, Bluestone JA, Segal DM (1985) Specific targeting of cytotoxic T cells by anti-T3 linked to anti-target cell antibody. Nature 316(6026):354–356

Polyak MJ, Deans JP (2002) Alanine-170 and proline-172 are critical determinants for extracellular CD20 epitopes; heterogeneity in the fine specificity of CD20 monoclonal antibodies is defined by additional requirements imposed by both amino acid sequence and quaternary structure. Blood 99(9):3256–3262

Qu Z, Goldenberg DM, Cardillo TM, Shi V, Hansen HJ, Chang C-H (2008) Bispecific anti-CD20/22 antibodies inhibit B-cell lymphoma proliferation by a unique mechanism of action. Blood 111(4):2211–2219

Rossi EA, Sharkey RM, McBride W, Karacay H, Zeng L, Hansen HJ, Goldenberg DM, Chang C-H (2003) Development of new multivalent-bispecific agents for pretargeting tumor localization and therapy. Clin Cancer Res 9(10 Pt 2):3886s–3896s

Rossi EA, Goldenberg DM, Cardillo TM, McBride WJ, Sharkey RM, Chang C-H (2006) Stably tethered multifunctional structures of defined composition made by the dock and lock method for use in cancer targeting. Proc Natl Acad Sci USA 103(18):6841–6846

Rossi EA, Goldenberg DM, Cardillo TM, Stein R, Wang Y, Chang C-H (2008) Novel designs of multivalent anti-CD20 humanized antibodies as improved lymphoma therapeutics. Cancer Res 68(20):8384–8392

Rossi EA, Goldenberg DM, Cardillo TM, Stein R, Chang C-H (2009) Hexavalent bispecific antibodies represent a new class of anticancer therapeutics. 1. Properties of anti-CD20/CD22 antibodies in lymphoma. Blood 113(24):6161–6171

Saga T, Neumann RD, Heya T, Sato J, Kinuya S, Le N, Paik CH, Weinstein JN (1995) Targeting cancer micrometastases with monoclonal antibodies: a binding-site barrier. Proc Natl Acad Sci USA 92(19):8999–9003

Schoffelen R, Sharkey RM, Goldenberg DM, Franssen G, McBride WJ, Rossi EA, Chang C-H, Laverman P, Disselhorst JA, Eek A, van der Graaf WT, Oyen WJ, Boerman OC (2010) Pretargeted immuno-positron emission tomography imaging of carcinoembryonic antigen-expressing tumors with a bispecific antibody and a 68Ga- and 18F-labeled hapten peptide in mice with human tumor xenografts. Mol Cancer Ther 9(4):1019–1027

Sharkey RM, Juweid M, Shevitz J, Behr T, Dunn R, Swayne LC, Wong GY, Blumenthal RD, Griffiths GL, Siegel JA, Leung S, Hansen HJ, Goldenberg DM (1995) Evaluation of a complementarity-determining region-grafted (humanized) anti-carcinoembryonic antigen monoclonal antibody in preclinical and clinical studies. Cancer Res 55(Suppl):5935s–5945s

Sharkey RM, Karacay H, Vallabhajosula S, McBride WJ, Rossi EA, Chang C-H, Goldsmith SJ, Goldenberg DM (2008) Metastatic human colonic carcinoma: molecular imaging with pretargeted SPECT and PET in a mouse model. Radiology 246(2):497–507

Sharkey RM, Karacay H, Johnson CR, Litwin S, Rossi EA, McBride WJ, Chang C-H, Goldenberg DM (2009) Pretargeted versus directly targeted radioimmunotherapy combined with anti-CD20 antibody consolidation therapy of non-Hodgkin lymphoma. J Nucl Med 50(3):444–453

Sharkey RM, Rossi EA, McBride WJ, Chang C-H, Goldenberg DM (2010) Recombinant bispecific monoclonal antibodies prepared by the dock-and-lock strategy for pretargeted radioimmunotherapy. Semin Nucl Med 40(3):190–203

Shen J, Vil MD, Jimenez X, Iacolina M, Zhang H, Zhu Z (2006) Single variable domain-IgG fusion. A novel recombinant approach to Fc domain-containing bispecific antibodies. J Biol Chem 281(16):10706–10714

Shen J, Vil MD, Jimenez X, Zhang H, Lacolina M, Mangalampalli V, Balderes P, Ludwig DL, Zhu Z (2007) Single variable domain antibody as a versatile building block for the construction of IgG-like bispecific antibodies. J Immunol Methods 318(1–2):65–74

Staerz UD, Bevan MJ (1986) Hybrid hybridoma producing a bispecific monoclonal antibody that can focus effector T-cell activity. Proc Nat Acad Sci USA 83(5):1453–1457

Teeling JL, French RR, Cragg MS, van den Brakel J, Pluyter M, Huang H, Chan C, Parren PWH, Hack CE, Dechant M, Valerius T, van der Winkel JGJ, Glennie MJ (2004) Characterization of new human CD20 monoclonal antibodies with potent cytolytic activity against non-Hodgkin lymphoma. Blood 104(6):1793–1800

Teeling JL, Mackus WJ, Wiegman LJ, van den Brakel J, Beers SA, French RR, van Meerten T, Ebeling S, Vink T, Slootsra JW, Parren PW, Glennie MJ, van de Winkel J (2006) The biological activity of human CD20 monoclonal antibodies is linked to unique epitopes on CD20. J Immunol 177(1):362–371

Unruh TL, Li H, Mutch CM, Shariat N, Grigoriou L, Sanyal R, Brown CB, Deans JP (2005) Cholesterol depletion inhibits src family kinase-dependent calcium mobilization and apoptosis induced by rituximab crosslinking. Immunology 116(2):223–232

van der Neut Kolfschoten M, Schuurman J, Losen M, Bleeker WK, Martinez-Martinex P, Vermeulen E, den Bleker TH, Wiegman L, Vink T, Aarden LA, De Baets MH, van der Winkel JG, Aalberse RC, Parren PW (2007) Anti-inflammatory activity of human IgG4 antibodies by dynamic Fab arm exchange. Science 317(5844):1554–1557

van Rij C, Franssen G, Sharkey R, Frielink C, McBride B, Oyen W, Goldenberg D, Boerman O (2010) Pretargeted immunoPET imaging and radioimmunotherapy (RIT) of prostate cancer with an anti-EGP1 x anti-HSG bispecific antibody (bsMAb). J Nucl Med 51(Supplement 2):501

Weinstein JN, van Osdol W (1992) Early intervention in cancer using monoclonal antibodies and other biological ligands: micropharmacology and the "binding site barrier". Cancer Res 52(9 Suppl):2747s–2751s

Wong W, Scott JD (2004) AKAP signaling complexes: focal points in space and time. Nat Rev Mol Cell Biol 5(12):959–970

Wu C, Ying H, Grinnell C, Bryant S, Miller R, Clabbers A, Bose S, McCarthy D, Zhu RR, Santora L, Davis-Taber R, Kunes Y, Fung E, Schwarz A, Sakorafas P, Gu J, Tarcsa E, Murtaza A, Ghayur T (2007) Simultaneous targeting of multiple disease mediators by a dual-variable-domain immunoglobulin. Nat Biotechnol 25(11):1290–1297

Chapter 13
Effector Cell Recruitment by Bispecific Antibodies

Christian Kellner, Matthias Peipp, and Thomas Valerius

13.1 Introduction

In recent years, therapeutic antibodies have been included in the standard treatment for a variety of malignant diseases (Oldham and Dillman 2008). Antibodies such as rituximab and trastuzumab, which are approved for the treatment of B-cell lymphomas and breast cancer, respectively, are generally well tolerated and have clearly improved the treatment outcome of respective patients. Based on these success stories, other antibodies have entered the clinic or are in various stages of development. However, despite these encouraging results, the success rate of antibody therapy is still unsatisfactory: antibodies do rarely cure as single agents, responses are often restricted to subgroups of patients, and chances of relapse or progress remain serious issues. In addition, antibody therapy often dramatically increases treatment costs while the clinical benefit remains limited (Reichert et al. 2005). Therefore, it is widely agreed that antibody therapy requires further improvement. A possibility to achieve this aim relies on understanding clinically important effector mechanism of therapeutic antibodies, and translating these insights into rational approaches to specifically enhance these mechanisms. Several observations indicated that recruitment of effector cells by engagement of Fcγ receptors (FcγR) critically determined the clinical efficacy of monoclonal antibodies – suggesting that enhancing the interaction of the antibody with FcR may improve effector cell recruitment and antibody therapy (Carter 2006; Presta 2008).

Bispecific antibodies have long been recognized as an elegant way to improve effector cell recruitment for antibody therapy (Staerz et al. 1985; Fanger et al. 1991; Segal et al. 1999). For tumor therapy, bispecific antibodies typically recognize a target antigen on tumor cells, whereas the second site engages a stimulatory trigger

C. Kellner • M. Peipp • T. Valerius (✉)
Division of Stem Cell Transplantation and Immunotherapy, 2nd Department of Medicine, Christian-Albrechts-University Kiel, Schittenhelmstr. 12, 24105 Kiel, Germany
e-mail: t.valerius@med2.uni-kiel.de

R.E. Kontermann (ed.), *Bispecific Antibodies*,
DOI 10.1007/978-3-642-20910-9_13,

molecule on effector cells. Bispecific antibodies are attractive as they may improve recruitment of FcR-expressing cells or allow redirection of T lymphocytes which lack FcR and therefore are not available as effector cells for conventional antibodies. Recently, promising clinical results were obtained with T cell-engaging bispecific antibodies. Thus, the trifunctional bispecific antibody catumaxumab targeting EpCAM was the first bispecific antibody receiving approval by the European Medicines Agency (EMEA) in 2009 (Zeidler et al. 1999). Blinatumumab targeting CD19 on B-cell lymphomas and leukemias is expected to follow soon (Loffler et al. 2000).

In this chapter, the scientific rationale for the use of bispecific antibodies in cancer immunotherapy will be outlined and different formats for their generation will be presented. Then different lymphocytic and myeloid effector cells and their candidate trigger molecules for bispecific approaches will be described, and pre-clinical and clinical experiences with individual bispecific antibodies will be summarized.

13.2 Rationale for Bispecific Antibodies

Clinical responses to therapeutic antibodies led to a renewed interest in the effector mechanisms of monoclonal antibodies, assuming that a better understanding of antibody function would lead to the development of improved antibody-based therapeutics. Different modes of action for monoclonal antibodies have been identified in vitro which are conceptually divided into direct and indirect mechanisms (Cragg et al. 1999). Direct mechanisms are mediated by binding of the "fragment variable" (Fv) to its target antigen and include blocking of receptor–ligand interactions and induction of apoptosis and growth arrest or tumor dormancy. In contrast, indirect mechanisms involve recruitment of immune effector functions which are mediated by the "fragment crystallizable" (Fc). Thus, the Fc portion interacts with components of the complement system resulting in induction of complement-dependent cytotoxicity (CDC) or induces effector cell-mediated functions such as antibody-dependent cell-mediated cytotoxicity (ADCC) or antibody-dependent cell-mediated phagocytosis (ADCP) by engagement of FcR. Furthermore, uptake of antibody-coated tumor cells by antigen-presenting cells such as dendritic cells may result in induction of adaptive immune responses and T cell immunity (Abes et al. 2010; Park et al. 2010). However, this differentiation into direct and indirect mechanisms is not absolute because direct antibody-induced apoptosis was enhanced in the presence of FcγR-expressing cells, which may serve as physiological cross-linkers (Shan et al. 1998).

The contribution of individual effector mechanisms to the clinical efficacy of different antibodies is still under intensive investigation, and may, for example, depend on the specific epitope recognized by individual CD20 antibodies (Glennie et al. 2007). Furthermore, also the anatomical site may significantly impact on the contribution of individual effector mechanisms (Gong et al. 2005). Nevertheless, several findings suggested a relevant role for interactions between therapeutic antibodies and FcγR (Nimmerjahn and Ravetch 2007). FcγR are grouped into activating and inhibitory receptors. Activating receptors mediate ADCC, ADCP, degranulation, cytokine release, and enhanced antigen presentation, whereas inhibitory receptors function as negative regulators. For IgG, three classes of FcγR exist in humans comprising FcγRI (CD64), FcγRII (CD32), and FcγRIII (CD16), which differ, for example, in their affinities for monomeric IgG and immune complexes (Bruhns et al. 2009). Various isoforms and genetic polymorphisms introduce further complexicity into this system (van de Winkel and Capel 1993; Kimberly et al. 2002). The importance of the interaction between the Fc portion and FcγR was demonstrated in several in vivo experiments. Thus, antibodies lost therapeutic activity in common γ-chain knock-out mice lacking expression of activating FcγR (Clynes et al. 2000). Furthermore, studies with class-switched antibody variants suggested that their therapeutic activities were determined by the ratio between the Fc-binding affinities for activating and inhibitory FcγR (A/I ratio; Nimmerjahn and Ravetch 2005). The impact of FcγR-signaling was further investigated in NoTAM mice which are deficient for the murine common FcRγ-chain but carry an ITAM-mutated human γ-chain as transgene (de Haij et al. 2010). The mutation prevents FcγR-signaling but allows cell surface FcγR expression comparable to that of wild type mice. Interestingly, CD20 antibodies lost therapeutic efficacy in these mice – supporting a substantial role for ADCC rather than antibody-induced apoptosis via FcγR-mediated crosslinking in vivo. Further evidence for the clinical importance of efficient FcγR-mediated effector cell recruitment was derived from analyses of FcγR polymorphisms in patients treated with rituximab, trastuzumab, or cetuximab. Expression of the CD16a-V158 allelic variant – binding IgG1 with greater affinity than the CD16a-F158 variant (van de Winkel and Capel 1993; Kimberly et al. 2002) – correlated with improved response rates to antibody therapy (Weng and Levy 2003; Musolino et al. 2008; Bibeau et al. 2009).

In conclusion, interactions between antibodies' Fc portion and FcγR on effector cells are important for monoclonal antibodies to elicit therapeutic effects – suggesting a major role for cell-mediated cytotoxicity. However, these interactions are often insufficient in vivo (Fanger et al. 1991). First antibodies will react with FcγR expressed on non-cytotoxic cells (e.g., CD32 on platelets and B cells) and with FcγR not triggering cytotoxicity (e.g., CD16b on granulocytes). Moreover, certain isoforms – such as CD32b on monocytes and macrophages – produce inhibitory signals in otherwise cytotoxic cells. Engagement of CD32b reduces the therapeutic efficacy of antibodies in vivo as revealed by the critical role for the A/I ratio and by elevated ADCC levels observed in CD32b-deficient mice in comparison to wildtype mice (Clynes et al. 2000; Nimmerjahn and Ravetch 2005). Second, therapeutic antibodies compete with large amounts of endogenous plasma immunoglobulins for binding to FcγR. This is especially crucial for CD64 which binds monomeric IgG with high affinity and which is blocked by endogenous immunoglobulins (Deo et al. 1997). Third, ADCC by monoclonal antibodies appeared to be compromised in the presence of components of the complement system (Wang et al. 2009). Moreover, polymorphisms of FcγR critically determine the efficacy of antibody therapy (Nimmerjahn and Ravetch 2007). All these factors may hamper therapeutic effects of monoclonal antibodies in vivo and thus strategies were developed to overcome these shortcomings. These include Fc engineering approaches to achieve higher affinities for activating FcγR (Carter 2006; Presta 2008), and the generation of bispecific antibodies (Fanger et al. 1991).

Bispecific antibodies allow recruitment of a more defined pool of effector cells by selectively engaging distinct trigger molecules and avoid binding to non-signaling or inhibitory FcR – thereby achieving optimal A/I ratios. Furthermore, FcR can be triggered by antigen-combining domains recognizing epitopes outside the Fc-binding site and avoiding competition with endogenous immunoglobulins. Therefore, bispecific antibodies may function in vivo when FcR are occupied. Additionally, the efficacy of bispecific antibodies should not be compromised by FcR-allelic variants – suggesting that especially patients homozygous for the CD16a-F158 allele may benefit from bispecific antibody therapy. Certain bispecific antibodies demonstrated higher cytotoxic activities in vitro than their corresponding parental monoclonal antibodies and were shown to operate at lower concentrations, which at least in part may be explained by their high affinities for the respective trigger antigens (Shahied et al. 2004). Accordingly, bispecific antibodies targeting activating FcγR may circumvent certain limitations of monoclonal antibodies – resulting in improved effector functions such as ADCC, ADCP, and induction of tumor-specific immune responses. Another appealing feature of bispecific antibodies is that their activities are not limited to FcR and FcR-expressing cells. The inclusion of binding sites for other trigger molecules into bispecific molecules also allows to attract cytotoxic cells lacking FcR such as T lymphocytes, which constitute attractive effector cells due to their high cytotoxic potential (Müller and Kontermann 2007).

13.3 Formats of Bispecific Antibodies

Bispecific antibodies have been designed by a variety of methods in different formats (Fig. 13.1; Müller and Kontermann 2007). Traditionally, bispecific antibodies were generated as quadroma antibodies or by chemical conjugation of antibodies or Fab-fragments. However, these first-generation bispecific antibodies were difficult to produce in sufficient quantity and quality, and many of them failed in clinical studies. Especially quadroma antibodies often could not be administered in sufficient doses to reach therapeutic efficacy because the agents induced severe side effects. Probably Fc-mediated interaction with FcR resulted in systemic activation of effector cells and triggering of cytokine release syndromes. Moreover, murine sequences were immunogenic and induced human anti-mouse antibody (HAMA) responses in patients (Segal et al. 1999).

Some of these shortcomings were addressed through development of new formats for bispecific IgG antibodies and the generation of recombinant bispecific molecules using single-chain fragments variable (scFv), Fabs, or domain antibodies as binding sites (Müller and Kontermann 2007). Production of bispecific IgG antibodies was improved by the generation of mouse IgG2a and rat IgG2b hybrid antibodies (triomabs) or by introducing knob-into-hole mutations into CH3

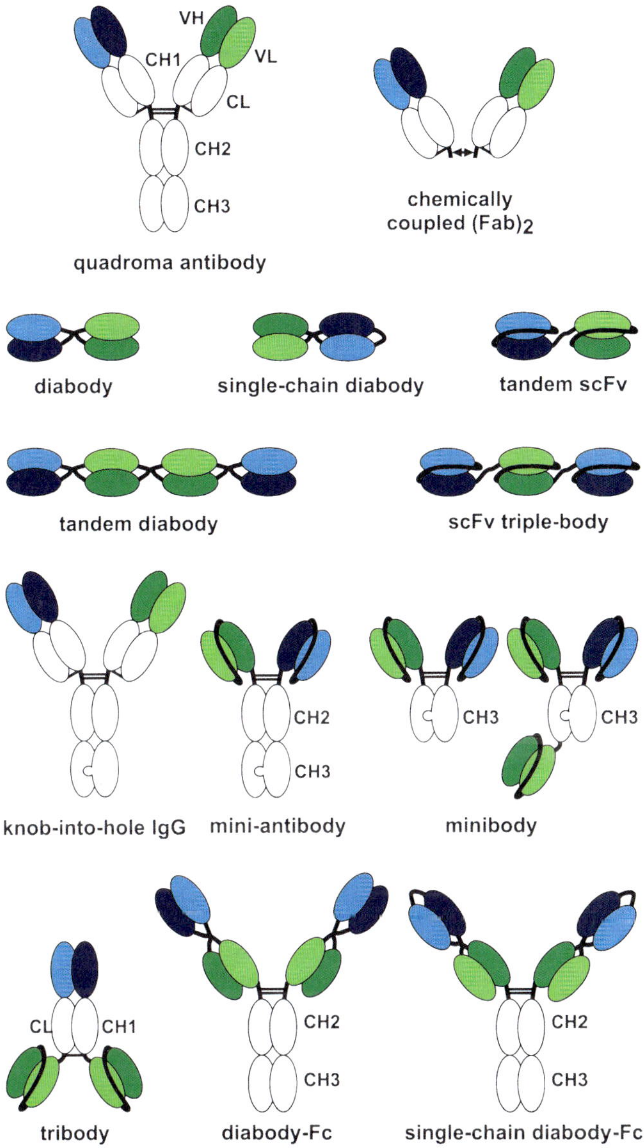

Fig. 13.1 Examples for the design of bispecific antibodies for recruitment of effector cells. VH, VL, antibody variable heavy and light domains (deep and light colored, respectively) from antibodies with different specificities as indicated in *blue* and *green*; CH1-3, antibody constant regions 1–3; CL antibody constant light chain region; *Fc* fragment crystalizable; *Fab* fragment antigen binding; *scFv* single-chain fragment variable

domains to force the formation of bispecific heterodimers (Merchant et al. 1998). The knob-into-hole technology was also employed to generate heterodimeric bispecific mini-antibodies and minibodies – fusion proteins of scFvs and Fc or

CH3 domains, respectively (Shahied et al. 2004; Xie et al. 2005). Bispecific scFv antibodies (bsscFvs) consist only of V-regions and lack antibody Fc portions. Formats include single-chain molecules such as tandem scFvs and single-chain diabodies as well as heterodimeric proteins such as diabodies or disulfide-stabilized variants called dual-affinity re-targeting proteins (DART; Müller and Kontermann 2007; Johnson et al. 2010). Due to their small size of approx. 50 kDa and the lack of Fc portions, they potentially exhibit higher tissue penetration, avoid uptake by various FcR and are supposed to be less immunogenic. However, bsscFvs do not interact with the neonatal FcR (FcRn) and are rapidly cleared from the bloodstream due to their small size – resulting in half-lifes of only a few hours. Therefore, plasma retention times were prolonged by increasing the size, modifying the glycosylation pattern, or introducing protein domains such as human serum albumin interacting with FcRn (Kontermann 2009). Also bispecific mini-antibodies containing an Fc domain may have more favorable pharmacokinetic properties (Xie et al. 2005). Another possibility to achieve sufficient steady-state plasma levels of bsscFvs is to apply them by continuous intravenous infusion (Bargou et al. 2008).

An important finding was the observation that the avidity and affinity of binding to the target antigen critically determined the cytotoxic potential of bispecific antibodies. Studies with a series of bsscFv variants binding the same target antigen with different affinity revealed that an increase in affinity was associated with enhanced cytotoxicity (McCall et al. 2001). Likewise, the cytotoxic potential of bispecific antibodies was enhanced by increasing the avidity for the tumor antigen (Shahied et al. 2004; Kellner et al. 2008). Meanwhile, several formats for bispecific proteins with increased valency have been reported (Fig. 13.1). However, according to studies in mice, an affinity optimum appears to exist for agents targeting solid tumors: scFvs with affinities beyond this optimum exhibited restricted tumor penetration (Adams et al. 2001). The impact of binding affinity to trigger molecules on the cytotoxic capacity of bispecific antibodies has not been addressed in detail. For CD3-directed molecules, strong cytotoxic effects were demonstrated even when affinities for CD3 were low with equilibrium binding constants in high nanomolar ranges (Dreier et al. 2002). Because improved CD3 binding sometimes led to lower cytotoxic abilities, it was argued that lower CD3 binders may favor serial killing by T cells (Bortoletto et al. 2002). Despite extensive studies performed with different formats, direct comparisons between them are rare and the ideal format appears not to be identified so far.

13.4 Effector Cells and Trigger Molecules for Bispecific Antibodies

13.4.1 Effector Cells for Bispecific Antibody Approaches

Bispecific approaches in cancer treatment involve five classes of effector cells selected from cell populations mediating tumor surveillance or participating in antibody-dependent destruction of tumors (de Gast et al. 1997). Appropriate cells which have been recruited to kill cancer cells besides T lymphocytes include natural killer (NK) cells, monocytes, macrophages, and granulocytes, each of which were able to eliminate malignant cells through cell-mediated cytotoxicity and/or phagocytosis (Fig. 13.2).

T and NK cells play key roles in the host defense against emerging tumor cells (Smyth et al. 2002). The infiltration of tumors by cytotoxic T cells correlated with good prognosis and overall survival. Cytotoxic T cells with specific T cell receptors (TCR) recognize tumor peptides presented by the malignant cells on MHC class I molecules. For full activation, an additional stimulus is necessary, which is usually provided by the interaction between CD28 on T cells with B7 molecules on the surface of antigen-presenting cells or tumor cells (Alegre et al. 2001). To escape T-cell recognition, some tumor cells down-regulate the expression of MHC class I, which renders them more sensitive for attack by NK cells (Smyth et al. 2002). NK cells express several stimulatory and inhibitory receptors which enable them to discriminate between normal and malignant cells. Inhibitory killer cell

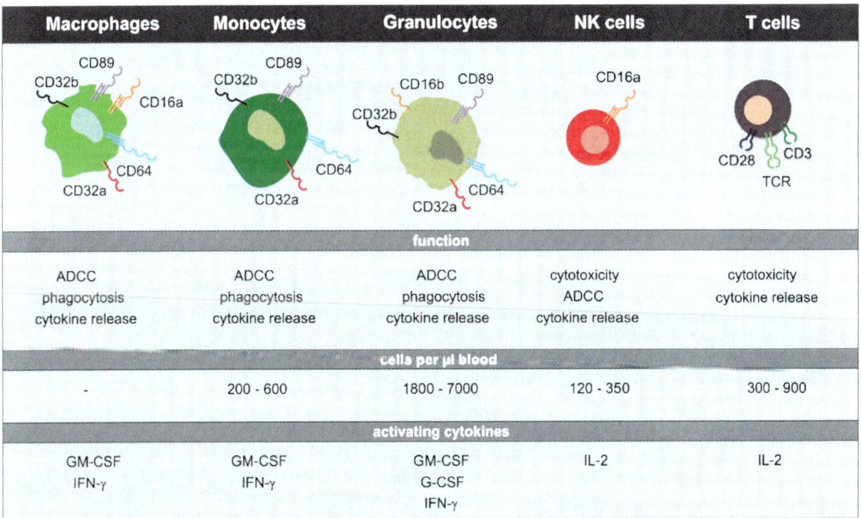

Fig. 13.2 Effector cell populations for bispecific antibodies. Each effector cell population expresses a defined pool of trigger molecules which is selected from the group of activating FcRs with CD16a being expressed on NK cells and macrophages, CD64 on monocytes, macrophages and granulocytes upon activation with G-CSF, and CD89 being displayed on monocytes, macrophages, and granulocytes. CD32a is not considered as an optimal trigger as CD32 is expressed also on non-cytotoxic cells. Granulocytes express CD16b, which is the non-signaling membrane-anchored isoform lacking any signaling domain and being unable to induce cytotoxic activities. CD32b is an inhibitory FcR on macrophages and monocytes with regulatory function. CD32b may be expressed by granulocytes from certain individuals. Cytotoxicity by FcR-negative T cells can be triggered through engagement of the TCR, CD28, and CD3. The functional state of different effector cell populations, but also the numbers of effector cells can be increased by immuno-stimulatory cytokines

immunoglobulin-like receptors recognize MHC class I molecules and block cytolytic activities against healthy host cells. Cytotoxicity is triggered when these inhibitory signals are lowered and/or stimulatory receptors are engaged by ligands expressed on malignant cells as danger signals. Moreover, the vast majority of NK cells express CD16a and are capable of killing tumor cells by ADCC. Importantly, recent findings suggested a major role for CD56-positive cells, mainly NK cells, in cetuximab-treated patients (Marechal et al. 2010). Killing mechanisms of T and NK cells are similar and include expression of ligands engaging death receptors on tumor cells and exocytosis of perforins and granzymes, which form pores into the outer target cell membrane and induce apoptosis through caspase cleavage (Trapani and Smyth 2002).

Myeloid effector cells lack specific tumor recognition receptors and therefore are unable to kill tumor cells in vitro unless sensitizing antibodies are present. Antibody opsonized tumor cells can be eliminated by ADCP and ADCC. Killing involves the expression of death receptor ligands, oxidative burst, and the release of cytotoxic mediators including nitric oxide, proteases, membrane-perforating agents (Di Carlo et al. 2001; Dale et al. 2008). Monocytes are circulating effector cells constitutively expressing CD64 and CD32a, but only a minor fraction displays surface CD16a (Gordon and Taylor 2005). For CD20 antibodies mouse models demonstrated monocytes to strongly contribute to antibody-induced B cell depletion (Uchida et al. 2004; Gong et al. 2005), and monocytes are considered to represent important effector cells also in humans. Macrophages are resident phagocytic cells which express three activating FcγR (CD64, CD32 and CD16a). Classically activated M1 macrophages exhibit anti-tumor activities and have been implicated in the immune surveillance of leukemic cells (Allavena et al. 2008; Jaiswal et al. 2010). Interestingly, high numbers of tumor infiltrating macrophages were associated with favorable outcome in follicular lymphoma patients treated with rituximab and chemotherapy (Taskinen et al. 2007). However, alternatively activated M2 macrophages are well known to play tumor-promoting roles. Granulocytes represent the most abundant cytotoxic cell population in the blood and have been suggested to contribute to tumor destruction (Di Carlo et al. 2001). They constitutively express the stimulatory receptors CD32a and FcαRI (CD89), and expression of CD64 can be induced by either interferon-γ or G-CSF. Although all these effector cells were demonstrated to mediate tumor cell killing with bispecific antibodies, it is still unclear which effector cell population will be most effective in eliminating tumor cells in patients. Eventually, their respective activities may depend on tumor localization and types as these factors may affect tumors' accessibility for different effector cells.

13.4.2 Trigger Molecules for Bispecific Antibodies

Different effector populations are characterized by a unique expression pattern of distinct sets of activating receptors which may serve as triggers for bispecific antibodies (Fig. 13.2). Ideally, trigger molecules should (1) be constitutively expressed on a defined pool of resting effector cells, (2) be capable of eliciting cellular cytotoxicity and/or phagocytosis without requirement for prior stimulation, and (3) induce the secretion of immunostimulatory cytokines attracting other effector cells.

Engaging T cells as effector cells for antibody-based therapy was one of the earliest applications for bispecific antibodies (Staerz et al. 1985). Several receptors (e.g., CD2, CD3, CD5 and CD28) expressed by T cells have been identified, which were demonstrated to induce cytokine release, proliferation, and cytotoxicity when triggered by bispecific antibodies (Müller and Kontermann 2007). Most experience is available with bispecific agents targeting CD3. CD3 is part of the T cell receptor (TCR) complex which signals through immunoreceptor tyrosine-based activation motifs (ITAMs) contained in the CD3 subunits and the associated TCR ζ chain (Samelson 2002). Most bispecific antibodies targeting CD3 required co-stimulatory signals to achieve full T cell activation, which were, for example, provided by CD28 antibodies (Bohlen et al. 1993). CD28 contains phosphorylated tyrosines and proline-rich signaling motifs and delivers co-stimulatory signals. However, CD28-specific superagonistic bispecific antibodies were shown to induce T-cell activation without ligation of the TCR/CD3 complex (Grosse-Hovest et al. 2003).

For recruitment of FcγR-expressing cells, all activating FcγR (CD64, CD32a and CD16a) can be considered as trigger molecules. FcR engagement can mediate multiple biological responses, including ADCP, ADCC, antigen presentation, oxidative burst, and cytokine release. Like CD3 on T cells, FcγR are associated with ITAMs (Nimmerjahn and Ravetch 2007). While CD32a itself contains an ITAM, CD16a and CD64 molecules associate with ITAM-bearing adapter proteins such as the common FcRγ and the TCR ζ chains. CD32a is not considered as a valuable trigger molecule since it is also expressed by non-cytotoxic platelets, which may initiate unwanted thrombotic complications after activation by antibodies. CD32b, an immunoreceptor tyrosine-based inhibition motif-associated FcγR on B cells, monocytes, and macrophages, even has inhibitory functions. CD16b, a GPI-linked FcγR on granulocytes, is incapable of transducing cytotoxic signals and is also shedded into the blood. Therefore, interactions with FcγRIIb or FcγRIIIb may reduce the therapeutic efficacy of bispecific antibodies.

In addition to FcγR, also the myeloid IgA Fc receptor (FcαRI, CD89) has been identified as an attractive trigger molecule for bispecific antibodies (van de Winkel and Capel 1993; Valerius et al. 1997; Monteiro and Van De Winkel 2003). Similar to activating FcγR, signaling through CD89 is coupled to the common FcR γ chain, and stimulation of CD89 has been shown to initiate ADCP and ADCC. Ligation of CD89 is expected to trigger a different pattern of immune responses compared to FcγR due to the antigen's unique expression on myeloid effector cells. CD89 does not bind IgG antibodies – the only class of antibodies that to date has been approved for clinical use. Bispecific molecules addressing CD89 are of therapeutic interest, but the full potential of this trigger molecule still remains to be determined.

13.5 Preclinical Data with Bispecific Antibodies

Initial results demonstrating biological activity of bispecific antibodies were obtained with quadroma antibodies or (Fab')$_2$ fragments. According to experience with conventional monoclonal antibodies, target antigens were chosen (1) to be absent from healthy stem cells, (2) to be abundantly expressed on tumor cells, and/or (3) to be essential for tumor cell survival. In contrast to IgG antibodies, which react with murine FcR and therefore are able to recruit murine effector cells, mouse models with bispecific antibodies have been more complicated. For example, most bispecific antibodies were specific for human target and trigger antigens and did not recognize murine homologues. To test the therapeutic efficacy of bispecific antibodies, immune-deficient mice were xenotransplanted with human tumor cells and the bispecific antibody was applied together with human effector cells. Mice transgenic for different human trigger molecules (CD3, CD16, CD64, and CD89) may offer more elegant systems to test bispecific antibodies in vivo (de la Hera et al. 1991; Heijnen and Van de Winkel 1995; Li et al. 1996; van Egmond et al. 1999). Furthermore, human cord blood cell-transplanted mice ("humanized mice") offer additional opportunities to establish more relevant animal models (Traggiai et al. 2004; Glorius et al. 2010). The next paragraph summarizes results from cell culture and animal experiments with bispecific antibodies addressing the most commonly used trigger molecules.

13.5.1 Bispecific Antibodies Recruiting T Cells by Engagement of CD3 or CD28

CD3-addressing bispecific antibodies have been developed targeting various tumor-associated antigens on different tumor types. These include molecules against the idiotype, HLA-DR, CD19, CD20, and CD30 on lymphomas and leukemias. Solid tumors were targeted by bispecific antibodies against, for example, Her-2, the epidermal growth factor receptor (EGFR), EpCAM, or EGP-2 (Müller and Kontermann 2007). Bispecific antibodies exhibited potent cytotoxic activity and demonstrated striking efficacy in animal models. For example, [CD19 × CD3] diabodies prolonged survival in mice xenotransplanted with human lymphoma cells when pre-activated human mononuclear cells (MNC) were co-transplanted (Cochlovius et al. 2000).

One particular class of bispecific antibodies was named "bispecific T cell engager" (BiTE), since these tandem scFvs were able to achieve full activation of T cells independent of other co-stimuli (see Chap. 15). Such BiTE molecules were highly efficacious in vitro and in vivo. For example, the [CD19 × CD3] bsscFv blinatumumab eliminated established human leukemic cells in mice and was capable of mediating B cell-depletion in chimpanzees (Dreier et al. 2003; Schlereth et al. 2006). Moreover, EGFR-specific BiTEs prevented growth of human colorectal cancer xenografts with mutated KRAS and BRAF proto-oncogenes in mice (Lutterbuese et al. 2010). These results were remarkable because recent analyses have demonstrated that colorectal cancer patients with mutations in these proto-oncogenes did not benefit from therapy with conventional EGFR antibodies (Karapetis et al. 2008), although KRAS-mutated tumor cells were susceptible to ADCC (Schlaeth et al. 2010).

Triomabs, which are generated through somatic hybridization of a CD3-specific rat IgG2b antibody with a tumor cell-directed mouse IgG2a antibody, represent another attractive class of T cell-recruiting bispecific antibodies (Lindhofer et al. 1995). Triomabs bind to tumor and T cells with their two Fv moieties, but are additionally capable of recruiting FcR-expressing cells such as NK cells, monocytes, macrophages, granulocytes, and dendritic cells through their Fc portions (Zeidler et al. 1999). Triomabs have been demonstrated to actually act as trifunctional molecules – inducing complexes of tumor cells, T lymphocytes, and accessory cells which provide co-stimulatory signals for T cells by secretion of cytokines or which directly eliminate tumor cells. Triomabs were produced against several tumor cell targets including EpCAM (prostate carcinomas, malignant ascites) and CD20 (Buhmann et al. 2009; Seimetz et al. 2010).

Promising results were also obtained with superagonistic bispecific antibodies addressing CD28. The tandem scFv rM28 targeting MCSP on malignant melanoma cells was shown to induce effective killing of tumor cells without stimulation of the TCR/CD3 complex in vitro and in vivo (Grosse-Hovest et al. 2003) and another similar molecule against CD20 mediated potent lysis of B-lymphoid tumor cells (Otz et al. 2009).

13.5.2 Bispecific Antibodies Recruiting CD16a-Positive NK Cells and Macrophages

Bispecific antibodies addressing CD16 eliminate tumor cells by recruitment of NK cells and macrophages both expressing the CD16a isoform and demonstrated promising cytotoxic activities. For treatment of solid tumors, bispecific (Fab')$_2$ conjugates targeting the EGFR mediated potent lysis of renal cancer cells and bispecific [Her2 × CD16] antibodies effectively killed ovarian and breast cancer cells (Stockmeyer et al. 1997; Elsasser et al. 1999). Furthermore, the bispecific Her2-targeting quadroma antibody 2B1 and a recombinant Her2-specific tribody proved therapeutic efficacy in tumor xenograft mouse models (Weiner et al. 1993a, b; Lu et al. 2008).

Tandem scFvs and scFv triple-bodies targeting HLA class II or CD19 on B-lineage lymphomas/leukemias as well as CD33 and CD123 on acute myeloid leukemias efficiently mediated killing of tumor cells in the presence of MNC effector cells (Bruenke et al. 2004; Kellner et al. 2008; Kugler et al. 2010). The DART molecule [CD32b × CD16] was highly potent in the elimination of xenotransplanted lymphoma cells in human CD16a transgenic mice (Johnson et al. 2010). Recently, a [CD20 × CD16] tribody efficiently depleted B cells in immune-deficient mice with a reconstituted human hematopoietic system by transplantation of hematopoietic progenitor cells from cord blood (Glorius et al. 2010). Interestingly, [CD19 × CD16] and [CD19 × CD3] diabodies acted synergistically in lymphoma mouse models indicating cooperation between NK and T cells (Kipriyanov et al. 2002). Moreover, [CD30 × CD16] diabodies induced regression of tumor growth in vivo in xenotransplanted Hodgkin lymphoma mouse models (Arndt et al. 1999).

13.5.3 Bispecific Antibodies Recruiting CD64-Positive Myeloid Effector Cells

Bispecific antibodies against CD64 recruit monocytes, macrophages as well as IFN-γ or G-CSF activated granulocytes. A [CD64 × Her2] bispecific antibody triggered killing of breast, ovarian, and lung cancer cells through ADCP and ADCC with monocytes, macrophages, and activated granulocytes (Keler et al. 1997; Stockmeyer et al. 1997). Interestingly, CD64-addressing molecules were more efficacious than respective [Her2 × CD16] and [Her2 × CD32] bispecific antibodies when whole blood from G-CSF treated patients was compared with blood from healthy donors as effector cell source. This difference was explained by higher monocyte and granulocyte counts, increased effector cell activation and higher expression levels of CD64 in G-CSF primed blood (Stockmeyer et al. 1997). Bispecific [CD64 × CD30] (Fab')$_2$ and diabodies induced ADCC and ADCP of Hodgkin lymphoma cells by monocytes and macrophages, respectively (Sundarapandiyan et al. 2001; Ranft et al. 2009). In a human CD64-transgenic mice model a set of bispecific (Fab')$_2$ antibodies against different lymphoma-associated antigens were tested for redirecting G-CSF-activated granulocytes (Honeychurch et al. 2000). The bispecific construct targeting the idiotype had the highest efficacy and provided long-term T cell tumor immunity. Interestingly, an unexpected antigen restriction for granulocyte-mediated lysis of B-lymphoid tumor cells was observed. Whereas monoclonal and CD64-addressing bispecific antibodies with specificities for HLA class II, the associated invariant chain CD74, or HLA class II variants effectively triggered granulocytes to kill tumor cells, molecules targeting other B cell differentiation antigens (e.g., CD19 and CD20) remained ineffective (Elsasser et al. 1996; Wurflein et al. 1998). This antigen restriction was not observed when monocytes or macrophages were used as effector cells, which efficiently phagocytosed and killed malignant B cells, for example, in the presence of [CD64 × CD19] bispecific (Fab')$_2$ fragments (Ely et al. 1996).

13.5.4 Bispecific Antibodies Triggering CD89 on Monocytes, Macrophages, and Granulocytes

CD89 is constitutively expressed on monocytes, macrophages, and granulocytes. The potency of CD89 as cytotoxic trigger molecule was demonstrated with bispecific

$(Fab')_2$ conjugates against Her2, which efficiently induced granulocyte-mediated killing of breast cancer and renal cell carcinoma cells (Valerius et al. 1997; Deo et al. 1998; Elsasser et al. 1999). An [EGFR × CD89] bispecific antibody was developed for potential treatment of renal cell carcinoma (Stadick et al. 2002). By comparing several bispecific antibodies addressing CD89, but targeting different tumor-associated antigens on lymphoma cells, HLA class II was not only identified as the target with the highest potency, but also a bispecific [CD20 × CD89] molecule-induced granulocyte-mediated killing. This was not observed with IgG1 or bispecific antibodies engaging FcγR (Stockmeyer et al. 1997). In whole blood assays, target cell lysis was increased when blood from patients during G-CSF or GM-CSF therapy was investigated. These growth factors dramatically increased the absolute numbers of CD89-positive effector cells in the blood by increasing total leukocyte counts, and at the same time activated individual cells, which both may account for the observed increased efficacy (Stockmeyer et al. 2001). Recently, a bispecific tandem scFv [HLAclassII × CD89] has been demonstrated to efficiently eliminate lymphoma cells with higher potency than a chimeric HLA class II antibody derived from the same hybridoma (Guettinger et al. 2010). Moreover, a bispecific [CD89 × CD30] $(Fab')_2$ conjugate was able to induce granulocyte- and monocyte-mediated killing of Hodgkin lymphoma cells (Sundarapandiyan et al. 2001). To the best of our knowledge, the in vivo potency of CD89-directed bispecific molecules in killing tumor cells has not been investigated up to now, but mice transgenic for human CD89 are available and may be attractive for testing such molecules in animal studies (van Egmond et al. 1999).

13.6 Clinical Experience with Bispecific Antibodies

Due to encouraging results from preclinical studies, several bispecific antibodies in different formats have entered clinical trials. To date, bispecific antibodies engaging CD3, CD28, CD16, or CD64 have been tested or are under current clinical evaluation (Table 13.1). The bispecific antibodies were either given alone or were combined with cytokines or chemotherapy. Whereas results with some bispecific antibodies were encouraging, others generated disappointing results – including unacceptable toxicity, or inconsistent tumor responses (Müller and Kontermann 2007). Results from selected clinical studies with bispecific antibodies are summarized in the next paragraphs.

13.6.1 Clinical Trials with Bispecific Antibodies Recruiting T Cells

Several clinical studies were performed with bispecific antibodies recruiting T cells with most of them targeting CD3. In initial clinical studies with bispecific

Table 13.1 Clinical trials with bispecific antibodies for cellular immunotherapy[a]

Bispecific antibody	Format	Indication	Stage	Reference or clinical trials.gov identifier
CD20Bi/ [CD20 × CD3]	Cross-linked IgG	Plasma cell neoplasms; NHL	Phase I	(NCT00938626; NCT00521261)
SHR-1/ [CD19 × CD3]	IgG	NHL	Phase I	de Gast et al. (1995)
OKT3x herceptin/ [CD3 × Her2]	IgG	Breast cancer	Phase I/II	(NCT00027807); Lum et al. (2003)
OKT3x 6A4/ [CD19 × CD3]	IgG	NHL	Phase I	Manzke et al. (2001)
OC/TR/[FR × CD3]	$F(ab')_2$	Ovarian cancer	Phase I/II	Canevari et al. (1995)
Bis-1/[EGP-2xCD3]	$F(ab')_2$	Renal carcinoma	Phase I	Kroesen et al. (1994)
Ertumaxumab/ [Her2 × CD3]	Triomab	Breast cancer	Phase II	(NCT00522457)
Bi20 (fBTA05)/ [CD20 × CD3]	Triomab	NHL	Phase I	Buhmann et al. (2009)
catumaxumab/ [EpCAM × CD3]	Triomab	EpCAM-positive tumors; malignant ascites	Approved	Seimetz et al. (2010)
blinatumumab/ [CD19 × CD3]	Tandem scFv	NHL; B ALL	Phase I and II	Bargou et al. (2008), Topp et al. (2009)
MT110/ [EpCAM × CD3]	Tandem scFv	EpCAM-positive tumors	Phase I	Fiedler et al. (2010)
rM28/ [MCSP × CD28]	Tandem scFv	Malignant melanoma	Phase I/II	(NCT00204594)
2B1/[Her2 × CD16]	IgG	Her2-positive tumors	Phase Ib/II	Borghaei et al. (2007)
HRS-3/A9/ [CD30 × CD16]	IgG	Hodgkin's lymphoma	Phase I/II	Hartmann et al. (1998)
H22xKi-4/ [CD30 × CD64]	$F(ab')_2$	Hodgkin's lymphoma	Phase I	Borchmann et al. (2002)
MDX-H210/ [Her2 × CD64]	$F(ab')_2$	Her2-positive tumors	Phase II	Posey et al. (1999), James et al. (2001)
MDX-447/ [EGFR × CD64]	$F(ab')_2$	EGFR-positive tumors and glioblastoma	Phase I	Fury et al. (2008); (NCT00014560)
[CD64 × CD33]	Cross-linked IgG	AML	Phase I	Chen et al. (1999)
[4g7xh22]/ [CD19 × CD64]	$F(ab')_2$	NHL; CLL	Phase I	(NCT00014560)

[a]*NHL* non-Hodgkin lymphoma; *B ALL* acute B-lymphoblastcic leukemia; *AML* acute myeloid leukemia; *CLL* chronic lymphocytic leukemia

quadroma antibodies or $(Fab')_2$ fragments, some clinical responses were reported, but problems occurred due to cytokine release and HAMA responses. Furthermore, antibody production problems prevented more extensive studies (Müller and Kontermann 2007). Meanwhile, different bispecific formats have been tested in patients with B-lymphoid malignancies or solid tumors (Table 13.1).

More recently, remarkable results were obtained with the BiTE molecule blinatumumab and the triomab trifunctional antibody catumaxumab targeting CD19 and EpCAM, respectively. Blinatumumab was tested in a phase I trial in patients with relapsed non-Hodgkin's lymphomas (Bargou et al. 2008). Thirty-eight patients received blinatumomab at daily doses from 0.0005 to 0.06 mg/m^2 by continuous intravenous infusions over periods from 4 to 8 weeks. Serum levels of blinatumumab were below 1 ng/ml. Common side effects included leukopenia/lymphopenia, chills, fever, pyrexia, and increase of C-reactive protein, which often normalized during continued treatment. Impressively, 11 major responses were observed, which included four complete and seven partial responses. Initial results from a phase II study in patients with minimal residual disease (MRD) of B-precursor acute lymphoblastic leukemia suggested that blinatumumab is capable of eliminating rare bone marrow-located MRD cells (Topp et al. 2009). Complete molecular responses were achieved in 13 of 16 evaluable patients during the first 4-week cycle at a dose of 0.015 mg/m^2 blinatumomab per day (see also Chap. 15).

The trifunctional bispecific triomab antibody catumaxumab targeting EpCAM received drug approval by the European Union in 2009 for the treatment of malignant ascites in patients with EpCAM-expressing carcinomas for whom no standard therapy is available (Seimetz et al. 2010). Catumaxumab treatment reduced ascites production accompanied by a robust elimination of EpCAM-positive tumor cells. As the majority of patients developed human anti-bispecific antibody responses (HABA), the possibility of repeated applications of catumaxomab in HABA-positive patients is currently investigated. Catumaxumab has also been tested in clinical trials in other indications including ovarian and gastric cancer. With ertumaxumab and Bi20 targeting Her2 and CD20, respectively, two other triomabs are under clinical investigation (Kiewe et al. 2006; Buhmann et al. 2009; see Chap. 16).

With the bispecific tandem scFv rM28 targeting MCSP, a CD28-addressing molecule has entered clinical trials and is currently tested in a phase I/II clinical study in patients with metastatic malignant melanoma (Grosse-Hovest et al. 2003). As a superagonistic CD28 antibody induced severe cytokine release in healthy volunteers (Suntharalingam et al. 2006), this approach requires particularly careful evaluation. However, rM28 triggered T cells in a strictly target cell-dependent manner – suggesting that similarly severe side effects may be prevented (Grosse-Hovest et al. 2005).

13.6.2 Clinical Experiences with Bispecific Antibodies Redirecting CD16a-Positive NK Cells and Macrophages

Clinical experiences with bispecific antibodies recruiting CD16a-positive effector cell populations are more limited, and up to now no recombinant formats have been clinically evaluated. HRS/A9, a bispecific IgG antibody targeting CD30 was analyzed in patients with refractory Hodgkin's disease (Hartmann et al. 1998). Fifteen patients received up to 64 mg/m^2 of HRS/A9 without dose-limiting toxicity in a multiple-dose phase I study. Responses to HRS/A9 included one complete, one partial, and three minor responses. After treatment, biological activity was evidenced by increased NK cell activity in most patients. Side effects occurred rarely and consisted of fever, pain in involved lymph nodes, and a maculopapulous rash. However, further treatment with this bispecific antibody was not possible due to the development of HAMA-responses in nine patients.

2B1 is a bispecific [Her2 × CD16] quadroma antibody (Weiner et al. 1993a, b). In a multiple-dose phase 1 clinical trial, 15 patients with Her2-expressing solid tumors (e.g., colon, kidney and breast carcinomas) received 2B1 at doses of up to 5 mg/m^2. 2B1 had a mean serum half life of 20 h. Patients experienced several non-dose-limiting toxicities such as fevers, chills, nausea, vomiting, and transient leukopenia, while thrombocytopenia was dose-limiting at 5 mg/m^2. HAMA responses developed in 14 patients. Strongly increased plasma levels of secondary cytokines such as TNFα, IL-6 and IL-8, and smaller increases in levels of GM-CSF and IFN-γ were reported. Only minor clinical responses in individual patients were observed. In a phase 1b/2 clinical trial, 20 women with metastatic breast cancer were treated with 2B1 (Borghaei et al. 2007). Due to dose-limiting systemic toxicities, which were observed in 3/8 patients, the dose was reduced from 2.5 to 1 mg/m^2. The toxicities were attributed to cross-linking of FcR by 2B1 – resulting in activation of leukocytes and cytokine induction. Although objective responses were not observed, patients who received maximum doses developed antibodies against both intracellular and extracellular domains of Her-2, indicating that 2B1 elicited humoral immune responses. Furthermore, increased infiltration of total leukocytes and NK cells into the tumor were reported.

13.6.3 Clinical Studies with Bispecific Antibodies Engaging CD64-Positive Myeloid Effector Cells

Clinical studies with CD64-targeting bispecific antibodies have been conducted with chemically linked (Fab')$_2$ or IgG conjugates, which were directed against EGFR or Her2 on solid tumors, or CD30 and CD33 on hematological malignancies (Curnow 1997 and Table 13.1). H22 × Ki-4, a (Fab')$_2$ conjugate against CD30, was analyzed in a phase 1 clinical study in 10 refractory Hodgkin's lymphoma patients (Borchmann et al. 2002). Patients intravenously received escalating doses of H22 × Ki-4 between 1 and 20 mg/m^2 on days 1, 3, 5, and 7. Side effects were transient and included hypotension, tachycardia, fatigue, and fever. The elimination half-life of H22 × Ki-4 was 11.1 h. After the second course, low levels of HABA were observed, that did neither affect serum concentrations of H22 × Ki-4 nor result in allergic responses. However, high HABA titers developed in one patient receiving four treatments. Remarkably, one complete and three partial remissions were reported, and additional four patients had stable disease.

The bispecific (Fab')$_2$ antibody MDX-447 targeting EGFR was tested in a two-armed phase I study in 64 patients with EGFR-positive solid tumors, mainly kidney and head and neck cancer (Fury et al. 2008). Patients received weekly infusions of increasing doses between 1 and 40 mg/m^2 of MDX-447 alone or in combination with G-CSF (3 μg/kg). Half-lives for MDX-447 ranged between 1.9 and 8.4 h.

MDX-447 alone was generally well tolerated with a maximum tolerated dose of 30 mg/m^2. In comparison, the combination of MDX-447 and G-CSF was less well tolerated, and dose escalation of MDX-447 beyond 3.5 mg/m^2 was not possible. In both arms hypotension was the dose-limiting toxicity. Among 58 patients evaluable for response from both groups, 35 had stable disease, while 23 patients experienced progression. No objective responses were observed in this study.

Several studies were performed with bispecific (Fab')$_2$ conjugates directed against Her2, MDX-210, and its partially humanized variant MDX-H210. MDX-210 was evaluated in a single-dose phase I study in 15 patients with advanced breast or ovarian cancer. Patients received infusions of increasing doses of MDX-210 between 0.35 and 10 mg/m^2, which were well tolerated with grade 1 and 2 fevers, malaise, and transient monocytopenia/lymphocytopenia being the most common side effects (Valone et al. 1995). The maximal tolerated dose was defined at 7.5 mg/m^2. Application of MDX-210 was accompanied by elevated plasma levels of TNF-α, IL-6, G-CSF, and neopterin. Among ten assessable patients, one partial and one mixed tumor response were reported. MDX-H210 was evaluated in combination with G-CSF in a single-dose phase I study (Repp et al. 2003). Thirty women with breast cancer received subcutaneous injections of G-CSF (5 mg/kg/day) for 8 days and a single infusion of MDX-210 at doses of 0.35–200 mg/m^2 on day 4 of the G-CSF cycle. Dose-limiting toxicity was not reached. Fever/chills and nausea/vomiting were the main side effects – being timely related to elevated plasma levels of TNF-α and IL-6 and responding to symptomatic therapy with anti-pyretics and anti-emetics. In patients treated with doses of 100 and 200 mg/m^2, MDX-H210 plasma levels exceeded 1 mg/ml – accompanied by 50% saturation of CD64 on peripherial blood myeloid cells until day 4 after the infusion. Tumor infiltration by monocytes and granulocytes was confirmed by immunohistochemistry. Although no clinical responses were observed, 11 patients had stable disease and many patients demonstrated a sustained decline in soluble HER-2 plasma levels. In a subsequent multiple-dose phase I study of MDX-210 plus G-CSF, 23 patients with metastatic breast cancer received weekly intravenous MDX-H210 infusions for three doses, followed by a 2-week break and another three doses (Pullarkat et al. 1999). Doses of MDX-210 ranged from 1 to 40 mg/m^2 per application without reaching dose-limiting toxicity, and were followed by subcutaneous G-CSF (5 mg/kg/day) for 3–5 days. Cytokine-release and side effects were similar to previous studies. The β-elimination half-life of MDX-210 varied from 4 to 9 h. By the third infusion, most patients developed significant HABA, and no objective clinical responses were observed in this group. Based on the capacity of CD64 to increase antigen presentation (Heijnen et al. 1996; van Vugt et al. 1999), MDX-H210 was combined with GM-CSF, which had demonstrated potent adjuvant functions in vaccine studies. In a pilot phase II trial, 13 patients with advanced HER-2-expressing tumors received once weekly infusions of MDX-210 at increasing doses (from 1 to 20 mg/m^2 per infusion) for 4 weeks after 4 days of priming with GM-CSF (250 mg/m^2) (Posey et al. 1999). Dose-limiting toxicity was not reached with similar side effects to other MDX-H210 trials. Notably, one patient had a 48% size reduction in an index tumor lesion, and six patients showed stable disease during treatment, while three had progressive disease.

To date, the most promising data with MDX-H210 were obtained in a clinical phase II study in prostate cancer patients (James et al. 2001). Twenty-four patients with hormone-refractory prostate cancer were treated with GM-CSF (5 mg/kg/day for 4 days) followed by 6 weekly infusions of MDX-H210 (15 mg/m^2). Side effects were as reported for other studies and generally mild. Among 20 assessable patients, seven had PSA declines exceeding 50% with a median duration of 128 days, and the relative PSA velocity decreased in 16 out of 18 evaluable patients. Importantly, 6 out of 12 patients with evaluable pain had improvements in their pain scores. However, an independent phase II trial showed that GM-CSF alone was efficacious in reducing PSA levels in prostate cancer patients rendering the results with MDXH-210 difficult to interpret (Small et al. 1999).

13.7 Conclusion

Recent clinical results with catumaxumab and blinatumumab have demonstrated the therapeutic potential of bispecific antibodies for tumor immunotherapy. Whereas these T-cell recruiting bispecific antibodies produced substantial clinical effects, no convincing clinical benefits have been described for bispecific antibodies engaging CD16 or CD64 on NK or myeloid effector cells, respectively. However, clinical experiences with these later constructs were limited to studies with first-generation bispecific IgG antibodies or F(ab)$_2$ fragments, which both may not represent ideal formats. Recombinant bispecific molecules targeting FcR demonstrated potent in vitro activity, but have not been clinically evaluated. As outlined above, these include formats with enhanced cytotoxic functions and improved pharmacokinetic characteristics, which are expected to translate also into improved clinical efficacy. Further studies with these novel molecules appear warranted.

References

Abes R, Gelize E, Fridman WH, Teillaud JL (2010) Long-lasting antitumor protection by anti-CD20 antibody through cellular immune response. Blood 116(6):926–934

Adams GP, Schier R, McCall AM, Simmons HH, Horak EM, Alpaugh RK, Marks JD, Weiner LM (2001) High affinity restricts the localization and tumor penetration of single-chain fv antibody molecules. Cancer Res 61(12):4750–4755

Alegre ML, Frauwirth KA, Thompson CB (2001) T-cell regulation by CD28 and CTLA-4. Nat Rev Immunol 1(3):220–228

Allavena P, Sica A, Garlanda C, Mantovani A (2008) The Yin-Yang of tumor-associated macrophages in neoplastic progression and immune surveillance. Immunol Rev 222:155–161

Arndt MA, Krauss J, Kipriyanov SM, Pfreundschuh M, Little M (1999) A bispecific diabody that mediates natural killer cell cytotoxicity against xenotransplantated human Hodgkin's tumors. Blood 94(8):2562–2568

Bargou R, Leo E, Zugmaier G, Klinger M, Goebeler M, Knop S, Noppeney R, Viardot A, Hess G, Schuler M, Einsele H, Brandl C, Wolf A, Kirchinger P, Klappers P, Schmidt M, Riethmuller G, Reinhardt C, Baeuerle PA, Kufer P (2008) Tumor regression in cancer patients by very low doses of a T cell-engaging antibody. Science 321(5891):974–977

Bibeau F, Lopez-Crapez E, Di Fiore F, Thezenas S, Ychou M, Blanchard F, Lamy A, Penault-Llorca F, Frebourg T, Michel P, Sabourin JC, Boissiere-Michot F (2009) Impact of Fc {gamma}RIIa-Fc{gamma}RIIIa polymorphisms and KRAS mutations on the clinical outcome of patients with metastatic colorectal cancer treated with cetuximab plus irinotecan. J Clin Oncol 27(7):1122–1129

Bohlen H, Hopff T, Manzke O, Engert A, Kube D, Wickramanayake PD, Diehl V, Tesch H (1993) Lysis of malignant B cells from patients with B-chronic lymphocytic leukemia by autologous T cells activated with CD3 × CD19 bispecific antibodies in combination with bivalent CD28 antibodies. Blood 82(6):1803–1812

Borchmann P, Schnell R, Fuss I, Manzke O, Davis T, Lewis LD, Behnke D, Wickenhauser C, Schiller P, Diehl V, Engert A (2002) Phase 1 trial of the novel bispecific molecule H22 × Ki-4 in patients with refractory Hodgkin lymphoma. Blood 100(9):3101–3107

Borghaei H, Alpaugh RK, Bernardo P, Palazzo IE, Dutcher JP, Venkatraj U, Wood WC, Goldstein L, Weiner LM (2007) Induction of adaptive Anti-HER2/neu immune responses in a Phase 1B/2 trial of 2B1 bispecific murine monoclonal antibody in metastatic breast cancer (E3194): a trial coordinated by the Eastern Cooperative Oncology Group. J Immunother 30(4):455–467

Bortoletto N, Scotet E, Myamoto Y, D'Oro U, Lanzavecchia A (2002) Optimizing anti-CD3 affinity for effective T cell targeting against tumor cells. Eur J Immunol 32(11):3102–3107

Bruenke J, Fischer B, Barbin K, Schreiter K, Wachter Y, Mahr K, Titgemeyer F, Niederweis M, Peipp M, Zunino SJ, Repp R, Valerius T, Fey GH (2004) A recombinant bispecific single-chain Fv antibody against HLA class II and FcgammaRIII (CD16) triggers effective lysis of lymphoma cells. Br J Haematol 125(2):167–179

Bruhns P, Iannascoli B, England P, Mancardi DA, Fernandez N, Jorieux S, Daeron M (2009) Specificity and affinity of human Fcgamma receptors and their polymorphic variants for human IgG subclasses. Blood 113(16):3716–3725

Buhmann R, Simoes B, Stanglmaier M, Yang T, Faltin M, Bund D, Lindhofer H, Kolb HJ (2009) Immunotherapy of recurrent B-cell malignancies after allo-SCT with Bi20 (FBTA05), a trifunctional anti-CD3 × anti-CD20 antibody and donor lymphocyte infusion. Bone Marrow Transplant 43(5):383–397

Canevari S, Stoter G, Arienti F, Bolis G, Colnaghi MI, Di Re EM, Eggermont AM, Goey SH, Gratama JW, Lamers CH et al (1995) Regression of advanced ovarian carcinoma by intraperitoneal treatment with autologous T lymphocytes retargeted by a bispecific monoclonal antibody. J Natl Cancer Inst 87(19):1463–1469

Carter PJ (2006) Potent antibody therapeutics by design. Nat Rev Immunol 6(5):343–357

Chen J, Bashey A, Holman P, Carrier E, Law P, Ball ED (1999) A phase I dose escalating study of infusion of a bispecific antibody for relapsed/refractory acute myeloid leukemia. Blood 94(suppl 1):227b

Clynes RA, Towers TL, Presta LG, Ravetch JV (2000) Inhibitory Fc receptors modulate in vivo cytoxicity against tumor targets. Nat Med 6(4):443–446

Cochlovius B, Kipriyanov SM, Stassar MJ, Christ O, Schuhmacher J, Strauss G, Moldenhauer G, Little M (2000) Treatment of human B cell lymphoma xenografts with a CD3 × CD19 diabody and T cells. J Immunol 165(2):888–895

Cragg MS, French RR, Glennie MJ (1999) Signaling antibodies in cancer therapy. Curr Opin Immunol 11(5):541–547

Curnow RT (1997) Clinical experience with CD64-directed immunotherapy. An overview. Cancer Immunol Immunother 45(3–4):210–215

Dale DC, Boxer L, Liles WC (2008) The phagocytes: neutrophils and monocytes. Blood 112(4):935–945 de Gast GC, Haagen IA, van Houten AA, Klein SC, Duits AJ, de Weger RA, Vroom TM, Clark MR, Phillips J, van Dijk AJ et al (1995) CD8 T cell activation after intravenous administration of CD3 × CD19 bispecific antibody in patients with non-Hodgkin lymphoma. Cancer Immunol Immunother 40(6):390–396

de Gast GC, van de Winkel JG, Bast BE (1997) Clinical perspectives of bispecific antibodies in cancer. Cancer Immunol Immunother 45(3–4):121–123

de Haij S, Jansen JH, Boross P, Beurskens FJ, Bakema JE, Bos DL, Martens A, Verbeek JS, Parren PW, van de Winkel JG, Leusen JH (2010) In vivo cytotoxicity of type I CD20 antibodies critically depends on Fc receptor ITAM signaling. Cancer Res 70(8):3209–3217

de la Hera A, Muller U, Olsson C, Isaaz S, Tunnacliffe A (1991) Structure of the T cell antigen receptor (TCR): two CD3 epsilon subunits in a functional TCR/CD3 complex. J Exp Med 173(1):7–17

Deo YM, Graziano RF, Repp R, van de Winkel JG (1997) Clinical significance of IgG Fc receptors and Fc gamma R-directed immunotherapies. Immunol Today 18(3):127–135

Deo YM, Sundarapandiyan K, Keler T, Wallace PK, Graziano RF (1998) Bispecific molecules directed to the Fc receptor for IgA (Fc alpha RI, CD89) and tumor antigens efficiently promote cell-mediated cytotoxicity of tumor targets in whole blood. J Immunol 160(4):1677–1686

Di Carlo E, Forni G, Lollini P, Colombo MP, Modesti A, Musiani P (2001) The intriguing role of polymorphonuclear neutrophils in antitumor reactions. Blood 97(2):339–345

Dreier T, Lorenczewski G, Brandl C, Hoffmann P, Syring U, Hanakam F, Kufer P, Riethmuller G, Bargou R, Baeuerle PA (2002) Extremely potent, rapid and costimulation-independent cytotoxic T-cell response against lymphoma cells catalyzed by a single-chain bispecific antibody. Int J Cancer 100(6):690–697

Dreier T, Baeuerle PA, Fichtner I, Grun M, Schlereth B, Lorenczewski G, Kufer P, Lutterbuse R, Riethmuller G, Gjorstrup P, Bargou RC (2003) T cell costimulus-independent and very efficacious inhibition of tumor growth in mice bearing subcutaneous or leukemic human B cell lymphoma xenografts by a CD19-/CD3-bispecific single-chain antibody construct. J Immunol 170(8):4397–4402

Elsasser D, Valerius T, Repp R, Weiner GJ, Deo Y, Kalden JR, van de Winkel JG, Stevenson GT, Glennie MJ, Gramatzki M (1996) HLA class II as potential target antigen on malignant B cells for therapy with bispecific antibodies in combination with granulocyte colony-stimulating factor. Blood 87(9):3803–3812

Elsasser D, Stadick H, Stark S, Van de Winkel JG, Gramatzki M, Schrott KM, Valerius T, Schafhauser W (1999) Preclinical studies combining bispecific antibodies with cytokine-stimulated effector cells for immunotherapy of renal cell carcinoma. Anticancer Res 19 (2C):1525–1528

Ely P, Wallace PK, Givan AL, Graziano RF, Guyre PM, Fanger MW (1996) Bispecific-armed, interferon gamma-primed macrophage-mediated phagocytosis of malignant non-Hodgkin's lymphoma. Blood 87(9):3813–3821

Fanger MW, Segal DM, Romet-Lemonne JL (1991) Bispecific antibodies and targeted cellular cytotoxicity. Immunol Today 12(2):51–54

Fiedler WM, Ritter B, Seggewiss R, Bokemeyer C, Fettes P, Klinger M, Vieser E, Ruettinger D, Kaubitzsch S, Wolf M (2010) Phase I safety and pharmacology study of the EpCAM/CD3-bispecific BiTE antibody MT110 in patients with metastatic colorectal, gastric, or lung cancer. J Clin Oncol, ASCO Annual Meeting Proceedings 28(15):2573

Fury MG, Lipton A, Smith KM, Winston CB, Pfister DG (2008) A phase-I trial of the epidermal growth factor receptor directed bispecific antibody MDX-447 without and with recombinant human granulocyte-colony stimulating factor in patients with advanced solid tumors. Cancer Immunol Immunother 57(2):155–163

Glennie MJ, French RR, Cragg MS, Taylor RP (2007) Mechanisms of killing by anti-CD20 monoclonal antibodies. Mol Immunol 44(16):3823–3837

Glorius P, Baerenwaldt A, Kellner C, Staudinger M, Parren PWHI, van de Winkel JGJ, Humpe A, Valerius T, Repp R, Nimmerjahn F, Gramatzki M, Peipp M (2010) The novel bispecific antibody [$(CD20)_2 \times$ CD16] efficiently triggers lysis of neoplastic B cells. Blood (ASH Annual Meeting Abstracts), 116(21):2846

Gong Q, Ou Q, Ye S, Lee WP, Cornelius J, Diehl L, Lin WY, Hu Z, Lu Y, Chen Y, Wu Y, Meng YG, Gribling P, Lin Z, Nguyen K, Tran T, Zhang Y, Rosen H, Martin F, Chan AC (2005) Importance of cellular microenvironment and circulatory dynamics in B cell immunotherapy. J Immunol 174(2):817–826

Gordon S, Taylor PR (2005) Monocyte and macrophage heterogeneity. Nat Rev Immunol 5(12):953–964

Grosse-Hovest L, Hartlapp I, Marwan W, Brem G, Rammensee HG, Jung G (2003) A recombinant bispecific single-chain antibody induces targeted, supra-agonistic CD28-stimulation and tumor cell killing. Eur J Immunol 33(5):1334–1340

Grosse-Hovest L, Wick W, Minoia R, Weller M, Rammensee HG, Brem G, Jung G (2005) Supraagonistic, bispecific single-chain antibody purified from the serum of cloned, transgenic cows induces T-cell-mediated killing of glioblastoma cells in vitro and in vivo. Int J Cancer 117(6):1060–1064

Guettinger Y, Barbin K, Peipp M, Bruenke J, Dechant M, Horner H, Thierschmidt D, Valerius T, Repp R, Fey GH, Stockmeyer B (2010) A recombinant bispecific single-chain fragment variable specific for HLA class II and Fc alpha RI (CD89) recruits polymorphonuclear neutrophils for efficient lysis of malignant B lymphoid cells. J Immunol 184(3):1210–1217

Hartmann F, Renner C, Jung W, Pfreundschuh M (1998) Anti-CD16/CD30 bispecific antibodies as possible treatment for refractory Hodgkin's disease. Leuk Lymphoma 31(3–4):385–392

Heijnen IA, Van de Winkel JG (1995) A human Fc gamma RI/CD64 transgenic model for in vivo analysis of (bispecific) antibody therapeutics. J Hematother 4(5):351–356

Heijnen IA, van Vugt MJ, Fanger NA, Graziano RF, de Wit TP, Hofhuis FM, Guyre PM, Capel PJ, Verbeek JS, van de Winkel JG (1996) Antigen targeting to myeloid-specific human Fc gamma RI/CD64 triggers enhanced antibody responses in transgenic mice. J Clin Invest 97(2):331–338

Honeychurch J, Tutt AL, Valerius T, Heijnen IA, Van De Winkel JG, Glennie MJ (2000) Therapeutic efficacy of FcgammaRI/CD64-directed bispecific antibodies in B-cell lymphoma. Blood 96(10):3544–3552

Jaiswal S, Chao MP, Majeti R, Weissman IL (2010) Macrophages as mediators of tumor immunosurveillance. Trends Immunol 31(6):212–219

James ND, Atherton PJ, Jones J, Howie AJ, Tchekmedyian S, Curnow RT (2001) A phase II study of the bispecific antibody MDX-H210 (anti-HER2 $\times$ CD64) with GM-CSF in HER2+ advanced prostate cancer. Br J Cancer 85(2):152–156

Johnson S, Burke S, Huang L, Gorlatov S, Li H, Wang W, Zhang W, Tuaillon N, Rainey J, Barat B, Yang Y, Jin L, Ciccarone V, Moore PA, Koenig S, Bonvini E (2010) Effector cell recruitment with novel Fv-based dual-affinity re-targeting protein leads to potent tumor cytolysis and in vivo B-cell depletion. J Mol Biol 399(3):436–449

Karapetis CS, Khambata-Ford S, Jonker DJ, O'Callaghan CJ, Tu D, Tebbutt NC, Simes RJ, Chalchal H, Shapiro JD, Robitaille S, Price TJ, Shepherd L, Au HJ, Langer C, Moore MJ, Zalcberg JR (2008) K-ras mutations and benefit from cetuximab in advanced colorectal cancer. N Engl J Med 359(17):1757–1765

Keler T, Graziano RF, Mandal A, Wallace PK, Fisher J, Guyre PM, Fanger MW, Deo YM (1997) Bispecific antibody-dependent cellular cytotoxicity of HER2/neu-overexpressing tumor cells by Fc gamma receptor type I-expressing effector cells. Cancer Res 57(18):4008–4014

Kellner C, Bruenke J, Stieglmaier J, Schwemmlein M, Schwenkert M, Singer H, Mentz K, Peipp M, Lang P, Oduncu F, Stockmeyer B, Fey GH (2008) A novel CD19-directed recombinant bispecific antibody derivative with enhanced immune effector functions for human leukemic cells. J Immunother 31(9):871–884

Kiewe P, Hasmuller S, Kahlert S, Heinrigs M, Rack B, Marme A, Korfel A, Jager M, Lindhofer H, Sommer H, Thiel E, Untch M (2006) Phase I trial of the trifunctional anti-HER2 × anti-CD3 antibody ertumaxomab in metastatic breast cancer. Clin Cancer Res 12(10):3085–3091

Kimberly RP, Wu J, Gibson AW, Su K, Qin H, Li X, Edberg JC (2002) Diversity and duplicity: human FCgamma receptors in host defense and autoimmunity. Immunol Res 26 (1–3):177–189

Kipriyanov SM, Cochlovius B, Schafer HJ, Moldenhauer G, Bahre A, Le Gall F, Knackmuss S, Little M (2002) Synergistic antitumor effect of bispecific CD19 × CD3 and CD19 × CD16 diabodies in a preclinical model of non-Hodgkin's lymphoma. J Immunol 169(1):137–144

Kontermann RE (2009) Strategies to extend plasma half-lives of recombinant antibodies. BioDrugs 23(2):93–109

Kroesen BJ, Buter J, Sleijfer DT, Janssen RA, van der Graaf WT, The TH, de Leij L, Mulder NH (1994) Phase I study of intravenously applied bispecific antibody in renal cell cancer patients receiving subcutaneous interleukin 2. Br J Cancer 70(4):652–661

Kugler M, Stein C, Kellner C, Mentz K, Saul D, Schwenkert M, Schubert I, Singer H, Oduncu F, Stockmeyer B, Mackensen A, Fey GH (2010) A recombinant trispecific single-chain Fv derivative directed against CD123 and CD33 mediates effective elimination of acute myeloid leukaemia cells by dual targeting. Br J Haematol 150(5):574–586

Li M, Wirthmueller U, Ravetch JV (1996) Reconstitution of human Fc gamma RIII cell type specificity in transgenic mice. J Exp Med 183(3):1259–1263

Lindhofer H, Mocikat R, Steipe B, Thierfelder S (1995) Preferential species-restricted heavy/light chain pairing in rat/mouse quadromas. Implications for a single-step purification of bispecific antibodies. J Immunol 155(1):219–225

Loffler A, Kufer P, Lutterbuse R, Zettl F, Daniel PT, Schwenkenbecher JM, Riethmuller G, Dorken B, Bargou RC (2000) A recombinant bispecific single-chain antibody, CD19 × CD3, induces rapid and high lymphoma-directed cytotoxicity by unstimulated T lymphocytes. Blood 95(6):2098–2103

Lu H, Shi M, Wang M, Xie Z, Hu M, Yu M, Shen B, Ma Y, Guo N (2008) In vitro and in vivo antitumor effect of a trivalent bispecific antibody targeting ErbB2 and CD16. Cancer Biol Ther 7(11):1744–1750

Lum LG, Rathore R, Cummings F, Colvin GA, Radie-Keane K, Maizel A, Quesenberry PJ, Elfenbein GJ (2003) Phase I/II study of treatment of stage IV breast cancer with OKT3 × trastuzumab-armed activated T cells. Clin Breast Cancer 4(3):212–217

Lutterbuese R, Raum T, Kischel R, Hoffmann P, Mangold S, Rattel B, Friedrich M, Thomas O, Lorenczewski G, Rau D, Schaller E, Herrmann I, Wolf A, Urbig T, Baeuerle PA, Kufer P (2010) T cell-engaging BiTE antibodies specific for EGFR potently eliminate KRAS- and BRAF-mutated colorectal cancer cells. Proc Natl Acad Sci USA 107(28):12605–12610

Manzke O, Tesch H, Borchmann P, Wolf J, Lackner K, Gossmann A, Diehl V, Bohlen H (2001) Locoregional treatment of low-grade B-cell lymphoma with CD3 × CD19 bispecific antibodies and CD28 costimulation. I. Clinical phase I evaluation. Int J Cancer 91(4):508–515

Marechal R, De Schutter J, Nagy N, Demetter P, Lemmers A, Deviere J, Salmon I, Tejpar S, Van Laethem JL (2010) Putative contribution of CD56 positive cells in cetuximab treatment efficacy in first-line metastatic colorectal cancer patients. BMC Cancer 10(1):340

McCall AM, Shahied L, Amoroso AR, Horak EM, Simmons HH, Nielson U, Adams GP, Schier R, Marks JD, Weiner LM (2001) Increasing the affinity for tumor antigen enhances bispecific antibody cytotoxicity. J Immunol 166(10):6112–6117

Merchant AM, Zhu Z, Yuan JQ, Goddard A, Adams CW, Presta LG, Carter P (1998) An efficient route to human bispecific IgG. Nat Biotechnol 16(7):677–681

Monteiro RC, Van De Winkel JG (2003) IgA Fc receptors. Annu Rev Immunol 21:177–204

Müller D, Kontermann RE (2007) Bispecific antibodies. In: Dübel S (ed) Handbook of therapeutic antibodies, vol 2. Wiley-VCH, Weinheim, pp 345–378

Musolino A, Naldi N, Bortesi B, Pezzuolo D, Capelletti M, Missale G, Laccabue D, Zerbini A, Camisa R, Bisagni G, Neri TM, Ardizzoni A (2008) Immunoglobulin G fragment C receptor polymorphisms and clinical efficacy of trastuzumab-based therapy in patients with HER-2/neu-positive metastatic breast cancer. J Clin Oncol 26(11):1789–1796

Nimmerjahn F, Ravetch JV (2005) Divergent immunoglobulin g subclass activity through selective Fc receptor binding. Science 310(5753):1510–1512

Nimmerjahn F, Ravetch JV (2007) Antibodies, Fc receptors and cancer. Curr Opin Immunol 19(2):239–245

Oldham RK, Dillman RO (2008) Monoclonal antibodies in cancer therapy: 25 years of progress. J Clin Oncol 26(11):1774–1777

Otz T, Grosse-Hovest L, Hofmann M, Rammensee HG, Jung G (2009) A bispecific single-chain antibody that mediates target cell-restricted, supra-agonistic CD28 stimulation and killing of lymphoma cells. Leukemia 23(1):71–77

Park S, Jiang Z, Mortenson ED, Deng L, Radkevich-Brown O, Yang X, Sattar H, Wang Y, Brown NK, Greene M, Liu Y, Tang J, Wang S, Fu YX (2010) The therapeutic effect of anti-HER2/neu antibody depends on both innate and adaptive immunity. Cancer Cell 18(2):160–170

Posey JA, Raspet R, Verma U, Deo YM, Keller T, Marshall JL, Hodgson J, Mazumder A, Hawkins MJ (1999) A pilot trial of GM-CSF and MDX-H210 in patients with erbB-2-positive advanced malignancies. J Immunother 22(4):371–379

Presta LG (2008) Molecular engineering and design of therapeutic antibodies. Curr Opin Immunol 20(4):460–470

Pullarkat V, Deo Y, Link J, Spears L, Marty V, Curnow R, Groshen S, Gee C, Weber JS (1999) A phase I study of a HER2/neu bispecific antibody with granulocyte-colony-stimulating factor in patients with metastatic breast cancer that overexpresses HER2/neu. Cancer Immunol Immunother 48(1):9–21

Ranft K, Thepen T, Fischer R, Barth S, Stocker M (2009) Recombinant bispecific single chain antibody fragments induce Fc gamma-receptor-mediated elimination of CD30+ lymphoma cells. Cancer Lett 282(2):187–194

Reichert JM, Rosensweig CJ, Faden LB, Dewitz MC (2005) Monoclonal antibody successes in the clinic. Nat Biotechnol 23(9):1073–1078

Repp R, van Ojik HH, Valerius T, Groenewegen G, Wieland G, Oetzel C, Stockmeyer B, Becker W, Eisenhut M, Steininger H, Deo YM, Blijham GH, Kalden JR, van de Winkel JG, Gramatzki M (2003) Phase I clinical trial of the bispecific antibody MDX-H210 (anti-FcgammaRI × anti-HER-2/neu) in combination with Filgrastim (G-CSF) for treatment of advanced breast cancer. Br J Cancer 89(12):2234–2243

Samelson LE (2002) Signal transduction mediated by the T cell antigen receptor: the role of adapter proteins. Annu Rev Immunol 20:371–394

Schlaeth M, Berger S, Derer S, Klausz K, Lohse S, Dechant M, Lazar GA, Schneider-Merck T, Peipp M, Valerius T (2010) Fc-engineered EGF-R antibodies mediate improved antibody-dependent cellular cytotoxicity (ADCC) against KRAS-mutated tumor cells. Cancer Sci 101(5):1080–1088

Schlereth B, Quadt C, Dreier T, Kufer P, Lorenczewski G, Prang N, Brandl C, Lippold S, Cobb K, Brasky K, Leo E, Bargou R, Murthy K, Baeuerle PA (2006) T-cell activation and B-cell depletion in chimpanzees treated with a bispecific anti-CD19/anti-CD3 single-chain antibody construct. Cancer Immunol Immunother 55(5):503–514

Segal DM, Weiner GJ, Weiner LM (1999) Bispecific antibodies in cancer therapy. Curr Opin Immunol 11(5):558–562

Seimetz D, Lindhofer H, Bokemeyer C (2010) Development and approval of the trifunctional antibody catumaxomab (anti-EpCAM × anti-CD3) as a targeted cancer immunotherapy. Cancer Treat Rev 36(6):458–467

Shahied LS, Tang Y, Alpaugh RK, Somer R, Greenspon D, Weiner LM (2004) Bispecific minibodies targeting HER2/neu and CD16 exhibit improved tumor lysis when placed in a divalent tumor antigen binding format. J Biol Chem 279(52):53907–53914

Shan D, Ledbetter JA, Press OW (1998) Apoptosis of malignant human B cells by ligation of CD20 with monoclonal antibodies. Blood 91(5):1644–1652

Small EJ, Reese DM, Um B, Whisenant S, Dixon SC, Figg WD (1999) Therapy of advanced prostate cancer with granulocyte macrophage colony-stimulating factor. Clin Cancer Res 5(7):1738–1744

Smyth MJ, Hayakawa Y, Takeda K, Yagita H (2002) New aspects of natural-killer-cell surveillance and therapy of cancer. Nat Rev Cancer 2(11):850–861

Stadick H, Stockmeyer B, Kuhn R, Schrott KM, Kalden JR, Glennie MJ, van de Winkel JG, Gramatzki M, Valerius T, Elsasser D (2002) Epidermal growth factor receptor and g250: useful target antigens for antibody mediated cellular cytotoxicity against renal cell carcinoma? J Urol 167(2 Pt 1):707–712

Staerz UD, Kanagawa O, Bevan MJ (1985) Hybrid antibodies can target sites for attack by T cells. Nature 314(6012):628–631

Stockmeyer B, Valerius T, Repp R, Heijnen IA, Buhring HJ, Deo YM, Kalden JR, Gramatzki M, van de Winkel JG (1997) Preclinical studies with Fc(gamma)R bispecific antibodies and granulocyte colony-stimulating factor-primed neutrophils as effector cells against HER-2/neu overexpressing breast cancer. Cancer Res 57(4):696–701

Stockmeyer B, Elsasser D, Dechant M, Repp R, Gramatzki M, Glennie MJ, van de Winkel JG, Valerius T (2001) Mechanisms of G-CSF- or GM-CSF-stimulated tumor cell killing by Fc receptor-directed bispecific antibodies. J Immunol Methods 248(1–2):103–111

Sundarapandiyan K, Keler T, Behnke D, Engert A, Barth S, Matthey B, Deo YM, Graziano RF (2001) Bispecific antibody-mediated destruction of Hodgkin's lymphoma cells. J Immunol Methods 248(1–2):113–123

Suntharalingam G, Perry MR, Ward S, Brett SJ, Castello-Cortes A, Brunner MD, Panoskaltsis N (2006) Cytokine storm in a phase 1 trial of the anti-CD28 monoclonal antibody TGN1412. N Engl J Med 355(10):1018–1028

Taskinen M, Karjalainen-Lindsberg ML, Nyman H, Eerola LM, Leppa S (2007) A high tumor-associated macrophage content predicts favorable outcome in follicular lymphoma patients treated with rituximab and cyclophosphamide-doxorubicin-vincristine-prednisone. Clin Cancer Res 13(19):5784–5789

Topp MS, Zugmaier G, Goekbuget N, Kufer P, Mariele Goebeler, Klinger M, Degenhard E, Baeuerle PA, Schmidt M, Nagorsen D, Neumann S, Horst HA, Raff T, Viardot A, Stelljes M, Schmid M, Ottmann OG, Burmeister T, Einsele H, Riethmueller G, Hoelzer D, Bargou RC (2009) Report of a phase II trial of single-agent BiTE® antibody blinatumomab in patients with minimal residual disease (MRD) positive B-precursor acute lymphoblastic leukemia (ALL). Blood 114:346 (abstract 840)

Traggiai E, Chicha L, Mazzucchelli L, Bronz L, Piffaretti JC, Lanzavecchia A, Manz MG (2004) Development of a human adaptive immune system in cord blood cell-transplanted mice. Science 304(5667):104–107

Trapani JA, Smyth MJ (2002) Functional significance of the perforin/granzyme cell death pathway. Nat Rev Immunol 2(10):735–747

Uchida J, Hamaguchi Y, Oliver JA, Ravetch JV, Poe JC, Haas KM, Tedder TF (2004) The innate mononuclear phagocyte network depletes B lymphocytes through Fc receptor-dependent mechanisms during anti-CD20 antibody immunotherapy. J Exp Med 199(12):1659–1669

Valerius T, Stockmeyer B, van Spriel AB, Graziano RF, van den Herik-Oudijk IE, Repp R, Deo YM, Lund J, Kalden JR, Gramatzki M, van de Winkel JG (1997) FcalphaRI (CD89) as a novel trigger molecule for bispecific antibody therapy. Blood 90(11):4485–4492

Valone FH, Kaufman PA, Guyre PM, Lewis LD, Memoli V, Deo Y, Graziano R, Fisher JL, Meyer L, Mrozek-Orlowski M et al (1995) Phase Ia/Ib trial of bispecific antibody MDX-210 in patients with advanced breast or ovarian cancer that overexpresses the proto-oncogene HER-2/neu. J Clin Oncol 13(9):2281–2292

van de Winkel JG, Capel PJ (1993) Human IgG Fc receptor heterogeneity: molecular aspects and clinical implications. Immunol Today 14(5):215–221 van Egmond M, van Vuuren AJ, Morton HC, van Spriel AB, Shen L, Hofhuis FM, Saito T, Mayadas TN, Verbeek JS, van de Winkel JG (1999) Human immunoglobulin A receptor (FcalphaRI, CD89) function in transgenic mice requires both FcR gamma chain and CR3 (CD11b/CD18). Blood 93(12):4387–4394

van Vugt MJ, Kleijmeer MJ, Keler T, Zeelenberg I, van Dijk MA, Leusen JH, Geuze HJ, van de Winkel JG (1999) The FcgammaRIa (CD64) ligand binding chain triggers major histocompatibility complex class II antigen presentation independently of its associated FcR gamma-chain. Blood 94(2):808–817

Wang SY, Veeramani S, Racila E, Cagley J, Fritzinger DC, Vogel CW, St John W, Weiner GJ (2009) Depletion of the C3 component of complement enhances the ability of rituximab-coated target cells to activate human NK cells and improves the efficacy of monoclonal antibody therapy in an in vivo model. Blood 114(26):5322–5330

Weiner LM, Holmes M, Adams GP, LaCreta F, Watts P, Garcia de Palazzo I (1993a) A human tumor xenograft model of therapy with a bispecific monoclonal antibody targeting c-erbB-2 and CD16. Cancer Res 53(1):94–100

Weiner LM, Holmes M, Richeson A, Godwin A, Adams GP, Hsieh-Ma ST, Ring DB, Alpaugh RK (1993b) Binding and cytotoxicity characteristics of the bispecific murine monoclonal antibody 2B1. J Immunol 151(5):2877–2886

Weng WK, Levy R (2003) Two immunoglobulin G fragment C receptor polymorphisms independently predict response to rituximab in patients with follicular lymphoma. J Clin Oncol 21(21):3940–3947

Wurflein D, Dechant M, Stockmeyer B, Tutt AL, Hu P, Repp R, Kalden JR, van de Winkel JG, Epstein AL, Valerius T, Glennie M, Gramatzki M (1998) Evaluating antibodies for their capacity to induce cell-mediated lysis of malignant B cells. Cancer Res 58(14):3051–3058

Xie Z, Guo N, Yu M, Hu M, Shen B (2005) A new format of bispecific antibody: highly efficient heterodimerization, expression and tumor cell lysis. J Immunol Methods 296(1–2):95–101

Zeidler R, Reisbach G, Wollenberg B, Lang S, Chaubal S, Schmitt B, Lindhofer H (1999) Simultaneous activation of T cells and accessory cells by a new class of intact bispecific antibody results in efficient tumor cell killing. J Immunol 163(3):1246–1252

Chapter 14
Bispecific Antibodies for Arming Activated T Cells and Other Effector Cells for Tumor Therapy

Lawrence G. Lum and Archana Thakur

14.1 Introduction

The advances in molecular biology, hybridoma technology, and protein engineering associated with effective methods for screening and identifying specific target antigens provide a unique opportunity for developing new strategies for producing bispecific antibodies (BiAb) to redirect T cells to cancer cells. This chapter focuses on the use of BiAbs to retarget T cells in vivo as a therapeutic platform. A bispecific antibody is a protein construct that binds to two different specificities. The common approach to produce BiAbs is to combine the variable domains of the selected monoclonal antibodies (mAbs) into a single construct with the desired structure and function. BiAbs were developed in preclinical murine models and the first clinical trials were performed nearly 20 years ago. The construct provides a unique opportunity to bring together an effector with a target (or tumor cells). There are numerous formats of BiAbs that include chemical heteroconjugation of two complete mAbs or fragments of mAbs, quadroma $F(ab)_2$, diabodies, tandem diabodies, single-chain variable fragment antibodies (scFv), and multi-valent-multifunctional dock and lock (DNL) (Dreier et al. 2002; Asano et al. 2007, 2010; Rossi et al. 2006). The formats and design are limited only by the engineer's imagination and specificity requirements. Engineering BiAbs range from using simple chemistry to elegant recombinant technology to produce BiAbs that target effector cells,

---

L.G. Lum (✉)
BMT and Immunotherapy Program, Department of Oncology, Wayne State University, Detroit, MI, USA

Barbara Ann Karmanos Cancer Institute, Rm 740.1 HWCRC, 4100 John R, Detroit, MI 48201, USA
e-mail: luml@karmanos.org

A. Thakur
BMT and Immunotherapy Program, Department of Oncology, Wayne State University, Detroit, MI, USA

R.E. Kontermann (ed.), *Bispecific Antibodies*,
DOI 10.1007/978-3-642-20910-9_14, © Springer-Verlag Berlin Heidelberg 2011

drugs, prodrugs, toxins, DNA, enzymes, anti-vascular agents, vectors, or radionuclides to tumor-associated antigens (TAA) on cancer cells. Unfortunately, early clinical studies using BiAbs were disappointing because of dose-limiting side effects and immunogenicity of the constructs. Fortunately, persistence of those working in protein engineering, molecular biology, and cell-based immunotherapy has led to the resurrection of BiAbs for clinical applications.

This chapter focuses on approaches that use BiAbs to redirect T cells, activated T cells (ATC), NK cells, or macrophages to target hematologic malignancies or solid tumors. We highlight concepts and technologies and their clinical translation involving BiAb binding and the redirection of non-MHC restricted cytotoxicity of T cells. The pros and cons of the BiAb redirection of T cells and ex vivo activated T cells are discussed. The advantages and disadvantages of a specific format for arming activated T cells will be incorporated into the discussion where apropos.

In the process of juxtaposing effector cells with tumor cells, selected BiAbs induce the upregulation of key activation molecules and/or bind to co-activation receptors. The binding and activating signals created by cross-linking the T cell receptor (CD3) induce cytotoxic and proliferative responses in T cells. The characteristics engineered into a BiAb construct depend upon the desired in vivo mechanism of action, pharmacokinetics, and clinical endpoint. BiAbs have the design flexibility that enables them for tissue penetration, serum stability, enhanced specificity for targeting and the desired high or low affinity for targeting effector cells to tumor cells. Specific BiAb engineering technologies and strategies have been extensively reviewed by others (Cao and Lam 2003; Segal et al. 2001) and are not reviewed in this chapter.

14.1.1 The Challenge and Principles of Adoptive Cellular Immune Therapy

Manipulating the patient's immune system in a non-toxic manner to breach tumor-induced tolerance by altering the cytokine balance toward a Th_1 cytokine pattern and to enhance anti-tumor effects is a challenging task. Although adoptively transferred lymphokine activated killer cells (LAK) (Grimm et al. 1982; Anderson et al. 1988), tumor infiltrating lymphocytes (TIL) (Rosenberg et al. 1986), anti-CD3 ATC (Uberti et al. 1994; Ueda et al. 1993), and anti-CD3/anti-CD28 co-activated T cells (COACTS) (Lum et al. 2001; Fowler et al. 2006; Garlie et al. 1999) have been successful in eliminating or reducing lethal tumor burden in preclinical murine models, translating these principles to the clinical setting have been problematic. These approaches were met with enthusiasm when there were dramatic clinical responses observed in patients with malignant melanoma (MM) or renal cell carcinoma (RCC) using TIL infusions (Rosenberg et al. 1986, 1988). However, no subsequent study has unequivocally shown improved clinical remission rates or overall survival. Since 1986, investigators have sought to develop methods or approaches to dissect the mechanisms and identify the key components of immune circuitry that would enhance Th_1 anti-tumor responses in hematologic malignancies and solid tumors. It has been a challenge to show that effector cell therapy per se provides a sustained memory anti-tumor or anti-leukemia effect. Furthermore, the strategies involving ex vivo expansion of TIL, ATC, or COACTS are limited by the numbers of functional cytotoxic T lymphocytes (CTL) that can be grown ex vivo. One major limitation of autologous T cell-based products is their impaired functional capacity due to intrinsic T cell defects caused by malignancy (Whiteside 1999), chemotherapy effects, or a combination of both. It is clear that patients who have cancer and/or have undergone chemotherapy for their cancer have impaired T cell and other immune components.

A prime example of successful T cell adoptive immunotherapy has been the allogeneic stem cell transplant (SCT) paradigm wherein donor T cells in the stem cell graft or donor lymphocyte infusions (DLIs) provide a graft-vs-leukemia or lymphoma (GVL) effect. In allogeneic stem cell transplantation, there is a clear anti-leukemia effect when DLIs are infused after relapse of chronic myelogenous leukemia (CML), acute myelogenous leukemia (AML), acute lymphocytic leukemia (ALL), and other hematologic malignancies (Deol 2010; Kolb et al. 1990). CML is so sensitive to immunotherapy that DLI can induce both cytogenetic (disappearance of the Philadelphia chromosome) and molecular (disappearance of bcr-abl) remissions (Kolb et al. 1990; Kolb and Holler 1997) whereas acute myelogenous leukemia, and acute lymphocyte leukemia are not as responsive to DLI.

Clinical responses were reported in patients with advanced stage malignant melanoma (MM) (Dudley et al. 2002). Cloned specific CTL were given to patients with MM during the nadir of their lymphocyte and neutrophil counts ($<20/mm^3$) 7 days after non-myeloablative chemotherapy consisting of cyclophosphamide (60 mg/kg $\times$ 2 days) and fludarabine (25 mg/m^2 $\times$ 5 days) (Dudley et al. 2002). A mean of 7.8×10^{10} ($2.3–13.7 \times 10^{10}$) anti-melanoma CTL were infused into MM patients with concomitant 720,000 IU of IL-2/kg bolus given every 8 h. Six of 13 (46%) patients had objective clinical responses, while 4 of 13 (30%) had mixed responses.

A GVL effect was seen in patients who developed EBV-driven lymphoproliferative disorders (LPD) after T cell-depleted allogeneic SCT (Liu et al. 2002). In patients with LPD, infusion of EBV-specific CTL induced clinical remissions in patients who had developed LPD after allogeneic SCT (Rooney et al. 1998). On the other hand, objective clinical responses were minimal or absent in patients with most solid tumors. The use of DLI was recently reviewed (Deol 2010). Poor clinical responses to immunotherapy in solid tumor patients pose a major clinical challenge for successful targeted immunotherapies for cancer.

14.1.2 Mechanisms of Tumor Resistance and Escape

Mechanisms that tumors use to escape immune surveillance include low expression of tumor or HLA antigens, low or no expression of co-stimulatory molecules, or defects (zeta chain defects) in patient T cell signal transduction pathways. All these immune mechanisms may confer "resistance" to T cell-based immunotherapies (Whiteside 1999). Tumors may also induce tolerance via secretion of suppressive cytokines that can induce myeloid and lymphoid regulatory cells. Potent T regulatory cells (Tregs), myeloid derived suppressor cells (MDSC), and suppressor monocytes can inhibit the development of anti-tumor CTL as well as specific antibody responses to tumor vaccines. Preclinical in vitro and in vivo studies suggest that strategies which take advantage of non-MHC restricted T cell cytotoxicity as an approach of targeting T cells to tumors may circumvent tumor escape mechanisms.

14.1.2.1 HLA-Mediated Tumor Resistance

Several studies show that decreased or altered levels of HLA antigen expression on tumors allow them to become resistant to lysis by Class I restricted antigen specific CTL (Rivoltini et al. 1995; Cohen et al. 1994). The phenomenon of altering HLA antigen expression has been reported in carcinomas of the breast (Concha et al. 1991), prostate (Blades et al. 1995), colon (Browning et al. 1996), lung (Redondo et al. 1991), pancreas (Torres et al. 1996), and in malignant melanoma (Ferrone and Marincola 1995).

14.1.2.2 Immunosuppressive Cytokines

Tumor cells subvert immune responses by secreting suppressive cytokines that inhibit T cell proliferation and function. TGF-β secreted by tumors down-regulate anti-tumor cytokines such as IL-2 and IFN-γ by secretion of IL-10 and IL-6. The suppression of IL-2 and IFN-γ not only inhibits T cell proliferation, but also inhibits production of perforin granules and granzyme B responsible for mediating T cell cytotoxicity (Smyth et al. 1991). Immunosuppressive cytokines have been reported to decrease clinical responses to treatment with IL-2 or IFN-α (Blay et al. 1992; Tartour et al. 1996).

Preclinical studies show that BiAb-armed activated T cells can lyse tumor targets not only in the presence, but also in the absence of IL-2 (Zeidler et al. 1999). TNF-α, IFN-γ, GM-CSF, IL-4, IL-6, and IL-10 are released during the process of BiAb-armed T cell mediated tumor lysis in the absence of IL-2 (Klein et al. 1997). Furthermore, analysis of serum cytokine levels in patients treated with BiAb-armed T cells showed increased levels of TNF-α and IFN-γ and no IL-4 (Klein et al. 1997). TNF-α and IFN-γ secreted during BiAb engagement of the TCR shift the in vivo tumor milieu towards a Th_1 anti-tumor environment. Furthermore, repeated reactivation of T cells armed with anti-CD3 × anti-CD19 or anti-CD3 × anti-Her2 may increase survival and enhance in vivo cytotoxicity (Klein et al. 1997; Grabert et al. 2006). Recent studies using anti-CD3 × anti-Her2/*neu* BiAb armed ATC showed not only induction of Th_1 cytokine secretion but also induction of GM-CSF, RANTES, and MIP1α (Grabert et al. 2006); which may enhance killing and provide an environment for immunizing the patient's endogenous immune system against the tumor antigens released from the cytotoxic event.

In addition to tumor escape and sabotage of immune responses, physical barriers of tumor architecture and vasculature provide a formidable if not impenetrable fortress that blocks the infiltration of immune effectors and antibodies. Effector cells and molecules are required to "swim upstream" against pressure gradients, or insinuate their way through a multi-layer defense perimeter created by the tumor to prevent contact with effector cells.

14.1.2.3 Armed T Cells with BiAb May Avoid Fratricide of Effector T Cells

Cytotoxic T cells bearing Fas ligand (FasL) are known to induce apoptosis in tumor cells that express Fas receptors (CD95). Antigen-dependent binding of T cells to tumors releases IFNγ which upregulates Fas on tumor targets (Mullbacher and Lobigs 2002). The FasL–Fas interaction between the T cells and tumor cells induces caspase proteolytic activity leading to apoptosis of tumor cells. Unfortunately, the tumor cells mount a counterattack inducing apoptosis in the effector T cells (Zeytun et al. 1997; Zaks et al. 1999; Walker et al. 1997). Colon, prostate, liver, lung, pancreatic, and malignant melanoma tumor cells have all been reported to express FasL. This FasL counterattack mechanism may eliminate effector T cells and sabotage the specific immune response before it begins (O'Connell et al. 1996). Furthermore, if there is no co-stimulation with CD28 during engagement of the TCR, activation-induced cell death (AICD) may occur (Daniel et al. 1997). The FasL on T cells can also react with Fas-sensitive T cells and induce effector cell apoptosis if co-stimulatory signals are insufficient to induce enough IL-2 secretion to protect T cells from AICD. It is noteworthy that TCR stimulation can protect CD8+ T cells from CD95-mediated apoptosis (Karas et al. 2001). BiAb-armed T cells that restimulate their TCRs upon binding with a target may be resistant to AICD since cross linking of the TCR and juxtapositioning of other signaling receptors via the BiAb complex may mimic or substitute for second signals, thereby preventing apoptosis (Grabert et al. 2003). In our recent study, we showed that armed ATC can repeatedly mediate cytotoxicity as well as proliferate after killing (Grabert et al. 2003). Therefore, targeting ATC with BiAbs may not only re-stimulate the armed ATC to proliferate and secrete Th_1 cytokines, but also enhance in vivo T cell survival by preventing apoptosis and/or self-induced AICD.

14.1.3 BiAb Retargeting of Effector Cells

14.1.3.1 Redirecting T Cells and NK Cells with BiAbs

It has been nearly three decades since the construction of the first BiAbs created by joining two mAbs (Raso and Griffin 1981; Titus et al. 1987; Perez et al. 1986; Segal et al. 1988). A critical element for BiAbs used to target effector cells is the ability of the anti-effector binding portion of the BiAb to trigger the desired effector cell function. In other words, simultaneous binding of BiAb to receptors on the effector cells and the targets would trigger functionality associated with activation of that particular cell type. For example, anti-CD3 binding of anti-CD3 × anti-Her2 bispecific antibody to T cell and Her2/*neu* on the breast cancer cell lines, respectively, triggers activation leading to cytotoxicity, cytokine secretion, and proliferation of the T cells only upon binding to tumor target.

The most pragmatic and straightforward approach for the use of BiAbs would be to design a BiAb that could be administered to the patient in vivo. In this situation, the infused BiAbs would activate and target effectors to the selected tumor target. The conventional approach for clinical application has been to administer the BiAb intravenously at concentrations that would optimize binding to both the effector and tumor cells. The majority of the clinical trials using BiAbs have infused BiAbs as "drugs". Details for engineering, clinical applications, side effects and clinical limitation, and clinical results are discussed and reviewed in the other chapters.

14.1.3.2 Retargeting of Ex Vivo Expanded T and NK Cells

Arming of ex vivo expanded ATC with anti-CD3 × anti-TAA -BiAb transforms non-specific cytotoxic T cells into tumor antigen specific CTL conferred by the mAb directed at the TAA (Renner et al. 1997a). This approach creates an antibody receptor on all T cells that target a TAA and kill tumor cells using the perforin/granzyme system.

There was a lot of excitement in the early 1990s for ex vivo activated T cells armed with BiAbs. This strategy would theoretically take advantage of multiple-fold short-term (10–14 days) ex vivo expansion of patient T cells and arming with BiAb. Multiple infusions of large numbers of armed ATC would significantly increase the precursor frequency of specific CTL in patients. Treatment with BiAb armed ATC should lead to binding and enrichment of T cells at tumor sites leading to augmented anti-tumor activity. Figure 14.1 shows a model of in situ cytotoxicity leading to local and systemic vaccination. Table 14.1 summarizes the preclinical studies involving arming of effector T and NK-T cells with BiAbs that contain anti-CD3 as one parent mAb and a second mAb directed at TAA. Table 14.2 summarizes the clinical trials that involve ex vivo or in vivo arming of effector cells. BiAbs that contain anti-CD3 can be infused to arm T cells in vivo and engage T cells and tumor cells simultaneously. On the other hand, ex vivo arming has the advantages and disadvantages of ex vivo expansion of ATC.

14.1.4 BiAbs for In Vivo Retargeting

A total of 22 mAbs are clinically approved. Their therapeutic applications include organ graft rejection, anti-platelet therapy, rheumatoid arthritis, respiratory syncytial virus infections, Crohn's disease, breast cancer, colon cancer, asthma, and

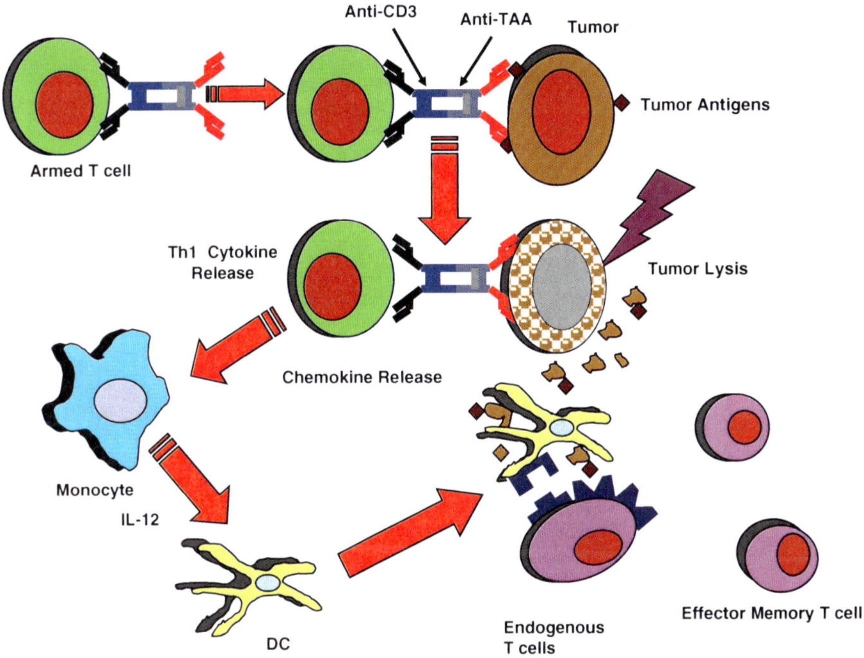

Fig. 14.1 BiAb targeting of T cells to tumor leads to in situ immunization of the endogenous immune system. Activated T cells armed with BiAb engage the tumor by targeting TAAs on the tumor. The process of targeting induces Th1 cytokine release from the T cells and TAA release from tumor lysis that are processed by dendritic cells (DCs) induced by IL-12 (produced by monocytes). The DCs process and present antigens to endogenous naïve T cells recruited to the tumor site by chemokine release leading to *in situ* vaccination and systemic immunization to the TAAs

Table 14.1 Anti-CD3 × anti-TAA combinations for preclinical studies

Target antigen	BiAb	Tumor target	References
Tenascin	Anti-CD3 × anti-tenascin	Human glioma	Davico et al. (1995)
Glioma	Anti-CD3 × anti-glioma	Human glioma	Nitta et al. (1990b)
CD13	Anti-CD3 × anti-CD13	$CD13^{+}$ acute myeloid leukemia	Kaneko et al. (1993)
MUC1	Anti-CD3 × anti-MUC1	Bile duct carcinoma	Katayose et al. (1996)
EpCAM	Anti-CD3 × anti-EpCAM	Epithelial cell adhesion molecule on adenocarcinomas	Riesenberg et al. (2001)
kDal K29	Anti-CD3 × anti-kDalK29	Renal cell carcinoma	Zhu et al. (1994)
G250	Anti-CD3 × anti-G250	Renal cell carcinoma	Luiten et al. (1996)
Transferrin receptors	Anti-CD3 × anti-transferrin receptor	transferrin receptors	Jost et al. (1996)

(continued)

Table 14.1 (continued)

Target antigen	BiAb	Tumor target	References
AMOC-31	Anti-CD3 × anti-AMOC-31	40-kDa membrane-bound glycoprotein expressed by carcinomas	Kroesen et al. (1993)
B cell idiotype	Anti-CD3 × anti-idiotype	Idiotype on BCL1 lymphoma in Balb/c mice	Demanet et al. (1996)
CD19	Anti-CD3 × anti-CD19	Leukemic B cells	Bohlen et al. (1993a) and Bohlen et al. (1993b)
CD19	Anti-CD3 × anti-CD19	Malignant B cells	Klein et al. (1997), Bohlen et al. (1993a), Anderson et al. (1992), Bejeck et al. (1995), de Gast et al. (1995b)
CD20	Anti-CD3 × anti-CD20	NHL	Gall et al. (2005)
CD20	Anti-CD3 × anti-CD20	Multiple myeloma	Gall et al. (2005)
tumor $(Fab')_2$	Anti-CD3 × anti-tumor $(Fab')_2$	For retargeting TIL	Chapoval et al. (1995)
CEA	Anti-CD3 × anti-CEA	Human carcinoembryonic Ag expressing cells	Kuwahara et al. (1997)
Her-2/*neu*	Anti-CD3 × anti-Her2/*neu*	HER2/*neu* receptor expressed by renal cell, colon, and breast, prostate carcinomas	Shalaby et al. (1992), Shalaby et al. (1995), Brossart et al. (1998), Sen et al. (2001), Lum et al. (2005), Davol et al. (2004a), Zhu et al. (1995)
CD30	Anti-CD3 × anti-CD30	Hodgkin's lymphoma	Renner et al. (1994), Renner et al. (1996), Pohl et al. (1993)
PSA	Anti-CD3 × anti-PSA	Prostate specific antigen expressed by prostate carcinomas	Katzenwadel et al. (2000)
CA19-9	Anti-CD3 × anti-CA19-9	CA 19-9	Hombach et al. (1997b)
HLA-DR beta chains	Anti-CD3 × anti-HLA-DR beta chains	Malignant B cells	Kostelny et al. (2001)
EGFR	Anti-CD3 × anti-EGFR	Glioma, neoplastic keratinocytes	Negri et al. (1995)
CA125	Anti-CD3 × anti-CA125	Ovarian carcinomas	Chan et al. (2006)
EGFR	Anti-CD3 × anti-EGFR	Colon, head and neck, and lung	Reusch et al. (2006)

Table 14.2 Clinical trials using bispecific antibodies

Targets	BiAb *INNs/Trade names (company)*	Effector cells	References
Human glioma	Anti-CD3 × anti-glioma	LAK	Nitta et al. (1990a)
Ovarian carcinoma	Anti-OCAA × anti-CD3 *OC/TR*	T-cells	Bolhuis et al. (1992)
Non-Hodgkin's lymphoma	Anti-CD3 × anti-CD19 *SHR-1*	T-cells	de Gast et al. (1995a)
Human carcinoembryonic Ag expressing cells	Anti-CD3 × anti-CEA	T-cells	Kuwahara et al. (1996)
Lung cancer, renal cell cancer	Anti-CD3 × Anti-EGP2 BIS-1	T-cells	Kroesen et al. (1997)
AML	CD64 × CD33 251-22	Monocyte/ macrophage	Chen et al. (1999)
CA19-9+ tumors	Anti-CD3 × anti-CA19-9 *OKT3/NSI19-9*	T-cells	Hombach et al. (1997b)
Ovarian, gastric, colon, breast (malignant ascites)	Anti-CD3 × anti-EpCAM *Removab*® *(Trion Pharma)*	T-cells	Burges et al. (2007) (Market Approved 2009)
Ovarian carcinoma	Anti-CD3 × anti-CEA	T-cells	Compte et al. (2007)
Gastric, lung, colorectal	Anti-CD3 × anti-EpCAM *Removab*® *(Trion Pharma)*	T-cells	Sebastian et al. (2007)
Her2+ tumors	Anti-CD16 × anti-HER2 *(2B1)*	NK	Borghaei et al. (2007), Weiner et al. (1995)
Hodgkin's disease	Anti-CD16 × anti-CD30 *HRS-3/A9*	NK	Hartmann et al. (1997)
Breast and prostate cancer	Anti-CD64 × anti-Her2 *(MDX-210; MDX-H210)*	Monocyte/ macrophage	James et al. (2001), Valone et al. (1995)
Solid tumors	Anti-CD64 × anti-EGFR *(MDX-447)*	Monocyte/ macrophage	Fury et al. (2008)
NHL	Anti-CD3 × anti-CD20 *CD20BiAb*	T-cells	ClinicalTrials (2010e)
Metastatic breast cancer	Anti-CD3 × anti-Her2 *Her2BiAb*	T-cells	ClinicalTrials (2010g) Phase II
Ovarian carcinoma	Anti-CD3 × anti-EpCAM *Removab (Trion Pharma)*	T-cells	ClinicalTrials (2010b) Phase II
Metastatic breast cancer	Anti-CD3 × anti-Her2 *Ertumaxomab (Trion Pharma)*	T-cells	ClinicalTrials (2010d) Phase 1/II
Metastatic melonoma	Anti-CD28 × MAPG *rM28*	T-cells	ClinicalTrials (2010a) Phase 1/II
NHL	Anti-CD3 × anti-CD19 *Blinatumomab (Micromet)*	T-cells	ClinicalTrials (2010c) Phase 1/II
B-precursor acute lymphoblastic leukemia	Anti-CD3 × anti-CD19 *Blinatumomab (Micromet)*	T-cells	ClinicalTrials (2010f) Phase 1/II

hematologic malignancies. Furthermore, new unconjugated, radioimmunoconjugated, and chemoimmunoconjugate mAbs have been approved based upon clinical efficacy and impact through specific targeting of CD20+ lymphomas, EGFR, and Her2/*neu*+ solid tumors. Table 14.2 summarizes the clinical trials using BiAbs for treatment of cancer. The ready availability of FDA-approved mAbs provides a unique opportunity to use any one of the clinically approved mAbs for heteroconjugation to create BiAbs against various tumor targets.

14.1.4.1 Clinical Trials Using BiAbs to Activate and Target T Cells In Vivo

Soon after the first BiAbs were produced in the early 1980s by hybridoma technology, the impact of clinical applications for targeting cancer was quickly recognized (Raso and Griffin 1981; Titus et al. 1987; Perez et al. 1985). The first phase I clinical trial using a BiAb infusion was conducted in non-Hodgkin's lymphoma (NHL) patients expressing CD19 using anti-CD3 × anti-CD19 (SHR-1) BiAb produced from a quadroma (de Gast et al. 1995a). The clinical strategy was to determine if SHR-1 infusions could arm endogenous T cells and retarget the T cells to lymphomas. Except for thrombocytopenia, doses of SHR-1 ranging from 5 to 10 mg were not toxic. Unfortunately, the favorable toxicity profile was not accompanied by clinical efficacy in chronic lymphocytic leukemia (CLL) patients. Failure to induce clinical responses in the phase I trials was thought to be related to rapid clearance of SHR-1. Since the pharmacokinetics of each BiAb construct can vary dramatically, customizing the engineering of each BiAb to improve persistence in the circulation, affinity to optimize T cell activation, and/or affinity to optimize tumor targeting to improve clinical efficacy is important.

BIS-1 (anti-CD3 × anti-EGP-2) was constructed to target T cells to carcinoma cells expressing the 38-kDa epithelial carcinoma-associated transmembrane glycoprotein, EGP-2 (clustered with CO17-1A, KS1/4, AUA-1, and MOC-31) (de Gast et al. 1995a). BIS was engineered to persist in vivo. The anti-EGP-2 mAb component of BIS was clinically effective for imaging tumors (Kosterink et al. 1995). In a Phase I clinical trial in renal cell cancer patients, intravenous infusions of BIS-1 in combination with IL-2 induced high levels of specific cytotoxicity associated with elevated serum levels of TNF-α and IFN-γ (de Gast et al. 1995a). The maximum tolerated dose (MTD) of 5 μg/kg was reached with dose limiting toxicities (DLTs) that included dyspnea, vasoconstriction, and fever without anti-tumor responses. This study illustrated that clinical toxicities in patients could not be predicted by lack of toxicity in the preclinical toxicology studies in rats.

In contrast to systemic injection, injection of autologous ATC and BIS-1 into carcinomatous ascites or pleural effusions did not cause systemic toxicities (Kroesen et al. 1997). The ex vivo ATC were produced by stimulating peripheral blood mononuclear cells (PBMC) for 2 days with IL-2. Seventy percent of the CD3+ cells were CD8+ cells and specific cytotoxicity directed at EGP-2 targets was 60–80% at an effector/target ratio (E/T) of 9. There was a strong anti-inflammatory response and anti-tumor activity only when both BIS-1 and ATC were infused together. Samples of ascites or pleural exudates showed aggregates of T cells and tumor cells with subsequent clearance of tumor cells from the ascites or pleural exudates. There was a decrease in local CEA levels, a marked granulocytosis, and an increase in TNF-α levels 24 h after starting the treatment. The expression of ICAM-1 increased on leukocytes and tumor cells, within 4 h after the infusion. There were no DLTs other than mild fever. The anti-tumor efficacy observed under local administration suggests that DLTs resulting from systemic administration of BIS-1 was a critical factor that limited efficacy. Injection of BIS-1 into ascites armed the injected ATC but would not arm circulating T cells thereby avoiding cytokine storm. On the other hand, systemic administration of BIS-1 could potentially activate as many as a trillion T cells via FcγR binding leading to cytokine storm as illustrated in Fig. 14.2. Therefore, if the anti-CD3 F(ab′) were highly effective at activating T cells, then it would not be surprising for systemic intravenous infusion of BIS-1 to cause cytokine storm similar to that reported for OKT3. Therefore, even if ex vivo ATC were armed with 100 ng/10^6 ATC and washed prior to infusion, the dose of BiAb that would be infused on ATC would be many logs less than the dose infused as free BIS-1. Figure 14.1 shows how anti-CD3 $\times$ anti-TAA armed T cells would target tumor, release cytokines, lyse tumor, and create tumor fragments that antigen presenting cells could process and present to endogenous T cells. Figure 14.2 shows how the free Fc portions of BiAbs can interact with FcγR on monocytes, NK cells, or mast cells leading to cytokine storm. Furthermore, FcγR-bearing cells may bind to the armed T cells triggering T cells to secrete cytokines. Anti-CD16 (anti-FcγRIII) $\times$ anti-TAA binding can be used to

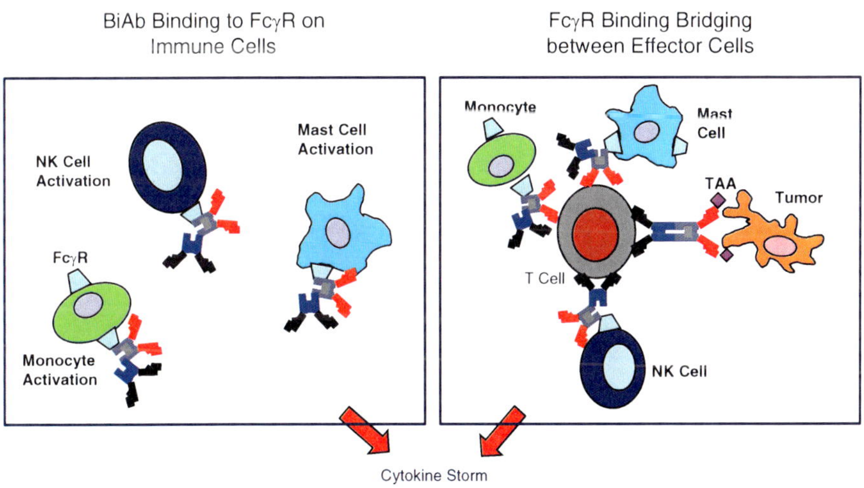

Fig. 14.2 Mechanism leading to cytokine storm. *Left Panel* shows how infusions of BiAbs alone can bind to FcγR on NK, mast, and monocytes and induce the release of cytokines/chemokines leading to the cytokine storm. *Right Panel* shows how the interaction between armed ATC can occur via available Fc portions of the BiAb on ATC

target NK cells and neutrophils whereas monocytes can be redirected with anti-CD64 (anti-FCγRI) to target TAA.

FcγRIII (CD16) × anti-HER2/*neu* (2B1) created from a quadroma targets NK cells to lyse HER2/*neu* + tumors. In a phase I clinical trial involving 15 patients, the non-dose limiting toxicities were fever, chills, nausea, vomiting, and leucopenia. The major DLT was thrombocytopenia at the 5 mg/m^2/dose. Unfortunately, only minor clinical responses were obtained. Intense cytokine storm was induced by cross linking of FcγRIII receptors on circulating leukocytes (Weiner et al. 1995). Treatment induced a 100-fold increase in circulating levels of TNFα, IL-2, and IL-8; with a slight increase in GM-CSF and IFNγ. The maximum tolerated dose (MTD) was 2.5 mg/m^2 and 14 of 15 patients developed human antibody to mouse antibody (HAMA) responses.

In a phase I/II trial using HRS-3/A9 (FcγRIII × anti-CD30) to treat patients with refractory Hodgkin's disease, there were encouraging clinical results with mild fevers as a side effect (Hartmann et al. 1997). In 15 patients, there were 1 complete response (CR), 1 partial response (PR), 3 minor responses, and 1 mixed response. The MTD was not reached at a dose of 64 mg/m^2. In a subsequent study involving 16 patients, there was 1 CR, 3 SD, and 3 PR. The most clinical experience was obtained with anti-CD64 (FCγRI) × anti-Her-2/*neu* (MDX-H210). MDX-H210 consists of chemically linked bsFab′ fragments of a humanized anti-CD64 Fab′ and a murine anti-HER2/*neu* Fab′. This second generation construct was engineered by deleting the Fc domains in an effort to decrease cytokine storm. In a phase I trial, patients with refractory breast and ovarian cancer were given intravenous MDX-H210 doses ranging from 0.35 to 18 mg/m^2 without DLTs (Valone et al. 1995). In the phase I trial, patients developed grade I and II "flu-like" symptoms, grade I/II chest pain, dyspnea, increased serum creatinine, transient grade IV thrombocytopenia, and grade III hypotension (Valone et al. 1995). One partial and one mixed tumor response was observed in ten patients. In a subsequent phase II study in prostate cancer patients, MDX-H210 was well tolerated at doses between 1 and 8 mg/m^2 with side effects of fever, chills, myalgias, and fatigue (Schwaab et al. 2001). No DLTs were observed. In five of six patients, prostate-specific antigen (PSA) levels over the 40-day course of therapy were stable while circulating HER-2/*neu* levels decreased by 80% at days 12 and 29 ($p = 0.03$ and 0.06).

In an encouraging phase II study, 25 patients with HER2/*neu*+ advanced prostate cancer were treated with MDX-H210 (5 μg/kg/day for 4 days/week every 6 weeks) and GM-CSF. In the 20 evaluable patients, 35% (7/20) had >50% decrease in their PSA levels with a median duration of 128 days (71–184). The PSA relative velocity decreased in 83% (15/18) of the evaluable patients (James et al. 2001). Pain scores decreased in 58% (7/12) of the patients who were evaluable for pain. Therapy was stopped in two patients who developed heart failure, dyspnea, and an allergic reaction and therapy was continued in nine who developed grade III events. There were no treatment-related deaths.

In a multi-dose trial, MDX-H210 (given day 4 weekly) and GM-CSF (250 μg/m^2 on days 1–4) were given for four consecutive weeks. In 13 patients treated at doses from 1 to 20 mg/m^2, there were no DLTs. Fever and chills developed up to 2 h post-infusion were associated with peak serum levels of TNF-α (median 88.2 pg/ml) and IL-6 (median 371 pg/ml). The patients became tachyphylactic to the infusions with decreased side effects by the fourth week. Five of 11 patients developed human anti-BiAb (HABA) levels that were 200–500× above baseline after 3 weeks of treatment. Ten patients completed treatment resulting in one patient having a 48% reduction in index lesions, six patients with SD, and three patients with PD before the fourth week (Posey et al. 1999).

MDX-447 (anti-CD64 × anti-EGFR) was tested in a phase I study in 64 patients to target renal cell carcinoma or head and neck cancer. MDX-447 was produced by cross linking F(ab′) fragments of H22 directed at FcγR and H425 directed at EGFR. Groups of three to six patients received MDX-447 (dose levels ranging from 1 to 40 mg/m^2/week) alone or in combination with G-CSF (3 μg/kg/day SC on days −3 to 1, days 4–8, and days 11–15). Hypotension was the DLT with grade III toxicities including fever, hypertension, arrhythmia, allergic reaction, dyspnea, and tumor pain (Fury et al. 2008).

A construct consisting of conjugated F(ab′) fragments (anti-CD64 × anti-CD30 [H22 × Ki-4]) was given IV to ten patients with refractory Hodgkin's disease in a phase I dose escalation study. Doses ranging from 1 to 20 mg/m^2/day on days 1, 5, and 7 were given with transient and mild side effects that included hypotension (4 of 10), tachycardia (6 of 10), fatigue (10 of 10), and fever (2 of 10). There was one patient with a CR, three patients with PR, and four patients with SD (Borchmann et al. 2002).

Recent studies using single chain scFv-based bispecific T-cell engager (BiTE) format (Micromet) have produced promising clinical results. BiTE antibodies have induced potent lysis of target antigen-expressing cells at pico- to femto-molar concentrations while associated with conditional T cell activation (Baeuerle et al. 2003; Brischwein et al. 2007). Details of this approach and clinical results are in Chap. 15 (Baeuerle). Phase I/II clinical trials are being performed with MT103 (anti-CD19 × anti-CD3, blintumomab) and MT110 (anti-EpCAM × anti-CD3) BiTE antibodies (Dreier et al. 2002, 2003; Loffler et al. 2000). Clinical results with MT103 from ongoing phase 1/2 studies strongly suggest that T cells can be engaged and redirected to lyse tumors by MT103 (Bargou et al. 2008). In 38 non-Hodgkin's lymphoma (NHL) patients with follicular lymphoma, mantle cell lymphoma, and CLL who received MT103 at doses from 0.0005 to 0.06 mg/m^2 per day, 11 patients had objective clinical responses. Four had a CR and seven had PR at doses of 0.015 mg/m^2 per day and higher. Seven of these patients who received 0.06 mg/m^2 per day had objective responses. MT103 at doses of 0.015 mg/m^2 per day and higher eliminated tumor cells in the blood, lymph nodes, spleen, and bone marrow. In 9 of 11 cases with bone marrow involvement, immunohistochemical staining and flow cytometry show complete (6/11) or partial (3/11) elimination of tumor cells (Bargou et al. 2008). In an ongoing phase II trial in patients with precursor B cell acute lymphoblastic leukemia (B-ALL), MT103 eradicated rare tumor cells in bone marrow that can only be detected by sensitive quantitative PCR methods (Topp 2008). These data suggest that MT103 recruits and redirects T cells against both bulky and minimal residual hematological malignancy (Baeuerle and Reinhardt 2009).

Another promising approach is the use of trifunctional antibodies (TriAb) to improve the clinical effectiveness of BiAb (Zeidler et al. 1999, 2000). Chapter 16 by Horst Lindhofer provides specific details on TriAbs. A phase I/II study using TriAb Removab® (anti-CD3 × anti-EpCAM, catumaxomab, Trion Pharma) was done by giving intraperitoneal injections of Removab® to 23 patients with malignant ascites from pretreated refractory ovarian cancer resulting in a 5 log reduction of EpCAM positive tumor cells in the ascites fluid with acceptable safety profiles (Burges et al. 2007). In 2009, Removab® received European Union approval for IP treatment of ascites in patients with EpCAM-positive carcinomas. Removab® is not only the first drug approved for malignant ascites, but also the first BiAb approved for clinical use.

Kiewe et al. (2006) reported a recent phase I trial in women with metastatic breast cancer using the TriFab ertumaxomab (anti-CD3 × anti-Her2/*neu* designed with a Fcγ type I/III receptor, ertumaxomab, Fresenius Biotech GmbH). Ertumaxomab establishes a tri-cell complex between T cells, Fc-receptor positive cells, and tumor cells (Kiewe et al. 2006) (Fig. 14.2). Fifteen of 17 patients completed the study with a MTD of 100 μg with mild and transient side effects with objective tumor responses in 5 out of 15 patients (1 CR, 2 PR, and 2 SD). There was a Th_1 cytokine pattern in the serum of patients.

In a phase I trial using Removab® for non-small cell lung cancer, Removab® was infused intravenously in combination with dexamethasone to reduce cytokine side effects. The MTD was 5 μg catumaxomab given after 50 mg of dexamethasone (Sebastian et al. 2007). The key observation is that steroids can be given to prevent DLT and still maintain clinical responses.

A potentially highly innovative and novel future application of Removab® would be to target T cells to solid tumors in patients undergoing allogeneic stem cell transplant [US Patent 2000601154810 entitled "Treating tumor Growth and metastasis by using TriFab antibodies to reduce the risk for GVHD in allogeneic anti-tumor cell therapy"]. If Removab® could induce a specific anti-tumor effect after allogeneic stem cell transplantation while preventing the development of acute graft-vs-host disease, it would be an extraordinary clinical advance for the treatment of solid tumors and hematologic malignancies.

In the past 15 years, there have been significant advances in the design and application of BiAbs for intravenous and local injections for the treatment of tumors. The toxicity and clinical efficacy profiles have progressed from BiAbs that caused cytokine storm limiting their clinical effectiveness to the more recent constructs that have improved clinical efficacy without inducing DLTs. The lessons learned from the series of BiAb trials can be summarized as follows: (1) design of the anti-effector activating/engaging portion (anti-CD3 or anti-FcγRIII or anti-FcγRI) is critical to the activating function of the BiAb; (2) deletion or modification of the Fc portions may decrease the side effects of cytokine storm; and (3) the concomitant use with immunomodulating drugs or chemotherapy may alter or improve the tolerability of BiAb infusions.

14.1.5 Clinical Trials Using Activated T Cells or Effector Cells Armed with BiAbs

The first clinical trial using ex vivo armed autologous LAK was reported in 1990 (Nitta et al. 1990a). LAK were co-injected with chemically heteroconjugated anti-CD3 × anti-glioma BiAb into ten patients. Both armed and unarmed LAK were given locally into the brain tumors. Of the ten patients who received armed LAK, four patients had tumor regression and four patients had improved overall survival (OS). OS and progression-free survival (PFS) appeared better than that seen for patients who received unarmed LAK.

In a subsequent phase I trial for glioma patients, anti-CD3 activated T cells (ATC) were co-injected into the surgical cavity of glioma patients in combination with two BiAbs (anti-CD3 × anti-EGFR and anti-CD28 × anti-EGFR) via an Ommaya reservoir. Two of ten patients with clinical responses received 70 and 250×10^6 ATC. It is quite remarkable that such a small number of cells led to clinical responses with only transient fever, nausea, headache, and aggravation of pre-existing neurologic deficits (Jung et al. 2001).

In 1992, 5×10^9 phytohemagglutinin (PHA)- or anti-CD3-activated T cells armed with anti-CD3 × anti-Mov28 (ovarian carcinoma-associated antigen) were injected intraperitoneally to treat five patients with ovarian carcinoma (Lamers et al. 1992). T cells were cultured for 14 and 16 days in media contaning 78% RPMI 1640, 20% AIM-V, and 2% human plasma. Both CD4 and CD8 T cells expanded, but the predominant population was CD8. Immunomodulation and anti-tumor effects were also noted in patients who received locally injected anti-CD3 × anti- folate receptor (OC/TR) armed T cells (Lamers et al. 1997).

In a phase II study in 28 advanced ovarian carcinoma patients who had undergone debulking laparatomies, intraperitoneal injections of ATC armed with OC/TR induced regressions of advanced ovarian carcinoma (Canevari et al. 1995). The patients were treated with two cycles of five daily IP doses consisting of OC/TR armed ATC and 0.6×10^6 IU of IL-2. Transient side effects from the IL-2 were mild to moderate. ATC were armed with 1 mg/10^9 BiAb (~1,000 ng/10^6 cells). There were no differences between the 17 patients who received armed T cells after removal of excess OC/TR and the 11 patients who received additional daily infusions of OC/TR and IL-2. The T cells were cultured in polyolefin gas permeable bags in RPMI, AIM-V, and a serum mix containing 600 IU of IL-2/ml with >50-fold expansion of lymphocytes over 2–3 weeks of culture. Ten patients received between 4 and 9×10^9 armed ATC (target dose was 10^{10}). Despite poor prognostic features, 27% (7 of 26) of the patients showed complete (4) or partial (3) responses. The most frequent side effects were mild to moderate fever, nausea, and emesis. HAMA responses developed in 84% (21 of 25) of patients between 4 and 5 weeks after therapy. Multiple metastatic lesions that were >2 cm in diameter disappeared after therapy.

In a pilot phase I dose escalation clinical trial in patients with squamous cell carcinoma of the head and neck, PBMC were collected by leukopheresis, incubated with Removab® for 24 h, washed free of the IFN-γ and TNF-α cytokines that were released after arming, and infused into four patients (Riechelmann et al. 2007). This strategy decreased the side effects of cytokine storm.

14.1.6 Evidence for In Situ Vaccination Using Armed Targeted T Cells

In our own preclinical and clinical studies, ATC armed with chemically heteroconjugated anti-CD3 $\times$ anti-HER2/*neu* BiAb (OKT3 $\times$ Herceptin®) [Her2Bi] was used. In clinical trials, our overall approach was to use polyclonal expansion of T cells (up to 320 $\times$ 10^9) and arming the ATC with BiAb to create a large population of TAA-specific CTL. Our preclinical studies show that Her2Bi armed ATC (1) bind to tumor cells and kill Her2/*neu* breast (Sen et al. 2001), prostate (Lum et al. 2005), and pancreatic (Lum et al. 2002) cancer cell lines; (2) secrete IFNγ, TNFα, GM-CSF, RANTES, and MIP-1α (Grabert et al. 2003); (3) can prevent the development of tumors in SCID/Beige mice or induce remissions when directly injected into established prostate PC-3 xenografts (Davol et al. 2004a); (4) traffic to PC-3 tumors in SCID/Beige mice and inhibit tumor growth (Davol et al. 2004a); and (5) kill tumor cells multiple times and proliferate after tumor cell killing without undergoing apoptosis (Grabert et al. 2003).

Our phase I clinical trials involve multiple infusions of Her2Bi-armed ATC for the treatment of patients with high risk stage II–III breast cancer ($\geq$4 positive nodes), stage IV breast cancer, stage IV hormone refractory prostate cancer, and locally advanced pancreatic cancer. ATC are expanded for 14 days in culture containing RPMI 1640 supplemented with 2% human serum and 100 IU of IL-2/ml. At the end of the culture, the ATC are harvested, armed with 50 ng Her2Bi/10^6 ATC, washed, aliquoted, and cryopreserved.

Our phase I dose escalation trials revealed no DLT at total dose levels ranging from 80 to 160 $\times$ 10^9 per patient. The patients received two infusions per week for 4 weeks for a total of eight infusions. The patients were also given low-dose IL-2 (300,000 IU/m^2 daily) and low-dose GM-CSF (250 μg/m^2 twice per week) starting 3 days prior to the first infusion and ending 1 week after the last infusion. Infusions up to 320 $\times$ 10^9 armed ATC were well tolerated and the toxicity profile, which included chills, fever, hypotension, and fatigue were not dose limiting. The chills and fever were easily managed with prophylactic antihistamines and antipyretics. The hypotension was managed with vigorous hydration. None of the patients developed dose-limiting allergic, cardiac, pulmonary, renal, or GI toxicities. A total of 19 women with metastatic breast cancer, 9 women with an adjuvant protocol for women with high risk ($\geq$4 positive nodes) breast cancer, 6 men with hormone refractory prostate cancer, and 1 patient with metastatic pancreatic cancer received Her2Bi-armed ATC. One patient died of cardiac toxicity mostly likely due to toxic digoxin levels and not due to immunotherapy. Decreased pain levels occurred in women with breast cancer and men with prostate cancer (Davol et al. 2004b). ATC bearing Her2Bi could be detected by flow cytometry up to several weeks after infusion (Lum et al. 2003). There were high circulating serum levels of Th_1 cytokines (GM-CSF, IFN-γ, TNF-α, IL-2), IL-12, RANTES, and MIP-1α and low levels of Th_2 cytokines (IL-10, IL-4). Targeting and lysis of tumor cells with local release of cytokines and chemokines may, in turn, recruit naive T cells and antigen-presenting cells to induce a local vaccine response developing into a robust systemic response (Fig. 14.1).

In a phase I clinical trial, up to 80 $\times$ 10^9 anti-CD3 $\times$ anti-C20 (OKT3 $\times$ Rituxan®, CD20Bi) armed autologous ATC were infused in two different infusion schedules after high-dose chemotherapy and peripheral blood stem cell transplantation. Both schedules were well tolerated and there were no DLTs (Lum et al. 2010). Infusions of 20 $\times$ 10^9 armed ATC as early as 4 days after peripheral blood stem cell transplant were well tolerated.

14.1.7 Co-activation and Targeting T Cell Strategies

Preclinical strategies that combine co-stimulation with anti-CD3/anti-CD28 and arming with BiAb have been reported using a combination of BiAbs. Co-stimulation with anti-CD28 alone or anti-CD28 $\times$ anti-TAA results in enhanced signaling (Alvarez-Vallina and Hawkins 1996; Renner et al. 1995), cytokine production (Mazzoni et al. 1996), enhanced cytotoxicity in lymphoma and leukemia models (Demanet et al. 1996; Renner et al. 1994; Bohlen et al. 1993a), and cytotoxicity of colon carcinoma cell lines (Hombach et al. 1997a). It has been reported that BiAb and CD28 co-stimulation induced Th_1 differentiation (Hombach et al. 1997a).

14.1.8 T Cells Expressing Chimeric Antibody Receptors to Target TAA

Transducing ATC with a gene for the expression of a chimeric single chain Fv domain receptor on their surface creates a "T body" or T cells with chimeric antibody receptors (CARs). Similar to ATC armed with BiAbs, this approach to deliver a lethal hit to tumor cells bypasses HLA restrictions for targeting tumor antigens (Fitzer-Attas and Eshhar 1998; Eshhar et al. 1993; Hwu et al. 1993, 1995; Altenschmidt et al. 1997). However, it remains a challenge to transduce enough

T cells with enough potency, under appropriate conditions to mount maximal anti-tumor effects. Recent studies clearly show that lentiviral vectors have been more effective than the earlier retroviral vectors in increasing the number of T cells that would express CARs. Without reviewing T-bodies, the construction and function of most CARs can be summarized to include heavy and light chain derived variable regions connected by peptide linkers and activating signaling chains that consist of a gamma chain and CD28 signaling chains (Sadelain et al. 2003). Early constructs involved a CAR containing an scFv directed at HER2/*neu* and the human zeta signaling chain where engagement of the anti-HER2 receptor with HER2/*neu* induced IL-2 secretion and specific lysis (Moritz et al. 1994). T-bodies homed to tumors and inhibited the growth of HER2/*neu* transformed tumors in athymic nude mice (Moritz et al. 1994). Studies were done using ovarian cancer TIL transduced with a T body receptor directed at a folate binding protein on most ovarian carcinomas with a FcR signaling chain (Hwu et al. 1993, 1995). Unfortunately, despite promising preclinical data, the phase I clinical trial did not induce clinical responses (Kershaw et al. 2006). Another strategy was the transduction of hematopoietic stem cells so that T cells and other progeny would express CARs after high-dose chemotherapy and stem cell infusions (Wang et al. 1998). T body immunotherapy remains a challenge because gene-transduced T cells are known to down-regulate expression of their transgenes (Plavec et al. 1997; Quinn et al. 1998). This obstacle has been recently overcome by adding a CD28 signaling chain to the gamma chain or zeta chain. Recent strategies using such a construct delivered by vectors include anti-CD19 (Cooper et al. 2005), anti-CEA (Emtage et al. 2008), and anti-PSMA (Ma et al. 2004). Preclinical studies were also done to provide proof of principle that IL-1β gene transduced T cells would secrete IL-1β. The transduced T cells armed with anti-CD3 × anti-carcinoembryonic antigen (CEA) would secrete high levels of IL-1β as a result of engagement of CEA on the colon carcinoma cell line LS174T (Trevor et al. 2000).

14.1.9 Why Use BiAb Armed T Cells?

The major advantage of this approach is that large numbers of T cells can be produced from a single leukopheresis product by polyclonal expansion in 2 weeks and that arming of the ATC can be done with any BiAb that contains an anti-CD3 component. Different T cell subsets may be selected based on the desired function. The key versatility is that any TAA can be targeted as long as the anti-TAA monoclonal antibody is available and the BiAb and ATC are produced under cGMP conditions. Although armed ATC kill repeatedly in vitro, long-term in vivo expansion of the infused armed ATC may not continue to mediate cytotoxicity because the BiAb would be diluted on the cell surface by cell division or shed from the T cells. In contrast, smaller numbers of T bodies could be stimulated to proliferate and expand in vivo to establish a long memory pool of antigen-specific effector memory T cells without the need for multiple infusions. It is relatively inexpensive to produce ATC and arm them with cGMP grade BiAbs. This simple approach may override the need for producing large expensive batches of vectors that need to be qualified for the transduction of T cells with the T body constructs, which may be oncogenic.

14.1.9.1 Immunologic Space

A number of recent studies show that the depletion of Treg cells or creating immunologic space using cyclophosphamide leads to improved anti-tumor activity (Wang et al. 2005; Gattinoni et al. 2005; Grinshtein et al. 2009; Salem 2009). Other studies clearly indicate that timing and dose of the vaccine can be critical to the success of the vaccine. Vaccines have been successful before or after chemotherapy in contrast to the original paradigm that suggested immune recovery must occur before vaccination. Although it is counterintuitive, it may be theorized that specifically differentiated T effector memory cells and antigen-specific B cells may be relatively resistant to chemotherapy and will be the first to expand into the space created by chemotherapy or lymphodepletion. In a preclinical murine model, administration of purified activated CD4+ cells inhibited tumor growth or prolonged disease-free survival when the cells were infused after creating immunologic space with cyclophosphamide (Saxton et al. 1997). A clinical study using this strategy showed that purified anti-CD3 activated CD4+ cells induced remissions in patients with solid tumors using doses of cyclophosphamide ranging from 500 to 1,000/m (Asano et al. 2010; Curti et al. 1993). The injection of cyclophosphamide 4 days prior to cell infusions appeared to be critical to the success of the immunotherapy.

14.1.9.2 Binding Sites and Affinity

All anti-CD3 mAbs are not equivalent for inducing a variety of T cell functions. Anti-CD3 engagement of the TCR is a critical element in BiAbs that target T cells while simultaneously restimulating T cell functions. The affinity, isotype, targeted epitope on CD3, and arming concentration may impact on the ability of anti-CD3 mAbs to signal and induce proliferation, cytokine synthesis, and cytotoxicity. Decreasing valency from two binding sites to a single binding site can markedly decrease the ability of the anti-CD3 based BiAb to induce specific T cell functions. In a recombinant BiAb (E3Bi) directed at the EpCAM antigen on tumor cells and CD3 on T cells, the E3Bi produced from clones of the scFv of OKT3 and scFv of GA733.2 connected via a 63 amino acid linker (CD8a Ig hinge-like domain) enhanced specific cytotoxicity of ATC directed at EpCAM tumor targets when E3Bi was added to co-cultures (Ren-Heidenreich et al. 2004), but E3Bi could not be detected on ATC by flow cytometry. On the other hand, increasing the affinity for the TAA can enhance cytotoxicity (McCall et al. 2001). In a series of anti-HER2/*neu* extracellular domain-directed scFvs with affinities ranging from 10^{-11} to 10^{-7}, increasing affinity resulted in increasing cytotoxicity at the same E/T (McCall et al. 2001).

14.1.9.3 Types of Activation, Coactivation, Time in Culture, and T Cell Subsets

The time in culture after activation of T cells does not affect the ability of BiAbs to bind to the effector ATC. Co-activation of resting T cells with anti-CD28 in combination with anti-CD3 × anti-CA19-9 induced a Th_1 cytokine pattern consisting of high amounts of IL-2 and IFNγ, no IL-4, and low amounts of IL-10 (Hombach et al. 1997a). Highly specific CTL were obtained 4–14 days after activation and expansion in low doses of IL-2 (ranging from 100 to 500 IU/ml). In ATC, purified CD4 and CD8 cells were cytotoxic with CD8 cells being more cytotoxic than T cells and CD4 cells being less cytotoxic than T cells (Sen et al. 2001). Others have demonstrated that CD8+CD45RO+ cells exhibit the highest levels of cytotoxicity and those levels correlate with high levels of mRNA for perforin and granzyme B (Renner et al. 1997a, b). Unactivated PBMC can be redirected by arming with BiAb to mediate lower and less consistent levels of cytotoxicity (Sen et al. 2001).

14.1.9.4 Arming of Ex Vivo Activated T Cells vs. In Vivo Arming of T Cells

In vitro activated and unactivated circulating T cells can be armed with BiAbs to target non-MHC restricted cytotoxicity to tumor targets (Haagen et al. 1994). Fresh T cells can be armed with BiAb, leading to proliferation and cytotoxicity (Haagen et al. 1994). The vast majority of investigators choose this strategy to induce anti-tumor cytotoxicity. The approach we have taken is to avoid in vivo activation of a very large number of T cells by using ex vivo arming of ATC prior to infusions (Sen et al. 2001).

Strategies that use ex vivo expanded T cells differ substantially from approaches that infuse BiAb. Considerably more BiAb (micrograms to milligrams per kg body weight) is infused into patients when the BiAbs are infused directly as opposed to arming (ng/million) ex vivo expanded ATC. The disadvantage of infusing BiAb is that the BiAb will bind to all circulating T cells, tumor targets, and Fc-receptor bearing cells shortly after infusion leading to activation of all cells binding enough BiAb to trigger activation. The cytokine storm observed in BiAb infusions are due in large measure to the binding and activation of large numbers of endogenous T cells as well as FcγR-related activation on Fc-bearing cells. The bridging and aggregation of effector cells and tumor cells will be dependent upon the affinity of each arm of the BiAb constructs. The presence of free BiAbs that bind to tumors prior to the arrival of armed T cells may block TAA binding sites on the tumor leading to diminished tumor cell lysis.

Another consideration for using armed effector cells is that the immobilized BiAb on the surface of T cells, NK cells, monocytes, or neutrophils may resist clearance from circulation better than circulating free BiAbs. BiAbs composed of single chains, diabodies, single chain diabodies, minibodies, leucine zippered, or knob into holes are cleared more rapidly than effector cells (Cao and Lam 2003). Smaller molecules have better tumor penetration but are cleared rapidly, whereas antibody molecules (large or small) immobilized on the surface of cells are more likely to persist. The concentration of BiAb on the T cell surface will be a function of the arming BiAb dose and the affinity of the anti-CD3 partner. Other important factors such as the optimal arming concentration of the BiAb and the behavior of the BiAb to re-activate T cells upon tumor engagement are critical. Overloading ATC with BiAb may trigger or induce AICD. It goes without saying that arming normal and patient ATC may markedly differ and, therefore, before clinical trials are performed with a specific BiAb construct, it would be important to perform in vitro testing as well as in vivo preclinical animal testing to characterize the armed T cell product.

14.2 Summary

The use of BiAbs for cancer therapy shows great promise. As the understanding of interactions between cancer stem cells and their immunologic environment improves, various types of customized BiAb constructs for optimal targeting will be engineered for in vivo or ex vivo retargeting of T cells to various tumors. The critical variables that need to be considered for successful manipulation of the immune system include the type and state of activation of the effector cells, the affinity of the anti-effector cell and the anti-TAA construct, the presence and type of regulator or suppressor cells, the type of BiAb (species, number of binding sites), the presence of decoy antigens, whether the TAA modulates after being engaged by antibody, the amount and type of tumor, and the "immunologic" state of the patient prior to the adoptive transfer of T cells.

Acknowledgement We thank the physicians, immunotherapy technical staff, clinical coordinators, nursing staff, administrative staff, and the leadership of the Barbara Ann Karmanos Cancer Institute for their efforts, and the support that have made it possible to perform the preclinical research, develop our clinical trials, and treat patients with cancer. Special thanks for the immunotherapy coordinators Wendy Young, Annette Olson, Lori Hall, Patricia Steele, and Karen Myers for their compassionate efforts to serve the immunotherapy patients. This investigation was supported in part by grants R01 CA 92344 and R01 CA 140314 from the NCI, DHHS, Translational Grants #6092-09 and #6066-06 from the Leukemia and Lymphoma Society, Susan G. Komen Foundation BCTR0707125, and Cancer Center Support Grant P30 CA022453-25, and startup funds from the Karmanos Cancer Institute. We thank Dr. Abhinav Deol and Vidya Kondadasula for their careful reading of this chapter.

References

Altenschmidt U, Moritz D, Groner B (1997) Specific cytotoxic T lymphocytes in gene therapy. J Mol Med 75:259–266

Alvarez-Vallina L, Hawkins RE (1996) Antigen-specific targeting of CD28-mediated T cell co-stimulation using chimeric single-chain antibody variable fragment-CD28 receptors. Eur J Immunol 26:2304–2309

Anderson PM, Bach FH, Ochoa AC (1988) Augmentation of cell number and LAK activity in peripheral blood mononuclear cells activated with anti-CD3 and interleukin-2. Preliminary results in children with acute lymphocytic leukemia and neuroblastoma. Cancer Immunol Immunother 27:82–88

Anderson PM, Crist W, Hasz D et al (1992) G19.4(αCD3) x B43(αCD19) monoclonal antibody heteroconjugate triggers CD19 antigen-specific lysis of t(4;11) acute lymphoblastic leukemia cells by activated CD3 antigen- positive cytotoxic T cells. Blood 80:2826–2834

Asano R, Watanabe Y, Kawaguchi H et al (2007) Highly effective recombinant format of a humanized IgG-like bispecific antibody for cancer immunotherapy with retargeting of lymphocytes to tumor cells. J Biol Chem 282:27659–27665

Asano R, Ikoma K, Kawaguchi H et al (2010) Application of the Fc fusion format to generate tag-free bi-specific diabodies. FEBS J 277:477–487

Baeuerle PA, Reinhardt C (2009) Bispecific T-cell engaging antibodies for cancer therapy. Cancer Res 69:4941–4944

Baeuerle PA, Kufer P, Lutterbuse R (2003) Bispecific antibodies for polyclonal T-cell engagement. Curr Opin Mol Ther 5:413–419

Bargou R, Leo E, Zugmaier G et al (2008) Tumor regression in cancer patients by very low doses of a T cell-engaging antibody. Science 321:974–977

Bejeck BE, Wang D, Berven E et al (1995) Development and characterization of three recombinant single chain antibody fragments (scFvs) directed against the CD19 antigen. Cancer Res 55:2346–2351

Blades RA, Keating PJ, McWilliam LJ, George NJ, Stern PL (1995) Loss of HLA class I expression in prostate cancer: implications for immunotherapy. Urology 46:681–686

Blay JY, Negrier S, Combaret V et al (1992) Serum level of interleukin 6 as a prognosis factor in metastatic renal cell carcinoma. Cancer Res 52:3317–3322

Bohlen H, Hopff T, Manzke O et al (1993a) Lysis of malignant B cells from patients with B-chronic lymphocytic leukemia by autologous T cells activated with CD3 x CD19 bispecific antibodies in combination with bivalent CD28 antibodies. Blood 82:1803–1812

Bohlen H, Manzke O, Patel B et al (1993b) Cytolysis of leukemic B-cells by T-cells activated via two bispecific antibodies. Cancer Res 43:4310–4314

Bolhuis RL, Lamers CH, Goey SH et al (1992) Adoptive immunotherapy of ovarian carcinoma with bs-MAb-targeted lymphocytes: a multicenter study. Int J Cancer Suppl 7:78–81

Borchmann P, Schnell R, Fuss I et al (2002) Phase 1 trial of the novel bispecific molecule H22xKi-4 in patients with refractory Hodgkin lymphoma. Blood 100:3101–3107

Borghaei H, Alpaugh RK, Bernardo P et al (2007) Induction of adaptive Anti-HER2/neu immune responses in a Phase 1B/2 trial of 2B1 bispecific murine monoclonal antibody in metastatic breast cancer (E3194): a trial coordinated by the Eastern Cooperative Oncology Group. J Immunother 30:455–467

Brischwein K, Parr L, Pflanz S et al (2007) Strictly target cell-dependent activation of T cells by bispecific single-chain antibody constructs of the BiTE class. J Immunother 30:798–807

Brossart P, Stuhler G, Flad T et al (1998) Her-2/neu-derived peptides are tumor-associated antigens expressed by human renal cell and colon carcinoma lines and are recognized by in vitro induced specific cytotoxic T lymphocytes. Cancer Res 58:732--736

Browning M, Petronzelli F, Bicknell D et al (1996) Mechanisms of loss of HLA class I expression on colorectal tumor cells. Tissue Antigens 47:364–371

Burges A, Wimberger P, Kumper C et al (2007) Effective relief of malignant ascites in patients with advanced ovarian cancer by a trifunctional anti-EpCAM x anti-CD3 antibody: a phase I/II study. Clin Cancer Res 13:3899–3905

Canevari S, Stoter G, Arienti F et al (1995) Regression of advanced ovarian carcinoma by intraperitoneal treatment with autologous T lymphocytes retargeted by a bispecific monoclonal antibody. J Natl Cancer Inst 87:1463–1469

Cao Y, Lam L (2003) Bispecific antibody conjugates in therapeutics. Adv Drug Deliv Rev 55:171–197

Chan JK, Hamilton CA, Cheung MK et al (2006) Enhanced killing of primary ovarian cancer by retargeting autologous cytokine-induced killer cells with bispecific antibodies: a preclinical study. Clin Cancer Res 12:1859–1867

Chapoval AI, Nelson H, Thibault C (1995) Anti-CD3 x anti-tumor $F(ab')_2$ bifunctional antibody activates and retargets tumor-infiltrating lymphocytes. J Immunol 155:1296–1303

Chen J, Bashey A, Holman P et al (1999) A phase I dose escalating study of infusion of a bispecific antibody (BsAb) for relapsed/refractory acute myeloid leukemia (AML). Blood 94:227B

ClinicalTrials.gov identifier: NC00204594. 2010a

ClinicalTrials.gov identifier: NCT00189345. 2010b

ClinicalTrials.gov identifier: NCT00274742. 2010c

ClinicalTrials.gov identifier: NCT00452140. 2010d

ClinicalTrials.gov identifier: NCT00521261. 2010e

ClinicalTrials.gov identifier: NCT00560794. 2010f

ClinicalTrials.gov identifier: NCT01022138. 2010g

Cohen SB, Katsikis PD, Feldmann M, Londei M (1994) IL-10 enhances expression of the IL-2 receptor alpha chain on T cells. Immunology 83:329–332

Compte M, Blanco B, Serrano F et al (2007) Inhibition of tumor growth in vivo by in situ secretion of bispecific anti-CEA x anti-CD3 diabodies from lentivirally transduced human lymphocytes. Cancer Gene Ther 14:380–388

Concha A, Cabrera T, Ruiz-Cabello F, Garrido F (1991) Can the HLA phenotype be used as a prognostic factor in breast carcinomas? Int J Cancer Suppl 6:146–154

Cooper LJ, Al-Kadhimi Z, Serrano LM et al (2005) Enhanced antilymphoma efficacy of CD19-redirected influenza MP1-specific CTLs by cotransfer of T cells modified to present influenza MP1. Blood 105:1622–1631

Curti BD, Longo DL, Ochoa AC et al (1993) Treatment of cancer patients with *ex vivo* anti-CD3-activated killer cells and interleukin-2. J Clin Oncol 11:652–660

Daniel PT, Kroidl A, Cayeux S et al (1997) Costimulatory Signals Through B7.1/CD28 Prevent T Cell Apoptosis During Target Cell Lysis. J Immunol 159:3808–3815

Davico BL, De Monte LB, Spagnoli GC et al (1995) Bispecific monoclonal antibody anti-CD3 x anti-tenascin: an immunotherapeutic agent for human glioma. Int J Cancer 61:509–515

Davol PA, Smith JA, Kouttab N, Elfenbein GJ, Lum LG (2004a) Anti-CD3 x Anti-HER2 bispecific antibody effectively redirects armed T cells to inhibit tumor development and growth in hormone-refractory prostate cancer-bearing SCID-Beige mice. Clin Prostate Cancer 3:112–121

Davol PA, Gall JM, Grabert RC et al (2004b) Infusions of T cells armed with anti-CD3 x anti-Her2/neu bispecific antibody modulate in vivo patient immune responses in phase I clinical trials for breast and hormone refractory prostate cancers [abstract]. Blood 104:379a

de Gast GC, van Houten AA, Haagen IA et al (1995a) Clinical experience with CD3 x CD19 bispecific antibodies in patients with B cell malignancies. J Hematother 4:433–437

de Gast GC, Haagen I-A, van Houten AA et al (1995b) CD8 T cell activation after intravenous administration of CD3 x CD19 bispecific antibody in patients with non-Hodgkin lymphoma. Cancer Immunol Immunother 40:390–396

Demanet C, Brissinck J, De Jong J, Thielemans K (1996) Bispecific antibody-mediated immunotherapy of the BCL_1 lymphoma: increased efficacy with multiple injections and CD28-induced costimulation. Blood 87:4390–4398

Deol A, Lum LG (2010) Role of donor lymphocyte infusions in relapsed hematological malignancies after stem cell transplantation revisited. Cancer Treat Rev 36:528–538

Dreier T, Lorenczewski G, Brandl C et al (2002) Extremely potent, rapid and costimulation-independent cytotoxic T-cell response against lymphoma cells catalyzed by a single-chain bispecific antibody. Int J Cancer 100:690–697

Dreier T, Baeuerle PA, Fichtner I et al (2003) T cell costimulus-independent and very efficacious inhibition of tumor growth in mice bearing subcutaneous or leukemic human B cell lymphoma xenografts by a CD19-/CD3- bispecific single-chain antibody construct. J Immunol 170: 4397–4402

Dudley ME, Wunderlich JR, Robbins PF et al (2002) Cancer regression and autoimmunity in patients after clonal repopulation with antitumor lymphocytes. Science 298:850–854

Emtage PC, Lo AS, Gomes EM et al (2008) Second-generation anti-carcinoembryonic antigen designer T cells resist activation-induced cell death, proliferate on tumor contact, secrete cytokines, and exhibit superior antitumor activity in vivo: a preclinical evaluation. Clin Cancer Res 14:8112–8122

Eshhar Z, Waks T, Gross G, Schindler DG (1993) Specific activation and targeting of cytotoxic lymphocytes through chimeric single chains consisting of antibody-binding domains and the gamma or zeta subunits of the immunoglobulin and T-cell receptors. Proc Natl Acad Sci USA 90:720–724

Ferrone S, Marincola FM (1995) Loss of HLA class I antigens by melanoma cells: molecular mechanisms, functional significance and clinical relevance. Immunol Today 16:487–494

Fitzer-Attas CJ, Eshhar Z (1998) Tyrosine kinase chimeras for antigen-selective T-body therapy. Adv Drug Deliv Rev 31:171–182

Fowler DH, Odom J, Steinberg SM et al (2006) Phase I clinical trial of costimulated, IL-4 polarized donor CD4+ T cells as augmentation of allogeneic hematopoietic cell transplantation. Biol Blood Marrow Transplant 12:1150–1160

Fury MG, Lipton A, Smith KM, Winston CB, Pfister DG (2008) A phase-I trial of the epidermal growth factor receptor directed bispecific antibody MDX-447 without and with recombinant human granulocyte-colony stimulating factor in patients with advanced solid tumors. Cancer Immunol Immunother 57:155–163

Gall JM, Davol PA, Grabert RC, Deaver M, Lum LG (2005) T cells armed with anti-CD3 x anti-CD20 bispecific antibody enhance killing of CD20+ malignant B cells and bypass complement-mediated rituximab resistance in vitro. Exp Hematol 33:452–459

Garlie NK, LeFever AV, Siebenlist RE et al (1999) T cells co-activated with immobilized anti-CD3 and anti-CD28 as potential immunotherapy for cancer. J Immunother 4:335–345

Gattinoni L, Finkelstein SE, Klebanoff CA et al (2005) Removal of homeostatic cytokine sinks by lymphodepletion enhances the efficacy of adoptively transferred tumor-specific CD8+ T cells. J Exp Med 202:907–912

Grabert RC, Smith J, Tiggs J, Kouttab N, Lum LG (2003) Anti-CD3 activated T cells armed with OKT3 x Herceptin bispecific antibody, survive and divide, and secrete cytokines and chemokines after multiple cycles of killing directed at Her2/neu+ tumor targets [abstract]. American Association for Cancer Research 44:656a

Grabert RC, Cousens LP, Smith JA et al (2006) Human T cells armed with Her2/neu bispecific antibodies divide, are cytotoxic, and secrete cytokines with repeated stimulation. Clin Cancer Res 12:569–576

Grimm EA, Mazumder A, Zhang HZ, Rosenberg SA (1982) Lymphokine-activated killer cell phenomenon. Lysis of natural killer-resistant fresh solid tumor cells by interleukin 2-activated autologous human peripheral blood lymphocytes. J Exp Med 155:1823–1841

Grinshtein N, Ventresca M, Margl R et al (2009) High-dose chemotherapy augments the efficacy of recombinant adenovirus vaccines and improves the therapeutic outcome. Cancer Gene Ther 16:338–350

Haagen IA, de Lau WB, Bast BJ et al (1994) Unprimed CD4+ and CD8+ T cells can be rapidly activated by a CD3 x CD19 bispecific antibody to proliferate and become cytotoxic. Cancer Immunol Immunother 39:391–396

Hartmann F, Renner C, Jung W et al (1997) Treatment of refractory Hodgkin's disease with an anti-CD16/Cd30 bispecific antibody [published erratum appears in Blood 1998, 91:1832]. Blood 89:2042–2047

Hombach A, Tillmann T, Jensen M et al (1997a) Specific activation of resting T cells against tumour cells by bispecific antibodies and CD28-mediated costimulation is accompanied by Th1 differentiation and recruitment of MHC-independent cytotoxicity. Clin Exp Immunol 108:352–357

Hombach A, Tillmann T, Jensen M et al (1997b) Specific activation of resting T cells against CA19-9+ tumor cells by an anti-CD3/CA19-9 bispecific antibody in combination with a costimulatory anti-CD28 antibody. J Immunother 20:325–333

Hwu P, Shafer GE, Treisman JS et al (1993) Lysis of ovarian cancer cells by human lymphocytes redirected with a chimeric gene composed of an antibody variable region and the Fc receptor gamma chain. J Exp Med 178:361–366

Hwu P, Yang JC, Cowherd R et al (1995) In vivo antitumor activity of T cells redirected with chimeric antibody/T-cell receptor genes. Cancer Res 55:3369–3373

James ND, Atherton PJ, Jones J et al (2001) A phase II study of the bispecific antibody MDX-H210 (anti-HER2 x CD64) with GM-CSF in HER2+ advanced prostate cancer. Br J Cancer 85:152–156

Jost CR, Titus JA, Kurucz I, Segal DM (1996) A single-chain bispecific Fv2 molecule produced in mammalian cells redirects lysis by activated CTL. Mol Immunol 33:211–219

Jung G, Brandl M, Eisner W et al (2001) Local immunotherapy of glioma patients with a combination of 2 bispecific antibody fragments and resting autologous lymphocytes: evidence for in situ T-cell activation and therapeutic efficacy. Int J Cancer 91:225–230

Kaneko T, Fusauchi Y, Kakui Y et al (1993) A bispecific antibody enhances cytokine-induced killer-mediated cytolysis of autologous acute myeloid leukemia cells. Blood 81:1333–1341

Karas M, Zaks TZ, Yakar S, Dudley ME, LeRoith D (2001) TCR stimulation protects CD8+ T cells from CD95 mediated apoptosis. Human Immunol 62:32–38

Katayose Y, Kudo T, Suzuki M et al (1996) MUC1-specific targeting immunotherapy with bispecific antibodies: inhibition of xenografted human bile duct carcinoma growth. Cancer Res 56:4205–4212

Katzenwadel A, Schleer H, Gierschner D, Wetterauer U, Elsasser-Beile U (2000) Construction and in vivo evaluation of an anti-PSA x anti-CD3 bispecific antibody for the immunotherapy of prostate cancer. Anticancer Res 20:1551–1555

Kershaw MH, Westwood JA, Parker LL et al (2006) A phase I study on adoptive immunotherapy using gene-modified T cells for ovarian cancer. Clin Cancer Res 12:6106–6115

Kiewe P, Hasmuller S, Kahlert S et al (2006) Phase I trial of the trifunctional anti-HER2 x anti-CD3 antibody ertumaxomab in metastatic breast cancer. Clin Cancer Res 12:3085–3091

Klein SC, Boer LH, de Weger RA, de Gast GC, Bast EJEG (1997) Release of cytokines and soluble cell surface molecules by PBMC after activation with the bispecific antibody CD3 x CD19. Scand J Immunol 46:452–458

Kolb HJ, Holler E (1997) Adoptive immunotherapy with donor lymphocyte transfusions [Review]. Curr Opin Oncol 9:139–145

Kolb HJ, Mittermuller J, Clemm C et al (1990) Donor leukocyte transfusions for treatment of recurrent chronic myelogenous leukemia in marrow transplant patients. Blood 76:2462–2465

Kostelny SA, Link BK, Tso JY et al (2001) Humanization and characterization of the anti-HLA-DR antibody 1D10. Int J Cancer 93:556–565

Kosterink JG, de Jonge MW, Smit EF et al (1995) Pharmacokinetics and scintigraphy of indium-111-DTPA-MOC-31 in small-cell lung carcinoma. J Nucl Med 36:2356–2362

Kroesen BJ, ter Haar A, Willemse P et al (1993) Local antitumor treatment in carcinoma patients with bispecific-monoclonal-antibody-redirected T cells. Cancer Immunol Immunother 37:401–407

Kroesen BJ, Nieken J, Sleijfer DT et al (1997) Approaches to lung cancer treatment using the CD3 x EGP-2-directed bispecific monoclonal antibody BIS-1. Cancer Immunol Immunother 45:203–206

Kuwahara M, Kuroki M, Arakawa F et al (1996) A mouse/human-chimeric bispecific antibody reactive with human carcinoembryonic antigen-expressing cells and human T-lymphocytes. Anticancer Res 16:2661–2667

Kuwahara M, Kuroki M, Arakawa F et al (1997) A mouse/human-chimeric bispecific antibody reactive with human carcinoembryonic antigen-expressing cells and human T-lymphocytes. Anticancer Res 16:2661–2668

Lamers CHJ, van de Griend RJ, Braakman E et al (1992) Optimization of culture conditions for activation and large-scale expansion of human T lymphocytes for bispecific antibody-directed cellular immunotherapy. Int J Cancer 51:973–979

Lamers CH, Bolhuis RL, Warnaar SO, Stoter G, Gratama JW (1997) Local but no systemic immunomodulation by intraperitoneal treatment of advanced ovarian cancer with autologous T lymphocytes re-targeted by a bi-specific monoclonal antibody. Int J Cancer 73:211–219

Liu Z, Savoldo B, Huls H et al (2002) Epstein-Barr virus (EBV)-specific cytotoxic T lymphocytes for the prevention and treatment of EBV-associated post-transplant lymphomas. Recent Results Cancer Res 159:123–133

Loffler A, Kufer P, Lutterbuse R et al (2000) A recombinant bispecific single-chain antibody, CD19 x CD3, induces rapid and high lymphoma-directed cytotoxicity by unstimulated T lymphocytes. Blood 95:2098–2103

Luiten RM, Coney LR, Fleuren GJ, Warnaar SO, Litvinov SV (1996) Generation of chimeric bispecific G250/anti-CD3 monoclonal antibody, a tool to combat renal cell carcinoma. Br J Cancer 74:735–744

Lum LG, LeFever AV, Treisman JS, Garlie NK, Hanson JP Jr (2001) Immune modulation in cancer patients after adoptive transfer of anti-CD3/anti-CD28-costimulated T cells-phase I clinical trial. J Immunother 24:408–419

Lum LG, Davol P, Grabert R et al. (2002) Targeting pancreatic cancer with armed activated T cells directed at Her2/neu receptors [abstract]. Exp Hematol Suppl 30:56

Lum LG, Rathore R, Colvin GA et al. (2003) Targeting HER2/neu tumor cells with anti-CD3 activated T cells: clinical trials and trafficking studies [abstract]. ASCO: meeting proceedings 22, pp 179

Lum HE, Miller M, Davol PA et al (2005) Preclinical studies comparing different bispecific antibodies for redirecting T cell cytotoxicity to extracellular antigens on prostate carcinomas. Anticancer Res 25:43–52

Lum LG, Thakur A, Al-Kadhimi Z, Abidi M, Ayash L, Cummings F, Rathore R, Uberti JP, Ratanatharathorn V (2010) Phase I Dose Esclation of Activated T Cells (ATC) Armed with Anti-CD3 x Anti-CD20 Bispecific Antibody (CD20Bi) after Stem Cell Transplant (SCT) In Non-Hodgkin's Lymphoma (NHL). The American Society of Hematology (ASH) December 4–7, 2010, Orlando, FL, Abstract # 488. 2010

Ma Q, Safar M, Holmes E et al (2004) Anti-prostate specific membrane antigen designer T cells for prostate cancer therapy. Prostate 61:12–25

Mazzoni A, Mezzanzanica D, Jung G et al (1996) CD3-CD28 costimulation as a means of avoiding T cell preactivation in bispecific monoclonal antibody-based treatment of ovarian carcinoma. Cancer Res 56:5443–5449

McCall AM, Shahied L, Amoroso AR et al (2001) Increasing the affinity for tumor antigen enhances bispecific antibody cytotoxicity. J Immunol 166:6112–6117

Moritz D, Wels W, Mattern J, Groner B (1994) Cytotoxic T lymphocytes with a grafted recognition specificity for ERBB2-expressing tumor cells. Proc Natl Acad Sci USA 91:4318–4322

Mullbacher A, Lobigs M, Tha Hla R et al (2002) Antigen-dependent release of IFN-gamma by cytotoxic T Cells up-regulates Fas on target cells and facilitates exocytosis-independent specific target cells lysis. J Immunol 169:145–150

Negri DR, Tosi E, Valota O et al (1995) In vitro and in vivo stability and anti-tumour efficacy of an anti- EGFR/anti-CD3 F(ab')2 bispecific monoclonal antibody. Br J Cancer 72:928–933

Nitta T, Sato K, Yagita H, Okumura K, Ishii S (1990a) Preliminary trial of specific targeting therapy against malignant glioma. Lancet 335:368–371

Nitta T, Sato K, Okumura K, Ishii S (1990b) Induction of cytotoxicity in human T cells coated with anti-glioma x anti-CD3 bispecific antibody against human glioma cells. J Neurosurg 72: 476–481

O'Connell J, O'Sullivan GC, Collins JK, Shanahan F (1996) The Fas Counterattack:Fas-Mediated T cell killing by colon cancer cells expressing Fas ligand. J Exp Med 184:1075–1082

Perez P, Hoffman RW, Shaw S (1985) Specific targeting of cytotoxic T cells by anti-T3 linked to anti-target cell antibody. Nature 316:354–356

Perez P, Titus JA, Lotze MT et al (1986) Specific lysis of human tumor cells by T cells coated with anti-T3 cross-linked to anti-tumor antibody. J Immunol 137:2069–2072

Plavec I, Agarwal M, Ho KE et al (1997) High transdominant RevM10 protein levels are required to inhibit HIV-1 replication in cell lines and primary T cells: implication for gene therapy. Gene Ther 4:128–139

Pohl C, Denfeld R, Renner C et al (1993) CD30-antigen-specific targeting and activation of T cells via murine bispecific monoclonal antibodies against CD3 and CD28: potential use for the treatment of Hodgkin's lymphoma. Int J Cancer 54:820–827

Posey JA, Raspet R, Verma U et al (1999) A pilot trial of GM-CSF and MDX-H210 in patients with erbB-2-positive advanced malignancies. J Immunother 22:371–379

Quinn ER, Lum LG, Trevor KT (1998) T cell activation modulates retrovirus-mediated gene expression. Hum Gene Ther 9:1457–1467

Raso V, Griffin T (1981) Hybrid antibody with dual specificity for the delivery of ricin to immunoglobin bearing target cells. Cancer Res 41:2073–2078

Redondo M, Concha A, Oldiviela R et al (1991) Expression of HLA class I and II antigens in bronchogenic carcinomas: its relationship to cellular DNA content and clinical-pathological parameters. Cancer Res 51:4948–4954

Ren-Heidenreich L, Davol PA, Kouttab NM, Elfenbein GJ, Lum LG (2004) Redirected T-cell cytotoxicity to EpCAM over-expressing adenocarcinomas by a novel recombinant antibody, E3Bi, in vitro and in an animal model. Cancer 100:1095–1103, Ref Type: Journal (Full)

Renner C, Jung W, Sahin U et al (1994) Cure of xenografted human tumors by bispecific monoclonal antibodies and human T cells. Science 264:833–835

Renner C, Jung W, Sahin U, van Lier R, Pfreundschuh M (1995) The role of lymphocyte subsets and adhesion molecules in T cell-dependent cytotoxicity mediated by CD3 and CD28 bispecific monoclonal antibodies. Eur J Immunol 25:2027–2033

Renner C, Bauer S, Sahin U et al (1996) Cure of disseminated xenografted human Hodgkin's tumors by bispecifc monoclonal antibodies and human T cells: the role of human T-cell subsets in a preclinical model. Blood 87:2930–2937

Renner C, Held G, Ohnesorge S et al (1997a) Role of perforin, granzymes and the proliferative state of the target cells in apoptosis and necrosis mediated by bispecific-antibody-activated cytotoxic T cells. Cancer Immunol Immunother 44:70–76

Renner C, Held G, Ohnesorge S et al (1997b) Role of naive and memory T cells in tumor cell lysis mediated by bi-specific antibodies. Immunobiology 197:122–132

Reusch U, Sundaram M, Davol PA et al (2006) Anti-CD3 x anti-EGFR bispecific antibody redirects T cell cytolytic activity to EGFR-positive cancers in vitro and in an animal model. Clin Cancer Res 12:183–190

Riechelmann H, Wiesneth M, Schauwecker P et al (2007) Adoptive therapy of head and neck squamous cell carcinoma with antibody coated immune cells: a pilot clinical trial. Cancer Immunol Immunother 56:1397–1406

Riesenberg R, Buchner A, Pohla H, Lindhofer H (2001) Lysis of prostate carcinoma cells by trifunctional bispecific antibodies (alpha EpCAM x alpha CD3). J Histochem Cytochem 49:911–917

Rivoltini L, Barracchini KC, Viggiano V et al (1995) Quantitative correlation between HLA class I allele expression and recognition of melanoma cells by antigen-specific cytotoxic T lymphocytes. Cancer Res 55:3149–3157

Rooney CM, Smith CA, Ng CY et al (1998) Infusion of cytotoxic T cells for the prevention and treatment of Epstein-Barr virus-induced lymphoma in allogeneic transplant recipients. Blood 92:1549–1555

Rosenberg SA, Spiess P, Lafreniere R (1986) A new approach to the adoptive immunotherapy of cancer with tumor-infiltrating lymphocytes. Science 233:1318–1321

Rosenberg SA, Packard BS, Aebersold PM et al (1988) Use of tumor-infiltrating lymphocytes and interleukin-2 in the immunotherapy of patients with metastatic melanoma. A preliminary report. N Engl J Med 319:1676–1680

Rossi EA, Goldenberg DM, Cardillo TM et al (2006) Stably tethered multifunctional structures of defined composition made by the dock and lock method for use in cancer targeting. Proc Natl Acad Sci USA 103:6841–6846

Sadelain M, Riviere I, Brentjens R (2003) Targeting tumours with genetically enhanced T lymphocytes. Nat Rev Cancer 3:35–45

Salem ML, Az-Montero CM, Al-Khami AA et al (2009) Recovery from cyclophosphamide-induced lymphopenia results in expansion of immature dendritic cells which can mediate enhanced prime-boost vaccination antitumor responses in vivo when stimulated with the TLR3 agonist poly(I:C). J Immunol 182:2030–2040

Saxton ML, Longo DL, Wetzel HE et al (1997) Adoptive transfer of anti-CD3-activated CD4$^+$ T cells plus cyclophosphamide and liposome-encapsulated interleukin-2 cure murine MC-38 and 3LL tumors and establish tumor-specific immunity. Blood 89:2529–2536

Schwaab T, Lewis LD, Cole BF et al (2001) Phase I pilot trial of the bispecific antibody MDXH210 (anti-Fc gamma RI X anti-HER-2/neu) in patients whose prostate cancer overexpresses HER-2/neu. J Immunother 24:79–87

Sebastian M, Passlick B, Friccius-Quecke H et al (2007) Treatment of non-small cell lung cancer patients with the trifunctional monoclonal antibody catumaxomab (anti-EpCAM x anti-CD3): a phase I study. Cancer Immunol Immunother 56:1637–1644

Segal DM, Garrido MA, Perez P et al (1988) Targeted cytotoxic cells as a novel form of cancer immunotherapy. Mol Immunol 25:1099–1103

Segal DM, Weiner GJ, Weiner LM (2001) Introduction: bispecific antibodies. J Immunol Methods 248:1–6

Sen M, Wankowski DM, Garlie NK et al (2001) Use of anti-CD3 x anti-HER2/neu bispecific antibody for redirecting cytotoxicity of activated T cells toward HER2/neu Tumors. Journal of Hematotherapy & Stem Cell Research 10:247–260

Shalaby MR, Shepard HM, Presta L et al (1992) Development of humanized bispecific antibodies reactive with cytotoxic lymphocytes and tumor cells overexpressing the *HER2* protooncogene. J Exp Med 175:217–225

Shalaby MR, Carter P, Maneval D, Giltinan D, Kotts C (1995) Bispecific Her2 x CD3 antibodies enhance T-cell cytotoxicity *in vitro* and localize to Her2-overexpressing xenograpfts in nude mice. Clin Immunol Immunopathol 74:185–192

Smyth MJ, Strobl SL, Young HA, Ortaldo JR, Ochoa AC (1991) Regulation of lymphokine-activated killer activity and pore-forming protein gene expression in human peripheral blood CD8+ T lymphocytes. Inhibition by transforming growth factor-β. J Immunol 146:3289–3297

Tartour E, Blay JY, Dorval T et al (1996) Predictors of clinical response to interleukin-2-based immunotherapy in melanoma patients: a French multiinstitutional study. J Clin Oncol 14: 1697–1703

Titus JA, Perez P, Kaubisch A, Garrido MA, Segal DM (1987) Human K/natural killer cells targeted with hetero-cross-linked antibodies specifically lyse tumor cells in vitro and prevent tumor growth in vivo. J Immunol 139:3153–3158

Topp MS (2008) Treatment with anti-CD19 BiTE antibody blinatumomab (MT103/MEDI-538) is able to eliminate minimal residual disease (MRD) in patients with B-Precursor acute lymphoblastic leukemia (ALL): first results of an ongoing phase II study. Blood 112:1926

Torres MJ, Ruiz-Cabello F, Skoudy A et al (1996) Loss of an HLA haplotype in pancreas cancer tissue and its corresponding tumor derived cell line. Tissue Antigens 47:372–381

Trevor KT, Quinn ER, Sen M et al (2000) Bispecific antibody reactivation of gene-transduced T cells: implications for cancer immunotherapy and gene therapy. Tumor Targeting 4:245–256

Uberti JP, Joshi I, Ueda M et al (1994) Preclinical studies using immobilized OKT3 to activate human T cells for adoptive immunotherapy: optimal conditions for the proliferation and induction of non-MHC restricted cytotoxicity. Clin Immunol Immunopathol 70:234–240

Ueda M, Joshi ID, Dan M et al (1993) Preclinical studies for adoptive immunotherapy in bone marrow transplantation: II. Generation of anti-CD3 activated cytotoxic T cells from normal donors and autologous bone marrow transplant candidates. Transplantation 56:351–356

Valone FH, Kaufman PA, Guyre PM et al (1995) Phase Ia/Ib trial of bispecific antibody MDX-210 in patients with advanced breast or ovarian cancer that overexpresses the proto-oncogene HER-2/neu. J Clin Oncol 13:2281–2292

Walker PR, Saas P, Dietrich PY (1997) Role of Fas ligand (CD95L) in immune escape: the tumor cell strikes back. J Immunol 158:4521–4524

Wang G, Chopra RK, Royal RE et al (1998) A T cell-dependent antitumor response in mice with bone marrow cells retrovirally transduced with an antibody/Fc-gamma chain chimeric receptor gene recognizing a human ovarian cancer antigen. Nat Med 4:168–172

Wang LX, Li R, Yang G et al (2005) Interleukin-7-dependent expansion and persistence of melanoma-specific T cells in lymphodepleted mice lead to tumor regression and editing. Cancer Res 65:10569–10577

Weiner LM, Clark JI, Davey M et al (1995) Phase I trial of 2B1, a bispecific monoclonal antibody targeting c-erbB-2 and Fc gamma RIII. Cancer Res 55:4586–4593

Whiteside TL (1999) Signaling defects in T lymphocytes of patients with malignancy. Cancer Immunol Immunother 48:346–352

Zaks TZ, Chappell DB, Rosenberg SA, Restifo NP (1999) Fas-mediated suicide of tumor-reactive T cells following activation by specific tumor: selective rescue by caspase inhibition. J Immunol 162:3273–3279

Zeidler R, Reisbach G, Wollenberg B et al (1999) Simultaneous activation of T cells and accessory cells by a new class of intact bispecific antibody results in efficient tumor cell killing. J Immunol 163:1246–1252

Zeidler R, Mysliwietz J, Csanady M et al (2000) The Fc-region of a new class of intact bispecific antibody mediates activation of accessory cells and NK cells and induces direct phagocytosis of tumour cells. Br J Cancer 83:261–266

Zeytun A, Hassuneh M, Nagarkatti M, Nagarkatti PS (1997) Fas-Fas ligand-based interactions between tumor cells and tumor-specific cytotoxic T lymphocytes: a lethal two-way street. Blood 90:1952–1959

Zhu Z, Ghose T, Lee SH et al (1994) Tumor localization and therapeutic potential of an antitumor-anti-CD3-heteroconjugate antibody in human renal cell carcinoma xenograft models. Cancer Lett 86:127–134

Zhu Z, Lewis GD, Carter P (1995) Engineering high affinity humanized anti-p185^{HER2}/anti-CD3 bispecific F(ab')$_2$ for efficient lysis of p185^{HER2} overexpressing tumor cells. Int J Cancer 62:319–324

Chapter 15 Bispecific T Cell Engager for Cancer Therapy

Patrick A. Baeuerle, Gerhard Zugmaier, and Dominik Rüttinger

15.1 Introduction

Treatment with monoclonal antibodies of the human IgG1 and IgG2 isotypes have become a mainstay in cancer therapy, particularly, in hematological malignancies (Ball and Broome 2010). Engagement of Fcγ receptor-bearing immune effector cells, complement fixation, induction of apoptosis, and neutralization of growth factors are major modes of action of conventional monoclonal antibodies. A limited efficacy of such antibodies in oncology as monotherapy has prompted intense efforts in academia and industry to improve their clinical activity. One approach is defucosylation or the introduction of point mutations in the Fcγ1 domain of antibodies, which significantly improves antibody-dependent cellular cytotoxicity (ADCC) (Kubota et al. 2009). Several new antibodies with improved Fcγ receptor-binding properties are now in early-stage clinical development. Another technology to boost antibody efficacy is coupling of toxic payloads, such as chemotherapeutics or bacterial toxins (Alley et al. 2010). While the US market approval for the anti-CD33/calicheamycin conjugate Mylotarg®, the first marketed drug of this kind, has been retracted in 2010, a novel generation of antibody drug conjugates (ADC) is well advanced in development using new technologies for more stably conjugating payloads. For instance, an anti-CD30 for treatment of Hodgkin's lymphoma called SGN-35, and an anti-Her-2/neu ADC called T-DM1 for treatment of metastatic breast cancer are in pivotal trials and have shown high response rates in patients exceeding those of the parental unconjugated antibodies.

A yet different approach for increasing antibody efficacy is bispecific antibodies capable of engaging cytotoxic T cells (Müller and Kontermann 2010). T cells lack

P.A. Baeuerle (✉) • G. Zugmaier • D. Rüttinger
Micromet, Inc., 9201 Corporate Boulevard, Suite 400, Rockville, MD 20850, USA
and
Micromet AG, Staffelseestr. 2, 81477 Munich, Germany
e-mail: Patrick.baeuerle@micromet.com

R.E. Kontermann (ed.), *Bispecific Antibodies*, 273
DOI 10.1007/978-3-642-20910-9_15,

antibody-binding Fcγ receptor which is why they cannot be engaged by regular IgG1 monoclonal antibodies. Combining in one antibody molecule the binding specificity for a tumor-associated surface antigen with that for a T cell receptor component holds the promise of not only temporarily connecting a cytotoxic T cell with a tumor cell but of activating the inherent killing potential most T cells have. If this is properly achieved by a bispecific antibody, the T cell will adopt a serial lysis mode and is induced to proliferate at the site where it has been activated by the bispecific antibody.

Many parameters apparently need to be optimized in order to achieve a proper T cell engagement by a bispecific antibody. These relate to (a) the selection of an appropriate surface antigen on cancer cells that is frequently and highly expressed but ideally absent or inaccessible on normal cells; and (b) the selection of an appropriate triggering molecule on the surface of T cells that potently activates T cells but only when a target cell is bound to the T cell. (c) Quite a number of structural requirements have to be met by a bispecific antibody including but not limited to fine-tuned binding affinities, selection of appropriate binding epitopes on the two antigens it is binding to, and the optimal distance and arrangement of the two distinct binding arms. (d) Finally, bispecific antibodies have to fulfil the high standards for a pharmaceutical drug candidate such as, for example, good productivity, high stability and homogeneity, high specificity, well-defined mechanism of action, and features allowing for its toxicological assessment in relevant animal species.

15.2 Diverse Approaches for T-Cell Engagement

T cells seem to be uniquely suited to fight cancer cells. A large number of nonclinical studies have shown that induction of T cell responses in animal models can lead to tumor rejection and long-term protection from recurrence (e.g., Renner et al. 1994; Altenschmidt et al. 1997; Schlereth et al. 2005; Carpenito et al. 2009). Adoptive transfer of tumor-derived T cells to melanoma patients has been effective in treating metastatic disease (Rosenberg et al. 2008), and many cancer vaccine trials have reported partial or complete responses albeit at rather low rates. There is an astounding correlation found between the presence of cytotoxic effector T cells inside tumors and the survival of colorectal (Pagès et al. 2005), ovarian (Zhang et al. 2003), and non-Hodgkin's lymphoma patients (Wahlin et al. 2007). If tumors contain many of such T cells, patients live with high statistical significance longer than those patients whose tumors are not populated by this particular T cell subtype. This indicates that cytotoxic T cells can control survival of cancer patients even if tumors are not eliminated.

We are currently seeing an increasing number of pharmaceuticals entering the market place that work by engaging T cells. Two examples are the prostate cancer vaccine sipuleucel-T (Provenge®), which is inducing a prostatic acidic phosphatase (PAP)-specific T cell response (Morse and Whelan 2010), and ipilimumab, an anti-CTLA-4-blocking human antibody, which can globally boost natural T cell responses against cancer cells (Weber 2009). Another example is an EpCAM/CD3-bispecific trifunctional antibody called catumaxomab (Removab®) (for structure, see Fig. 15.1), which gained market approval in Europe in 2009 for treatment of malignant ascites in cancer patients with EpCAM-expressing tumors (Shen and Zhu 2008; Linke et al. 2010). This antibody not only introduced EpCAM as a novel target antigen for treatment of cancer to the market, but is the first T cell-engaging antibody that achieved registration.

Several other T cell-engaging antibodies are under clinical (Fig. 15.1) and preclinical development (Fig. 15.2). Apart from "bispecific T cell engager" (BiTE), which is the subject of this review article, several bispecific trifunctional antibodies, a bacterial superantigen/Fab fusion protein, and two kinds of chemically crosslinked monoclonal antibodies are currently in clinical development. The bacterial superantigen/anti-5T4 Fab fusion protein called naptumomab estafenatox (Anyara®) has completed a large phase 3 trial in renal cell cancer that is awaiting efficacy assessment (Robinson et al. 2010). By chemically crosslinking commercial antibodies anti-Her-2/neu Herceptin® or anti-CD20 Rituxan® with anti-CD3 murine antibody Orthoclone® (OKT-3), conjugates can be produced for ex-vivo "arming" of T cells followed by their adoptive transfer to patients (Lum et al. 2006). Several phase 1 and 2 studies with this approach are ongoing. In preclinical development are

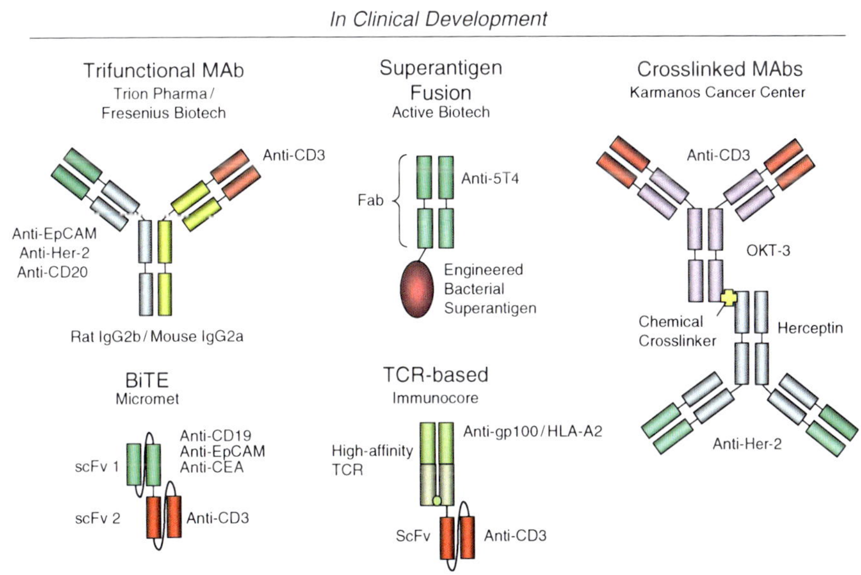

Fig. 15.1 T-cell engaging antibodies under current clinical development. The design of various antibodies and antibody-derived constructs is shown. *Colored boxes* correspond to immunoglobulin domains. *Red*: T-cell engaging part; *green*: target antigen-binding part. Details can be found in references in the text. Companies and academic institutions pursuing the respective approaches are listed. *MAb* monoclonal antibody

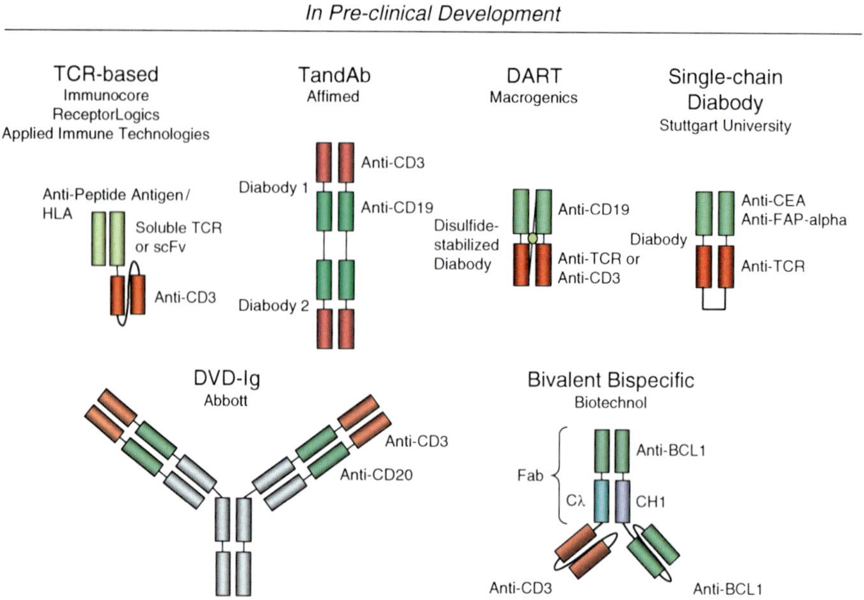

Fig. 15.2 T-cell engaging antibodies under current pre-clinical development. For BiTE antibodies under pre-clinical development, see Table 15.1. The design of various antibodies and antibody-derived constructs is shown. This is only selection. Many more antibodies may be under pre-clinical development but are not disclosed. *Colored boxes* correspond to immunoglobulin domains. *Red*: T-cell engaging part; *green*: target antigen-binding part. Details can be found in references in the text, and in respective web pages of listed companies and institutions. The TCR-based approach uses either soluble single-chain T cell receptors (TCRs; Immunocore, Receptor Logics) or single-chain antibodies with specificity for peptide/MHC I complexes (Applied Immune Technologies). *TandAb* tandem diabody; *DART* dual affinity re-targeting; *DVD* dual variable domain

so-called tandem diabodies (TandAbs) (Reusch et al. 2004), disulfide-stabilized dual affinity re-targeting diabodies (DARTs) (Moore et al. 2010 AACR), single-chain diabodies (Stork et al. 2009), dual variable domain (DVD) antibodies (Wu et al. 2009), and bispecific constructs based on Fab fragments (Willems et al. 2005), each one using a different structural principle for engaging T cells, as depicted in Fig. 15.2. A yet different approach is targeting cancer cells by single-chain T-cell receptors or single-chain antibodies recognizing MHC class I/peptide antigens, both in combination with an anti-CD3 antibody (Fig. 15.2). One such molecule, called IMCgp100 by Immunocore, has begun clinical phase 1 testing in late 2010 (see Fig. 15.1). Like vaccines, this class of bispecific antibodies is still restricted by specific MHC class I complexes presenting tumor-associated peptide antigens. Several companies and academic groups are pursuing these approaches (Denkberg and Reiter 2006).

Very few studies have attempted to side-by-side compare different bispecific antibody formats (Molhoj et al. 2007). A great variety of parameters can obviously influence the activity of bispecific antibodies in cell culture assays including source of donor T cells, T cell pre-conditioning, cell culture conditions, assay duration, effector-to-target ratio, and target cell properties. Likewise, many parameters can impact the activity of bispecific antibodies in mouse xenograft studies. It is therefore very difficult to draw firm conclusions about the potency and other properties of bispecific antibodies solely from comparing published data.

There are two obvious advantages of engaging T cells with bispecific antibodies. One is that any cytotoxic T cell – regardless of its T cell receptor specificity – can be engaged for tumor cell lysis. This obviates the need of a specific T cell response, the generation of which is frequently hampered by a variety of immune escape mechanisms (Rabinovich et al. 2007). The other is that certain kinds of bispecific antibodies allow T cells to recognize complete surface antigens as have been validated by monoclonal antibody therapies in clinical settings.

15.3 Bispecific T Cell Engager

A well-characterized class of bispecific T cell-engaging antibodies are "bispecific T cell engaging" or BiTE antibodies (Baeuerle et al. 2009). Their structure is based on two single-chain antibodies that are covalently linked by a peptide linker. This way, a total of four variable domains encoded by four distinct genes are aligned on a single polypeptide chain encoded by just one gene. BiTE antibodies can use a great variety of tumor-associated surface antigens for engagement of T cells (Table 15.1). Best studied are BiTE antibodies binding to CD19, as expressed by most human B cell malignancies (Dreier et al. 2002, 2003), and to EpCAM (Brischwein et al. 2006a), which is expressed on the surface of many solid tumors (Munz et al. 2009), and has recently been validated by Removab® (Linke et al. 2010), as described above. Pre-clinical studies have shown that BiTE antibodies can also specifically target tumor cells expressing carcinoembryonal antigen (CEA) (Lutterbuese et al. 2009), tyrosine kinase EphA2 (Hammond et al. 2007), epidermal growth factor receptor (EGFR) (Lutterbuese et al. 2010), and a large melanoma-associated proteoglycan called MCSP (Bluemel et al. 2010). Presentations at annual meetings of the American Association for Cancer Research have reported on novel BiTE antibodies that mediate redirected specific lysis of cancer cells expressing CD33, insulin-like growth factor receptor type 1, c-Met, fibroblast-activating protein alpha, prostate stem cell antigen, and Her-2/neu (Kischel et al. 2008; Raum et al. 2010). The investigated target antigens for BiTE antibodies compiled in Table 15.1 allow the conclusion that BiTE antibodies are able to use most if not all surface target antigens that have been pursued by conventional monoclonal antibody therapies. Further BiTE programs are in joint development with three large biopharmaceutical companies, which have not disclosed their respective target antigens.

In the case of the EGFR target, a very recent study has shown that the variable domains of commercial antibodies cetuximab (Erbitux®) and panitumumab (Vectibix®) can be used to generate highly potent BiTE antibodies that lyse tumor cells at very low pg/ml (i.e., sub-picomolar) concentrations using either

Table 15.1 Surface target antigens thus far tested for BiTE antibodies

BiTE target (development partner)	Indication/target tissue	Development status
CD19	B cell malignancies and disorders	Pivotal, phase 2 and phase 1 clinical studies
EpCAM	EpCAM$^+$ solid tumors	Clinical phase 1
CEA (MedImmune)	CEA$^+$ solid tumors	Pre-clinical
N.d. (Bayer Schering Pharma)	Solid tumors	Pre-clinical
N.d. (Sanofi-aventis)	Solid tumors	Pre-clinical
N.d. (Boehringer Ingelheim)	Multiple myeloma	Pre-clinical
EGFR	EGFR$^+$ solid tumors	In-vivo PoC (monkey, mouse)
CD33	AML, CML, MDS	In-vivo PoC (monkey, mouse)
MCSP	Melanoma	In-vivo PoC (monkey, mouse)
EphA2	EphA2$^+$ solid tumors	In-vivo efficacy (mouse)
PSCA	Prostate cancer	In-vitro activity
FAP-alpha	Sarcoma, stromal fibroblasts	In-vitro activity
IGF-1R	IGF-1R$^+$ solid tumors	In-vitro activity
Her-2/neu	Breast and gastric cancer	In-vitro activity
Endoslialin	Neovasculature	In-vitro activity
Carboanhydrase IX	Renal cancer	In-vitro activity
cMet	cMet$^+$ solid tumors	In-vitro activity

Abbrevations used are: *EpCAM* epithelial cell adhesion molecule; *CEA* carcinoembryonal antigen (also called CD66e; CEACAM5); *N.d.* not disclosed; *EGFR* epidermal growth factor receptor (also called HER-1); *MCSP* melanoma-associated chondroitin sulfate proteoglycan (also called HMW-MAA, CSPG4, NG2); *EphA2* ephrin type-A receptor tyrosine kinase; *PSCA* prostate stem cell antigen; *FAP-alpha* fibroblast-activating protease-alpha; *IGF-1R* insulin-like growth factor 1 receptor; *Her-2/neu* epidermal growth factor receptor 2; *cMet* hepatocyte growth factor receptor tyrosine kinase; *PoC* proof of concept. For references, see text

unstimulated or pre-stimulated T cells (Lutterbuese et al. 2010). Unlike the parental monoclonal antibodies, the BiTE antibodies were highly active against cancer cell lines harboring mutations in K-RAS and B-RAF genes. This relates to the fact that BiTE antibodies do not rely for their activity on receptor inhibition as the monoclonal antibodies but do solely work by engagement of T cells via the EGFR surface receptor, a mode of action not affected by activating mutations in downstream signaling proteins K-RAS and B-RAF. Therefore, BiTE antibodies have the potential to overcome limitations of conventional antibodies while still using their well-defined target-binding specificities.

15.4 Properties and Mode of Action of BiTE Antibodies

BiTE antibodies have been studied in great detail for their in-vitro and in-vivo activities, and for their mode of action (Kufer et al. 2004; Baeuerle et al. 2008, 2009; Baeuerle and Reinhardt 2009). Because BiTE characteristics have been regularly reviewed, we will here only briefly summarize their properties. Key features of BiTE antibodies include (a) high potency of redirected target cell lysis, (b) strictly target cell-dependent polyclonal activation of $CD8^+$ and $CD4^+$ T cells, (c) induction of serial lysis mode by T cells, and (d) the induction of T-cell proliferation. The high potency of redirected target cell lysis of most BiTE antibodies in the pico- to femtomolar range may relate to a highly efficient activation of the T-cell receptor (TCR) complex (Brandl et al. 2007; Brischwein et al. 2007a). Activation of only a single digit number of receptors of a specific cytotoxic T cell by MHC I/peptide complexes is sufficient for a full-blown cytotoxic reaction against an appropriate target cell (Purbhoo et al. 2004), showing a high degree of signal amplification. A few BiTE antibodies bound to the surface of a target cell therefore appear as efficient as a few MHC class I/peptide complexes for fully activating a transiently connected T cell. Both stimuli lead to formation of a cytolytic synapse (Offner et al. 2006). Triggering of the TCR by either MHC class I/peptide complexes or an anti-CD3 antibody requires clustering of a finite small number of receptors (Minguet and Schamel 2008), This may explain why single-armed binding of BiTE antibodies to CD3 does not fully activate T cells unless BiTE antibodies are presented to the T cell as an "activation matrix" on the surface of a target cell (Brischwein et al. 2007a).

Once activated by a BiTE antibody, T cells can adopt a serial lysis mode through which they can sequentially eliminate target cells (Hoffmann et al. 2005). This is accompanied by upregulation of granzyme and perforin expression (Haas et al. 2009), as needed for re-loading secretory granules of cytotoxic T cells. Experimentally, serial lysis can be demonstrated by complete target cell lysis at low effector-to-target cell ratios or by following the activity of individual T cells in cell culture using video-assisted microscopy. BiTE stimulation induces proliferation of T cells with one cell cycle occurring approximately every 24 h (Brischwein et al. 2006). This reaction will increase available T cells in target tissue, which may greatly contribute to the velocity of tumor eradication.

The mode of BiTE action involves formation of a cytolytic synapse (Offner et al. 2006), and critically relies on the pore-forming activity of the protein perforin (Haas et al. 2009). When pore formation is blocked by chelation of extracellular calcium, no target cell lysis is observed. Perforin is therefore important for both perforation of the target cell membrane as well as for delivery of granzymes into the cytoplasm of target cells (Kurschus et al. 2008). Among the various granzymes stored in secretory granules of cytotoxic T cells, granzyme B appears most potent in activating pro-caspases and inducing programmed cell death (apoptosis) in target cells (Kurschus and Jenne 2010). A pan-caspase inhibitor can, however, not prevent target cell lysis by BiTE antibodies, indicating that perforin and perhaps other granzymes, which have different protease substrate specificities, are sufficient for lysis. Following synapse formation with BiTE-activated T cells, the cytoplasmic enzyme adenylate kinase is leaking from damaged target cells into the culture medium and, in parallel, target cells show various hallmarks of apoptosis, including pro-caspase activation, cleavage of poly-ADP ribose polymerase and DNA fragmentation (Haas et al. 2009). Therefore, it appears that multiple mechanisms simultaneously contribute to target cell lysis by BiTE-activated T cells such that there is very little opportunity for target cells to evade T cell-mediated damages.

BiTE antibodies can engage various T cell subpopulations for redirected lysis (Kischel et al. 2009). Most effective are so-called effector memory T cells of either CD4 or CD8 phenotype. Effector memory T cells were also found to most strongly expand in peripheral blood of NHL patients in response to BiTE treatment (Bargou et al. 2008). Central memory, effector memory T cells expressing CD45RA, and naïve and regulatory T cells in peripheral blood did not show a comparable expansion. CD19-, EpCAM-, and CEA-specific BiTE antibodies have also been found to engage T cells present in patient-derived mononuclear cell preparations for lysis of autologous cancer cells (Löffler et al. 2003; Wimberger et al. 2003; Osada et al. 2010). These in-vitro experiments also demonstrated that the T cells left after chemotherapy in patients can still lyse cancer cells from the same patient.

15.5 Clinical Experience with BiTE Antibodies

Clinical experience has been obtained in patients with CD19/CD3- and EpCAM/CD3-bispecific BiTE antibodies (Nagorsen et al. 2009a). The CD19/CD3-bispecific BiTE antibody blinatumomab (MT103) has completed a phase 2 clinical trial in patients with minimal residual disease (MRD) of B-lineage acute lymphoblastic leukemia (ALL) and is evaluated in an ongoing phase 1 study in patients with various forms of non-Hodgkin lymphoma (NHL), mostly of the follicular and mantle cell type. The EpCAM/CD3-bispecific antibody MT110 is currently being tested in a dose-escalating phase 1 study in patients with gastrointestinal (colorectal, gastric, and gastro-esophageal junction), lung (adenocarcinoma and small cell), breast, ovarian, or prostate cancer. Extensive immunopharmacological analyses have been performed in all studies to date in order to investigate the influence of BiTE antibodies on T cell counts, subpopulations, distribution, adhesiveness, and activation state. Due to their short serum half life of several hours, BiTE antibodies are being administered by continuous i.v. (c.i.v.) infusion using portable minipumps and port systems. C.i.v. administration has the advantage of achieving very predictable serum levels, which can be maintained for prolonged periods of time. Initial treatment lengths of 4–8 weeks are currently used. Efficacious serum levels of blinatumomab for treatment of ALL and NHL were found in the range of 1–3 ng/ml, corresponding to 18–53 pM.

Blinatumomab has shown astounding objective response rates in patients with both NHL and ALL. At constant dosing with 60 μg/m^2/day, eight out of eight evaluable NHL patients with relapsed or refractory follicular or mantle cell lymphoma showed either partial or complete remission, including response duration of ≥2 years in four patients and ongoing beyond 30 months in the patient with longest follow-up (Goebeler et al. 2010). Anti-tumor activity of blinatumomab was evident in blood, lymph nodes, spleen, liver, and bone marrow of patients with NHL (Bargou et al. 2008). In patients with ALL having persistent or re-appearing tumor cell presence in the bone marrow despite high dose chemotherapy, a dose level of 15 μg blinatumomab/m^2/day was sufficient to elicit in 16 out of 20 evaluable patients (80%) a complete MRD response as determined by PCR-based assays of bone marrow biopsies (Topp et al. 2009). The median relapse free survival has not been reached after 480 days indicative for long-lasting responses of ALL patients to blinatumomab treatment (Bargou et al. 2010).

Objective responses in patients with lymphoma were not seen before a dose level of 15 μg/m^2/day was reached (Bargou et al. 2008). It is interesting to note that clinical activity of the bispecific antibody required serum levels of blinatumomab in patients that closely matched EC_{90} values for redirected lysis in cell culture assays. This close correlation is not evident for conventional monoclonal antibodies where efficacious serum levels in patients (μg/ml) typically exceed in-vitro EC_{90} values (ng/ml) by several orders of magnitude.

Upon infusion of BiTE antibodies, T cells disappear within 1 h from peripheral blood and reappear within 1–2 days. The initial disappearance correlates with an increase in their affinity for soluble ICAM-1, indicating that T cells enhance their adhesiveness to vessel walls in response to BiTE stimulation. Although BiTE antibodies have the potential to activate any T cell by virtue of binding to the invariant CD3 component of the TCR, no overt or sustained release of pro-inflammatory cytokine in serum from activated T cells has been noted (Bargou et al. 2008, 2010). In fact, only transient and minute cytokine peaks are measured at the start of BiTE infusion with interleukin-10 being the most prominent cytokine. This observation is consistent with cell culture experiments where target antigen-expressing cells are needed to fully activate T cells and cause a transient release of cytokines, including IL-10 (Brandl et al. 2007). Many of the adverse events occurring within the first few days following start of BiTE infusion, such as fever, chills, headache, C reactive protein increase, and lymphopenia can be explained by an initial inflammatory response of BiTE-activated T cells.

T cells have the intrinsic property of quickly adapting to an initial stimulation. Despite continued presence of an activating stimulus, T cells will cease to release cytokines while they continue to conduct target cell lysis. This has been verified in mouse models with a murine EpCAM/ murine CD3 specific BiTE antibody, called muS110 (Amann et al. 2008; 2009a, b). The adaptive behavior of T cells would well explain why adverse events upon BiTE infusion cumulate within the first days and drop in number and intensity during continued treatment.

A peculiar adverse event of the CD19-specific BiTE antibody blinatumomab are transient dysfunctions of the central nervous system (CNS), such as apraxia, aphasia/dysarthria, tremor, encephalopathy, cerebellar syndrome, and seizure. These events can occur within the first days of after start of infusion, or upon intra-patient dose escalation, and may therefore result from a first T-cell activation in the CNS. With a low peripheral ratio of B to T cells, a biomarker has been identified in patients with NHL, which seems to identify patients at risk of developing CNS adverse events leading to treatment discontinuation (Nagorsen et al. 2009b). Either patients with a low B:T ratio can be excluded from treatment until their B cell counts have recovered, or they can be treated with a step-dosing regimen. The latter involves an intra-patient dose escalation from a low starting to the target dose for a certain time period that allows for adaptation of T cells even in the presence of low numbers of peripheral target cells (Goebeler et al. 2010). Blinatumomab is now advancing into the clinical development phase aiming at registration/marketing authorization for patients with B-lineage ALL.

The ongoing phase 1 trial with the EpCAM-/CD3-bispecific BiTE antibody MT110 in patients with solid tumors has reached a dose level of 24 μg/day, which translates into a constant serum level of ca. 1 ng/ml (18 pM) (Fiedler et al. 2010). In cell culture experiments, 1 ng/ml MT110 would result in half-maximal lysis of target cancer cells (Brischwein et al. 2006a). Thus far, cytotoxic T-cell infiltration into tumor tissue has been observed in biopsies after therapy, but apart from disease stabilization, no responses according to RECIST criteria were yet apparent. The situation in solid tumors is known to be different from that of blood-bourne tumors in that limited penetration by drugs, immunosuppressive microenvironment, and aberrant vasculature pose extra barriers to treatment. Of note, CNS adverse events have so far not been observed in 28 patients treated with MT110 (Fiedler et al. 2010). A noteworthy biochemical adverse event of MT110 was an initial, sharp increase of liver enzymes at the start of infusion or upon intra-patient dose escalation, which in the majority of patients normalized under continued therapy or after treatment stop. A localized release of cytokines from newly activated T cells in liver tissue is being discussed as a possible cause. From mouse models it is known that TNF-alpha and interferon-gamma released by lectin-activated T cells in the liver is toxic to hepatocytes (Küsters et al. 1996, 1997). Otherwise, from a clinical perspective, MT110 at doses explored to date has shown a very benign safety profile with low-grade pyrexia, nausea, vomiting, diarrhea, and fatigue being observed most frequently.

The CEA/CD3-bispecific BiTE antibody MT111/MEDI-565 has entered clinical phase 1 testing patients with gastrointestinal cancers in the beginning of 2011.

15.6 Outlook

The BiTE technology is being intensely evaluated in ongoing clinical trials and by non-clinical experiments. Apart from two clinical stage BiTE antibodies, a total of three more BiTE antibodies currently are in pre-clinical development, two of which target surface antigens on solid tumors, and one on multiple myeloma cells (see Table 15.1). Development of BiTE antibodies targeting distinct antigens will mitigate risks associated with a particular solid tumor target related, for instance, to therapeutic window, limited expression, or potential loss from cancer cells under therapeutic pressure.

The challenges for blinatumomab are to now optimally prepare the way to registration, and to leverage the large potential it may have in treating human B cell malignancies. In fact, CD19 is expressed on the majority of B cell-derived hematological malignancies with the sole exception of multiple myeloma. Blinatumomab has already shown outstanding efficacy in several indications and a manageable safety profile.

The BiTE platform has matured into a robust technology, which has thus far attracted four biopharmaceutical companies as development partners. Considerable knowhow has been accumulated regarding BiTE lead candidate generation and selection, cell clone generation and process development, safety and efficacy assessment in diverse animal models, bioassay development, and clinical development of BiTE antibodies. Existing monoclonal antibody therapies can be converted into highly active BiTE antibodies (Lutterbuese et al. 2010) and more than a dozen target antigens have to date been shown suitable for potent T cell engagement by BiTE antibodies. It therefore appears that BiTE antibodies can be generated for treatment of a large variety of cancer indications.

References

Alley SC, Okeley NM, Senter PD (2010) Antibody-drug conjugates: targeted drug delivery for cancer. Curr Opin Chem Biol 14(4):529–537

Altenschmidt U, Klundt E, Groner B (1997) Adoptive transfer of in-vitro-targeted, activated lymphocytes results in total tumor regression. J Immunol 159(11):5509–5515

Amann M, Brischwein K, Lutterbuese P, Parr L, Petersen L, Lorenczewski G, Krinner E, Bruckmeier S, Lippold S, Kischel R, Lutterbuese R, Kufer P, Baeuerle PA, Schlereth B (2008) Therapeutic window of MuS110, a single-chain antibody construct bispecific for murine EpCAM and murine CD3. Cancer Res 68(1):143–151

Amann M, Friedrich M, Lutterbuese P, Vieser E, Lorenczewski G, Petersen L, Brischwein K, Kufer P, Kischel R, Baeuerle PA, Schlereth B (2009a) Therapeutic window of an EpCAM/CD3-specific BiTE antibody in mice is determined by a subpopulation of EpCAM-expressing lymphocytes that is absent in humans. Cancer Immunol Immunother 58(1):95–109

Amann M, D'Argouges S, Lorenczewski G, Brischwein K, Kischel R, Lutterbuese R, Mangold S, Rau D, Volkland J, Pflanz S, Raum T, Münz M, Kufer P, Schlereth B, Baeuerle PA, Friedrich M (2009b) Antitumor activity of an EpCAM/CD3-bispecific BiTE antibody during long-term treatment of mice in the absence of T-cell anergy and sustained cytokine release. J Immunother 32(5):452–464

Baeuerle PA, Reinhardt C, Kufer P (2008) BiTE – a new class of antibodies that recruit T cells. Drugs of the Future 33:137–147

Baeuerle PA, Reinhardt C (2009) Bispecific T cell engaging antibodies for cancer therapy. Cancer Res 69(12):4941–4944

Baeuerle PA, Kufer P, Bargou R (2009) BiTE: teaching antibodies to engage T cells. Curr Opin Mol Ther 11(1):22–30

Ball ED, Broome HE (2010) Monoclonal antibodies in the treatment of hemtologic malignancy. Best Pract Res Clin Haematol 23(3):403–416

Bargou R, Leo E, Zugmaier G, Goebeler M, Knop S, Noppeney R, Viardot A, Hess G, Schuler M, Einsele H, Riethmüller G, Brandl C, Wolf A, Kirchinger P, Klappers P, Schmidt M, Klinger M, Reinhardt C, Baeuerle PA, Kufer P (2008) Tumor regression in cancer patients by very low doses of a T cell-engaging antibody. Science 321(5891):974–977

Bargou R, Zugmaier G, Goekbuget N, Neumann S, Horst H-A, Raff T, Brueggemann M, Kneba M, Viardot A, Schmid M, Schaich M, Stelljes M, Goebeler M, Pfeifer H, Ottmann O, Burmeister T, Degenhard E, Schmidt M, Scheele J, Einsele H, Kufer P, Klinger M, Nagorsen D, Hoelzer D, Topp M (2010) Prolonged leukemia free survival following blinatumomab (anti-CD19 BiTE) treatment of patients with minimal residual disease (MRD) of B precursor ALL: updated results of a phase II study. Annual meeting of the European Hematology Association, Barcelona 2010, abstract no 598

Bluemel C, Hausmann S, Fluhr P, Sriskandarajah M, Stallcup WB, Baeuerle PA, Kufer P (2010) Epitope distance to the target cell membrane and antigen size determine the potency of T cell-mediated lysis by BiTE antibodies specific for a large melanoma surface antigen. Cancer Immunol Immunother 59(8):1197–1209

Brandl C, Haas C, d' Argouges S, Fisch T, Kufer P, Brischwein K, Prang N, Bargou R, Suzich J, Baeuerle PA, Hofmeister R (2007) The effect of dexamethasone on polyclonal T cell activation and redirected target cell lysis as induced by a CD19/CD3-bispecific single-chain antibody construct. Cancer Immunol Immunother 56(10):1551–1563

Brischwein K, Schlereth B, Guller B, Steiger C, Wolf A, Lutterbuese R, Offner S, Locher M, Urbig T, Raum T, Kleindienst P, Kimmig R, Fichtner I, Kufer P, da Silva AJ, Baeuerle PA (2006) MT110: a novel bispecific single-chain antibody construct with high efficacy in eradicating solid tumors. Mol Immunol 43(8):1129–1143

Brischwein K, Parr L, Pflanz S, Volkland J, Lumsden J, Klinger M, Locher M, Hammond SA, Kiener P, Kufer P, Schlereth B, Baeuerle PA (2007) Strictly target cell-dependent activation of T cells by bispecific single-chain antibody constructs of the BiTE class. J Immunother 30(8):798–807

Carpenito C, Milone MC, Hassan R, Simonet JC, Lakhal M, Suhoski MM, Varela-Rohena A, Haines KM, Heitjan KM, Albelda SM, Carroll RG, Riley JL, Pastan I, June CH (2009) Control of large, established tumor xenografts with genetically engineered human T cells containing CD28 and CD137 domains. Proc Natl Acad Sci USA 106(9):3360–3365

Denkberg G, Reiter Y (2006) Recombinant antibodies with T-cell receptor-like specificity: novel tools to study MHC class I penetration. Autoimmun Rev 5(4):252–257

Dreier T, Lorenczewski G, Brandl C, Hoffmann P, Syring U, Hanakam F, Kufer P, Riethmuller G, Bargou R, Baeuerle PA (2002) Extremely potent, rapid and costimulation-independent cytotoxic T-cell response against lymphoma cells catalyzed by a single-chain bispecific antibody. Int J Cancer 100(6):690–697

Dreier T, Baeuerle PA, Fichtner I, Grün M, Schlereth B, Lorenczewski G, Kufer P, Lutterbüse R, Riethmüller G, Gjorstrup P, Bargou RC (2003) T cell costimulus-independent and very efficacious inhibition of tumor growth in mice bearing subcutaneous or leukemic human B cell lymphoma xenografts by a CD19-/CD3-bispecific single-chain antibody construct. J Immunol 170(8):4397–4402

Fiedler WM, Ritter B, Seggewiss R, Bokemeyer C, Fettes P, Klinger M, Vieser E, Ruettinger D, Kaubitzsch S, Wolf M (2010) Phase I safety and pharmacology study of the EpCAM/CD3-bispecific BiTE antibody MT110 in patients with metastatic colorectal, gastric, or lung cancer. J Clin Oncol 28:15s, abstract no 2573

Goebeler M, Viardot A, Knop S, Topp M, Sökeler M, Kanz L, Krause S, Mackensen A, Noppeney R, Nagorsen D, Zugmaier G, Klinger M, Schmidt M, Scheele J, Klappers P, Kufer P, Einsele H, Bargou R (2010) Confirmation of safety, efficacy and response duration in non-Hodgkin lymphoma patients treated with 60 μg/m/d of BiTE antibody blinatumomab. Annual meeting of the European Hematology Association, Barcelona 2010, abstract no 559

Haas C, Krinner E, Brischwein K, Hoffmann P, Lutterbüse R, Schlereth B, Kufer P, Baeuerle PA (2009) Mode of cytotoxic action of T cell-engaging BiTE antibody MT110. Immunobiology 214(6):441–453

Hammond SA, Lutterbuese R, Roff S, Lutterbuese P, Schlereth B, Bruckheimer E, Kinch MS, Coats S, Baeuerle PA, Kufer P, Kiener PA (2007) Selective targeting and potent control of tumor growth using an EphA2/CD3-bispecific single-chain antibody construct. Cancer Res 67(8):3927–3935

Hoffmann P, Hofmeister R, Brischwein K, Brandl C, Crommer S, Bargou R, Itin C, Prang N, Baeuerle PA (2005) Serial killing of tumor cells by cytotoxic T cells redirected with a CD19-/CD3-bispecific single-chain antibody construct. Int J Cancer 115(1):98–104

Kischel R, Raum T, Hausmann S, Rau D, Bluemel C, Mangold S, Meier P, Ebert E, Kvesic M, Hoffman P, Baeuerle PA, Kufer P (2008) Characterization in primates of MCSP- and CD33-

specific human BiTE antibodies for treatment of melanoma and AML. Proc Am Assoc Cancer Res 99: Abstract no 2404

Kischel R, Hausmann S, Klinger M, Baeuerle PA, Kufer P (2009) Effector memory T cells make a major contribution to redirected lysis by T cell-engaging BiTE antibody MT110. 100th meeting of the American Association for Cancer Research 2009, abstract no 3252

Kubota T, Niwa R, Satoh M, Akinaga S, Shitara K, hanai N (2009) Engineered therapeutic antibodies with improved effector functions. Cancer Sci 100(9):1566–1572

Kufer P, Lutterbuese R, Baeuerle PA (2004) A revival of bispecific antibodies. Trends Biotechnol 22:238–244

Kurschus FC, Jenne D (2010) Delivery and therapeutic potential of human granzyme B. Immunol Rev 235(1):159–171

Kurschus FC, Fellows E, Stegmann E, Jenne DE (2008) Granzyme B delivery via perforin is restricted by size, but not by heparan sulfate-dependent endocytosis. Proc Natl Acad Sci USA 105(37):13799–13804

Küsters S, Gantner F, Künstle G, Tiegs G (1996) Interferon gamma plays a critical role in T cell-dependent liver injury in mice initiated by concanavalin A. Gastroenterology 111(2):462–471

Küsters S, Tiegs G, Alexopoulo L, Pasparakis M, Douni E, Künstle G, Bluethmann H, Wendel A, Pfizenmaier K, Kollias G, Grell M (1997) In vivo evidence for a functional role of both tumor necrosis factor (TNF) receptors and transmembrane TNF in experimental hepatitis. Eur J Immunol 27(11):2870–2905

Linke R, Klein A, Seimetz D (2010) Catumaxomab: clinical development and future directions. MAbs 2(2):129–136

Löffler A, Gruen M, Wuchter C, Schriever F, Kufer P, Dreier T, Hanakam F, Baeuerle PA, Bommert K, Karawajew L, Dörken B, Bargou RC (2003) Efficient elimination of chronic lymphocytic leukaemia B cells by autologous T cells with a bispecific anti-CD19/anti-CD3 single-chain antibody construct. Leukemia 17(5):900–909

Lum LG, Davol PA, Lee RJ (2006) The new face of bispecific antibodies: targeting cancer and much more. Exp Hematol 34(1):1–6

Lutterbuese R, Raum T, Kischel R, Lutterbuese P, Schlereth B, Schaller E, Mangold S, Rau D, Meier P, Kiener PA, Mulgrew K, Oberst MD, Hammond SA, Baeuerle PA, Kufer P (2009) Potent control of tumor growth by CEA/CD3-bispecific single-chain antibody constructs that are not competitively inhibited by soluble CEA. J Immunother 32(4):341–352

Lutterbuese R, Raum T, Kischel R, Hoffmann P, Mangold S, Rattel B, Friedrich M, Thomas O, Lorenczewski G, Rau D, Schaller E, Herrmann I, Wolf A, Urbig T, Baeuerle PA, Kufer P (2010) T cell-engaging BiTE antibodies specific for EGFR potently eliminate KRAS- and BRAF-mutated colorectal cancer cells. Proc Natl Acad Sci USA 107(28):12605–12610

Minguet S, Schamel WW (2008) A permissive geometry model for TCR-CD3 activation. Trends Biochem Sci 33(2):51–57

Molhoj M, Crommer S, Rau D, Sriskandarajah M, Kufer P, Brischwein K, Hofmeister R, Baeuerle PA (2007) CD19/CD3-bispecific antibody of the BiTE class is far superior to tandem diabody with respect to redirected tumor cell lysis. Mol Immunol 44(8):1945–1953

Moore PA, Zhang W, Rainey J, Burke S, Huang L, Li H, Veri C, Yang Y, He L, Ciccarone V, Gorlatov S, Aggarwal S, Alderson R, Koenig S, Bonvini E, Johnson S (2010) Application of a novel bispecific antibody-based scaffold for optimal redirected T-cell killing of cancer cells. 101st Annual meeting of the American Association for Cancer Research 2010, abstract no 5629

Morse MA, Whelan M (2010) A year of successful cancer vaccines points to a path forward. Curr Opin Mol Ther 12:11–13

Müller D, Kontermann RE (2007) Recombinant bispecific antibodies for cellular cancer immunotherapy. Curr Opin Mol Ther 9(4):319–326

Müller D, Kontermann RE (2010) Bispecific antibodies for cancer immunotherapy: current perspectives. BioDrugs 24(2):89–98

Munz M, Baeuerle PA, Gires O (2009) The emerging role of EpCAM in cancer and stem cell signaling. Cancer Res 69(14):5627–5629

Nagorsen D, Bargou R, Ruettinger D, Kufer P, Baeuerle PA, Zugmaier G (2009a) Immunotherapy of lymphoma and leukemia with T cell-engaging BiTE antibody blinatumomab. Leuk Lymphoma 50(6):1–6

Nagorsen D, Zugmaier G, Viardot A, Goebeler M, Noppeney R, Schmidt M, Klappers P, Baeuerle PA, Kufer P, Bargou RC (2009b) Confirmation of safety, efficacy and response duration in non-Hodgkin lymphoma patients treated with 60 $\mu g/m^2/d$ of BiTE® antibody blinatumomab. Blood 114:4793 (ASH Annual Meeting Abstracts)

Offner S, Hofmeister R, Romaniuk A, Kufer P, Baeuerle PA (2006) Induction of regular cytolytic T cell synapses by bispecific single-chain antibody constructs on MHC class I-negative tumor cells. Mol Immunol 43(6):763–771

Osada T, Hsu D, Hammond S, Hobeika A, Devi G, Clay TM, Lyerly HK, Morse MA (2010) Metastatic colorectal cancer cells from patients previously treated with chemotherapy are sensitive to T-cell killing mediated by CEA/CD3-bispecific T-cell-engaging BiTE antibody. Br J Cancer 102(1):124–133

Pagès F, Berger A, Camus M, Sanchez-Cabo F, Costes A, Molidor R, Mlecnik B, Kirilovsky A, Nilsson M, Damotte D, Meatchi T, Bruneval P, Cugnenc PH, Trajanoski Z, Fridman WH, Galon J (2005) Effector memory T cells, early metastasis, and survival in colorectal cancer. N Engl J Med 353(25):2654–2666

Purbhoo MA, Irvine DJ, Huppa JB, Davis MM (2004) T cell killing does not require the formation of a stable mature immunological synapse. Nat Immunol 5(5):524–530

Rabinovich GA, Gabrilovich D, Sotomayor EM (2007) Immunosuppressive strategies that are mediated by tumor cells. Annu Rev Immunol 25:267–296

Raum R, Rau D, Kischel R, Bluemel C, Hausmann S, Steiger C, Meier P, Ebert E, Mangold S, Hoffmann P, Strasser S, Cierpka R, Kvesic M, Baeuerle PA (2010) Novel primate-crossreactive BiTE antibodies that eliminate cancer cells expressing cMet, IGFR-1, FAP-alpha, PSCA, Endosialin, CAIX or Her2/neu. 101st Annual meeting of the American Association for Cancer Research 2010, abstract no 2434

Renner C, Jung W, Sahin U, Denfeld R, Pohl C, Trümper L, Hartmann F, Diehl V, van Lier R, Pfreundschuh M (1994) Cure of xenografted human tumors by bispecific monoclonal antibodies and human T cells. Science 264(5160):833–835

Reusch U, Le Gall F, Hensel M, Moldenhauer G, Ho AD, Little M, Kipriyanov SM (2004) Effect of tetravalent bispecific CD19 × CD3 recombinant antibody construct and CD28 costimulation on lysis of malignant B cells from patients with chronic lymphocytic leukemia by autologous T cells. Int J Cancer 112(3):509–518

Robinson MK, Alpaugh RK, Borghaei H (2010) Naptumomab estafenatox: a new immunoconjugate. Expert Opin Biol Ther 10(2):273–279

Roman K, Raum T, Hausmann S, Rau D, Bluemel C, Mangold S, Meier P, Ebert E, Kvesic M, Hoffman P, Baeuerle PA, Kufer P (2008) Characterization of novel CD33- and MCSP-specific BiTE antibodies for the treatment of acute myeloid leukemia and melanoma, respectively, that are fully human in sequence. 99th meeting of the American Association for Cancer Research 2008, abstract no 2404

Rosenberg SA, Restifo N, Yang JC, Morgan RA, Dudley ME (2008) Adoptive cell transfer: a clinical path to effective cancer immunotherapy. Nat Rev Cancer 8(4):299–308

Schlereth B, Fichtner I, Lorenczewski G, Kleindienst P, Brischwein K, Kufer P, Lutterbüse R, Wimberger P, Kimmig R, Baeuerle PA (2005) Eradication of tumors from a human colon cancer cell line and primary ovarian cancer metastases in immunodeficient mice by a single-chain Ep-CAM-/CD3-bispecific antibody construct. Cancer Res 65(7):2882–2889

Shen J, Zhu Z (2008) Catumaxomab a rat/murine hybrid trifunctional bispecific monoclonal antibody for the treatment of cancer. Curr Opin Mol Ther 10(3):273–284

Stork R, Campigna A, Robert B, Müller D, Kontermann RE (2009) Biodistribution of a bispecific single-chain diabody and its half-life extended derivatives. J Biol Chem 284 (38):25612–25619

Topp MS, Zugmaier G, Goekbuet N, Kufer P, Goebeler M, Klinger M, Degenhard E, Baeuerle PA, Schmidt M, Nagorsen D, Neumann S, Horst HA, Raff T, Viardot A, Stelljes M, Schmid M, Ottmann OG, Burmeister T, Einsele H, Riethmueller G, Hoelzer D, Bargou RC (2009) Report of a phase II trial of single-agent BiTE antibody blinatumomab in patients with minimal residual disease (MRD) positive B-precursor acute lymphocytic leukemia (ALL). Blood 114:840 (ASH Annual Meeting Abstract)

Wahlin BE, Sander B, Christensson B, Kimby E (2007) $CD8^+$ T-cell content in diagnostic lymph nodes measured by flow cytometry is a predictor of survival in follicular lymphoma. Clin Cancer Res 13(2):388–397

Weber J (2009) Ipilimumab: controversies in its development, utility and autoimmune adverse events. Cancer Immunol Immunother 58(8):823–830

Willems A, Schoonooghe S, Eeckhout D, De Jaeger G, Grooten J, Mertens N (2005) CD3 × CD28 cross-interacting bispecific antibodies improve tumor cell dependent T-cell activation. Cancer Immunol Immunother 54:1059–1071

Wimberger P, Xiang W, Mayr D, Diebold J, Dreier T, Baeuerle PA, Kimmig R (2003) Efficient tumor cell lysis by autologous, tumor-resident T lymphocytes in primary ovarian cancer samples by an EP-CAM-/CD3-bispecific antibody. Int J Cancer 105(2):241–248

Wu C, Ying H, Bose S, Miller R, Medina L, Santora L, Ghayur T (2009) Molecular construction and optimization of anti-human IL-1alpha/beta dual variable domain immunoglobulin (DVD-Ig) molecules. MAbs 1(4):339–347

Zhang L, Conejo-Garcia JR, Katsaros D, Gimotty PA, Massobrio M, Regnani G, Makrigiannakis A, Gray H, Schlienger K, Liebman MN, Rubin SC, Coukos G (2003) Intratumoral T cells, recurrence, and survival in epithelial ovarian cancer. N Engl J Med 348(3):203–213

Chapter 16
Trifunctional Triomab® Antibodies for Cancer Therapy

Horst Lindhofer, Juergen Hess, and Peter Ruf

16.1 History of Bispecific Antibodies

In the 1980s, two groups described in parallel and for the first time T-cell-redirecting bispecific antibodies (bsAbs) for the elimination of tumor cells (Perez et al. 1985; Staerz et al. 1985). In vitro, this approach was extremely potent, highlighting the superiority of T-cell-induced compared to accessory cell-mediated killing. But two major drawbacks hampered the rapid transition of this new approach from preclinical investigations to successful clinical trials and market approval.

First, at the beginning of the 1990s, production and purification of bsAbs was a real challenge. Bispecific antibodies were produced using quadroma technology, i.e., the fusion of two different hybridoma cell lines. However, quadroma cells derived from two different mouse hybridomas produced mainly nonfunctional mismatched heavy (H)/light (L) chain variants (De Lau et al. 1991). Purification of the functional bsAbs from among the many mismatch variants was difficult and hardly manageable for clinical application. Second, regulation of T cells is complex, requiring two independent stimulation signals for physiological activation. Obviously, this outcome could hardly be accomplished using a single bispecific molecule delivering only one signal, e.g., via CD3 binding to the T cell (Jung et al. 1991).

H. Lindhofer (✉)
TRION Pharma GmbH, Frankfurter Ring 193a, 80807 Munich, Germany

TRION Research GmbH, Am Klopferspitz 19, 82152 Martinsried, Germany
e-mail: horst.lindhofer@trionpharma.de

J. Hess
TRION Pharma GmbH, Frankfurter Ring 193a, 80807 Munich, Germany

P. Ruf
TRION Research GmbH, Am Klopferspitz 19, 82152 Martinsried, Germany

R.E. Kontermann (ed.), *Bispecific Antibodies*,
DOI 10.1007/978-3-642-20910-9_16,

To address this situation, we developed rat/mouse bsAbs with species-restricted H/L chain pairing, reducing the mismatch variants from about 70% to <10% (Lindhofer et al. 1995). Moreover, because of the different binding affinities of rat and mouse IgG isotypes for protein A matrices, we found a solution to the production and purification problem. Of importance, we also discovered a preferential binding of the hybrid Fc portion to human-activating Fc-gamma (γ) receptors (R) I/IIa and III when we used the isotype combination of rat IgG2b and mouse IgG2a (see Sect. 16.3.1). This characteristic led to a solution for the second drawback of T-cell-redirecting bsAbs, the lack of a co-stimulatory signal. With activation of FcγR-positive cells such as macrophages, dendritic cells (DCs), or natural killer (NK) cells, co-stimulatory molecules such as CD40 or CD80/CD86 are upregulated on the surface, allowing delivery of the important second stimulation signal to the T cell (see Sect. 16.3).

Others suggested a different approach, such as Weiner et al. (1994), who presented pre-clinical data demonstrating strong release of cytokines with whole IgG T-cell-engaging bsAbs. In their conclusion, they recommended removal of the Fc region to reduce the risk of cytokine-mediated negative side effects. Various groups adopted this approach, which was promoted by the advent of recombinant antibody technology. As a result, a plethora of different bsAb formats lacking the Fc region have been developed (Müller and Kontermann 2010). In contrast, our group pursued the whole-IgG approach, developing the concept of trifunctional Triomab® antibodies as outlined in detail in the following sections.

16.2 Purification of Quadroma Supernatants

The quadroma technology, i.e., the somatic hybridization of two different hybridomas to hybrid-hybridoma (quadroma) cells, represents one of the first techniques for the production of bsAbs. However, single-species quadroma cell lines such as mouse/mouse or human/human hybrid-hybridomas often showed random H/L chain pairing with up to ten possible assembling variants (Smith et al. 1992; Tiebout et al. 1987). As a result, the output of functional bsAbs was low and the purification process complex. In contrast, Triomab® antibodies are produced using hybrid mouse/rat quadroma cell lines. Here, we present an overview of the manufacturing and purification processes, identifying three essential features in the biology of rat/mouse quadromas and in the structural nature of produced Triomab® antibodies.

First, stable mouse/rat quadroma cell lines secrete Triomab® antibodies into the culture supernatant. Cells can be grown in defined protein-free medium with sufficient post-production stability using cell factory systems and batch fermentation. Then, antibodies are purified from the cell-free culture supernatant by means of protein A affinity and ion-exchange chromatography, followed by diafiltration, nanofiltration, and final formulation (Fig. 16.1). Structural and functional characterization achieved by applying liquid chromatography, mass spectrometry and

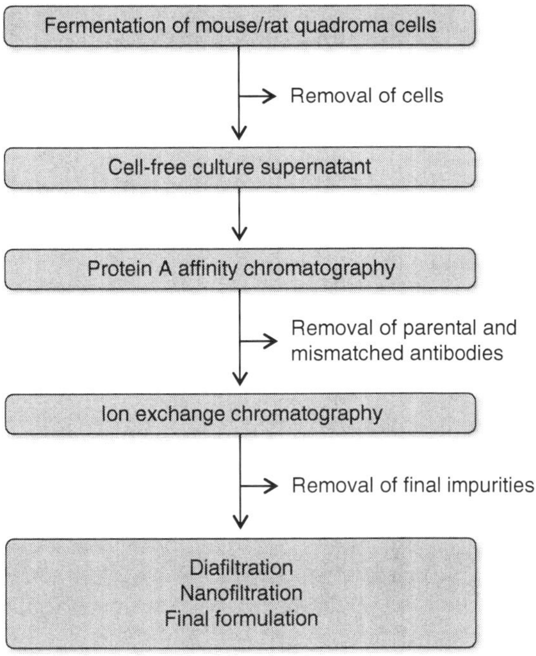

Fig. 16.1 Overview of the manufacturing and purification process of Triomab® antibodies

various immunological methods and bioassays confirmed that Triomab® antibodies of high integrity, purity, and quality result from this downstream processing technology (Chelius et al. 2010).

The high yield of functional Triomab® antibodies and the simple purification process are based on special characteristics of mouse IgG2a/rat IgG2b quadroma cell lines distinct from conventional single-species hybrid-hybridomas.

16.2.1 Preferential Intra-species H/L Chain Pairing

In a comprehensive study, De Lau et al. (1991) investigated H and L chain pairing in eight different mouse/mouse quadromas. Surprisingly, in only 4 of 16 combinations did the L chains associate preferentially with their homologous H chain partners. In other words, the majority of mouse/mouse quadromas displayed a random association of H and L chains with the consequence of a low yield of functional bsAbs. Theoretically, only 12.5% of produced antibodies are of a bispecific nature if all H and L chains combine stochastically (Milstein and Cuello 1983). In contrast, Triomab® antibodies produced by mouse/rat quadromas show a preferential pairing of the mouse H with the mouse L chain on the one side and of the rat H with the rat L chain on the other side. With this strategy, the amount of correctly paired bsAbs is significantly increased, as we have demonstrated

(Lindhofer et al. 1995). In three different mouse IgG2a/rat IgG2b and in one mouse IgG2a/rat IgG2a quadromas analyzed, the bsAbs fractions accounted for 30–49% of whole produced IgG. The major impurities detected were parental mouse and parental rat IgG, whereas H/L mismatch variants generally occurred to a minor degree (4–10%).

16.2.2 *Efficient Heterologous H/H Chain Pairing*

Because of an approximately 25% difference in the amino acid sequence of the Fc region between mouse IgG2a and rat IgG2b (http://imgt.cines.fr), a preferential homologous H/H chain pairing might be expected; however, this pairing has not been observed. In contrast to H/L chain pairing, H/H chain assembly is apparently not species restricted. Otherwise, the high yield of up to 49% heterologous bsAbs, which approaches the theoretically expected proportion of 50%, cannot be explained. However, a closer look at the hinge region of mouse IgG2a and rat IgG2b reveals a crucial homology. This flexible inter-domain region located between C_{H1} and C_{H2} is considered most important for H chain assembly because of the formation of inter-chain disulfide bonds. Interestingly, the hinge regions of mouse IgG2a and rat IgG2b are of comparable lengths, and three matched cysteine residues easily allow the covalent interaction of the H chains. In this way, inter-species H chain pairing between mouse IgG2a and rat IgG2b may not be hindered but instead be equally effective.

16.2.3 *Different Affinities of Mouse and Rat IgG for Protein A*

Whole IgG antibodies can be simply purified by affinity chromatography using protein A, which binds to the C_{H2}–C_{H3} interface of the Fc region (Deisenhofer 1981). However, different IgG isotypes bind with different affinities. In contrast to mouse IgG2a, which is a strong binder, rat IgG2b shows only a weak affinity for protein A (Burton 1985). This fact was exploited for the efficient separation of heterologous mouse IgG2a/rat IgG2b from homologous parental IgG antibody variants: With application of a decreasing pH gradient, the IgG produced by a mouse/rat quadroma cell line was eluted in three main peaks, as demonstrated in Fig. 16.2. In accordance with the increasing binding affinity for protein A, the peak analysis revealed mainly parental rat IgG2b in the first fraction, rat/mouse bispecific IgG in the second fraction, and parental mouse IgG2a in the third fraction. Peak integration indicated about 35% bsAbs produced. Because some bsAbs had already co-eluted with parental rat IgG in the first peak, the real production rate was even higher.

In summary, the combination of mouse IgG2a and rat IgG2b is characterized by intra-species restricted H/L chain pairing and by non-restricted H/H chain

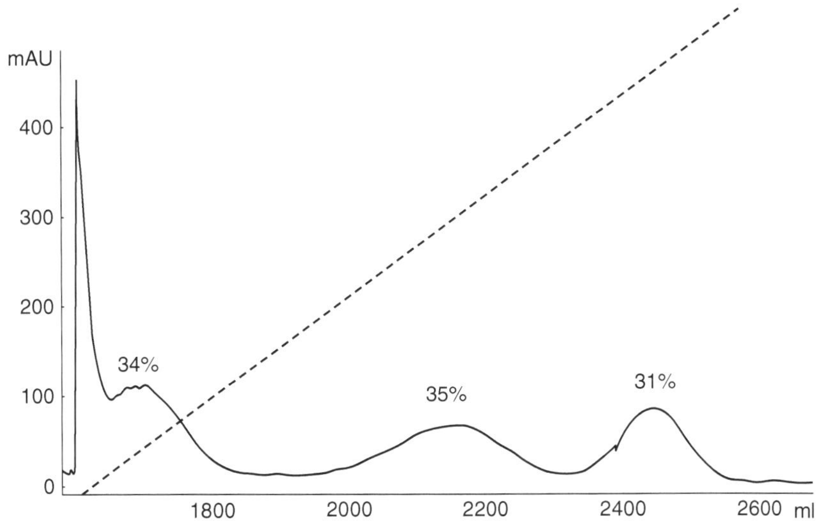

Fig. 16.2 Purification of Triomab® antibody by protein A affinity chromatography. A total of 500 ml of cell-free culture supernatant were loaded onto a 19.7-ml protein A column (MabSelectSure, GE Healthcare), and bound antibodies were eluted by decreasing the pH gradient (pH 8.5 to 3.5 in 75 column volumes), indicated by the *dotted line*. The UV_{280nm} elution profile showed three main peaks with 34%, 35%, and 31% area under the curve distribution. The first fraction contained mainly parental rat IgG, the second one bispecific rat/mouse IgG, and the third parental mouse IgG as analyzed by isoelectric focusing, mass spectrometry, and flow cytometry (data not shown). The first sharp peak was excluded from peak integration because it contained no protein but was the result of an absorbing media component

assembly, resulting in a high yield of 30–50% of correctly paired functional Triomab® antibodies. The main impurities consist of parental rat and mouse IgG, which can be efficiently removed by sequential pH elution/protein A affinity chromatography. Based on these features, TRION Pharma (Munich, Germany) developed a global unique platform technology for the cost-effective and cGMP-compliant production of Triomab® antibodies, a prerequisite for clinical development and commercialization.

16.3 The Mode of Action of Triomab® Antibodies in Tumor Therapy

Triomab® antibodies are bispecific and trifunctional, consisting of a tumor-associated antigen (TAA)-specific binding arm, a second binding arm specific for CD3 expressed on T cells, and a chimeric mouse IgG2a × rat IgG2b Fc region that preferentially binds to activating Fcγ receptors present on accessory cells such as macrophages, DCs, or NK cells. More than 12 years of intensive work has led to the proposed mode of action of Triomab® antibodies in tumor therapy. The first crucial step in this mode of action is thought to be redirection of T cells to the tumor via the bispecific Triomab® antibody-mediated crosslink of a TAA with CD3 (Fig. 16.3, upper). Antibody-mediated engagement of CD3 as a component of the T-cell receptor complex is a powerful first stimulus to activate T cells in a major histocompatibility complex (MHC)-independent manner, accompanied by TNF-α and IFN-γ secretion (Jacobs et al. 1997). However, the physiological activation of T cells requires a second signal. Attracted by opsonized T cells and tumor cells as well as proinflammatory cytokines, FcγR-positive immune cells are additionally engaged via the Fc region of Triomab® antibodies. A cluster of different immune cell types is formed at the tumor cell (Fig. 16.3, upper). This tri-cell complex formation consisting of tumor cells, T cells, and FcγR-positive accessory immune cells suggests several important consequences: First, there is mutual stimulation of accessory immune cells and T cells. Triomab® antibody-triggered interaction of T cells and CD14-positive monocytes results in the upregulation of CD83, CD86, and CD40 (Riechelmann et al. 2007; Stanglmaier et al. 2008; Zeidler et al. 2000). Thus, T cells receive a second co-stimulatory signal in the form of CD40/CD40L or CD80-CD86/CD28 interaction. As a consequence, they are profoundly and physiologically activated, as characterized by high secretion of IL-2 and strong proliferation with detection of the proliferation marker Ki-67 (Riesenberg et al. 2001; Ruf et al. 2004; Zeidler et al. 1999). Additionally, the T-cell activation markers CD25 and CD69 are upregulated (Riechelmann et al. 2007). Conversely, accessory immune cells are stimulated by interaction with T cells and the FcγR crosslink. This stimulation is manifested as high levels of proinflammatory cytokines such as IL-6 and IL-12 are measured, which are mainly secreted by accessory cells (Hirschhaeuser et al. 2009; Jäger et al. 2009b; Zeidler et al. 1999). Furthermore, the cross-talk between accessory and T cells is indicated by the release of Th1-biased cytokines, especially IL-2 and IFN-γ (Hirschhaeuser et al. 2009; Jäger et al. 2009b; Zeidler et al. 1999). Finally, the targeted tumor cells are efficiently destroyed by the concerted attack of different types of immune effector cells, as shown in allogeneic settings as well as in autologous human ex vivo systems (Gronau et al. 2005). Necrotic and apoptotic tumor cells and particles are phagocytosed (Riesenberg et al. 2001; Zeidler et al. 2000) and may be processed and presented by professional antigen-presenting cells in a stimulatory context, the ideal prerequisite for anti-tumor immunization (Fig. 16.3, *bottom*). In this regard, the elevated expression of the DC-specific cytokine DC-CK1 in the presence of DCs, targeted tumor cells, and Triomab® antibodies was an important finding that indicated the specific activation of professional antigen-presenting cells (Zeidler et al. 1999).

The observed Fc-mediated nonspecific activation of immune cells off-site of the tumor was regarded as a major obstacle to previous clinical approaches with intact bispecific antibodies. Uncontrolled systemic release of cytokines that causes severe adverse events is a serious concern (Weiner et al. 1994). Therefore, it was an important finding that Triomab® antibodies reach their full activating efficacy only at the tumor cell when all three binding partners are present (Ruf et al. 2004). The relevance of this feature holds especially true for the production of IL-2, the most

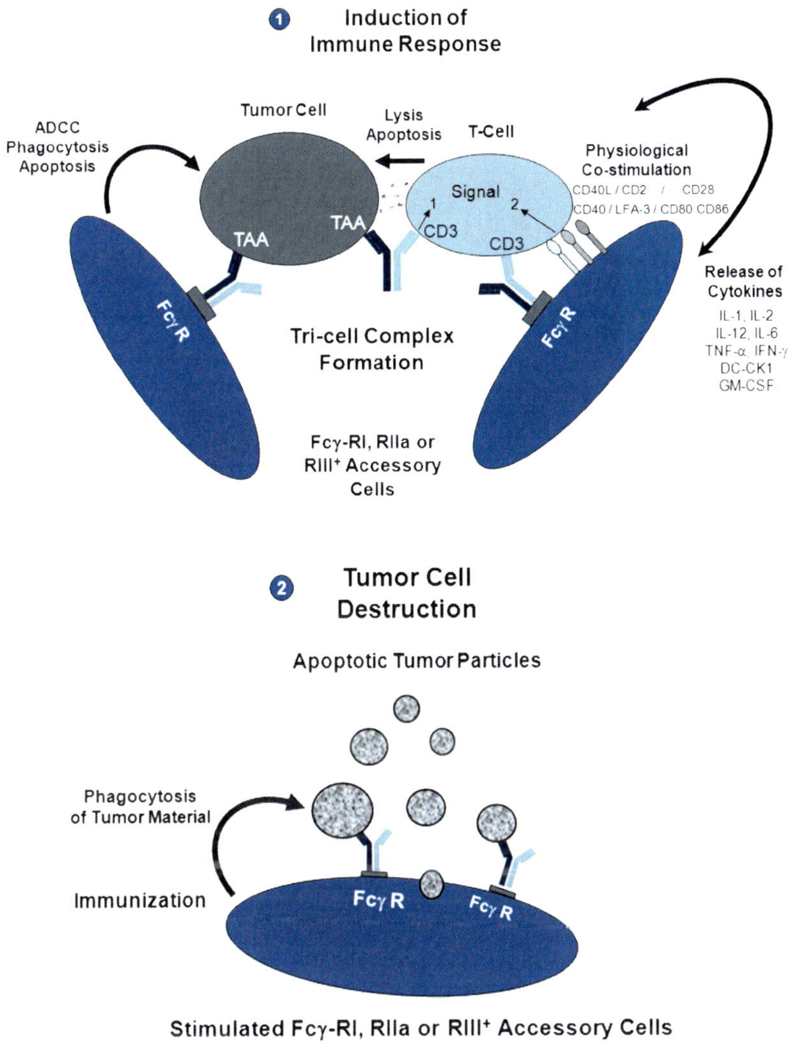

Fig. 16.3 Stepwise mode of action of Triomab® antibodies. In a first step for this process, (1) T cells are redirected to and activated at the tumor cells in an MHC-independent manner by Triomab® antibody-mediated crosslink of TAA with CD3 (Fig. 16.3, *upper*; signal 1). Simultaneously or subsequently, accessory cells can be additionally bound via the Fc region of the trifunctional antibody (Fig. 16.3, *upper*). A process of mutual cross-talk between T cells and accessory immune cells (macrophages, DCs and NK cells) is initiated. T cells are strongly activated in a physiological manner, receiving a second co-stimulatory signal (signal 2), while accessory immune cells are stimulated via FcγR crosslinking leading to the release of proinflammatory cytokines. As a result, tumor cells are effectively destroyed by a concerted attack of T cells and accessory immune cells applying different killer mechanisms like antibody-dependent cellular cytotoxicity (ADCC), phagocytosis, or perforin/granzyme-mediated lysis and apoptosis induction. Finally, T cell proliferation occurs as well as necrotic or apoptotic tumor particles are phagocytosed, processed, and presented by stimulated professional antigen-presenting cells (macrophages, DCs), a prerequisite for the induction of long-term anti-tumor immunization (2) (Fig. 16.3, *bottom*)

important autocrine growth factor for T cells. Significant amounts of IL-2 are induced only when trifunctional binding is enabled (Stanglmaier et al. 2008; Zeidler et al. 1999; Ruf et al. 2004). It should be emphasized that IL-2 is absolutely needed for self-supporting proliferation of T cells upon activation (Robb et al. 1981).

Triomab® antibodies cannot be substituted by a combination of both parental, monospecific antibodies because the antigen-binding functions must be closely linked to achieve the immunostimulatory capacity required. In different in vitro and in vivo models, cytokine secretion and tumor cell elimination are significantly less efficient with an equimolar combination of both parental antibodies (Ruf and Lindhofer 2001; Zeidler et al. 1999). Thus, the excellent efficacy of Triomab® antibodies in the destruction of tumor cells allows the identification of a therapeutic window with very low (microgram range) but effective drug concentrations and an acceptable safety profile. This possibility has been impressively demonstrated in 2009 by the first approval of a trifunctional Triomab® antibody (catumaxomab, Removab®) for the intraperitoneal (i.p.) treatment of malignant ascites (MA) in the European Union (Heiss et al. 2010; Linke et al. 2010). Several clinical studies involving different Triomab® candidates and showing promising results indicate that both locoregional (Heiss et al. 2010; Sebastian et al. 2009a) and systemic Triomab® antibody administration are feasible (Buhmann et al. 2009; Kiewe et al. 2006; Sebastian et al. 2007; Stemmler et al. 2005).

16.3.1 The Role of the Fc Region

The Fc region is of exceptional importance for the therapeutic concept and mode of action of Triomab® antibodies. Cross-talk between and activation of different types of immune effector cells in close proximity to the tumor cells can be accomplished only by intact Fc-containing bsAbs. Recent findings have recalled the important role of the Fc region of conventional monospecific antibodies in tumor therapy. Fc-mediated killing mechanisms such as ADCC, antibody-dependent cellular phagocytosis (ADCP), and complement-dependent cytotoxicity (CDC) are essential in tumor models and in the clinic. Prominent examples are trastuzumab (Herceptin®) and rituximab (Rituxan®) (Clynes et al. 2000; Racila et al. 2008; Weng and Levy 2003). Consequently, much effort has focused on improving therapeutic monoclonal antibodies (mAbs) by means of Fc engineering (Desjarlais et al. 2007; Moore et al. 2010). The selectively enhanced binding of mAbs to the activating isoform CD32a but not to its inhibitory counterpart CD32b of FcγRII (CD32) turned out to be of special interest: An improved CD32a/CD32b binding ratio strongly increases ADCP (Richards et al. 2008) and favors activation and maturation of DCs with a positive effect on the induction of tumor immunity (Boruchov et al. 2005; Kalergis and Ravetch 2002).

The Fc region of Triomab® antibodies consists of mouse IgG2a and rat IgG2b half antibodies, each mediating strong effector functions such as ADCC as homodimers (Chassoux et al. 1988; Larson et al. 1988). Moreover, it is worth mentioning that the constant Fc regions of mouse and rat H chain isotypes IgG2a and IgG2b have an amino acid homology of only approximately 75%, giving rise to the truly complex FcγR-binding behavior of this heterodimeric entity with its underlying immunostimulatory features. Interestingly, the IgG2a and IgG2b isotypes of mouse and rat also demonstrate efficacious activation of human effector cells such as monocytes via FcγRI and IIa interaction (Haagen et al. 1995; Koolwijk et al. 1991). Thus, a chimeric mouse IgG2a × rat IgG2b Fc region might interact equally well with human FcγR expressed on different immune effector cells. This idea has been investigated in detail, as follows. First, the binding of Triomab® antibodies to purified NK cells ($CD56a^+/CD3^-$) verified recognition of FcγRIII (CD16), the only FcγR expressed on NK cells (Zeidler et al. 2000). Moreover, interaction with the high-affinity FcγRI (CD64) was analyzed by surface plasmon resonance. With this approach, the Triomab® antibody catumaxomab showed binding kinetics comparable to the humanized alemtuzumab (hIgG1) with Ka values of 2.46×10^7 and 3.18×10^7 [1/M], respectively (Table 16.1).

Finally, preferential binding to the activating FcγRIIa (R131) in comparison to the inhibitory FcγRIIb has been observed (Table 16.2). Monomeric catumaxomab bound about 2.1 times more strongly to FcγRIIa than to FcγRIIb. However, when pre-aggregated catumaxomab was used instead of the monomeric form, the FcγRIIa/IIb binding ratio increased up to 19 times (Fig. 16.4). This increase resulted from

Table 16.1 Characterization of the binding of Triomab® antibody catumaxomab to human FcγRI by surface plasmon resonance

Antibody	Fc	ka[a] [1/Ms]	kd[b] [1/s]	Ka[a] [1/M]	Kd[b] [M]
Catumaxomab	mIgG2a × rIgG2b chimeric	$2.35 \pm 0.096 \times 10^4$	$9.59 \pm 0.936 \times 10^{-4}$	$2.46 \pm 0.146 \times 10^7$	$4.08 \pm 0.252 \times 10^{-8}$
Alemtuzumab	Human IgG1	3.5×10^4	1.08×10^{-3}	3.18×10^7	3.26×10^{-8}

The experiments were performed by BioAnaLab (Oxford, UK) Abbreviations: *m* mouse; *r* rat
[a]Association rates (ka) and constants (Ka) as well as [b]dissociation rates (kd) and constants (Kd) (± standard deviation), respectively, were determined by Biacore analysis using Langmuir 1:1 binding model fits. Mean values of three individual catumaxomab and two alemtuzumab batch runs are shown. $F(ab')_2$ fragment controls of catumaxomab displayed no specific binding

Table 16.2 Binding of monomeric and aggregated catumaxomab to human FcγRI, FcγRIIa, and FcγRIIb

Catumaxomab status	FcγRI[a]	FcγRIIa[a] R131	FcγRIIb[a]	IIa/IIb[b] Ratio
Monomeric	0.34 ± 0.01	26.56 ± 1.48	54.45 ± 5.22	2.1
Aggregated	–	1.66 ± 0.04	31.79 ± 1.66	19.2
Fold	–	16	1.7	–

Abbreviation: *EC50* effective concentration at half maximal binding signals
[a]EC50 values indicated in μg/ml (± standard error) were obtained from GraphPadPrism one site specific binding with Hill slope fit of ELISA data. Mean data of two independent experiments are shown
[b]IIa/IIb ratio = EC50 (FcγRIIb)/EC50 (FcγRIIa)
[c]Fold equals EC50 (FcγRII, catumaxomab monomeric)/EC50 (FcγRII, catumaxomab aggregated)

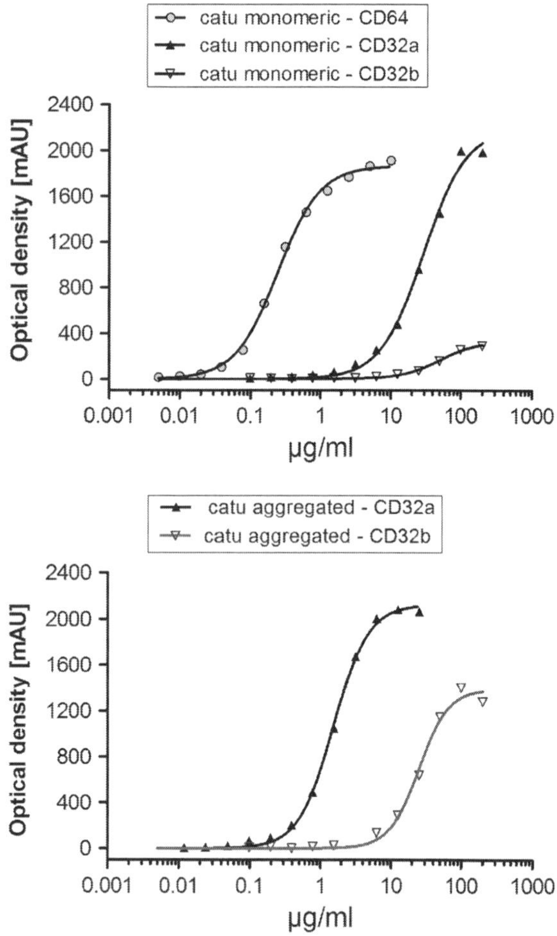

Fig. 16.4 Binding of monomeric catumaxomab to FcγRI, IIa (R131), and IIb (*upper*), and of aggregated catumaxomab to FcγRIIa (R131) and IIb (*bottom*). Binding was measured by ELISA, coating the corresponding FcγRs on the plate and detecting bound catumaxomab using a mixture of biotinylated F(ab′) 2-fragments of goat anti-mouse and anti-rat IgG, both F(ab′)2-fragment specific. For complex formation, catumaxomab was cross-linked by equimolar amounts of biotin-labeled anti-mouse and anti-rat IgG beforehand. Nonspecific binding of the $F(ab')_2$ fragment of catumaxomab was subtracted

the fact that the avidity of aggregated catumaxomab for FcγRIIa increased greatly in comparison to the binding to FcγRIIb (Table 16.2). Of importance, the use of aggregated antibodies clearly better resembles the physiological situation of an immune-complexed or cell-bound antibody while monomeric antibodies will hardly bind to the low-affinity FcγRII at therapeutic concentrations. In summary, the chimeric mouse IgG2a × rat IgG2b Fc region of Triomab® antibodies preferentially binds to activating FcγRI, IIa, and III. Thus, different immune effector cells are involved and activated by trifunctional Triomab® antibodies. Taken together, these findings indicate that by chance, the heterodimeric structure of Triomab® antibodies may already fulfill the balanced interplay with activating Fcγ receptors that is considered to be the next important step in optimizing therapeutic antibody properties by means of Fc engineering.

16.3.2 *The Different Killer Mechanisms of Involved Immune Cells*

According to the postulated mode of action, Triomab® antibodies induce a concerted attack of different types of immune effector cells against the tumor. Based on this orchestrated tumor defense, Triomab® antibodies are superior to conventional therapeutic antibodies. This superiority also holds true for recognition of tumor cells showing only low surface expression of the relevant target antigen, such as CD20 or HER2/neu (Jäger et al. 2009b; Stanglmaier et al. 2008). This important hallmark of Triomab® antibodies has been shown, for instance, by the capacity of anti-HER2/neu × anti-CD3 ertumaxomab to mediate in vitro killing of tumor target cells that express only low HER2/neu levels (scored 1+) (Jäger et al. 2009b). In contrast, trastuzumab absolutely failed to eliminate low HER2/neu-expressing tumor cell lines in the presence of peripheral blood mononuclear cells (PBMCs), reflecting the disadvantage of the monospecific recognition mode in the case of rare surface antigen expression. Below, the different cytotoxic mechanisms of recruited immune cells are described.

16.3.2.1 T Lymphocytes

T cells are the most potent effector cells of the immune system with regard to tumor therapy. Bispecific antibodies that crosslink TAA with the trigger molecule CD3 recruit cytotoxic T cells while bypassing natural T cell receptor specificity. In this way, all mature CD3-positive T cells are in principle available for an anti-tumor attack, although different T-cell subpopulations demonstrate unequal cytotoxic potential: CD8 T cells of the memory phenotype (RO^+) expand rapidly and are most efficacious with a combination of anti-CD3/CD30 and anti-CD28/CD30 bispecific antibodies for T cell stimulation (Renner et al. 1997). In this context, it should be noted that cytotoxic CD4 T effector cells may also contribute to tumor cell cytotoxicity mediated by bispecific antibodies with anti-CD3 stimulation function, as others have shown (Porakishvili et al. 2004).

T cells targeted by Triomab® antibodies are physiologically stimulated within the tri-cell complex, receiving anti-CD3 and additional co-stimulatory signals from accessory immune cells. Along this stimulation pathway, the activation markers CD25 and CD69 are upregulated, and cytokines such as IFN-γ and TNF-α that indicate T-cell activation are secreted in vitro (Riechelmann et al. 2007; Ruf et al. 2004) and in vivo (Heiss et al. 2005). Especially, significant IL-2 production combined with increased expression of the IL-2R α-chain (CD25) exemplifies a kind of self-supporting system characterized by strong T-cell proliferation and most efficacious eradication of targeted tumor cells (Stanglmaier et al. 2008; Zeidler et al. 1999). Remarkably, Triomab® antibody-mediated tumor cell killing requires no additional co-stimulation of immune effector cells, as other bispecific antibodies do. Once bound, T cells are induced to secrete pore-forming perforin at

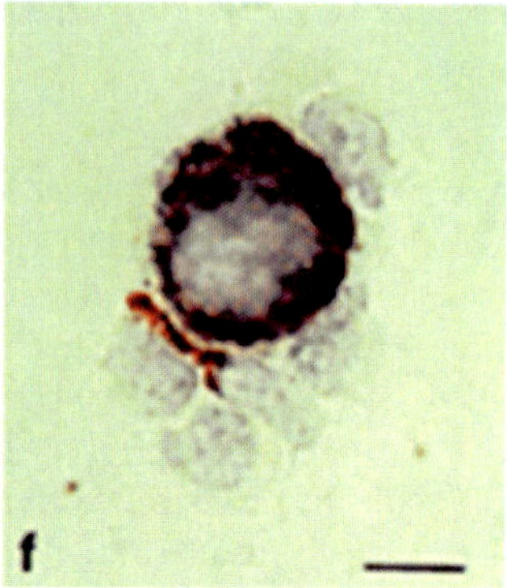

Fig. 16.5 Triomab® antibody-induced pore-forming perforin attack against a targeted tumor cell. PBMC (*blue*) partly expressing pore-forming perforin proteins (*brown*) concentrated toward the targeted tumor cell (LNCaP, *black*). Bar = 10 μm. Reproduced with permission from Riesenberg et al. (2001)

the membrane contact side to the targeted tumor cells, resulting in the osmotic lysis of the target (Riesenberg et al. 2001, Fig. 16.5). In vitro, tumor cells burst 9–17 h after incubation initiation, as monitored by computerized sequential video microscopy. Interestingly, increased granzyme B secretion by PBMCs indicates an additional contribution of apoptotic processes to Triomab®-mediated cytotoxic efficacy (Riechelmann et al. 2007). Thus, along with necrosis for tumor cell killing, the induction of apoptosis is likely another T-cell-mediated killer mechanism that occurs in the context of Triomab® antibody therapies.

16.3.2.2 Monocytes/Macrophages and DCs

As Zeidler et al. (1999) have shown, monocytes and macrophages clearly contribute to Triomab® antibody-mediated tumor cell elimination. Further studies have revealed enhanced phagocytosis of tumor cells by CD14-positive monocytes after addition of Triomab® antibodies (Zeidler et al. 2000). The profound activation of monocytes is indicated by high IL-6 secretion and by the upregulation of CD25 and CD40 (Stanglmaier et al. 2008). Thus, the interaction of monocytes with T cells is essential. Cross-talk of T cells and monocytes at the tumor cell results in mutual effector cell stimulation and enhanced tumor destruction. Accordingly, Triomab® antibody-induced tumor cell killing is much greater when the whole PBMC population is used instead of purified T cells (Zeidler et al. 1999). The preferential binding of the chimeric mouse IgG2a × rat IgG2b Fc region of Triomab® antibodies to activating FcγRI, IIa, and III on monocytes/macrophages and DCs may be the key to understanding this strong stimulation effect. As other groups have shown, selective blockade of the inhibitory FcγRIIb (Dhodapkar et al. 2007) or selective binding to the activating FcgRIIa by Fc engineering (Moore et al. 2010) leads to activation of DCs and monocytes and enhanced ADCP. The essential role of monocytes and macrophages in the destruction of tumor cells by Triomab® antibodies is further confirmed by experiments using multicellular tumor spheroids (MCTS) as a model for micrometastases. This work showed that co-cultures of

PBMC effector cells and Triomab® antibody-treated MCTS of about 0.9–1.8 mm diameter completely disintegrated into single cells and debris within 11 days (Hirschhaeuser et al. 2009). Further, at 6 days after therapy, clonogenic survival assays showed that around 98% of tumor cells were destroyed. Of importance, destruction of MCTS was accompanied by massive infiltration of CD45-positive leukocytes. So far, FcγRI-positive accessory immune cells have been identified as the major infiltrating cell population (Walz et al. 2004). Thus, reciprocal cross-talk and stimulation between T cells and monocytes/macrophages potentiates anti-tumor efficacy, resulting in the destruction of both the tumor cell monolayer and the three-dimensional tumor spheroids. Therefore, not only single disseminated tumor cells but also micrometastases or even solid tumors of patients might be attacked by Triomab® antibody-directed immune cells. Partial and complete remissions of metastases observed in clinical studies with Triomab® antibodies confirm this conclusion (Kiewe et al. 2006; Riechelmann et al. 2007).

16.3.2.3 NK Cells

Classical ADCC is mainly executed by NK cells that express the low affinity but activating FcγRIIIa. The binding of Triomab® antibodies to this receptor has been shown using $CD56^{+}CD3^{-}$ purified NK cells. Antibody binding was followed by increased expression of the NK cell-activation marker CD95 (APO-1, Fas) (Zeidler et al. 2000). Finally, high tumor cell lysis was observed, demonstrating that Triomab® antibodies can recruit cytotoxic NK cells. Thus, it is evident that NK cells can contribute to Triomab® antibody-mediated tumor cell killing. However, in consideration of their low frequency in regard to lymphocytes (2–13%), NK cells may add to a lesser extent to direct tumor cell killing in comparison to T cells and monocytes or macrophages.

16.4 The Tumor Cell Is a Smart Immunologist

Cancer cells can develop a number of resistance- and tolerance-inducing mechanisms during their evolution in the patient's body. These mechanisms are based on mutations in single tumor cells at different timepoints and are responsible for the heterogeneity of tumor cells or, for example, differences between the primary tumor and metastases. Thus, the patient's immune system must confront not only a single phenotype of tumor but a variety that can in total contribute different tumorigenic factors. Thus, tumor cells can, for example, secret immune modulators such as vascular endothelial growth factor (VEGF) (improving tumor vasculature) or IL-6 (a growth factor for lymphomas) and immunosuppressing factors such as IL-10 or TGF-β (Del Prete et al. 1993; de Souza and Bonorino 2009). Moreover, tumor cells obviously can raise peripheral tolerance against themselves by inducing regulatory T cells (Onishi et al. 2010). Curiel et al. (2004) demonstrated that in MA, about 10–17% of T cells have a regulatory $CD4^+$ $CD25^+$ $FOXP3^+$ T cell phenotype that can inhibit cytotoxic T cells and the release of proinflammatory cytokines. In this context, it was of interest to determine whether Triomab® antibodies can overcome this highly immunosuppressive environment after locoregional application. Lindhofer et al. (2009) and Jäger et al. (2009a) showed that this is indeed possible: Analysis of ascites samples before, during, and after i.p. therapy with the therapeutic Triomab® antibody catumaxomab (anti-EpCAM × anti-CD3) demonstrated (a) elimination of $EpCAM^+$ tumor cells and $CD133^+$/$EpCAM^+$ double-positive putative cancer stem cells (CSCs) (Lindhofer et al. 2009) and (b) the in vivo activation of $CD4^+$ and $CD8^+$ T cells as well as a significant decrease in VEGF levels (Burges et al. 2007; Jäger et al. 2009a, 2010; Heiss et al. 2005).

16.5 From In Vitro to In Vivo

The important features of the Triomab® antibody treatment concept have already been demonstrated in preclinical and clinical settings using anti-EpCAM catumaxomab or anti-HER2/neu ertumaxomab or surrogate antibodies such as BiLu (Burges et al. 2007; Heiss et al. 2005, 2010; Kiewe et al. 2006; Ruf and Lindhofer 2001). Table 16.3 summarizes the important hallmarks for the clinical development program of these antibodies.

The efficient dose–response profile of the Triomab® antibody format [e.g., catumaxomab and ertumaxomab given intravenously (i.v.) up to a 7.5 μg per single dose (Sebastian et al. 2007) or at a triple dosing of 10–200–200 μg at weekly intervals, respectively (Kiewe et al. 2006), is characterized by its fine-tuned balance among (a) the binding affinity for the tumor target antigen (e.g., EpCAM or HER2/neu); (b) the FcγR binding capacity of the constant Fc region (e.g., on NK cells, monocytes/macrophages, DCs) without considerable engagement of the inhibitory FcγRIIb; (c) the additional polyclonal CD3-specific T cell recruitment and

Table 16.3 Clinical studies with Triomab® antibodies

Name	Target	Cancer type	Trial stage	References
Catumaxomab	EpCAM × CD3	Malignant ascites	Phase I/II	Burges et al. (2007)
Catumaxomab	EpCAM × CD3	Malignant ascites	Phase II/III	Heiss et al. (2010)
Catumaxomab	EpCAM × CD3	Peritoneal carcinomatosis	Phase I	Ströhlein et al. (2009)
Catumaxomab	EpCAM × CD3	Pleural effusion	Phase I/II	Sebastian et al. (2009a, b)
Catumaxomab	EpCAM × CD3	Non-small cell lung cancer	Phase I	Sebastian et al. (2007)
Ertumaxomab	HER2/neu × CD3	Metastatic breast cancer	Phase I	Kiewe et al. (2006)
FBTA05	CD20 × CD3	CLL and NHL	Proof-of-concept	Buhmann et al. (2009)

activation; (d) the IL-2-mediated self-supporting maintenance of the immune response without co- or restimulation (e.g., w/o additional administration of IL-2); (e) the scientifically tenable pharmacokinetic and toxicological features (e.g., slow renal clearance rate and avoidance of severe adverse events such as "cytokine storm"); and (f) a convenient treatment window prior to the occurrence of anti-antibody-based immune responses.

16.5.1 Clinical Studies with Triomab® Antibodies

16.5.1.1 Catumaxomab

Based on these structural and underlying therapeutic properties, the Triomab® family member catumaxomab (Removab®) represents the first bispecific anti-tumor antibody worldwide (Sebastian et al. 2009a, b; Shen and Zhu 2008) that has clearly met its expected clinical endpoints in a pivotal phase II/III study and was therefore approved for the treatment of MA in 2009 in European Union (Heiss et al. 2010). Malignant ascites is a manifestation of an advanced disease in a variety of indications including ovarian, breast, or gastric cancer, and is characterized by strong fluid influx and tumor growth into the peritoneal cavity. The onset and progression of MA is associated with significant morbidity and deterioration in quality of life. The prognosis of MA patients is generally poor, with a median survival following diagnosis of ascites that varies from 75 days (non-ovarian) up to 7–12 months (ovarian) (Parsons et al. 1996).

Briefly, 258 patients suffering from epithelial ovarian cancer or non-ovarian cancer with confirmed $EpCAM^+$ MA were enrolled and randomized to catumaxomab treatment (one paracentesis plus four subsequent intraperitoneal (i.p.) catumaxomab infusions of 10, 20, 50, and 150 μg within 11 days) or to the control intervention (paracentesis alone as the best supportive care). Here, it should be noted that the unusual i.p. route of antibody administration represents the best delivery mode for attacking tumor cells that reside within the peritoneum. Moreover, in a pharmacokinetic analysis with ten MA patients, it was determined that catumaxomab became increasingly concentrated in ascites during the course of treatment, attaining effective concentrations in the ng/ml range (Ruf et al. 2010). Median antibody concentrations in the ascites arose from 552 to over 1,721 and then to 6,121 pg/ml, in proportion to the dosing scheme of 10, 20, and 50 μg. However, due to the established peritoneal carcinomatosis, the catumaxomab transfer from the peritoneal cavity into the systemic circulation was low (<1%); a median plasma concentration of 403 pg/ml was achieved with an antibody half-life of 2.13 days using the i.p. infusion route for MA treatment. Interestingly, catumaxomab showed a high in vivo stability, thus retaining its cytotoxic potential even after several days in the systemic circulation (Ruf et al. 2010).

By means of i.p. application, catumaxomab therapy reached a statistically significant puncture-free survival of 46 compared to 11 days for control treatment ($P < 0.0001$), respectively, underlining the potency of the catumaxomab treatment even in MA patients with such a poor prognosis. Moreover, median puncture-free time was 77 versus 13 days ($P < 0.0001$), indicating a clear gain in life quality. Although treatment benefits with regard to overall survival (OS) are unlikely in such advanced disease, the Kaplan–Meier curves indicate that catumaxomab prolonged OS in ~50% of patients (Heiss et al. 2010). In particular, a prospectively planned subgroup analysis of gastric cancer patients, demonstrated a statistically significant prolongation of OS ($n = 66$; 71 versus 44 days; $p = 0.0313$). One reason for this observation could be that nonovarian cancer patients (including gastric cancer) had fewer previous chemotherapies (median of one) compared to ovarian cancer patients (median of three) resulting in a less impaired immune system (see also Sect. 16.5.4).

Of greatest importance, tumor cells were almost completely eliminated from ascites fluid, suggesting the induction of prominent catumaxomab-mediated anti-tumor responses. Pyrexia, nausea, vomiting, and a transient elevation of liver enzymes as the most frequent adverse events were generally mild to moderate and reversible.

Interestingly, in a preliminary monitoring study, putative $CD133^+/EpCAM^+$ CSCs were present in peritoneal fluids of 62% of analyzed MA patients with different underlying primary tumor entities (Lindhofer et al. 2009) before therapeutic intervention. After four i.p. catumaxomab infusions, these presumed CSCs were completely eliminated from the peritoneal fluids of all tested MA patients. Thus, even under enhanced immunosuppressive conditions (Curiel et al. 2004, see also Sect. 16.4), catumaxomab efficiently destroyed $CD133^+/EpCAM^+$ CSCs within the peritoneal fluids of MA patients. Therefore, catumaxomab-based therapeutic measures may offer an additional treatment opportunity to eliminate CSCs in $EpCAM^+$ malignancies.

Taken together, results suggest that patients with $EpCAM^+$ MA do benefit from catumaxomab therapy by the prolongation of puncture-free survival and the reduction of ascites symptoms in comparison with MA patients receiving palliative paracentesis only. Of note, two further studies targeting treatment of peritoneal carcinomatosis (phase I) and pleural effusion (phase I/II) have been performed with acceptable safety and promising efficacy results (Sebastian et al. 2009a, b; Ströhlein and Heiss 2009). Moreover, because EpCAM is frequently over-expressed, as in prostate, breast, lung, colon, ovarian, or gastric carcinomas, catumaxomab may offer novel future treatment opportunities for a variety of cancers provided that the trifunctional antibody will be administered via the systemic route, which will be investigated in future trials.

16.5.1.2 Ertumaxomab

Ertumaxomab is a Triomab® antibody targeting CD3 and HER2/neu, the latter representing a well-characterized breast tumor marker that is also central to trastuzumab (Herceptin®) therapy. In vitro ertumaxomab has been compared to trastuzumab for its ability to kill tumor cells expressing various levels of HER2/neu. Exposure to ertumaxomab led to the efficient lysis of cells expressing high or very low amounts of tumor antigen, whereas trastuzumab was completely ineffective even at high concentrations at low HER2/neu expression (Jäger et al. 2009b). As already outlined in Sect. 16.3.2, this difference can be probably explained by the mode of action of these antibodies; trastuzumab triggers NK-mediated ADCC, while ertumaxomab mainly relies on T-cell-mediated killing and the interaction between T cells and accessory cells, as demonstrated in a phase I trial by the release of proinflammatory Th1-biased cytokines such as IFN-γ. To summarize the study briefly, 15 metastatic breast cancer patients who completed the study received i.v. three ascending doses of ertumaxomab (10–200 μg) on days 1, 7, and 14. Adverse events (e.g., fever, rigors, headache, nausea, and vomiting) related to ertumaxomab were generally mild, transient, and fully reversible. An anti-tumor response was seen in 5 out of 15 evaluable patients (one with a complete response, two with a partial response, and two with stable disease) (Kiewe et al. 2006).

16.5.2 Autologous In Situ Immunization Mediated by Triomab® Antibodies

The world's first therapeutic cancer vaccine, which was recently approved by the U. S. Food and Drug Administration, represents a major breakthrough in harnessing the immune system to fight tumors. Moreover, this therapeutic vaccine approval symbolizes a "proof-of-concept" development in cancer immunotherapy in general. The novel vaccine, Provenge (sipuleucel-T) from Dendreon (USA), will be used to treat men for advanced prostate cancer in an individual-targeted manner. In view of this successful cancer vaccine development, it should be noted that trifunctional IgG-shaped antibodies with anti-CD3 recognition of T cells can mediate vaccine-like adaptive immune responses against tumors. As already outlined, Triomab® antibodies have the potential to target and activate DCs in situ while initiating phagocytosis of tumor cells via opsonization (see also Sect. 16.3.1). After the destruction of tumor cells phagocytosed by DCs, subsequent antigen trafficking along MHC class I and class II processing and presentation pathways leads to the stimulation of tumor-specific CD8 or CD4 T cells, respectively. Data taken from preclinical experiments and preliminary clinical observations have indicated the induction of long-lasting CD4 and CD8 T cell responses (Ruf and Lindhofer 2001). If comprehensive immunomonitoring programs confirm these interesting findings (e.g., long-term immune and memory responses against tumors) in parallel with a beneficial therapeutic outcome, Triomab® antibodies may give rise to the induction of adaptive immunity, bridging the gap between active and passive immunization. Chames and Baty (2009) have described this mode of action as the "Holy Grail" of antibody-based cancer therapy.

In fact, in the A20 mouse lymphoma model, it has been demonstrated that the trifunctional antibody BiLu (anti-human EpCAM × anti-mouse CD3) could be used as an immunizing agent for anti-tumor vaccination. The immunization efficacy was dependent on the presence of the chimeric mouse/rat Fc-region, as the F(ab′)$_2$ fragment of the same trifunctional antibody did not elicit long-term protection (Ruf and Lindhofer 2001). Furthermore, a strong correlation has been observed between the induction of a humoral immune response with tumor-reactive antibodies and mouse survival.

Moreover, case observations with patients suffering from progressive peritoneal carcinomatosis from gastric and ovarian cancer were immunomonitored for long-lasting cell-mediated immune responses as a consequence of Triomab® therapy (Ströhlein et al. 2009). Briefly, after chemotherapy, the patient received three escalating doses of catumaxomab (10 μg–20 μg–40 μg) given i.p. The already primed immune responses were restimulated 30 days later after the end of catumaxomab therapy by means of an intradermal antigenic boost immunization via PBMCs loaded with 100-Gy irradiated autologous tumor cells and Triomab® antibody. As demonstrated by IFN-γ secretion assays, activation of tumor-reactive T cells was measured from day 39 up to day 110 at least after restimulation (Ströhlein et al. 2009). For these reasons, we have initiated for the next 2 years a clinical immunomonitoring program addressing the detection of memory responses. We are confident that important immunological parameters will be identified that will clearly mark the beginning of a new era in antibody therapy when the Triomab® concept will become a part of vaccinology.

16.5.3 Avoiding Graft-Versus-Host Disease While Redirecting Graft Versus Leukemia After Allogeneic Transplantation by Means of Triomab® FBTA05

With an annual incidence of four new cases per 100,000 population, chronic lymphocytic leukemia (CLL) represents the most frequent leukemia in the western world. In addition, about 19 new non-Hodgkin's lymphoma (NHL) cases per 100,000 population annually are currently diagnosed in Europe or the USA. CLL is not curable by means of the usual chemotherapeutic regimes even in combination with antibody application (e.g., anti-CD52 alemtuzumab and/or anti-CD20 rituximab). Although evidence for a graft-versus-lymphoma (GVL) effect of allogeneic stem cell transplantation (SCT) in highly malignant lymphoma is limited and controversial, allogeneic SCT represents the only curative option for CLL patients (Grigg and Ritchie 2004). However, several studies and observations report a GVL effect by donor lymphocyte infusions (DLI) following allogeneic hematopoietic SCT (Porter et al. 2006). Nevertheless, the use of allogeneic cell therapy in clinical practice can be frequently accompanied by life-threatening graft-versus-host disease (GVHD), which is difficult to control effectively with current treatment protocols. Thus, more sophisticated treatment regimes are required that allow optimal GVL effects while reducing the frequency and severity of devastating GVHD. Monoclonal antibodies such as rituximab have been used with increasing success in the treatment of patients with indolent and aggressive lymphoma as well as with CLL. The application of rituximab has improved anti-tumor response and survival of patients, but the mechanism of action is not entirely clear; ADCC, CDC, and induction of apoptosis have been demonstrated in in vitro experiments (Cartron et al. 2004).

Bispecific antibodies may represent a highly attractive therapeutic option for directing T cells efficiently toward tumor cells. In this context, the trifunctional anti-CD20 × anti-CD3 Triomab® antibody FBTA05 may improve the targeting of tumor cells by immune allogeneic effector cells while simultaneously reducing the risk of undesirable reactivity against normal host cells. In other words, FBTA05 may maximize GVL effects by simultaneously decreasing the incidence and severity of GVHD. Morecki and co-workers (2006, 2008) provided the first evidence that the hallmark of autologous vaccination also holds true for the concerted interaction of allogeneic lymphocytes and a trifunctional BiLu antibody (anti-EpCAM × anti-murine CD3) in an experimental B16-EpCAM tumor model. By means of this BiLu antibody-targeted DLI, long-lasting anti-tumor responses were induced that protected treated mice against a second B16-EpCAM tumor challenge (Morecki et al. 2008). Of importance, the use of this trifunctional antibody BiLu prevented mice from GVHD induced by the application of allogeneic haplo-identical lymphocytes. These preclinical findings represent the basic principle for the combination of DLI and trifunctional antibodies for CLL or NHL treatment. Based on the preclinical success described, a proof-of-concept clinical study was performed with a combination of the trifunctional anti-CD20 × anti-CD3 Triomab® antibody FBTA05 and DLI as therapeutic measures against B cell malignancies (Buhmann et al. 2009). This therapeutic setting showed FBTA05-induced anti-tumor responses in extensively pre-treated, p53-mutated, alemtuzumab- and rituximab-refractory patients, indicating the therapeutic capacity of this biopharmaceutical. It should be noted that FBTA05 antibodies can mediate effective killing of lymphoma cells even at very low surface expression levels of CD20 in vitro (Stanglmaier et al. 2008). Taken together, these findings suggest that the trifunctional antibody FBTA05 may offer novel treatment options, such as SCT followed by DLI plus FBTA05 antibodies, for therapy-resistant low-grade lymphoma (including CLL) and NHL patients in relapse.

16.5.4 HAMA Response as a Predictive Marker

Because Triomab® antibodies are considered "foreign" to the human immune system, it is not surprising that human anti-mouse (HAMA) and anti-rat antibodies (HARA) are induced in patients with preceding antibody applications. Nevertheless, results from clinical studies have demonstrated that Triomab® antibodies administered within a given treatment window, e.g., prior to the induction of neutralizing HAMA/HARA responses, may be sufficient for the therapeutic benefit in many cancer indications (Heiss et al. 2005, 2010; Kiewe et al. 2006; Ruf et al. 2010).

On the other hand, HAMA responses could be a predictive measure for the early onset of antigen-specific immune responses. In fact, these HAMA responses have been monitored through a pivotal phase II/III study with catumaxomab against MA and represented a predictive measure in that study. Briefly, 76% of the tested MA patients in the pivotal phase II/III trial were HAMA positive 8 days after completion of catumaxomab treatment (Ott et al. 2010). In contrast to HAMA-negative patients, they showed a significantly longer puncture-free survival, time to puncture, and especially overall survival. In fact, median overall survival in HAMA-positive patients was more than twofold increased compared to patients in the control group or HAMA-negative patients in the pivotal trial (129 days versus 64 days [HAMA neg.], $p = 0.0003$; Ott et al. 2010). Catumaxomab therapy was also beneficial for HAMA-negative patients in terms of time to first puncture; however, catumaxomab failed to improve overall survival in HAMA-negative patients. In summary, only early positive HAMA responses were strongly correlated with increased overall survival data with catumaxomab treatment. Thus, the beneficial clinical outcome may also be attributed to the contribution of acquired anti-tumor effector mechanisms mediated by catumaxomab, especially distinguished through its T-cell-recruiting capacity. Moreover, HAMA data may additionally reflect immunological fitness and antigen responsiveness with emphasis on acquired immunity. Therefore, the development of HAMA may represent an early biomarker and predictor for beneficial tumor therapy with Triomab® antibodies.

16.6 Future Perspectives

After the first approval in Europe for the treatment of MA, catumaxomab will be intensely developed against $EpCAM^+$ carcinomas including cancer stem cells and for alternative routes of administration. Moreover, ertumaxomab treatment of HER2/neu-expressing breast cancer patients, FBTA05 against B cell malignancies, and novel target antigens such as GD2-gangliosides will be further investigated and implemented into the Triomab® technology platform. Another interesting focus of research will be the aspect of active immunization. Therefore, immunomonitoring will be intensified to analyze possibly induced tumor-specific T cells as well as humoral anti-tumor responses.

References

Boruchov AM, Heller G, Veri MC, Bonvini E, Ravetch JV, Young JW (2005) Activating and inhibitory IgG Fc receptors on human DCs mediate opposing functions. J Clin Invest 115:2914–2923

Buhmann R, Simoes B, Stanglmaier M, Yang T, Faltin M, Bund D, Lindhofer H, Kolb HJ (2009) Immunotherapy of recurrent B-cell malignancies after allo-SCT with Bi20 (FBTA05), a trifunctional anti-CD3 × anti-CD20 antibody and donor lymphocyte infusion. Bone Marrow Transplant 43:383–397

Burges A, Wimberger P, Kümper C, Gorbounova V, Sommer H, Schmalfeldt B, Pfisterer J, Lichinitser M, Makhson A, Moiseyenko V, Lahr A, Schulze E, Jäger M, Ströhlein MA, Heiss MM, Gottwald T, Lindhofer H, Kimmig R (2007) Effective relief of malignant ascites in patients with advanced ovarian cancer by a trifunctional anti-EpCAM × anti-CD3 antibody: a phase I/II study. Clin Cancer Res 13(13):3899–3905

Burton DR (1985) Immunoglobulin G: functional sites. Mol Immunol 22:161–206

Cartron G, Watier H, Golay J, Solal-Celigny P (2004) From the bench to the bedside: ways to improve rituximab efficacy. Blood 104(9):2635–2642

Chames P, Baty D (2009) Bispecific antibodies for cancer therapy. Curr Opin Drug Discov Devel 12(2):276–283

Chassoux DM, Linares-Cruz LG, Bazin H, Stanislawski M (1988) K-cell-mediated cytotoxicity induced with rat monoclonal antibodies. I. Antibodies of various isotypes differ in their ability to induce cytotoxicity mediated by rat and human effectors. Immunology 65:623–628

Chelius D, Ruf P, Gruber P, Ploscher M, Liedtke R, Gansberger E, Hess J, Wasiliu M, Lindhofer H (2010) Structural and functional characterization of the trifunctional antibody catumaxomab. MAbs 2(3):309–319

Clynes RA, Towers TL, Presta LG, Ravetch JV (2000) Inhibitory Fc receptors modulate in vivo cytoxicity against tumor targets. Nat Med 6:443–446

Curiel TJ, Coukos G, Zou L, Alvarez X, Cheng P, Mottram P, Evdemon-Hogan M, Conejo-Garcia JR, Zhang L, Burow M, Zhu Y, Wei S, Kryczek I, Daniel B, Gordon A, Myers L, Lackner A, Disis ML, Knutson KL, Chen L, Zou W (2004) Specific recruitment of regulatory T cells in ovarian carcinoma fosters immune privilege and predicts reduced survival. Nat Med 10(9):942–949

De Lau WB, Heije K, Neefjes JJ, Oosterwegel M, Rozemuller E, Bast BJ (1991) Absence of preferential homologous H/L chain association in hybrid hybridomas. J Immunol 146:906–914

de Souza AP, Bonorino C (2009) Tumor immunosuppressive environment: effects on tumor-specific and nontumor antigen immune responses. Expert Rev Anticancer Ther 9(9):1317–1332

Deisenhofer J (1981) Crystallographic refinement and atomic models of a human Fc fragment and its complex with fragment B of protein A from *Staphylococcus aureus* at 2.9- and 2.8-A resolution. Biochemistry 20:2361–2370

Del Prete G, de Carli M, Almerigogna F, Giudizi MG, Biagiotti R, Romagnani S (1993) Human IL-10 is produced by both type 1 helper (Th1) and type 2 helper (Th2) T cell clones and inhibits their antigen-specific proliferation and cytokine production. J Immunol 150:353–360

Desjarlais JR, Lazar GA, Zhukovsky EA, Chu SY (2007) Optimizing engagement of the immune system by anti-tumor antibodies: an engineer's perspective. Drug Discov Today 12:898–910

Dhodapkar KM, Banerjee D, Connolly J, Kukreja A, Matayeva E, Veri MC, Ravetch JV, Steinman RM, Dhodapkar MV (2007) Selective blockade of the inhibitory Fcgamma receptor (FcgammaRIIB) in human dendritic cells and monocytes induces a type I interferon response program. J Exp Med 204:1359–1369

Grigg A, Ritchie D (2004) Graft-versus-lymphoma effects: clinical review, policy proposals, and immunobiology. Biol Blood Marrow Transplant 10(9):579–590

Gronau SS, Schmitt M, Thess B, Reinhardt P, Wiesneth M, Schmitt A, Riechelmann H (2005) Trifunctional bispecific antibody-induced tumor cell lysis of squamous cell carcinomas of the upper aerodigestive tract. Head Neck 27:376–382

Haagen IA, Geerars AJ, Clark MR, van de Winkel JG (1995) Interaction of human monocyte Fc gamma receptors with rat IgG2b. A new indicator for the Fc gamma RIIa (R-H131) polymorphism. J Immunol 154:1852–1860

Heiss MM, Strohlein MA, Jager M, Kimmig R, Burges A, Schoberth A, Jauch KW, Schildberg FW, Lindhofer H (2005) Immunotherapy of malignant ascites with trifunctional antibodies. Int J Cancer 117:435–443

Heiss MM, Murawa P, Koralewski P, Kutarska E, Kolesnik OO, Ivanchenko VV, Dudnichenko AS, Aleknaviciene B, Razbadauskas A, Gore M, Ganea-Motan E, Ciuleanu T, Wimberger P, Schmittel A, Schmalfeldt B, Burges A, Bokemeyer C, Lindhofer H, Lahr A, Parsons SL (2010) The trifunctional antibody catumaxomab for the treatment of malignant ascites due to epithelial cancer: results of a prospective randomized phase II/III trial. Int J Cancer 27(9):2209–2221

Hirschhaeuser F, Leidig T, Rodday B, Lindemann C, Mueller-Klieser W (2009) Test system for trifunctional antibodies in 3D MCTS culture. J Biomol Screen 14:980–990

Jacobs N, Mazzoni A, Mezzanzanica D, Negri DR, Valota O, Colnaghi MI, Moutschen MP, Boniver J, Canevari S (1997) Efficiency of T cell triggering by anti-CD3 monoclonal antibodies (mAb) with potential usefulness in bispecific mAb generation. Cancer Immunol Immunother 44:257–264

Jäger M, Schoberth A, Theissen B, Hess J, Friccius-Quecke H, Lindhofer H (2009a) Decrease of VEGF within malignant ascites during catumaxomab treatment: results from a pivotal phase II/III study. J Clin Oncol 27:15s (suppl; abstr 3029)

Jäger M, Schoberth A, Ruf P, Hess J, Lindhofer H (2009b) The trifunctional antibody ertumaxomab destroys tumor cells that express low levels of human epidermal growth factor receptor 2. Cancer Res 69:4270–4276

Jäger M, Schoberth A, Hennig M, Burges A, Heiss M, Wimberger P, Schmalfeldt B, Lindhofer H (2010) The trifunctional antibody catumaxomab (anti-EpCAM × anti-CD3) in patients with malignant ascites: immunomonitoring results of a pivotal phase II/III study (pooled population). J Clin Oncol 28:15s, suppl; abstr 2521)

Jung G, Freimann U, Von Marschall Z, Reisfeld RA, Wilmanns W (1991) Target cell-induced T cell activation with bi- and trispecific antibody fragments. Eur J Immunol 21(10):2431–2435

Kalergis AM, Ravetch JV (2002) Inducing tumor immunity through the selective engagement of activating Fcgamma receptors on dendritic cells. J Exp Med 195:1653–1659

Kiewe P, Hasmuller S, Kahlert S, Heinrigs M, Rack B, Marme A, Korfel A, Jager M, Lindhofer H, Sommer H, Thiel E, Untch M (2006) Phase I trial of the trifunctional anti-HER2 × anti-CD3 antibody ertumaxomab in metastatic breast cancer. Clin Cancer Res 12:3085–3091

Koolwijk P, van de Winkel JG, Pfefferkorn LC, Jacobs CW, Otten I, Spierenburg GT, Bast BJ (1991) Induction of intracellular Ca^{2+} mobilization and cytotoxicity by hybrid mouse monoclonal antibodies. Fc gamma RII regulation of Fc gamma RI-triggered functions or signaling? J Immunol 147:595–602

Larson LN, Johansson C, Lindholm L, Holmgren J (1988) Mouse monoclonal antibodies for experimental immunotherapy promotes killing of tumor cells. Int J Cancer 42:877–882

Lindhofer H, Mocikat R, Steipe B, Thierfelder S (1995) Preferential species-restricted heavy/light chain pairing in rat/mouse quadromas. Implications for a single-step purification of bispecific antibodies. J Immunol 155:219–225

Lindhofer H, Schobert A, Pelster D, Hess J, Herold J, Jäger M (2009) Elimination of cancer stem cells (CD133+/EpCAM+) from malignant ascites by the trifunctional antibody catumaxomab: results from a pivotal trial. J Clin Oncol 27:15s (suppl; abst 3014)

Linke R, Klein A, Seimetz D (2010) Catumaxomab: clinical development and future directions. MAbs 2(2):129–136

Milstein C, Cuello AC (1983) Hybrid hybridomas and their use in immunohistochemistry. Nature 305:537–540

Moore GL, Chen H, Karki S, Lazar GA (2010) Engineered Fc variant antibodies with enhanced ability to recruit complement and mediate effector functions. MAbs 2(2):181–189

Morecki S, Lindhofer H, Yacovlev E, Gelfand Y, Slavin S (2006) Use of trifunctional bispecific antibodies to prevent graft versus host disease induced by allogeneic lymphocytes. Blood 107(4):1564–1569

Morecki S, Lindhofer H, Yacovlev E, Gelfand Y, Ruf P, Slavin S (2008) Induction of long-lasting antitumor immunity by concomitant cell therapy with allogeneic lymphocytes and trifunctional bispecific antibody. Exp Hematol 36(8):997–1003

Müller D, Kontermann RE (2010) Bispecific antibodies for cancer immunotherapy. BioDrugs 24(2):89–98

Onishi H, Wada J, Suzuki H, Yamasaki A, Nagai S, Morisaki T, Katano M (2010) Effect of TGF-β 1 expressed on tumor-derived exosomes on survival and function of regulatory T cells. J Clin Oncol 28:15s (suppl; abstr 2567)

Ott MG, Lindhofer H, Linke RG, Hennig M, Martinius H, Klein A, Seimetz D (2010) The trifunctional antibody catumaxomab: correlation between immunological response and clinical outcome – new analysis of a pivotal phase II/III study. J Clin Oncol 28:15s (suppl; abstr 2551)

Parsons SL, Lang MW, Steele RJ (1996) Malignant ascites: a 2-year review from a teaching hospital. Eur J Surg Oncol 22:237–239

Perez P, Hoffman RW, Shaw S, Bluestone JA, Segal DM (1985) Specific targeting of cytotoxic T cells by anti-T3 linked to anti-target cell antibody. Nature 316:354–356

Porakishvili N, Kardava L, Jewell AP, Yong K, Glennie MJ, Akbar A, Lydyard PM (2004) Cytotoxic CD4+ T cells in patients with B cell chronic lymphocytic leukemia kill via a perforin-mediated pathway. Haematologica 89(4):435–443

Porter DL, Levine BL, Bunin N, Stadtmauer EA, Luger SM, Goldstein S, Loren A, Phillips J, Nasta S, Perl A, Schuster S, Tsai D, Sohal A, Veloso E, Emerson S, June CH (2006) A phase 1 trial of donor lymphocyte infusions expanded and activated ex vivo via CD3/CD28 costimulation. Blood 107(4):1325–1331

Racila E, Link BK, Weng WK, Witzig TE, Ansell S, Maurer MJ, Huang J, Dahle C, Halwani A, Levy R, Weiner GJ (2008) A polymorphism in the complement component C1qA correlates with prolonged response following rituximab therapy of follicular lymphoma. Clin Cancer Res 14:6697–6703

Renner C, Held G, Ohnesorge S, Bauer S, Gerlach K, Pfitzenmeier JP, Pfreundschuh M (1997) Role of naive and memory T cells in tumor cell lysis mediated by bi-specific antibodies. Immunobiology 197:122–132

Richards JO, Karki S, Lazar GA, Chen H, Dang W, Desjarlais JR (2008) Optimization of antibody binding to FcgammaRIIa enhances macrophage phagocytosis of tumor cells. Mol Cancer Ther 7:2517–2527

Riechelmann H, Wiesneth M, Schauwecker P, Reinhardt P, Gronau S, Schmitt A, Schroen C, Atz J, Schmitt M (2007) Adoptive therapy of head and neck squamous cell carcinoma with antibody coated immune cells: a pilot clinical trial. Cancer Immunol Immunother 56:1397–1406

Riesenberg R, Buchner A, Pohla H, Lindhofer H (2001) Lysis of prostate carcinoma cells by trifunctional bispecific antibodies (alpha EpCAM × alpha CD3). J Histochem Cytochem 49:911–917

Robb RJ, Munck A, Smith KA (1981) T cell growth factor receptors. Quantitation, specificity, and biological relevance. J Exp Med 154:1455–1474

Ruf P, Lindhofer H (2001) Induction of a long-lasting antitumor immunity by a trifunctional bispecific antibody. Blood 98:2526–2534

Ruf P, Jager M, Ellwart J, Wosch S, Kusterer E, Lindhofer H (2004) Two new trifunctional antibodies for the therapy of human malignant melanoma. Int J Cancer 108:725–732

Ruf P, Kluge M, Jäger M, Burges A, Volovat C, Heiss MM, Hess J, Wimberger P, Brandt B, Lindhofer H (2010) Pharmacokinetics, immunogenicity and bioactivity of the therapeutic antibody catumaxomab intraperitoneally administered to cancer patients. Br J Clin Pharmacol 69(6):617–625

Sebastian M, Passlick B, Friccius-Quecke H, Jager M, Lindhofer H, Kanniess F, Wiewrodt R, Thiel E, Buhl R, Schmittel A (2007) Treatment of non-small cell lung cancer patients with the trifunctional monoclonal antibody catumaxomab (anti-EpCAM × anti-CD3): a phase I study. Cancer Immunol Immunother 56:1637–1644

Sebastian M, Kiewe P, Schuette W, Brust D, Peschel C, Schneller F, Ruhle KH, Nilius G, Ewert R, Lodziewski S, Passlick B, Sienel W, Wiewrodt R, Jager M, Lindhofer H, Friccius-Quecke H, Schmittel A (2009a) Treatment of malignant pleural effusion with the trifunctional antibody catumaxomab (Removab) (anti-EpCAM × Anti-CD3): results of a phase 1/2 study. J Immunother 32:195–202

Sebastian M, Kuemmel A, Schmidt M, Schmittel A (2009b) Catumaxomab: a bispecific trifunctional antibody. Drugs Today 45(8):589–597

Shen J, Zhu Z (2008) Catumaxomab, a rat/murine hybrid trifunctional bispecific monoclonal antibody for the treatment of cancer. Curr Opin Mol Ther 10(3):273–284

Smith W, Jarrett AL, Beattie RE, Corvalan JR (1992) Immunoglobulins secreted by a hybrid-hybridoma: analysis of chain assemblies. Hybridoma 11:87–98

Staerz UD, Kanagawa O, Bevan MJ (1985) Hybrid antibodies can target sites for attack by T cells. Nature 314:628–631

Stanglmaier M, Faltin M, Ruf P, Bodenhausen A, Schroder P, Lindhofer H (2008) Bi20 (FBTA05), a novel trifunctional bispecific antibody (anti-CD20 × anti-CD3), mediates efficient killing of B-cell lymphoma cells even with very low CD20 expression levels. Int J Cancer 123:1181–1189

Stemmler HJ, Salat C, Lindhofer H, Menzel H, Untch M, Kahlert S, Konecny G, Sauer H, Ledderose G, Heinemann V, Kolb HJ (2005) Combined treatment of metastatic breast cancer (MBC) by high-dose chemotherapy (HDCT) and bispecific antibodies: a pilot study. Anticancer Res 25:3047–3054

Ströhlein MA, Heiss MM (2009) Intraperitoneal immunotherapy to prevent peritoneal carcinomatosis in patients with advanced gastrointestinal malignancies. J Surg Oncol 100(4):329–330

Ströhlein MA, Siegel R, Jäger M, Lindhofer H, Jauch KW, Heiss MM (2009) Induction of anti-tumor immunity by trifunctional antibodies in patients with peritoneal carcinomatosis. J Exp Clin Cancer Res 28:18

Tiebout RF, van Boxtel-Oosterhof F, Stricker EA, Zeijlemaker WP (1987) A human hybrid hybridoma. J Immunol 139:3402–3405

Walz A, Mack B, Schmitt B, Gires O, Wollenberg B, Zeidler R (2004) Antitumour effects of a bispecific trivalent antibody in multicellular tumour spheroids. Anticancer Res 24:887–893

Weiner GJ, Kostelny SA, Hillstrom JR, Cole MS, Link BK, Wang SL, Tso JY (1994) The role of T cell activation in anti-CD3 × antitumor bispecific antibody therapy. J Immunol 152:2385–2392

Weng WK, Levy R (2003) Two immunoglobulin G fragment C receptor polymorphisms independently predict response to rituximab in patients with follicular lymphoma. J Clin Oncol 21:3940–3947

Zeidler R, Reisbach G, Wollenberg B, Lang S, Chaubal S, Schmitt B, Lindhofer H (1999) Simultaneous activation of T cells and accessory cells by a new class of intact bispecific antibody results in efficient tumor cell killing. J Immunol 163:1246–1252

Zeidler R, Mysliwietz J, Csanady M, Walz A, Ziegler I, Schmitt B, Wollenberg B, Lindhofer H (2000) The Fc-region of a new class of intact bispecific antibody mediates activation of accessory cells and NK cells and induces direct phagocytosis of tumour cells. Br J Cancer 83:261–266

Chapter 17
Bispecific Antibodies for the Retargeting of Cytokines

Bruno Robert, Christel Larbouret, David Azria, Jean-Pierre Mach, and André Pèlegrin

17.1 Introduction

The use of cytokines as adjuvant for different forms of cancer therapy represents one of the most promising areas of applied cancer research (Seruga et al. 2008). As examples, Interleukin-2 is now been a classical treatment for renal cell carcinoma (FDA approved in 1992) and always evaluated for other cancer immunotherapy, interferon α for melanoma therapy (FDA approved in 1995) and others, such as IL-15 emerged as good candidate since IL-15 was ranked 1st/124 immunomodulatory drugs in clinical and preclinical trials to treat cancer by an NCI consortium (Cheever 2008).

Despite some success, the usually narrow window between toxicity and efficacy has limited the use of cytokines as cancer treatment.

Among cytokines, the pleiotropic TNFα has focused a lot of interest since it has been shown in experimental animal models to have a potent antitumor activity, through its proapoptotic activity on certain tumors, as well as on the endothelial cells of the tumor vasculature (Brouckaert et al. 1986; Carswell et al. 1975; Balkwill et al. 1986; Asher et al. 1991; Creasey et al. 1986; Havell et al. 1988, Lejeune and Eggermont 2007). Thanks to the latter property, this cytokine has gained the name of "Tumor Necrosis Factor." However, it was rapidly found that TNF is a central immunomodulator of inflammation and, when produced in excess by macrophages during acute infection, a key factor responsible for septic shock (Tracey et al. 1987). That is the reason why cancer therapy clinical trials using systemic injection of TNFα were disappointing since patients were found to have significantly lower maximum

B. Robert (✉) • C. Larbouret • D. Azria • A. Pèlegrin
IRCM-Institut de Recherche en Cancérologie de Montpellier, Université MontpellierI, CRLC Val d'Aurelle-Paul Lamarque, 208 Rue des Apothicaires, 34298 Montpellier, Cedex 5, France
e-mail: bruno.robert@inserm.fr

J.-P. Mach
Department of Biochemistry, University of Lausanne, Chemin des Boveresses 155, CH-1066 Epalinges, Switzerland

R.E. Kontermann (ed.), *Bispecific Antibodies*,
DOI 10.1007/978-3-642-20910-9_17,

tolerated dose [about 10 μg/kg; (Moritz et al. 1989; Abbruzzese et al. 1989)] than mice [400 μg/kg; (Asher et al. 1991; Havell et al. 1988)] and the low doses injectable to patients had no antitumor effect. Thus, toxicity induced by TNFα prevented its systemic use in cancer patients.

To overcome these problems, a strategy involving isolated limb perfusion (ILP) (Lienard et al. 1992) of TNFα has demonstrated its potential for cancer therapy, but only when a high enough therapeutic concentration of TNFα was obtained in the tumor with a nontoxic systemic concentration. To date, TNFα is only used by ILP in sarcoma therapy in association with Melphalan and Hyperthermia with a good rate of success for the treatment of tumors limited to the perfused limb (Lejeune and Eggermont 2007). Interestingly, in the course of these attempts of cancer therapy by TNF injection, a relatively new vasoactive property of TNF was discovered, that is the capacity of small doses of this cytokine to increase the tumor blood flow as well as the permeability of tumor neovascularization, before the potential subsequent collapse of these vessels (Folli et al. 1993; Ten Hagen et al. 2000). Furthermore, the antitumor efficiency of TNF injection through ILP was in great part attributed to these new vasoactive properties of TNF, leading to a better accessibility to the tumor cells of the chemotherapeutic drug, always coinjected with TNF (de Wilt et al. 2000; Lejeune 2002). As we will see in this chapter, these TNF properties of increasing tumor blood flow and vascular permeability are likely to be responsible for the enhancement of the efficiency of external beam radiotherapy by TNF that we have demonstrated (Azria et al. 2003a, and b; Larbouret et al. 2007).

17.2 Different Way to Target Cytokine

In this context, different other strategies have been proposed to target specifically cytokines in tumors. One obvious approach was to directly couple cytokines with antitumor antibody using either chemical linkage (LeBerthon et al. 1991; Khawli et al. 1994; Rosenblum et al. 1991) or genetic fusion proteins between the antibody or its fragment directed against tumor-associated antigen (TAA) and the cytokines (Yang et al. 1995; Hoogenboom et al. 1991; Dela Cruz et al. 2004). These conjugates or fusion proteins were called immunocytokine (ICK). Different studies have shown the ability of some ICK to target tumor and to induce a tumor regression in mice (Kaspar et al. 2007; Dela Cruz et al. 2000). These promising results led to different phase I, II clinical trials (King et al. 2004; Ribas et al. 2009).

Another reported approach was to use gene therapy to transfect plasmids coding for different cytokines into tumor cells (Asher et al. 1991; Mullen et al. 1992), which involve invasive protocol and the known present limitations associated with gene therapy. Recently, new idea emerged with the design of mutated cytokines, such as TNFα with increased affinity to theirs receptors or mutant cytokines fused to vasculature localization peptide (Curnis et al. 2000).

In our group, we developed an original strategy consisting in the design of a bispecific antibody with one arm directed against a TAA and the other against a cytokine, in particular, TNFα (Robert et al. 1996). We hypothesized that this bispecific antibody could first target the cytokine specifically into tumors, and second increase the otherwise rapid circulating half-life of the cytokine, while, as a third property, the bispecific antibody could decrease the systemic toxicity observed following the intravenous injection of free cytokine. Results obtained with this last strategy is discussed in this chapter.

17.3 Bispecific Antibody to Target TNF

Based on previous reports, using bispecific antibody to target vinca alkaloides or saporin to tumor cells (Corvalan et al. 1987; French et al. 1995), as well as on the demonstration that bispecific antibody could target at the tumor site a radiolabeled hapten (Chatal et al. 1995), we developed a bispecific antibody conjugate having the capacity to target the CEA expressing tumor cells and to bind the human TNFα cytokine. Thus, this bispecific antibody could have the combined properties to target and concentrate TNFα at the tumor site, thus increasing the local availability of the cytokine and to decrease the cytokine systemic toxicity.

To produce this bispecific antibody, we used anti-carcinoembryonic antigen (CEA) monoclonal antibody 35A7 which already demonstrated efficient tumor targeting capacity both in xenografted mice model and in colon carcinoma patients in the form of $F(ab')_2$ or Fab fragments (Delaloye et al. 1986). Anti-CEA/anti-human TNFα bispecific antibody conjugates were produced using a chemical thioether link between Fab' fragments from anti-CEA 35A7 (Girardet et al. 1986) and the anti-human TNFα tnf18 mAb (provided by Dr. M. Brockhaus, Hoffmann-La Roche AG, Basel) using the bismaleimide method (Glennie et al. 1987). Figure 17.1, show a schematic diagram of the proposed mechanism of TNF tumor targeting through the bispecific antibody and its potential release to the local tumor or endothelial cells.

In contrast to immunocytokine in which cytokine is classically fused at the C-terminal of the Fc part of the mAb or chemically linked to mAbs, we hypothesized in a bispecific antibody tumor targeting strategy that when the TNF-loaded bispecific antibody reaches the tumor site the cytokines molecules can "jump" to their specific receptors present on tumor cells and endothelial cells from the tumor neovasculature. The "jump" of the cytokine to TNF receptor could be due to the higher affinity of the TNF receptor for its ligand, when compared with that of the mAb binding paratope.

The increased concentration of free TNFα at the tumor site should generate all the pleiotropic proprieties of this cytokine, such as increased local blood flow and vascular permeability and in some cases apoptosis/necrosis of tumor cells. Furthermore, the increased tumor vasoactivity and, the oxidative damage produced by TNFα can enhance cellular damage produced by ionizing radiation. Our group and others observed a synergistical antitumor effect of the combination of TNFα with ionizing radiation in vitro and demonstrated an enhancement of external beam irradiation antitumor effect in association with TNFα in some murine tumors and human tumor xenografts in vivo.

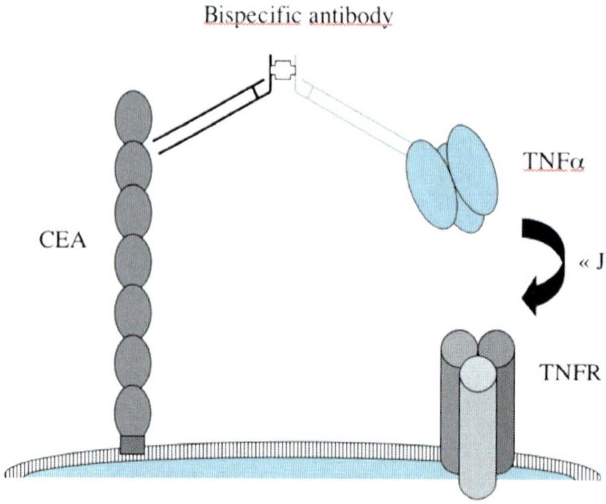

Fig. 17.1 Scheme representing membrane anchored carcinoembryonic antigen (CEA) and tumor necrosis factor receptor (TNFR). Bispecific antibody bind CEA and concentrate tumor necrosis factor α (TNFα) in the tumor at the vicinity of TNFR which attract ("JUMP") TNFα due to its higher affinity for the cytokine

One of the critical and limiting point for developing bispecific antibody using chemical linker as described here, is the selection of a parental mAbs showing high enough affinity to their targets even in the Fab monovalent form. Different anti-TNFα mAbs were tested and showed a massive loss of affinity after being reduced in Fab' fragments by pepsin digestion and mild reduction. Our final choice was the tnf18 mAb which present high affinity for TNFα in the monovalent Fab' format. The same point was verified with the evaluation of different anti-CEA mAbs, leading to the selection of hybridoma 35A7. The use of thioether chemical linkage between the two Fab' generated a good flexibility between the two Fab' fragments and has allowed to obtain a bispecific antibody with high affinity to both CEA tumor target and TNFα cytokine, as determined by BIAcore measurement, in comparison with intact parental $F(ab')_2$ (Table 17.1).

Biodistribution of the bispecific antibody is one of the most crucial characteristic to be taken in account for cytokine targeting. Thank to the high affinity of monovalent anti-CEA 35A7 Fab' used, the tumor localization and the biodistribution of the bispecific antibody was similar, at different time after injection, to those obtained with the dimeric parental anti-CEA 35A7 $F(ab')_2$, in a model of nude mice xenografted by a CEA expressing human colon carcinoma (Fig. 17.2). Furthermore, we observed relatively high amounts, 17% of the injected dose of the bispecific antibody per gram of tumor remaining at 48 h, which were similar to those obtained with the parental antitumor $F(ab')_2$ (Robert et al. 1996).

As previously described for other cytokines, TNFα showed a rapid blood and tissue elimination after i.v. injection in mice (Fig. 17.3).

The blood disappearance curves of TNFα fitted a biexponential model with values of 10 min and 5 h for the α and β half-lives, respectively. This rapid blood

Table 17.1 Affinity constants of different soluble proteins for a CEA or TNFα sensor chip

	Protein immobilized on sensor chips	
Soluble proteins	CEA	TNFα
BAb	1.7×10^{9a}	9.4×10^{8}
TNFα on BAb[b]	6.6×10^{8}	ND
CEA on BAb[b]	ND	2.5×10^{8}
35A7	9.7×10^{8}	ND
35A7 F(ab')$_2$	2.2×10^{8}	No fixation
tnf18	No fixation	1.8×10^{8}
tnf18F(ab')$_2$	ND	2.5×10^{8}
CEA	ND	No fixation
TNFα	No fixation	ND

ND not done[a]The *Ka* (k_{on}/k_{off}) values (M^{-1}) for the interaction of different soluble proteins with CEA or TNFα immobilized on different sensor chips were determined using BIAcore and BIAevaluation (Pharmacia)[b]For the determination of the *Ka* values of CEA or TNFα on BAb, BAb was first exposed to a TNFα- or CEA-bearing sensor chip, and when equilibrium was reached, CEA or TNFα was then injected

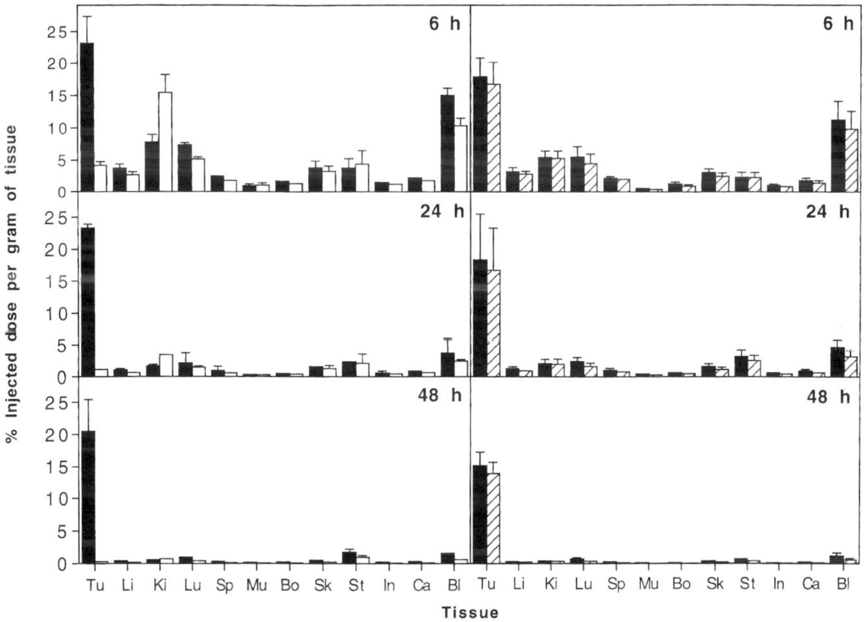

Fig. 17.2 Biodistribution of bispecific antibody and parental F(ab')$_2$ fragments in nude mice bearing human colon carcinoma. Iodine-125 radiolabeled bispecific antibody (*black bar*) was intravenously injected with iodine-131 parental F(ab')$_2$ (anti-TNF; *white bar* or anti-CEA; *hatched bar*) in mice which were dissected 6, 24, or 48 later. Results are expressed as the mean of percentages of injected dose per gram of tissue. *Tu* tumor; *Li* liver; *Ki* kidneys; *Lu* lung; *Sp* spleen; *Mu* muscle; *Bo* bone; *Sk* skin; *St* stomach; *In* intestine; *Ca* carcass; *Bl* blood (adapted from Robert et al. 1996)

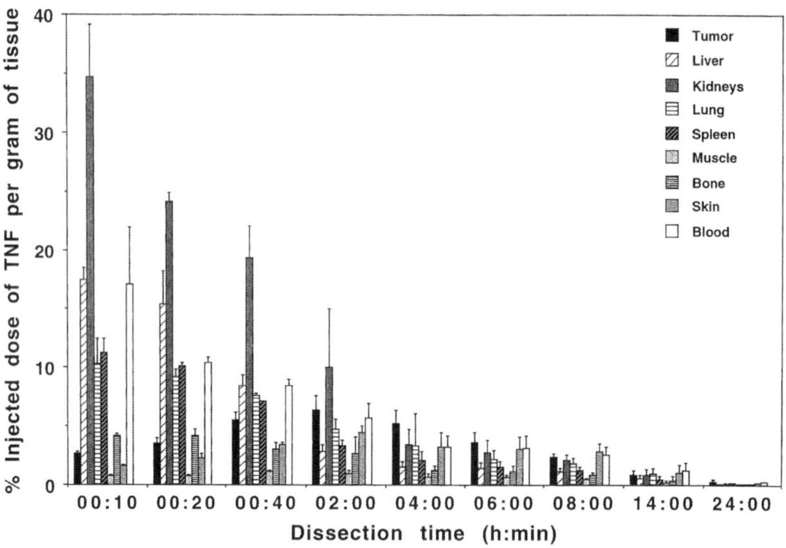

Fig. 17.3 Biodistribution of intravenous injection iodine-125 radiolabeled TNFα in nude mice bearing human colon carcinoma. Mice were dissected at different time and results are expressed as percentages of injected dose per gram of tissue (adapted from Robert et al. 1996)

elimination is due to a renal clearance since high-radiolabeled TNFα was recovered in this organ already 10 min after i.v. injection. As expected, we observed an increase of systemic half-life of TNFα when it was preincubated with the bispecific antibody before intravenous injection.

Near than twofold increase of blood concentration of TNFα was measured 6 h postinjection (Fig. 17.4), but the biodistribution of TNFα mixed with bispecific antibody was not the same as that of the bispecific antibody. At such time specified, the bispecific antibody showed between 10 and 15% ID/g in blood, in contrast to 4% ID/g of TNFα, also in blood when injected with bispecific antibody (Figs. 17.2 and 17.3). This observation strengthen the hypothesis of a possible "jump" of cytokine from the bispecific antibody to its high-affinity TNF receptor.

Furthermore, when injected i.v. alone, TNFα did not show any preferential tumor retention (Fig. 17.2). In contrast, when radiolabeled TNFα was injected with the bispecific antibody a high increase of the cytokine tumor localization was observed over time with a limited increase in other organs. The tumor uptake enhancement of TNFα, injected with the bispecific antibody, when compared with TNFα alone, was 3.25-fold, at 6 h and 8.15-fold at 14 h, after injection (Fig. 17.4).

By using radiolabeled TNFα, we also confirmed that the bispecific antibody is able to increase the retention of TNFα in the mice whole body. Indeed, 14 h after injection two times more TNFα was recovered from mice injected with bispecific antibody and TNFα than in mice injected with TNFα alone. Interestingly, this increased amounts of TNFα was in great part specifically localized in the tumor by bispecific antibody, since a 3.2% increase of tumor concentration was observed, between 6 and 14 h, when compared with the tumor concentration observed in control mice injected with TNFα alone between.

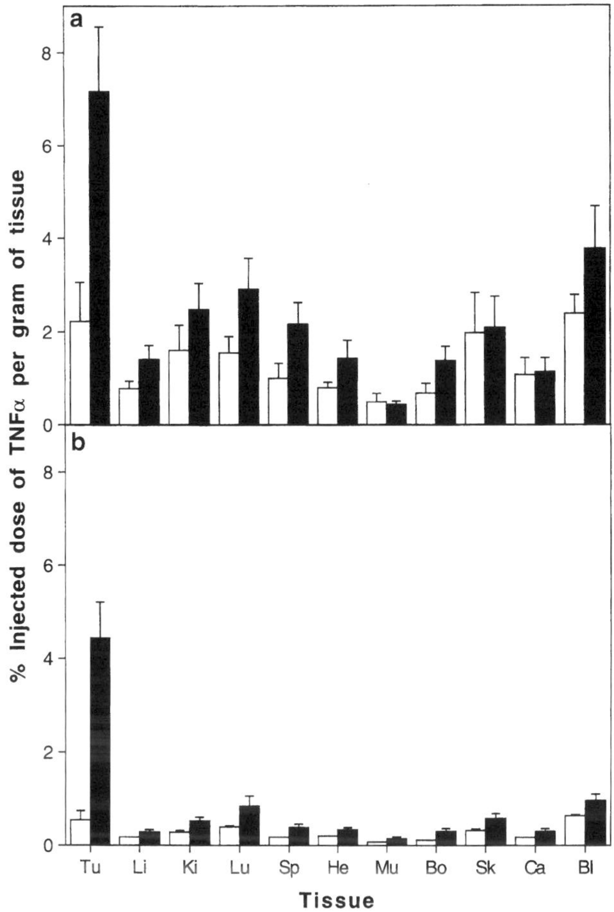

Fig. 17.4 Tumor targeting of TNFα when premixed with bispecific antibody in nude mice bearing human colon carcinoma. Mice were intravenously injected with iodine-131-labeled TNFα (*white bar*) alone or with bispecific antibody (*black bar*). Mice are dissected 6 (A) or 14 h (B) and results are expressed as percentages of injected dose per gram of tissue. *Tu* tumor; *Li* liver; *Ki* kidneys; *Lu* lung; *Sp* spleen; *Mu* muscle; *Bo* bone; *Sk* skin; *St* stomach; *In* intestine; *Ca* carcass; *Bl* blood (adapted from Robert et al. Cancer Res. 1996)

17.4 Tumor Therapy Using Cytokine Targeting and External Beam Radiotherapy

In vitro and in vivo results have reported an additive or supra-additive effect of TNFα, when associated with external beam radiation (Hallahan et al. 1990; Azria et al. 2003a, 2004). In view of the tumor targeting properties of the bispecific antibody strategy, we explored the therapeutic effect of TNFα targeted by bispecific antibody, in association with external beam radiotherapy in colon and pancreatic carcinomas models.

In a model of human colon carcinoma, relatively resistant to radiotherapy, xenografted in nude mice (Azria et al. 2003b), we demonstrated that TNFα targeted to the tumor by the anti-CEA/anti-TNF bispecific antibody in association with external radiotherapy enhanced significantly the tumor growth delay and the mice survival in comparison with radiotherapy alone or radiotherapy in combination with free TNFα (Figs. 17.5 and 17.6).

This association of tumor-targeted TNFα with external radiotherapy was also successful in another type of human carcinoma xenograft, the human pancreatic carcinoma BX-PC3 which express CEA (Azria et al. 2003a). As shown in Fig. 17.7, mice bearing well-established BX-PC3 xenografts, treated with TNF/bispecific antibody in association with irradiation showed better survival (time for tumors to reach 1,500 mm^3) than mice receiving each treatment modality alone (Fig. 17.7).

In terms of toxicity, at the end of all treatment, no significant differences were found in mouse body weight between control and treated group. No diarrhea was observed in any group, suggesting the absence of digestive toxicity. No other signs of toxicity (fluid retention, respiratory distress, etc.) were observed in any of the animals during the course of the therapeutic study.

However, the effect of TNFα on the immune system must also be taken into consideration, thus our results needed to be confirmed in an immunocompetent model in which the tumoricidal action of TNFα could be evaluated mainly by the immunologic and nonimmunologic mechanisms, such as damage to the tumor vasculature. We used the model of syngenic murine colon carcinoma MC-38 transfected with human CEA (Clarke et al. 1998) and transplanted in CEA transgenic

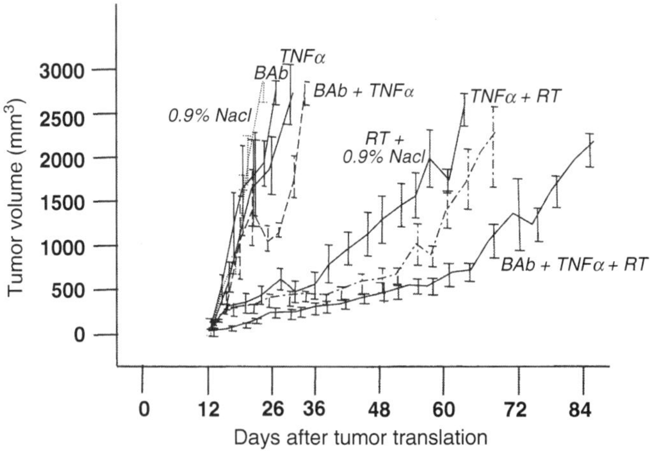

Fig. 17.5 Tumor volume as a function of time after LS174T cell transplantation and treatment with TNFα and/or bispecific antibody (BAb) and/or irradiation. Mice are randomly divided into seven groups as follows: Group 1, i.v. injection of NaCl; Group 2, i.v. injection of TNFα; Group 3, i.v. injection of BAb; Group 4, i.v. injection of BAb + TNFα; Group 5, local radiation; Group 6, local radiation + i.v. injection of TNFα 3 h before irradiation; Group 7, local radiation + i.v. injection of BAb + TNFα. $n = 5$ mice per group (adapted from Azria et al. 2003b)

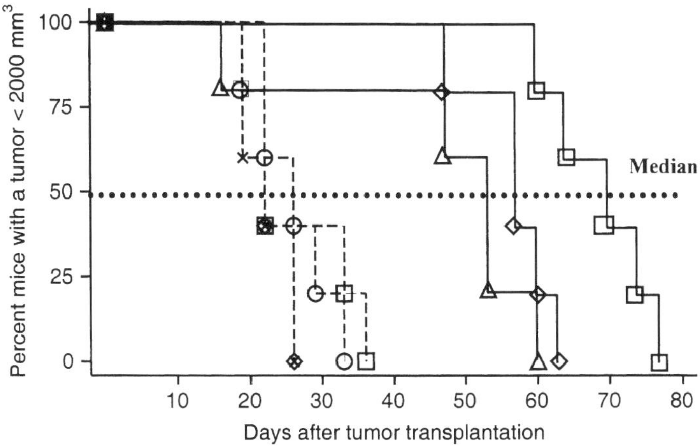

Fig. 17.6 Kaplan–Meier survival curves obtained as a function of time for all groups: Group 1, *dotted line* (X) no treatment; Group 2, *dotted line* (*open diamond*) TNFα alone; Group 3, *dotted line* (*open circle*) bispecific antibody (BAb) alone; Group 4, *dotted line* (*open square*) BAb + TNFα; Group 5, *solid line* (*open triangle*) local irradiation (RT) alone; Group 6, *solid line* (*open diamond*) RT + TNFα; Group 7, *solid line* (*open square*) RT + TNFα + BAb (adapted from Azria et al. 2003b)

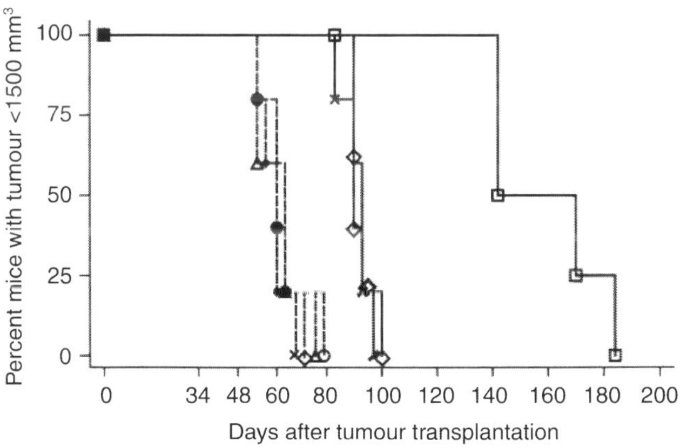

Fig. 17.7 Kaplan–Meier survival curves obtained as a function of time for all groups: Group 1, *dotted line* (*circle*) no treatment; Group 2, *dotted line* (*diamond*) TNFα; Group 3, *dotted line* (*cross*) bispecific antibody (BAb) alone; Group 4, *dotted line* (*triangle*) BAb + TNFα; Group 5, *solid line* (*cross*) local irradiation (RT) alone; Group 6, *solid line* (*diamond*) RT + TNFα; Group 7, *solid line* (*square*) RT + BAb + TNFα (adapted from Azria et al. 2003a)

C57/Bl6 mice). In these immunocompetent mice, we confirmed the advantage of our combination therapy strategy and observed a significantly longer tumor growth delay when TNFα/bispecific antibody conjugates were injected in association with radiotherapy, when compared with radiotherapy alone or radiotherapy combined with free TNFα (Fig. 17.8).

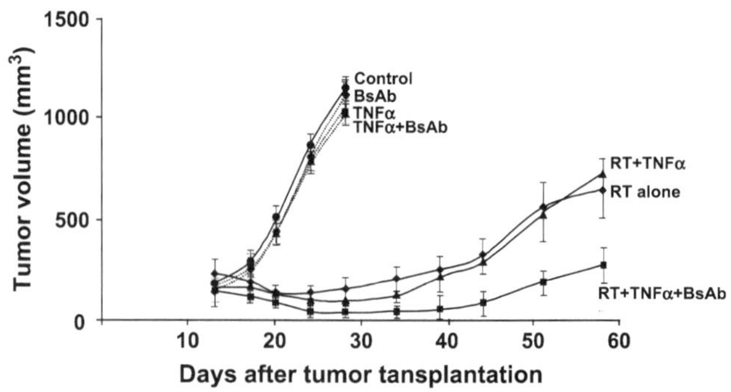

Fig. 17.8 Tumor volume as function of time of syngeneic MC38-CEA+ murine colon carcinoma growth in CEA-transgenic mice. Seven groups of mice were treated. All groups detailed on graphs after removing all cured mice from each group. Values represent mean tumor size of two consecutive experiments. *BsAb* bispecific antibody; *TNFα* tumor necrosis factor α; *RT* radiotherapy (adapted from Larbouret et al. 2007)

Indeed, in this model of aggressive murine colon carcinoma in syngenic and immunocompetent C57/Bl6 mice, we demonstrated a radiocurability of 50% of mice in two different experiences in the group injected with TNFα/bispecific antibody in association with radiotherapy, when compared with 18 and 20% cured mice in the groups receiving radiotherapy alone or i.v. injection of free TNFα in association with radiotherapy, respectively.

As shown in Fig. 17.8, the injection of TNFα/bispecific antibody conjugates with radiotherapy induced an increase in the median delay to reach a tumor volume of 1,000 mm^3, with a gain of 22 and 40 days, when compared with the groups receiving radiotherapy alone or radiotherapy + TNFα, respectively (Larbouret et al. 2007). Histologically, we could observed a dramatic increase in tumor necrosis when TNF was tumor targeted (Fig. 17.9).

The key in vivo results presented in this study are the correlation between the high number of complete tumor responses and the increase of tumor growth delay. To these two factors can be added, the observation of the subsequent detection of a high level of endogenous TNFα mRNA and a high percentage of necrosis by histopathologic examination (Fig. 17.9) in the tumor from the mice treated by radiotherapy in combination with bispecific antibody/TNFα complex. The increase of endogenous TNFα mRNA could be explained by cellular exposure to ionizing radiation (Hallahan et al. 1989).

17.5 Conclusion

All these preclinical in vivo data illustrate the therapeutic potential of the use of bispecific antibody to redirect and concentrate a cytokine such as TNFα at the tumor site. Furthermore, the results demonstrate that the association of external

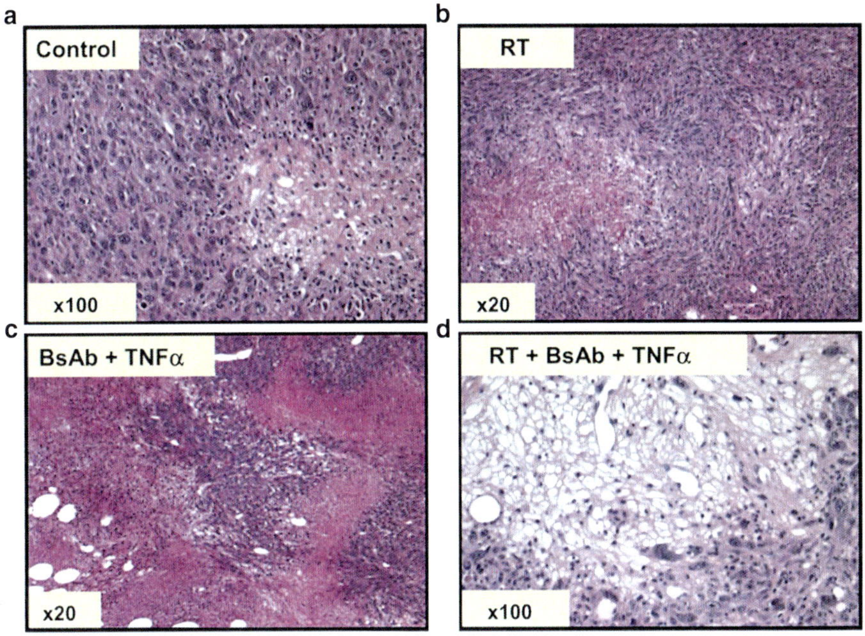

Fig. 17.9 Effect of local radiotherapy (RT) and tumor necrosis factor (TNF)-α combination on histologic parameters of explanted MC38-CEA+ tumors. Female transgenic mice with established tumor treated with (**a**) vehicle, (**b**) RT, (**c**) bispecific antibody (BsAb) plus TNFα, and (**d**) BsAb + TNFα + RT. After first week of treatment, tumors were removed and sections of fixed samples embedded in paraffin were stained with hematoxylin and eosin. We then performed two dedicated experiments to histologic analyses with one mouse per treatment group (adapted from Larbouret et al. 2007)

beam irradiation with bispecific antibody-targeted TNFα can significantly increase the curability of tumor in immunocompetent mice.

Based on these results, it is tempting to speculate that this association of bispecific antibody-targeted TNFα and radiotherapy could be directly applied to the clinic for the improvement of the radiotherapy of CEA expressing carcinomas. However, two practical problems should be resolved before planning a clinical trial.

The first point concerns the bispecific antibody. Although some clinical trials have been performed with bispecific antibody conjugates generated by chemical linkage (Curnow 1997) as used in our studies, the majority of clinical studies have used recombinant bispecific antibodies obtained by gene fusion which can be produced in large amounts under GMP conditions. Thus, we will have to produce our own recombinant anti-TNF/anti-CEA fusion protein (Müller et al. 2007; Osada et al. 2010). The second problem is that the toxicity of TNF is known to be much higher in human than in mice, thus very careful toxicity studies will be required, taking advantage of the fact that the anti-TNF binding site of our bispecific antibody will decrease its toxicity.

References

Abbruzzese JL et al (1989) Phase I trial of recombinant human gamma-interferon and recombinant human tumor necrosis factor in patients with advanced gastrointestinal cancer. Cancer Res 49(14):4057–4061

Asher AL et al (1991) Murine tumor cells transduced with the gene for tumor necrosis factor-alpha. Evidence for paracrine immune effects of tumor necrosis factor against tumors. J Immunol 146(9):3227–3234

Azria D et al (2003a) Potentiation of ionising radiation by targeting tumour necrosis factor alpha using a bispecific antibody in human pancreatic cancer. Br J Cancer 89(10):1987–1994

Azria D et al (2003b) Enhancement of radiation therapy by tumor necrosis factor alpha in human colon cancer using a bispecific antibody. Int J Radiat Oncol Biol Phys 55(5):1363–1373

Azria D et al (2004) A bispecific antibody to enhance radiotherapy by tumor necrosis factor-alpha in human CEA-expressing digestive tumors. Int J Radiat Oncol Biol Phys 58(2):580–588

Balkwill FR et al (1986) Human tumor xenografts treated with recombinant human tumor necrosis factor alone or in combination with interferons. Cancer Res 46(8):3990–3993

Brouckaert PG et al (1986) In vivo anti-tumour activity of recombinant human and murine TNF, alone and in combination with murine IFN-gamma, on a syngeneic murine melanoma. Int J Cancer 38(5):763–769

Carswell EA et al (1975) An endotoxin-induced serum factor that causes necrosis of tumors. Proc Natl Acad Sci USA 72(9):3666–3670

Chatal JF et al (1995) Bifunctional antibodies for radioimmunotherapy. Hybridoma 14(2):125–128

Cheever MA (2008) Twelve immunotherapy drugs that could cure cancers. Immunol Rev 222:357–368

Clarke P et al (1998) Mice transgenic for human carcinoembryonic antigen as a model for immunotherapy. Cancer Res 58(7):1469–1477

Corvalan JR et al (1987) Specific in vitro and in vivo drug localisation to tumour cells using a hybrid-hybrid monoclonal antibody recognising both carcinoembryonic antigen (CEA) and vinca alkaloids. Cancer Immunol Immunother 24(2):133–137

Creasey AA, Reynolds MT, Laird W (1986) Cures and partial regression of murine and human tumors by recombinant human tumor necrosis factor. Cancer Res 46(11):5687–5690

Curnis F et al (2000) Enhancement of tumor necrosis factor alpha antitumor immunotherapeutic properties by targeted delivery to aminopeptidase N (CD13). Nat Biotechnol 18(11): 1185–1190

Curnow RT (1997) Clinical experience with CD64-directed immunotherapy. An overview. Cancer Immunol Immunother 45(3–4):210–215

de Wilt JH et al (2000) Tumour necrosis factor alpha increases melphalan concentration in tumour tissue after isolated limb perfusion. Br J Cancer 82(5):1000–1003

Dela Cruz JS et al (2000) Recombinant anti-human HER2/neu IgG3-(GM-CSF) fusion protein retains antigen specificity and cytokine function and demonstrates antitumor activity. J Immunol 165(9):5112–5121

Dela Cruz JS et al (2004) Antibody-cytokine fusion proteins: innovative weapons in the war against cancer. Clin Exp Med 4(2):57–64

Delaloye B et al (1986) Detection of colorectal carcinoma by emission-computerized tomography after injection of 123I-labeled Fab or $F(ab')_2$ fragments from monoclonal anti-carcinoembryonic antigen antibodies. J Clin Invest 77(1):301–311

Folli S et al (1993) Tumor-necrosis factor can enhance radio-antibody uptake in human colon carcinoma xenografts by increasing vascular permeability. Int J Cancer 53(5):829–836

French RR et al (1995) Delivery of the ribosome-inactivating protein, gelonin, to lymphoma cells via CD22 and CD38 using bispecific antibodies. Br J Cancer 71(5):986–994

Girardet C et al (1986) Immunochemical characterization of two antigens recognized by new monoclonal antibodies against human colon carcinoma. J Immunol 136(4):1497–1503

Glennie MJ et al (1987) Preparation and performance of bispecific F(ab' gamma)2 antibody containing thioether-linked Fab' gamma fragments. J Immunol 139(7):2367–2375

Hallahan DE et al (1989) Increased tumor necrosis factor alpha mRNA after cellular exposure to ionizing radiation. Proc Natl Acad Sci USA 86(24):10104–10107

Hallahan DE et al (1990) The interaction between recombinant human tumor necrosis factor and radiation in 13 human tumor cell lines. Int J Radiat Oncol Biol Phys 19(1):69–74

Havell EA, Fiers W, North RJ (1988) The antitumor function of tumor necrosis factor (TNF), I. Therapeutic action of TNF against an established murine sarcoma is indirect, immunologically dependent, and limited by severe toxicity. J Exp Med 167(3):1067–1085

Hoogenboom HR, Raus JC, Volckaert G (1991) Targeting of tumor necrosis factor to tumor cells: secretion by myeloma cells of a genetically engineered antibody-tumor necrosis factor hybrid molecule. Biochim Biophys Acta 1096(4):345–354

Kaspar M, Trachsel E, Neri D (2007) The antibody-mediated targeted delivery of interleukin-15 and GM-CSF to the tumor neovasculature inhibits tumor growth and metastasis. Cancer Res 67(10):4940–4948

Khawli LA, Miller GK, Epstein AL (1994) Effect of seven new vasoactive immunoconjugates on the enhancement of monoclonal antibody uptake in tumors. Cancer 73(3 Suppl):824–831

King DM et al (2004) Phase I clinical trial of the immunocytokine EMD 273063 in melanoma patients. J Clin Oncol 22(22):4463–4473

Larbouret C et al (2007) Radiocurability by targeting tumor necrosis factor-alpha using a bispecific antibody in carcinoembryonic antigen transgenic mice. Int J Radiat Oncol Biol Phys 69(4):1231–1237

LeBerthon B et al (1991) Enhanced tumor uptake of macromolecules induced by a novel vasoactive interleukin 2 immunoconjugate. Cancer Res 51(10):2694–2698

Lejeune FJ (2002) Clinical use of TNF revisited: improving penetration of anti-cancer agents by increasing vascular permeability. J Clin Invest 110(4):433–435

Lejeune FJ, Eggermont AM (2007) Hyperthermic isolated limb perfusion with tumor necrosis factor is a useful therapy for advanced melanoma of the limbs. J Clin Oncol 25(11):1449–1450

Lienard D et al (1992) High-dose recombinant tumor necrosis factor alpha in combination with interferon gamma and melphalan in isolation perfusion of the limbs for melanoma and sarcoma. J Clin Oncol 10(1):52–60

Moritz T et al (1989) Phase I study of recombinant human tumor necrosis factor alpha in advanced malignant disease. Cancer Immunol Immunother 29(2):144–150

Mullen CA et al (1992) Fibrosarcoma cells transduced with the IL-6 gene exhibited reduced tumorigenicity, increased immunogenicity, and decreased metastatic potential. Cancer Res 52(21):6020–6024

Müller D et al (2007) Improved pharmacokinetics of recombinant bispecific antibody molecules by fusion to human serum albumin. J Biol Chem 282(17):12650–12660

Osada T et al (2010) Metastatic colorectal cancer cells from patients previously treated with chemotherapy are sensitive to T-cell killing mediated by CEA/CD3-bispecific T-cell-engaging BiTE antibody. Br J Cancer 102(1):124–133

Ribas A et al (2009) Phase I/II open-label study of the biologic effects of the interleukin-2 immunocytokine EMD 273063 (hu14.18-IL2) in patients with metastatic malignant melanoma. J Transl Med 7:68

Robert B et al (1996) Cytokine targeting in tumors using a bispecific antibody directed against carcinoembryonic antigen and tumor necrosis factor alpha. Cancer Res 56(20):4758–4765

Rosenblum MG et al (1991) Antibody-mediated delivery of tumor necrosis factor (TNF-alpha): improvement of cytotoxicity and reduction of cellular resistance. Cancer Commun 3(1):21–27

Seruga B et al (2008) Cytokines and their relationship to the symptoms and outcome of cancer. Nat Rev Cancer 8(11):887–899

Ten Hagen TL et al (2000) Low-dose tumor necrosis factor-alpha augments antitumor activity of stealth liposomal doxorubicin (DOXIL) in soft tissue sarcoma-bearing rats. Int J Cancer 87(6):829–837

Tracey KJ et al (1987) Anti-cachectin/TNF monoclonal antibodies prevent septic shock during lethal bacteraemia. Nature 330(6149):662–664

Yang J, Moyana T, Xiang J (1995) A genetically engineered single-chain FV/TNF molecule possesses the anti-tumor immunoreactivity of FV as well as the cytotoxic activity of tumor necrosis factor. Mol Immunol 32(12):873–881

Chapter 18
Bispecific Antibodies and Gene Therapy

Dirk M. Nettelbeck

18.1 Introduction

18.1.1 Gene Therapy

Gene therapy is the transfer of genes into patients' cells for therapeutic purposes (Fig. 18.1). Gene therapy was originally envisioned as a cure for inherited (monogenetic) diseases by gene correction, i.e., by replacing or complementing the causative mutated gene with a functional copy. In recent decades, however, gene therapy has been intensively investigated for treatment of many diseases by transfer of diverse classes of therapeutic genes from various species (Table 18.1). Examples are genes encoding pathogen antigens for prevention or treatment of infectious diseases (genetic vaccination); genes encoding agonists or antagonists of vascular growth factors for treatment of cardiovascular diseases; or genes that directly or indirectly mediate tumor cell killing for cancer treatment. Gene therapy drugs consist of the therapeutic gene, which defines the mode of therapeutic action, and the gene transfer vector, which needs to facilitate appropriate stability, delivery, and expression of the therapeutic gene (Fig. 18.1). Indeed, major efforts in gene therapy research focus on vector development, since the delivery of therapeutic genes is complex and critically determines treatment efficacy. Since the 1990s a multitude of gene therapy clinical trials have been performed with thousands of patients and therapeutic efficacy was demonstrated recently. Examples are the restoration of immunity in SCID patients, restoration of some degree of vision in childhood blindness or inhibition of neurodegeneration (Kohn 2010; Roy et al. 2010; Cartier et al. 2009). However, most gene therapy approaches necessitate

D.M. Nettelbeck (✉)
Helmholtz-University Group Oncolytic Adenoviruses, German Cancer Research Center (DKFZ) and Department of Dermatology, Heidelberg University Hospital, Im Neuenheimer Feld 242, 69120 Heidelberg, Germany
e-mail: d.nettelbeck@dkfz.de

R.E. Kontermann (ed.), *Bispecific Antibodies*,
DOI 10.1007/978-3-642-20910-9_18,

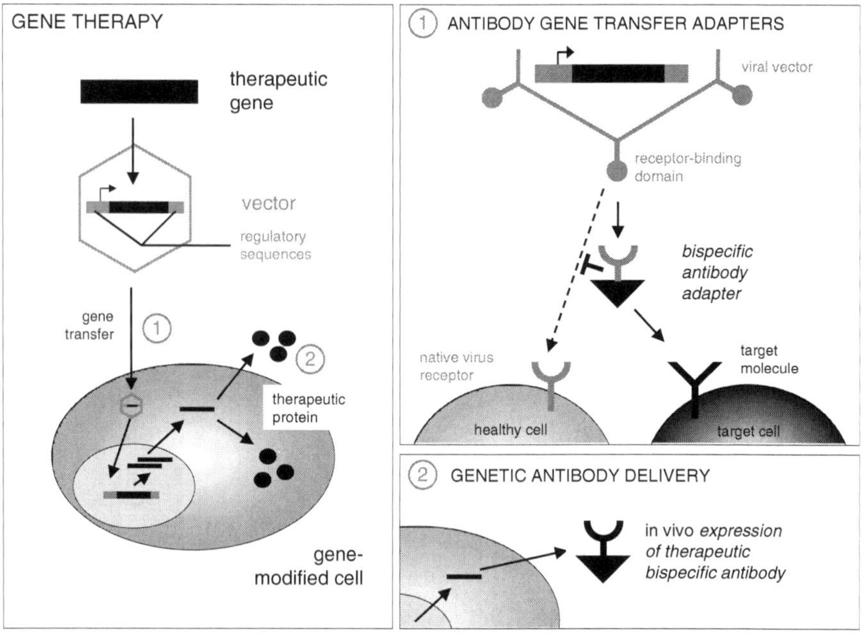

Fig. 18.1 Bispecific antibodies and gene therapy. *Left panel*: For gene therapy, a therapeutic gene is delivered into the patient's cell, where the gene product is expressed and mediates therapeutic activity. Examples are the complementation of the patient's genetic defects or the killing of cancer cells. For delivery and expression, the therapeutic gene is incorporated into a gene transfer vector containing regulatory sequences (e.g., for transcription start and termination). Frequently, the vector is also a means for efficient gene delivery into the patient's cells (e.g., replication-deficient viruses). *Right panels*: Bispecific antibodies can be either a tool for targeting gene transfer vectors to specific cell types (1) or gene transfer can be exploited as a tool for antibody therapy by antibody gene transfer and subsequent synthesis of the antibody in the patient (2)

improved efficacy or selectivity of gene transfer in order to facilitate successful applications in patients.

18.1.2 Gene Therapy Vectors

To ensure proper expression of the therapeutic gene in the patients' cells, a gene therapy vector contains a promoter and a transcription termination/polyadenylation signal (Fig. 18.1). Further regulatory elements can be exploited, for example, to achieve enhanced (introns) or bicistronic (internal ribosome entry sites, IRES) gene expression. Importantly, regulatory elements can be exploited for spatial or temporal control of gene expression. Examples are inducible or cell type-specific promoters or sequences differentially regulating mRNA stability or translation efficiency (Goverdhana et al. 2005; Dorer and Nettelbeck 2009; Brown and Naldini 2009). To improve stability of the therapeutic DNA, these eukaryotic expression

Table 18.1 Therapeutic genes used in gene therapy

Therapeutic gene	Disease	Activity	References
Correct copy of mutated gene encoding – Cytokine receptor – Retinal pigment epithelium protein – Blood coagulation factor – Transporter *ABCD1*	Monogenetic disease – Immunodeficiency – Childhood blindness – Hemophilia – Neurodegenerative disease	Gene correction	Kohn (2010), Roy et al. (2010), Margaritis and High (2010), Cartier et al. (2009)
Vascular growth factor gene; soluble vascular growth factor receptor gene	Vascular diseases, cancer	Angiogenesis, anti-angiogenesis	Karvinen and Yla-Herttuala (2010), Brandwijk et al. (2007)
Cytokine gene, co-stimulatory genes, gene encoding T cell receptor	Infectious diseases, cancer	Immunostimulation	Liu et al. (2010), Dotti et al. (2009)
Antibody-encoding gene	Infectious diseases, cancer	Antibody-dependent immune activation, neutralization of pathogens, receptor agonism or antagonism	Watanabe et al. (2010)
Gene encoding antigen – Pathogen-derived – Tumor-associated	– Infectious diseases – Cancer	Genetic vaccination	Smits et al. (2009), Rice et al. (2008)
Pro-apoptotic gene – Fas, TRAIL	Cancer	Cytotoxic; para- and autokrine apoptosis induction	Griffith et al. (2009)
Suicide gene (encodes prodrug-activating enzyme) – HSV thymidine kinase – Yeast cytosine deaminase	Cancer	Cytocidal in combination with prodrug – Ganciclovir – 5-FC	Portsmouth et al. (2007)
Gene encoding shRNA, miRNA, ribozyme	Various	Inhibition of target gene expression	Singer and Verma (2008)

cassettes are inserted either into circular plasmids, which might be further packaged by non-viral vectors, or into genomes of replication-deficient viral vectors (Table 18.2). Delivery of the resulting gene drug can be achieved by physical methods, for example, plasmid delivery into the skin by a gene gun for genetic vaccination. Alternatively, delivery can be mediated by the vector, for example, by penetration of virus vectors into the target tissue, uptake into cells and transfer of the recombinant virus genome containing the therapeutic gene into the nucleus (viral transduction). To ensure safety, viruses used as gene transfer vectors have been crippled by gene deletions in order to prevent virus replication. Gene transfer vectors determine the fate of the therapeutic gene: retroviral vectors insert their genome and thus the therapeutic gene into the host cell chromosome, whereas plasmids or adenoviral vectors result in episomal gene transfer. Thus, retroviruses facilitate stable gene transfer of transduced cells and their progeny. This is required for corrective treatment of inherited diseases, for example, by ex vivo gene transfer

Table 18.2 Gene transfer vectors used in gene therapy

Gene transfer vector	Delivery	Features	References
Naked plasmid	Injection, pressure-perfusion, gene gun, ultrasound, electroporation,	Regulatory DNA sequences, easy to produce, episomal gene transfer, often inefficient gene transfer	Wells (2004)
Plasmid + non-viral vectors (cationic lipids, liposomes, cationic polymers)	Non-specific uptake or ligand-mediated cell entry	See plasmid, varying efficiency, specific gene transfer possible	Wolff and Rozema (2008)
Retrovirus	Viral transduction, targeting ligand-mediated cell entry possible	Regulatory DNA sequences, stable gene transfer (chromosomal integration, problem of insertional mutagenesis), genetic vector engineering for improved gene transfer	Baum et al. (2006)
Adeno-associated virus (AAV)	Viral transduction, targeting ligand-mediated cell entry possible	Regulatory DNA sequences, episomal gene transfer, long-term gene transfer in postmitotic cells, genetic vector engineering for improved gene transfer	Buning et al. (2008)
Adenovirus	Viral transduction, targeting ligand-mediated cell entry possible	Regulatory DNA sequences, episomal gene transfer, high capacity for foreign DNA, genetic vector engineering for improved gene transfer	McConnell and Imperiale (2004)

into hematopoietic stem cells for treatment of inherited immunodeficiencies (Kohn 2010). In contrast, transient gene transfer is usually sufficient for genetic vaccination or cytoablative cancer therapy. For the latter, however, efficient gene transfer is pivotal and thus vector choice is determined by transduction efficiency. In this regard, conditionally replication-competent viral vectors have been recently engineered allowing for vector spread in tumors and thus amplified gene transfer (Parato et al. 2005; Cody and Douglas 2009). Such replication-competent vectors also mediate tumor cell lysis by virus replication, termed oncolysis or virotherapy. Hence, from the perspective of the virotherapist, insertion of therapeutic genes into the genome of oncolytic viruses is a strategy to complement oncolysis with gene therapy ("arming" of oncolytic viruses).

18.1.3 Targeting Gene Transfer

Many gene therapy applications require the restriction of gene transfer to specific cells. This is obvious for cytoablative gene therapy and for replication-competent vectors. Also effective genetic vaccination can depend on gene transfer into appropriate immune cells, as antigen expression in the wrong cells can trigger tolerance rather than immunity. Consequently, vector targeting is a major challenge for gene therapy research. Targeted gene therapy (or viral replication) can be achieved by inserting cell-binding ligands into the gene transfer vector for targeted cell entry, or by post-entry regulation of therapeutic gene expression using appropriate regulatory sequences, as mentioned above.

18.1.4 Bispecific Antibodies and Gene Therapy

Bispecific antibodies and gene therapy are connected in two ways. First, bispecific antibodies have been developed as promising tools for targeting cell entry of gene transfer vectors: as adapter molecules they link the vector to a marker molecule (specifically) expressed on the target cell surface (Fig. 18.1). Second, gene therapy can be an alternative means for delivery of therapeutic antibodies to patients, i.e., by antibody production in the patients' cells (genetic antibody therapy, Fig. 18.1). Besides genetic delivery of (established) soluble antibodies, such antibody gene transfer can also facilitate new applications for (bispecific) antibody therapy, for example by expression of membrane-bound or intracellular derivatives. Certainly combination therapies of bispecific antibodies and gene therapy can also be envisioned.

18.2 Bispecific Antibodies Are Tools for Gene Therapy

18.2.1 Adapters for Targeting Cell Entry of Viral Gene Therapy Vectors

Bispecific antibodies have been exploited in gene therapy as tools to direct viral gene transfer vectors to diseased cells. Therefore, an antibody with specificity for a viral surface protein is linked to a second antibody that binds to a cell surface molecule of interest, thus implementing an adapter molecule that binds the gene transfer vector to the target cell (Figs. 18.1 and 18.2). Such modification of virus tropism is required when virus receptor expression is lacking on target cells, preventing gene transfer, or when widespread expression of the native virus receptor on healthy cells leads to adverse side effects and vector sequestration. For the latter, either the viral attachment proteins have been mutated without losing their affinity for the adapter, or the receptor-binding domain of the virus attachment protein is shielded by the adapter. The resulting loss of virus tropism for healthy cells is termed de-targeting. Binding of and entry into target cells in both cases is mediated by the target of the cell-binding moiety of the adapter (re-targeting). Important advantages of the adapter strategy are (a) it does not require modifications to the virus structure, which might well turn out to be detrimental for

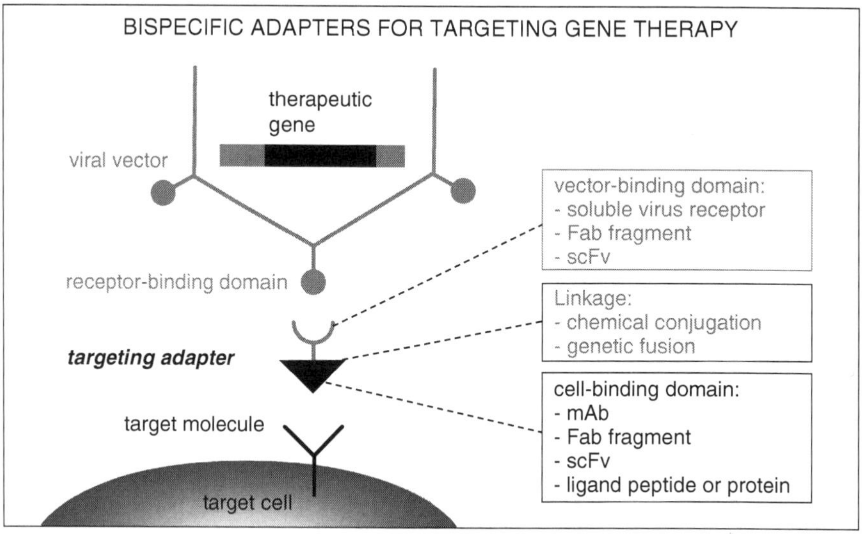

Fig. 18.2 Bispecific antibodies as tools for targeting gene therapy. Bispecific adapters binding to the gene transfer vector via one specificity and to a cell surface molecule with the other are used for delivering therapeutic genes to specific cell types. This strategy is of interest to gene therapy in order to ensure targeted therapy and avoid side effects. These bispecific adapters might contain one antibody or antibody fragment (for either vector or cell binding). Alternatively, they can be bispecific antibodies: chemical conjugate, diabody or tandem scFv

vector assembly, stability, or activity; (b) it is flexible as vector binding to any target molecule, to which an antibody can be raised, is possible by exchange of the adapter's cell-binding moiety and (c) once an effective adapter for a specific vector has been generated, it can be used for transfer of any therapeutic gene by corresponding derivatives of this vector.

18.2.2 Gene Therapy with Adenoviral Gene Transfer Vectors

Bispecific antibodies as adapters for targeting gene transfer have been most intensively investigated with adenoviral vectors. Adenoviruses (Ads, McConnell and Imperiale 2004) possess a double-stranded linear DNA genome covered by a protein capsid, but not a lipid envelope. The receptor-binding spike of the adenoviral capsid, made of the trimeric fiber protein, is responsible for attachment to host cells by binding to the virus receptor, which is the coxsackie-adenovirus receptor (CAR) for the mostly used human Ad serotype 5 (HAdV-5). Virus internalization into the host cell is then mediated by a secondary interaction of a different virus capsid protein, the penton base, with cellular integrins. By separating cell binding from entry, this two-step mechanism facilitates a high degree of flexibility for the nature of initial attachment of Ad vectors to cells. After entry of the vector into the cell, the virus genome is transferred to the nucleus, where viral genes are expressed from the episomal genome. Likewise, therapeutic genes are expressed after transfer of Ad vector genomes into patients' cells. Therefore, essential viral genes are replaced with the therapeutic gene, rendering the vector replication-deficient. More recently, therapeutic genes have been inserted into replication-competent Ads (McConnell and Imperiale 2004). Ads represent prominent gene therapy vectors, as they are stable, can be produced at high titers, possess an effective gene transfer machinery, and are only mildly pathogenic (HAdV-5 causes common cold symptoms) (McConnell and Imperiale 2004). They have been the most frequently used viral vectors in clinical gene therapy trials (Journal of Gene Medicine Clinical Trials Database). These trials have revealed a favorable safety profile of Ad vectors in patients. Cancer gene therapy and genetic vaccination are the regimens where Ad vectors are widely used. One therapeutic approach in cancer gene therapy is molecular chemotherapy, also termed gene-dependent enzyme prodrug therapy (GDEPT). This strategy is based on transfer of a gene encoding a prodrug-activating enzyme, which activates a harmless prodrug into an effective chemotherapeutic drug (Portsmouth et al. 2007). The rationale for this strategy is that tumor-restricted prodrug activation should facilitate effective concentrations of the chemotherapeutic drug in the tumor, which cannot be achieved by conventional systemic infusion of the drug due to dose-limiting side effects. GDEPT and other cytoablative cancer gene therapies depend on tumor-selective gene transfer which is not provided by unmodified HAdV-5 or other Ad serotypes due to widespread expression of Ad receptors also on healthy cells. Ads are also frequently used as vectors for genetic vaccination, which is most efficient when the antigen gene is transduced into professional antigen-presenting cells (APCs), which provide the proper signals for activation of immune effector cells. Dendritic cells (DCs) are the most effective APCs, but are difficult to transduce. Though Ads are the most effective gene transfer vectors for DCs, high vector titers are required for efficient DC transduction because of low expression of CAR.

18.2.3 Antibodies for Targeting Adenoviral Gene Transfer

Antibodies are attractive binding molecules for targeting gene transfer vectors based on their high affinity, specificity, and the opportunity to generate antibodies with specificity for virtually any cell surface target molecule. Three strategies have been pursued for insertion of targeting ligands into viral gene transfer vectors: genetic fusion to viral capsid or envelope proteins, complexing with bispecific adapters, or chemical linkage. Major drawbacks of the genetic and chemical strategies are that they are tedious and often interfere with viral functions. Moreover, genetic insertion of antibodies into the adenoviral capsid is hampered by the incompatibility of biosynthesis of capsid and antibody molecules. Ad capsid proteins are synthesized in the cytosol and transferred to the nucleus where viral particle assembly takes place, whereas antibodies are produced via the secretory pathway, which ensures their proper folding. Consequently, genetic fusion of antibodies to Ad capsid proteins has been inefficient and limited to a few cytosolically stable antibody fragments (Hedley et al. 2006; Vellinga et al. 2007; Poulin et al. 2010). In contrast, synthesis of adapter molecules can be separated from virus production. Moreover, the insertion of cell-binding antibodies into adapter molecules is less tedious than the engineering of a complete new virus genome and resulting adapters can be linked to any Ad vector, allowing for better flexibility. For production of adapters, antibody fragments binding to Ad capsid proteins, mostly the fiber, and antibodies or antibody fragments binding to cell surface target molecules of interest have been used (Fig. 18.2). They have been linked by chemical conjugation (see also: Chap. 3) or by genetic fusion, the latter generating tandem scFvs or scDbs (see also: Chap. 5). As an alternative to bispecific antibodies, adapters have been generated by linking virus-binding antibody fragments to cell-binding proteins or peptides, or by linking cell-binding antibody fragments to the soluble adenovirus receptor CAR.

18.2.4 Antibody-Derived Chemical Conjugates as Bispecific Adapter Molecules for Targeting Adenoviral Gene Transfer

The adapter strategy for targeting cell entry of Ad vectors has been pioneered by Douglas and co-workers for targeting of folate receptor overexpressing tumor cells (Douglas et al. 1996). To this end, they chemically conjugated folate to the

Fab fragment of a neutralizing anti-Ad fiber monoclonal antibody (MAb). A Fab fragment was used to avoid agglutination of Ad vectors by bivalent antibodies. After complexation to the respective Ad vector, the adapter mediated folate-dependent transfer of a reporter gene or of cytoablative genetic prodrug activation to target cells. The Fab fragment alone inhibited adenoviral transduction, which was expected as it was derived from a neutralizing antibody. Thus the Fab-folate adapter realized targeted gene transfer by both ablating virus binding to the native virus receptor and directing virus attachment to a novel cell surface molecule (Fig. 18.1). Wickham et al. described a bispecific antibody for directing Ad cell binding to integrins. This adapter consisted of the integrin-binding Mab chemically linked to a second MAb with specificity for a peptide tag, which was engineered into the Ad penton base (Wickham et al. 1996). The conjugate mediated enhanced adenoviral transduction of human smooth muscle and endothelial cells, which were only modestly tranduced by unmodified HAdV5 vectors. Subsequently, various bispecific antibody conjugates were reported, that consist of a fiber-binding Fab fragment covalently linked to a cell-binding antibody or antibody fragment. Such bispecific antibody conjugates have been reported to re-direct Ad gene transfer to various cell types via binding to different cell surface molecules, including CD40, EpCAM, Tag72, CD70, and ACE (Tillman et al. 1999, 2000; De Gruijl et al. 2002; Brandao et al. 2003; Miller et al. 1998; Haisma et al. 1999; Israel et al. 2001; Reynolds et al. 2000, 2001). These reports confirm the high flexibility of the adapter approach. For example, DCs, as professional antigen-presenting cells, represent targets of interest for gene transfer aiming at genetic vaccination for infectious or malignant diseases. Conjugates of α-fiber Fab and MAbs binding to the DC surface molecule CD40 allowed for efficient Ad gene transfer into mouse and human DCs (Tillman et al. 1999, 2000). With this adapter, improved efficiency and selectivity of Ad gene transfer to DCs was also achieved in situ using human skin explants (De Gruijl et al. 2002). In addition to targeting Ad entry, the α-fiber Fab/αCD40 mAb adapter triggered DC activation, as required for efficient induction of immune responses, via its CD40-binding activity. Accordingly, the adapter increased the efficiency of tumor vaccination with Ad vector transduced DCs in an animal model (Tillman et al. 2000). Adapter targeting of Ad vectors to cancer cells was demonstrated in cell culture studies with an EGFR-binding α-fiber Fab/MAb conjugate for squamous cell carcinoma, glioblastoma, and osteosarcoma (Miller et al. 1998; Blackwell et al. 1999; Barnett et al. 2002); with an EpCAM-binding Fab/Fab conjugate for various adenocarcinomas (Haisma et al. 1999; Heideman et al. 2001); with a TAG-72-binding α-fiber Fab/MAb conjugate for ovarian cancer (Kelly et al. 2000); and with a CD70-binding α-fiber Fab/MAb conjugate for B cell lymphomas (Israel et al. 2001). As CAR-expression varies on cancer cells, adapters frequently mediated markedly enhanced transduction of cancer cells. Yet another type of antibody-based adapter conjugate has been generated by linking α-fiber Fab fragments to basic fibroblast growth factor for targeting of various cancer cells (FGF2, Goldman et al. 1997; Rogers et al. 1997; Rancourt et al. 1998) to a synthetic lung-homing peptide (Trepel et al. 2000), or to the Hc-fragment of tetanus toxin for targeting neuronal cells (Schneider et al. 2000).

18.2.5 Recombinant Antibody-Derived Bispecific Adapter Molecules for Targeting Adenoviral Gene Transfer

Recombinant bispecific adapter molecules possess attributes that are advantageous for application in vector targeting when compared with chemical conjugates. Foremost, they can be produced by standardized procedures of prokaryotic or eukaryotic expression yielding well-defined molecules. Both tandem single chain variable fragments (tandem scFvs, see also: Chap. 5) and single chain diabodies (scDbs, see also: Chap. 5) have been used for targeting Ad gene transfer. Haisma and co-workers demonstrated that Ad transduction of glioblastoma and carcinoma cells can be increased by complexing the virus with a recombinant tandem α-fiber/α-EGFR scFv (Haisma et al. 2000). Our group reported in 2001 that a scDb with specificities for the Ad fiber and Endoglin, which is expressed on proliferative endothelium, facilitated targeted transduction of endothelial cells (Nettelbeck et al. 2001). In contrast to the tandem scFv, which was expressed in eukaryotic cells, the scDb was produced in bacteria. Ad transduction was also targeted to gastric cancer cells with a tandem scFv adapter binding to EpCAM (Heideman et al. 2002), to DCs with a CD40-binding tandem scFv adapter (Brandao et al. 2003), to melanoma cells using a scDb adapter binding the melanoma surface antigen HMWMAA (Nettelbeck et al. 2004), or to breast cancer cells with either a tandem scFv or a scDb binding to CEA (Korn et al. 2004). For improved de-targeting, "receptor-blind" Ad mutants were combined with tandem scFv or scDb adapters that were derived from α-fiber scFvs that retained binding to mutant fibers (van Beusechem et al. 2002; Nettelbeck et al. 2004; Carette et al. 2007). These Ad vectors could not bind CAR, even when individual fiber molecules were not protected after complexation with adapters. In consequence, this strategy of combined genetic/immunological tropism-modification implements a further increase in selectivity of gene transfer. Recombinant antibody-derived adapters for targeting adenoviral transduction were also obtained by fusion of α-fiber scFv to ligand proteins (EGF or uPAR, Watkins et al. 1997; Harvey et al. 2010) or to ligand peptides (Nicklin et al. 2000). Alternatively, cell-binding scFvs (α-c-erbB2, α-CD40 or α-FcγRI) were fused to monomeric or trimeric soluble CAR (Kashentseva et al. 2002; Pereboev et al. 2002; Kim et al. 2002; Sapinoro et al. 2007). Such sCAR-derived adapters offer the advantage of improving affinity to fiber by sCAR trimerization; however, they naturally cannot bind to "receptor-blind" fiber-mutant viruses. These strategies also demonstrated that the adapter, besides targeting gene transfer, might also influence the outcome of gene therapy in different ways: adenoviral gene transfer to DCs by CD40-binding adapters, but not by the FcγRI-binding adapter resulted in DC activation, thus influencing the type of immune response (immunization versus tolerization, Tillman et al. 1999; Sapinoro et al. 2007).

18.2.6 Toward Applications of Bispecific Adapter Molecules in Gene Therapy

In vitro studies with adapter molecules, including various bispecific antibodies, have clearly proven that viral cell entry can be re-directed via novel cell surface receptors, thus reprogramming virus tropism. This has been demonstrated in established cell cultures, freshly purified normal and tumor cells and in tissue explants, as for the demonstration of DC-targeted gene transfer in skin explants (de Gruijl et al. 2002). What are possible applications of bispecific antibodies and other antibody-derived gene transfer adapters? First, due to their modular composition and the opportunity to rapidly (in comparison with genetically engineered viruses) produce new adapters by chemical or genetic means, they facilitate the analysis, comparison, and screening of cell surface molecules for their feasibility as targets for viral gene transfer. Second, applications of adapters for ex vivo gene therapy are of interest. An example is genetic vaccination of cancer or infectious diseases by ex vivo gene transfer into DCs isolated from patients. Gene therapy of inherited diseases by ex vivo gene transfer into hematopoietic (stem) cells is a further application. Here, however, retroviral vectors are preferred over Ad vectors, as they facilitate stable gene transfer and thus prolonged gene correction or replacement (Table 18.2). Of note, adapter-targeting of retroviral gene transfer has been demonstrated recently (see below). Most gene therapy applications, however, require in vivo gene transfer. For establishing adapters for targeting gene transfer, in vivo extensive studies on the stability, efficiency, and selectivity of adapter-vector complexes after in vivo application are needed. Whereas rigorous studies for the evaluation of pharmacologic and therapeutic parameters of adapter-targeted gene transfer are still to be done, initial studies have shown efficacy of adapter-targeting in vivo. In an effort to facilitate gene therapy of pulmonary vascular disease, Reynolds and colleagues investigated a Fab-mAb conjugate adapter that binds angiotensin-converting enzyme (ACE) for targeting of Ad gene transfer to the lung endothelium in rats (Reynolds et al. 2000, 2001). By systemic application of adapter-bound or uncomplexed Ad vector, it was shown that this adapter increased gene transfer to the lung by more than 20-fold. Importantly, gene transfer was directed to endothelial cells. Moreover, gene transfer to the liver, the organ responsible for most Ad-induced side effects, was reduced more than 80%. Hence, this study demonstrated both systemic stability of the adapter-vector complex and adapter-dependent vector de- and re-targeting in vivo. For the Fab-FGF2 adapter, several studies in mice showed adapter-dependent reduction of liver transgene expression after systemic injection of Ad vectors and reduced toxicity of Ad-mediated genetic prodrug activation therapy. Furthermore, this adapter increased therapeutic activity of Ad-mediated genetic prodrug activation of peritoneal malignancies, when the Ad vectors were injected intraperitoneally (Rancourt et al. 1998; Gu et al. 1999;

Printz et al. 2000). In vivo stability of adaptor-vector complexes has also been demonstrated for recombinant proteins. Trimeric, but not monomeric sCAR significantly blocked liver gene transfer by Ads after systemic application of the sCAR-Ad vector complex into mice (Kim et al. 2002). However, in a different study, a sCAR-scFv adapter targeting CEA also reduced liver transduction by Ad vectors after systemic injection of adapter-virus complexes into mice (Li et al. 2007). After systemic injection, this adapter also increased adenoviral transduction of CEA-positive, but reduced transduction of CEA-negative tumors that were grafted to mouse livers. Furthermore, a trimeric derivative of the sCAR-CEA adapter mediated improved targeting of adenoviral gene transfer in vitro and in vivo. In combination with transcriptional targeting using the cox-2 promoter, this trimeric adapter increased therapeutic activity and at the same time reduced liver toxicity of genetic prodrug activation therapy with HSV-tk/GCV (Li et al. 2009). Studies with sCAR-EGF and trimeric sCAR-mCD40L confirm the re-targeting properties of sCAR-derived adapters in vivo (Liang et al. 2004; Huang et al. 2007). In addition to facilitating selective gene transfer, targeting adenoviral cell binding and entry is of interest also for improving oncolytic Ads. Toward this end, adapter molecules are of interest to re-direct the injected virus to target tumors. To also allow for targeting of progeny viruses of oncolytic Ads produced in patients' tumors, genes encoding recombinant bispecific adapters have been inserted into the genome of these viruses. Using a tandem scFv with specificity for the Ad fiber and EGFR, van Beusechem and co-workers demonstrated increased viral spread and oncolysis in two- and three-dimensional tumor cell cultures (van Beusechem et al. 2003; Carette et al. 2007).

18.2.7 Adapter Targeting is Feasible for Several Viral Gene Transfer Vectors

Although most widely investigated for Ad vectors, adapters have been also shown to facilitate targeted cell entry of other viruses. Adeno-associated viruses (AAV) are small non-enveloped viruses that are frequently used for diverse gene therapy applications (Buning et al. 2008). Tropism-modification of AAV vectors was achieved with a bispecific Fab/Fab antibody conjugate. The adapter with specificity for the virus capsid and for integrins facilitated gene transfer into megakaryocytes, which are not permissive to unmodified AAV vectors (Bartlett et al. 1999). For non-human coronaviruses, enveloped RNA viruses that naturally do not enter human cells, infectivity for human cancer cells was established with a bispecific tandem scFv with specificities for a coronavirus surface glycoprotein and EGFR (Wurdinger et al. 2005a, b). The idea of this approach was to selectively kill tumor cells by lytic virus infection rather than viral gene transfer. Similar results were obtained using a recombinant adapter built of soluble coronavirus receptor fused to the EGFR-binding scFv (Wurdinger et al. 2005a, b). Newcastle disease virus, which is in development for viral oncolysis and gene therapy, has been re-targeted using a recombinant adapter built of a virus-binding scFv and IL-2 (Bian et al. 2005, 2006). Retroviruses are enveloped RNA viruses, which insert their genome after reverse transcription into the chromosome of infected cells. Therefore, retroviral vectors facilitate long-term gene transfer which is especially suitable for gene correction therapy of monogenetic diseases. Adapter targeting of retrovirus cell entry was reported for recombinant proteins built of the virus receptor extracellular domain fused to EGF, VEGF, or an EGFR-specific scFv (Snitkovsky and Young 1998; Boerger et al. 1999; Snitkovsky et al. 2000, 2001).

18.3 Gene Transfer as a Tool for Antibody Therapy: Genetic Antibody Delivery

18.3.1 Genetic Antibody Delivery

Gene therapy can be exploited for expressing antibodies in patients, which might be advantageous for achieving sustained and/or efficient antibody concentrations and/or a favorable antibody biodistribution by local expression. Thus, gene therapy is a tool of interest to overcome rapid antibody clearance or poor access to tumors as reported for antibodies that are injected as proteins. Genetic antibody therapy can be implemented by in vivo or ex vivo gene transfer (Fig. 18.3), i.e., by direct injection of the gene transfer vector into patients or by gene transfer in cultures of previously isolated cells followed by injection of the resulting genetically engineered cells, respectively. Dependent on the design of the gene transfer vector, genetic antibody application can be transient or permanent, constitutive or inducible, targeted or ubiquitous. For example, retroviral vectors allow for stable gene transfer, inducible promoters facilitate control of antibody expression, and targeted vectors can direct gene transfer to specific cell types (see Sects. 18.1.2 and 18.1.3). Therefore, gene therapy possesses high potential and flexibility for implementing improved antibody delivery for specific applications. However, this area of research is still in its infancy and more widespread investigations are warranted.

With the advent of recombinant DNA technology it became possible to establish novel strategies for antibody production and to engineer antibody properties (for example affinity maturation and humanization), formats (single chain fragments), and fusion proteins (immunotoxins). Recombinant antibodies have been frequently produced in bacteria, but gene transfer into eukaryotic cells has also been utilized for in vitro production of immunoglobulins, antibody fragments or antibody fusion proteins. Having established the engineering of recombinant gene constructs for eukaryotic antibody expression, also the in vivo production of antibodies became feasible. Examples are the expression of functional recombinant MAbs in mice

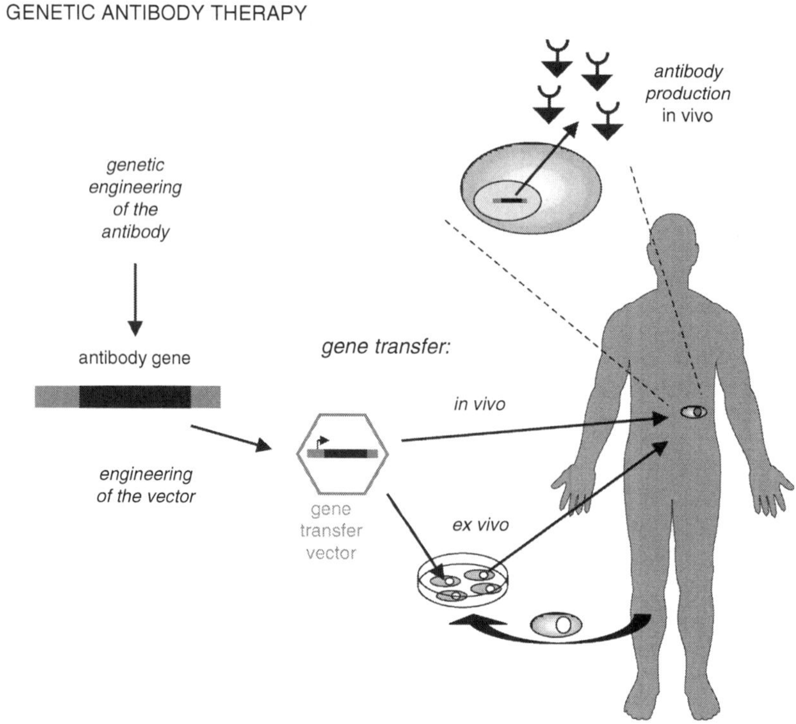

Fig. 18.3 Gene therapy as a tool for antibody delivery: Genetic antibody therapy. For genetic antibody delivery antibody genes, which can be engineered to match specific purposes, are incorporated into gene transfer vectors. These vectors are either directly injected into patients (in vivo gene therapy) or are used for gene transfer into cells previously purified from a patient followed by re-injection of the engineered cells into the patient (ex vivo gene therapy). The antibodies are produced in the patient from cells genetically modified by in vivo or ex vivo gene transfer. Dependent on the vector design, antibody production can be transient or prolonged, constitutive or inducible and show local or systemic activity

after transfer of genetically engineered cells (Noel et al. 1997) or after in vivo gene transfer with an adenoviral or AAV vector (Noel et al. 2002; Jiang et al. 2006; Watanabe et al. 2009; Lewis et al. 2002; Fang et al. 2005, 2007; Skaricic et al. 2008; De et al. 2008; Ho et al. 2009). Toward this end, Fang and coworkers optimized antibody production: they expressed the heavy and light chains of the MAb at equal amounts from a single open reading frame using a "ribosomal skip" sequence. Thereby, serum levels of >1 mg/ml antibody for extended time periods were obtained in mice after injection of a single dose of AAV vector. In a subsequent study, the same group demonstrated that by using an inducible promoter, serum antibody levels after in vivo gene transfer can be repeatedly shut off and on (Fang et al. 2007). This represents a promising strategy to increase safety and/or facilitate dose adaptation in potential future clinical applications of genetic antibody delivery. De and co-workers combined genetic delivery of a MAb gene by AAV and Ad vectors to achieve both rapid (Ad) and persistent (AAV) antibody production (De et al. 2008).

Functional expression in vivo was also demonstrated for recombinant antibody fragments or fusion proteins that contain such fragments after adenoviral gene transfer (Whittington et al. 1998; Arafat et al. 2002; Afanasieva et al. 2003; Kasuya et al. 2005; Liu et al. 2010). The expression of chimeric antigen receptors by T cells and subsequent adoptive T cell therapy is another important application of genetic antibody delivery.

18.3.2 In Vivo Expression of Bispecific Antibodies

Álvarez-Vallina and team have developed genetic delivery of bispecific antibodies by engineered cells. In 2003, they reported anti-tumor activity for a bispecific diabody expressed in vivo from irradiated, genetically engineered 293T cells (Blanco et al. 2003). They produced stably transfected 293T cells secreting a diabody with specificity for both CEA and CD3. A second cell line additionally secreted a bivalent CEA-specific diabody fused to the extracellular domain of B7-1. After co-injection with CEA-positive tumor cells into mice, these genetically engineered cells showed anti-tumor activity compared with co-injection of control 293T cells. Subsequent to this proof-of-principle study, the same group engineered a lentiviral gene transfer vector encoding the CEA-CD3 diabody (Compte et al. 2007). This vector facilitated the transduction of different types of hematopoietic cells that showed prolonged secretion of active diabody in vitro and antitumor activity in vivo. In a follow-up study, the group demonstrated that also the implantation of lentivirally transduced endothelial cells into mice resulted in prolonged production of the CEA/CD3 diabody with therapeutic activity (Compte et al. 2010). This study aims at a therapeutic regimen that allows for the production of therapeutic antibodies from neovessels that have incorporated ex vivo engineered endothelial cells.

Genetic delivery of bispecific antibodies has also been reported for intracellular applications: cell surface localization of two membrane proteins, VEGFR2 and Tie-2, could be blocked by expression of a corresponding bispecific, tetravalent antibody targeted to the endoplasmic reticulum (Jendreyko et al. 2003). This intracellular bispecific antibody showed anti-angiogenic activity in vitro, which was superior to monovalent control antibodies. A similar construct with specificity for VEGFR2 and Tie-2 mediated anti-angiogenic and anti-tumor activity in vivo after adenoviral gene transfer (Jendreyko et al. 2005).

18.4 Conclusions

Proof of principle has been demonstrated in several cell culture studies and animal models for both the utility of bispecific antibodies for targeting gene therapies and the feasibility of gene transfer for delivering recombinant bispecific antibodies.

Based on this fundamental work, bispecific antibody adapters and gene transfer technologies should now be considered for improving therapeutic regimens in gene therapy and antibody therapy, respectively. Cooperation between antibody engineers and gene therapists are warranted to further develop bispecific antibodies and gene transfer vectors for this purpose.

References

Afanasieva TA, Wittmer M, Vitaliti A, Ajmo M, Neri D, Klemenz R (2003) Single-chain antibody and its derivatives directed against vascular endothelial growth factor: application for antiangiogenic gene therapy. Gene Ther 10(21):1850–1859

Arafat WO, Gomez-Navarro J, Buchsbaum DJ, Xiang J, Wang M, Casado E, Barker SD, Mahasreshti PJ, Haisma HJ, Barnes MN, Siegal GP, Alvarez RD, Hemminki A, Nettelbeck DM, Curiel DT (2002) Effective single chain antibody (scFv) concentrations in vivo via adenoviral vector mediated expression of secretory scFv. Gene Ther 9(4):256–262

Barnett BG, Tillman BW, Curiel DT, Douglas JT (2002) Dual targeting of adenoviral vectors at the levels of transduction and transcription enhances the specificity of gene expression in cancer cells. Mol Ther 6(3):377–385

Bartlett JS, Kleinschmidt J, Boucher RC, Samulski RJ (1999) Targeted adeno-associated virus vector transduction of nonpermissive cells mediated by a bispecific F(ab'gamma)2 antibody. Nat Biotechnol 17(2):181–186

Baum C, Schambach A, Bohne J, Galla M (2006) Retrovirus vectors: toward the plentivirus? Mol Ther 13(6):1050–1063

Bian H, Fournier P, Moormann R, Peeters B, Schirrmacher V (2005) Selective gene transfer in vitro to tumor cells via recombinant Newcastle disease virus. Cancer Gene Ther 12(3):295–303

Bian H, Wilden H, Fournier P, Peeters B, Schirrmacher V (2006) In vivo efficacy of systemic tumor targeting of a viral RNA vector with oncolytic properties using a bispecific adapter protein. Int J Oncol 29(6):1359–1369

Blackwell JL, Miller CR, Douglas JT, Li H, Reynolds PN, Carroll WR, Peters GE, Strong TV, Curiel DT (1999) Retargeting to EGFR enhances adenovirus infection efficiency of squamous cell carcinoma. Arch Otolaryngol Head Neck Surg 125(8):856–863

Blanco B, Holliger P, Vile RG, Alvarez-Vallina L (2003) Induction of human T lymphocyte cytotoxicity and inhibition of tumor growth by tumor-specific diabody-based molecules secreted from gene-modified bystander cells. J Immunol 171(2):1070–1077

Boerger AL, Snitkovsky S, Young JA (1999) Retroviral vectors preloaded with a viral receptor-ligand bridge protein are targeted to specific cell types. Proc Natl Acad Sci USA 96(17):9867–9872

Brandao JG, Scheper RJ, Lougheed SM, Curiel DT, Tillman BW, Gerritsen WR, van den Eertwegh AJ, Pinedo HM, Haisma HJ, de Gruijl TD (2003) CD40-targeted adenoviral gene transfer to dendritic cells through the use of a novel bispecific single-chain Fv antibody enhances cytotoxic T cell activation. Vaccine 21(19–20):2268–2272

Brandwijk RJ, Griffioen AW, Thijssen VL (2007) Targeted gene-delivery strategies for angiostatic cancer treatment. Trends Mol Med 13(5):200–209

Brown BD, Naldini L (2009) Exploiting and antagonizing microRNA regulation for therapeutic and experimental applications. Nat Rev Genet 10(8):578–585

Buning H, Perabo L, Coutelle O, Quadt-Humme S, Hallek M (2008) Recent developments in adeno-associated virus vector technology. J Gene Med 10(7):717–733

Carette JE, Graat HC, Schagen FH, Mastenbroek DC, Rots MG, Haisma HJ, Groothuis GM, Schaap GR, Bras J, Kaspers GJ, Wuisman PI, Gerritsen WR, van Beusechem VW (2007) A conditionally replicating adenovirus with strict selectivity in killing cells expressing epidermal growth factor receptor. Virology 361(1):56–67

Cartier N, Hacein-Bey-Abina S, Bartholomae CC, Veres G, Schmidt M, Kutschera I, Vidaud M, Abel U, Dal-Cortivo L, Caccavelli L, Mahlaoui N, Kiermer V, Mittelstaedt D, Bellesme C, Lahlou N, Lefrere F, Blanche S, Audit M, Payen E, Leboulch P, L'Homme B, Bougneres P, Von Kalle C, Fischer A, Cavazzana-Calvo M, Aubourg P (2009) Hematopoietic stem cell gene therapy with a lentiviral vector in X-linked adrenoleukodystrophy. Science 326 (5954):818–823

Cody JJ, Douglas JT (2009) Armed replicating adenoviruses for cancer virotherapy. Cancer Gene Ther 16(6):473–488

Compte M, Blanco B, Serrano F, Cuesta AM, Sanz L, Bernad A, Holliger P, Alvarez-Vallina L (2007) Inhibition of tumor growth in vivo by in situ secretion of bispecific anti-CEA x anti-CD3 diabodies from lentivirally transduced human lymphocytes. Cancer Gene Ther 14(4):380–388

Compte M, Alonso-Camino V, Santos-Valle P, Cuesta AM, Sanchez-Martin D, Lopez MR, Vicario JL, Salas C, Sanz L, Alvarez-Vallina L (2010) Factory neovessels: engineered human blood vessels secreting therapeutic proteins as a new drug delivery system. Gene Ther 17(6):745–751

de Gruijl TD, Luykx-de Bakker SA, Tillman BW, van den Eertwegh AJ, Buter J, Lougheed SM, van der Bij GJ, Safer AM, Haisma HJ, Curiel DT, Scheper RJ, Pinedo HM, Gerritsen WR (2002) Prolonged maturation and enhanced transduction of dendritic cells migrated from human skin explants after in situ delivery of CD40-targeted adenoviral vectors. J Immunol 169(9):5322–5331

De BP, Hackett NR, Crystal RG, Boyer JL (2008) Rapid/sustained anti-anthrax passive immunity mediated by co-administration of Ad/AAV. Mol Ther 16(1):203–209

Dorer DE, Nettelbeck DM (2009) Targeting cancer by transcriptional control in cancer gene therapy and viral oncolysis. Adv Drug Deliv Rev 61(7–8):554–571

Dotti G, Savoldo B, Brenner M (2009) Fifteen years of gene therapy based on chimeric antigen receptors: "are we nearly there yet?". Hum Gene Ther 20(11):1229–1239

Douglas JT, Rogers BE, Rosenfeld ME, Michael SI, Feng M, Curiel DT (1996) Targeted gene delivery by tropism-modified adenoviral vectors. Nat Biotechnol 14(11):1574–1578

Fang J, Qian JJ, Yi S, Harding TC, Tu GH, VanRoey M, Jooss K (2005) Stable antibody expression at therapeutic levels using the 2A peptide. Nat Biotechnol 23(5):584–590

Fang J, Yi S, Simmons A, Tu GH, Nguyen M, Harding TC, VanRoey M, Jooss K (2007) An antibody delivery system for regulated expression of therapeutic levels of monoclonal antibodies in vivo. Mol Ther 15(6):1153–1159

Goldman CK, Rogers BE, Douglas JT, Sosnowski BA, Ying W, Siegal GP, Baird A, Campain JA, Curiel DT (1997) Targeted gene delivery to Kaposi's sarcoma cells via the fibroblast growth factor receptor. Cancer Res 57(8):1447–1451

Goverdhana S, Puntel M, Xiong W, Zirger JM, Barcia C, Curtin JF, Soffer EB, Mondkar S, King GD, Hu J, Sciascia SA, Candolfi M, Greengold DS, Lowenstein PR, Castro MG (2005) Regulatable gene expression systems for gene therapy applications: progress and future challenges. Mol Ther 12(2):189–211

Griffith TS, Stokes B, Kucaba TA, Earel JK Jr, VanOosten RL, Brincks EL, Norian LA (2009) TRAIL gene therapy: from preclinical development to clinical application. Curr Gene Ther 9(1):9–19

Gu DL, Gonzalez AM, Printz MA, Doukas J, Ying W, D'Andrea M, Hoganson DK, Curiel DT. Douglas JT, Sosnowski BA, Baird A, Aukerman SL, Pierce GF (1999) Fibroblast growth factor 2 retargeted adenovirus has redirected cellular tropism: evidence for reduced toxicity and enhanced antitumor activity in mice. Cancer Res 59:2608–14

Haisma HJ, Pinedo HM, Rijswijk A, der Meulen-Muileman I, Sosnowski BA, Ying W, Beusechem VW, Tillman BW, Gerritsen WR, Curiel DT (1999) Tumor-specific gene transfer via an adenoviral vector targeted to the pan-carcinoma antigen EpCAM. Gene Ther 6(8):1469–1474

Haisma HJ, Grill J, Curiel DT, Hoogeland S, van Beusechem VW, Pinedo HM, Gerritsen WR (2000) Targeting of adenoviral vectors through a bispecific single-chain antibody. Cancer Gene Ther 7(6):901–904

Harvey TJ, Burdon D, Steele L, Ingram N, Hall GD, Selby PJ, Vile RG, Cooper PA, Shnyder SD, Chester JD (2010) Retargeted adenoviral cancer gene therapy for tumour cells overexpressing epidermal growth factor receptor or urokinase-type plasminogen activator receptor. Gene Ther 17(8):1000–1010

Hedley SJ, Auf der Maur A, Hohn S, Escher D, Barberis A, Glasgow JN, Douglas JT, Korokhov N, Curiel DT (2006) An adenovirus vector with a chimeric fiber incorporating stabilized single chain antibody achieves targeted gene delivery. Gene Ther 13(1):88–94

Heideman DA, Snijders PJ, Craanen ME, Bloemena E, Meijer CJ, Meuwissen SG, van Beusechem VW, Pinedo HM, Curiel DT, Haisma HJ, Gerritsen WR (2001) Selective gene delivery toward gastric and esophageal adenocarcinoma cells via EpCAM-targeted adenoviral vectors. Cancer Gene Ther 8(5):342–351

Heideman DA, van Beusechem VW, Offerhaus GJ, Wickham TJ, Roelvink PW, Craanen ME, Pinedo HM, Meijer CJ, Gerritsen WR (2002) Selective gene transfer into primary human gastric tumors using epithelial cell adhesion molecule-targeted adenoviral vectors with ablated native tropism. Hum Gene Ther 13(14):1677–1685

Ho DT, Wykoff-Clary S, Gross CS, Schneider D, Jin F, Kretschmer PJ, Hermiston TW (2009) Growth inhibition of an established A431 xenograft tumor by a full-length anti-EGFR antibody following gene delivery by AAV. Cancer Gene Ther 16(2):184–194

Israel BF, Pickles RJ, Segal DM, Gerard RD, Kenney SC (2001) Enhancement of adenovirus vector entry into CD70-positive B-cell Lines by using a bispecific CD70-adenovirus fiber antibody. J Virol 75(11):5215–5221

Jendreyko N, Popkov M, Beerli RR, Chung J, McGavern DB, Rader C, Barbas CF 3rd (2003) Intradiabodies, bispecific, tetravalent antibodies for the simultaneous functional knockout of two cell surface receptors. J Biol Chem 278(48):47812–47819

Jendreyko N, Popkov M, Rader C, Barbas CF 3rd (2005) Phenotypic knockout of VEGF-R2 and Tie-2 with an intradiabody reduces tumor growth and angiogenesis in vivo. Proc Natl Acad Sci USA 102(23):8293–8298

Jiang M, Shi W, Zhang Q, Wang X, Guo M, Cui Z, Su C, Yang Q, Li Y, Sham J, Liu X, Wu M, Qian Q (2006) Gene therapy using adenovirus-mediated full-length anti-HER-2 antibody for HER-2 overexpression cancers. Clin Cancer Res 12(20 Pt 1):6179–6185

Karvinen H, Yla-Herttuala S (2010) New aspects in vascular gene therapy. Curr Opin Pharmacol 10(2):208–211

Kashentseva EA, Seki T, Curiel DT, Dmitriev IP (2002) Adenovirus targeting to c-erbB-2 oncoprotein by single-chain antibody fused to trimeric form of adenovirus receptor ectodomain. Cancer Res 62(2):609–616

Kasuya K, Boyer JL, Tan Y, Alipui DO, Hackett NR, Crystal RG (2005) Passive immunotherapy for anthrax toxin mediated by an adenovirus expressing an anti-protective antigen single-chain antibody. Mol Ther 11(2):237–244

Kelly FJ, Miller CR, Buchsbaum DJ, Gomez-Navarro J, Barnes MN, Alvarez RD, Curiel DT (2000) Selectivity of TAG-72-targeted adenovirus gene transfer to primary ovarian carcinoma cells versus autologous mesothelial cells in vitro. Clin Cancer Res 6(11):4323–4333

Kim J, Smith T, Idamakanti N, Mulgrew K, Kaloss M, Kylefjord H, Ryan PC, Kaleko M, Stevenson SC (2002) Targeting adenoviral vectors by using the extracellular domain of the coxsackie-adenovirus receptor: improved potency via trimerization. J Virol 76(4):1892–1903

Kohn DB (2010) Update on gene therapy for immunodeficiencies. Clin Immunol 135(2):247–254

Korn T, Nettelbeck DM, Volkel T, Muller R, Kontermann RE (2004) Recombinant bispecific antibodies for the targeting of adenoviruses to CEA-expressing tumour cells: a comparative analysis of bacterially expressed single-chain diabody and tandem scFv. J Gene Med 6(6):642–651

Lewis AD, Chen R, Montefiori DC, Johnson PR, Clark KR (2002) Generation of neutralizing activity against human immunodeficiency virus type 1 in serum by antibody gene transfer. J Virol 76(17):8769–8775

Li HJ, Everts M, Pereboeva L, Komarova S, Idan A, Curiel DT, Herschman HR (2007) Adenovirus tumor targeting and hepatic untargeting by a coxsackie/adenovirus receptor ectodomain anti-carcinoembryonic antigen bispecific adapter. Cancer Res 67(11):5354–5361

Li HJ, Everts M, Yamamoto M, Curiel DT, Herschman HR (2009) Combined transductional untargeting/retargeting and transcriptional restriction enhances adenovirus gene targeting and therapy for hepatic colorectal cancer tumors. Cancer Res 69(2):554–564

Liang Q, Dmitriev I, Kashentseva E, Curiel DT, Herschman HR (2004) Noninvasive of adenovirus tumor retargeting in living subjects by a soluble adenovirus receptor-epidermal growth factor (sCAR-EGF) fusion protein. Mol Imaging Biol 6(6):385–394

Liu L, Wang S, Shan B, Sang M, Liu S, Wang G (2010) Advances in viral-vector systemic cytokine gene therapy against cancer. Vaccine 28(23):3883–3887

Margaritis P, High KA (2010) Gene therapy in haemophilia – going for cure? Haemophilia 16 (Suppl 3):24–28

McConnell MJ, Imperiale MJ (2004) Biology of adenovirus and its use as a vector for gene therapy. Hum Gene Ther 15(11):1022–1033

Miller CR, Buchsbaum DJ, Reynolds PN, Douglas JT, Gillespie GY, Mayo MS, Raben D, Curiel DT (1998) Differential susceptibility of primary and established human glioma cells to adenovirus infection: targeting via the epidermal growth factor receptor achieves fiber receptor-independent gene transfer. Cancer Res 58(24):5738–5748

Nettelbeck DM, Miller DW, Jerome V, Zuzarte M, Watkins SJ, Hawkins RE, Muller R, Kontermann RE (2001) Targeting of adenovirus to endothelial cells by a bispecific single-chain diabody directed against the adenovirus fiber knob domain and human endoglin (CD105). Mol Ther 3(6):882–891

Nettelbeck DM, Rivera AA, Kupsch J, Dieckmann D, Douglas JT, Kontermann RE, Alemany R, Curiel DT (2004) Retargeting of adenoviral infection to melanoma: combining genetic ablation of native tropism with a recombinant bispecific single-chain diabody (scDb) adapter that binds to fiber knob and HMWMAA. Int J Cancer 108(1):136–145

Nicklin SA, White SJ, Watkins SJ, Hawkins RE, Baker AH (2000) Selective targeting of gene transfer to vascular endothelial cells by use of peptides isolated by phage display. Circulation 102(2):231–237

Noel D, Pelegrin M, Marin M, Biard-Piechaczyk M, Ourlin JC, Mani JC, Piechaczyk M (1997) In vitro and in vivo secretion of cloned antibodies by genetically modified myogenic cells. Hum Gene Ther 8(10):1219–1229

Noel D, Pelegrin M, Kramer S, Jacquet C, Skander N, Piechaczyk M (2002) High in vivo production of a model monoclonal antibody on adenoviral gene transfer. Hum Gene Ther 13(12):1483–1493

Parato KA, Senger D, Forsyth PA, Bell JC (2005) Recent progress in the battle between oncolytic viruses and tumours. Nat Rev Cancer 5(12):965–976

Pereboev AV, Asiedu CK, Kawakami Y, Dong SS, Blackwell JL, Kashentseva EA, Triozzi PL, Aldrich WA, Curiel DT, Thomas JM, Dmitriev IP (2002) Coxsackievirus-adenovirus receptor genetically fused to anti-human CD40 scFv enhances adenoviral transduction of dendritic cells. Gene Ther 9(17):1189–1193

Portsmouth D, Hlavaty J, Renner M (2007) Suicide genes for cancer therapy. Mol Aspects Med 28(1):4–41

Poulin KL, Lanthier RM, Smith AC, Christou C, Quiroz MR, Powell KL, O'Meara RW, Kothary R, Lorimer IA, Parks RJ (2010) Retargeting of adenovirus vectors through genetic fusion of a single-chain or single-domain antibody to capsid protein IX. J Virol 84(19):10074–10086

Printz MA, Gonzalez AM, Cunningham M, Gu DL, Ong M, Pierce GF, Aukerman SL (2000) Fibroblast growth factor 2-retargeted adenoviral vectors exhibit a modified biolocalization pattern and display reduced toxicity relative to native adenoviral vectors. Hum Gene Ther 11:191–204

Rancourt C, Rogers BE, Sosnowski BA, Wang M, Piche A, Pierce GF, Alvarez RD, Siegal GP, Douglas JT, Curiel DT (1998) Basic fibroblast growth factor enhancement of adenovirus-mediated delivery of the herpes simplex virus thymidine kinase gene results in augmented therapeutic benefit in a murine model of ovarian cancer. Clin Cancer Res 4(10):2455–2461

Reynolds PN, Zinn KR, Gavrilyuk VD, Balyasnikova IV, Rogers BE, Buchsbaum DJ, Wang MH, Miletich DJ, Grizzle WE, Douglas JT, Danilov SM, Curiel DT (2000) A targetable, injectable adenoviral vector for selective gene delivery to pulmonary endothelium in vivo. Mol Ther 2 (6):562–578

Reynolds PN, Nicklin SA, Kaliberova L, Boatman BG, Grizzle WE, Balyasnikova IV, Baker AH, Danilov SM, Curiel DT (2001) Combined transductional and transcriptional targeting improves the specificity of transgene expression in vivo. Nat Biotechnol 19(9):838–842

Rice J, Ottensmeier CH, Stevenson FK (2008) DNA vaccines: precision tools for activating effective immunity against cancer. Nat Rev Cancer 8(2):108–120

Rogers BE, Douglas JT, Ahlem C, Buchsbaum DJ, Frincke J, Curiel DT (1997) Use of a novel cross-linking method to modify adenovirus tropism. Gene Ther 4(12):1387–1392

Roy K, Stein L, Kaushal S (2010) Ocular gene therapy: an evaluation of recombinant adeno-associated virus-mediated gene therapy interventions for the treatment of ocular disease. Hum Gene Ther 21(8):915–927

Sapinoro R, Maguire CA, Burgess A, Dewhurst S (2007) Enhanced transduction of dendritic cells by FcgammaRI-targeted adenovirus vectors. J Gene Med 9(12):1033–1045

Schneider H, Groves M, Muhle C, Reynolds PN, Knight A, Themis M, Carvajal J, Scaravilli F, Curiel DT, Fairweather NF, Coutelle C (2000) Retargeting of adenoviral vectors to neurons using the Hc fragment of tetanus toxin. Gene Ther 7(18):1584–1592

Singer O, Verma IM (2008) Applications of lentiviral vectors for shRNA delivery and transgenesis. Curr Gene Ther 8(6):483–488

Skaricic D, Traube C, De B, Joh J, Boyer J, Crystal RG, Worgall S (2008) Genetic delivery of an anti-RSV antibody to protect against pulmonary infection with RSV. Virology 378(1):79–85

Smits EL, Anguille S, Cools N, Berneman ZN, Van Tendeloo VF (2009) Dendritic cell-based cancer gene therapy. Hum Gene Ther 20(10):1106–1118

Snitkovsky S, Young JA (1998) Cell-specific viral targeting mediated by a soluble retroviral receptor-ligand fusion protein. Proc Natl Acad Sci USA 95(12):7063–7068

Snitkovsky S, Niederman TM, Carter BS, Mulligan RC, Young JA (2000) A TVA-single-chain antibody fusion protein mediates specific targeting of a subgroup A avian leukosis virus vector to cells expressing a tumor-specific form of epidermal growth factor receptor. J Virol 74(20):9540–9545

Snitkovsky S, Niederman TM, Mulligan RC, Young JA (2001) Targeting avian leukosis virus subgroup A vectors by using a TVA-VEGF bridge protein. J Virol 75(3):1571–1575

Tillman BW, de Gruijl TD, Luykx-de Bakker SA, Scheper RJ, Pinedo HM, Curiel TJ, Gerritsen WR, Curiel DT (1999) Maturation of dendritic cells accompanies high-efficiency gene transfer by a CD40-targeted adenoviral vector. J Immunol 162(11):6378–6383

Tillman BW, Hayes TL, DeGruijl TD, Douglas JT, Curiel DT (2000) Adenoviral vectors targeted to CD40 enhance the efficacy of dendritic cell-based vaccination against human papillomavirus 16-induced tumor cells in a murine model. Cancer Res 60(19):5456–5463

Trepel M, Grifman M, Weitzman MD, Pasqualini R (2000) Molecular adaptors for vascular-targeted adenoviral gene delivery. Hum Gene Ther 11(14):1971–1981

van Beusechem VW, Grill J, Mastenbroek DC, Wickham TJ, Roelvink PW, Haisma HJ, Lamfers ML, Dirven CM, Pinedo HM, Gerritsen WR (2002) Efficient and selective gene transfer into primary human brain tumors by using single-chain antibody-targeted adenoviral vectors with native tropism abolished. J Virol 76(6):2753–2762

van Beusechem VW, Mastenbroek DC, van den Doel PB, Lamfers ML, Grill J, Wurdinger T, Haisma HJ, Pinedo HM, Gerritsen WR (2003) Conditionally replicative adenovirus expressing a targeting adapter molecule exhibits enhanced oncolytic potency on CAR-deficient tumors. Gene Ther 10(23):1982–1991

Vellinga J, de Vrij J, Myhre S, Uil T, Martineau P, Lindholm L, Hoeben RC (2007) Efficient incorporation of a functional hyper-stable single-chain antibody fragment protein-IX fusion in the adenovirus capsid. Gene Ther 14(8):664–670

Watanabe M, Boyer JL, Crystal RG (2009) Genetic delivery of bevacizumab to suppress vascular endothelial growth factor-induced high-permeability pulmonary edema. Hum Gene Ther 20(6):598–610

Watanabe M, Boyer JL, Crystal RG (2010) AAVrh.10-mediated genetic delivery of bevacizumab to the pleura to provide local anti-VEGF to suppress growth of metastatic lung tumors. Gene Ther 17(8):1042–1051

Watkins SJ, Mesyanzhinov VV, Kurochkina LP, Hawkins RE (1997) The ‘adenobody’ approach to viral targeting: specific and enhanced adenoviral gene delivery. Gene Ther 4(10):1004–1012

Wells DJ (2004) Gene therapy progress and prospects: electroporation and other physical methods. Gene Ther 11(18):1363–1369

Whittington HA, Ashworth LJ, Hawkins RE (1998) Recombinant adenoviral delivery for in vivo expression of scFv antibody fusion proteins. Gene Ther 5(6):770–777

Wickham TJ, Segal DM, Roelvink PW, Carrion ME, Lizonova A, Lee GM, Kovesdi I (1996) Targeted adenovirus gene transfer to endothelial and smooth muscle cells by using bispecific antibodies. J Virol 70(10):6831–6838

Wolff JA, Rozema DB (2008) Breaking the bonds: non-viral vectors become chemically dynamic. Mol Ther 16(1):8–15

Wurdinger T, Verheije MH, Broen K, Bosch BJ, Haijema BJ, de Haan CA, van Beusechem VW, Gerritsen WR, Rottier PJ (2005a) Soluble receptor-mediated targeting of mouse hepatitis coronavirus to the human epidermal growth factor receptor. J Virol 79(24):15314–15322

Wurdinger T, Verheije MH, Raaben M, Bosch BJ, de Haan CA, van Beusechem VW, Rottier PJ, Gerritsen WR (2005b) Targeting non-human coronaviruses to human cancer cells using a bispecific single-chain antibody. Gene Ther 12(18):1394–1404

Chapter 19
Bispecific Antibodies for Diagnostic Applications

Archana Parashar, Susmita Sarkar, Advaita Ganguly, Sai Kiran Sharma, and Mavanur R. Suresh

19.1 Introduction

Bispecific monoclonal antibodies (BsMAb) are uniquely engineered hetero-bifunctional macromolecules with two distinct binding specificities within a single molecule (Cao and Suresh 1998; Suresh et al. 1986b). They represent the second generation of monoclonal antibodies, and although structurally bivalent, they are functionally univalent for binding each molecule. Typically, BsMAb have one paratope specific for a particular protein or an antigen and another paratope specific for a detection moiety such as an enzyme (Fig. 19.1). BsMAb can therefore function as versatile hetero-bifunctional cross-linkers. The binding of BsMAb to both molecules is dependent on the affinity of epitope–paratope interaction and is highly specific. BsMAb can be produced by three methods: (a) chemical conjugation involving chemical cross-linking (the original method for BsMAb production); (b) somatic hybridization involving hybridoma technology; and (c) genetic engineering involving recombinant DNA technology (Das and Suresh 2005). Production of BsMAb by hybridoma technology involving bifunctional fusion (Fig. 19.2) of two different hybridomas that produce distinct monoclonal antibodies is more common compared to the other BsMAb production methods (Kohler and Milstein 1975). BsMAb have been exploited for a variety of in vivo and in vitro applications ranging from diagnostics to therapeutics (Cao and Suresh 1998). The use of BsMAb for different therapeutic applications and gene therapy is described in detail in earlier chapters. This chapter focuses only on the diagnostic applications of BsMAb.

BsMAb function as excellent diagnostic immunoprobes due to their intrinsic binding sites to any two molecules in a pre-determined order. They offer distinct

A. Parashar • S. Sarkar • A. Ganguly • S. Sharma • M.R. Suresh (✉)
Faculty of Pharmacy and Pharmaceutical Sciences, University of Alberta, Edmonton, AB T6G 2N8, Canada
e-mail: msuresh@pharmacy.ualberta.ca

R.E. Kontermann (ed.), *Bispecific Antibodies*,
DOI 10.1007/978-3-642-20910-9_19, © Springer-Verlag Berlin Heidelberg 2011

Fig. 19.1 Diagrammatic representation of a bispecific antibody. One paratope of the Bispecific antibody is directed against a specific target antigen of a tumor, virus, bacteria, or a parasite. The second paratope of the Bispecific antibody is directed against an effector molecule, which may be a detection moiety (such as an enzyme), toxin, drug, liposome, radioisotope, cytokines or immune cells. The diagram is adapted from Cao and Suresh, Bioconjugate Chemistry, 1998

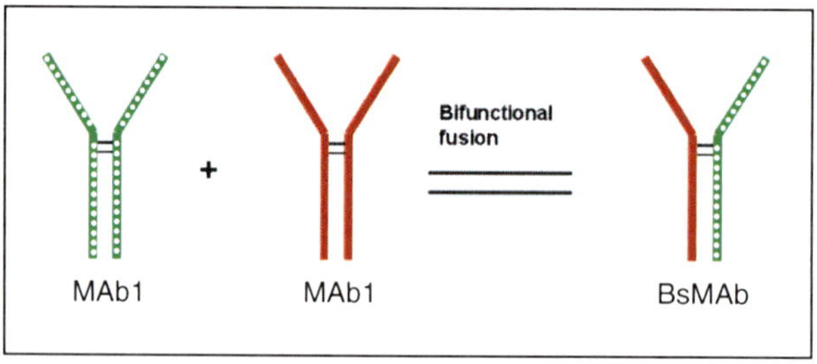

Fig. 19.2 Schematic representation of bifunctional fusion. Two different monoclonal antibodies (MAb1 and MAb2) each directed against a specific epitope are fused to generate a bispecific monoclonal antibody (BsMAb)

advantages over other conventional probes used in diagnostic assays due to their monovalency for one antigen and simultaneous binding to a marker enzyme. One epitope of BsMAb is usually directed against a marker enzyme such as horse radish peroxidase (HRPO), alkaline phosphatase, or β-galactosidase, which enables signal generation in immunodiagnostic assays. BsMAb have been used in a number of diagnostic applications such as immunoassays, immunohistochemistry, and radio-immunodiagnosis (Kreutz and Suresh 1997; Suresh 1991; Suresh et al. 1986a). Their use in diagnostic assays has revolutionized the field of immunodiagnostics due to the higher sensitivity and reproducibility that can be achieved in the assays.

19.2 Advantages of Using BsMAb in Diagnostics

The main advantage of using BsMAb in immunodiagnostics is that maximum specific activity and functional efficiency can be achieved since every antibody molecule is associated with a signal generating molecule (Kreutz and Suresh 1997). In addition to making the diagnostic assays rapid, BsMAb enhance sensitivity and specificity of detection. They significantly reduce or even eliminate false positive reactions especially in assays such as ELISA. In the case of immunohistochemistry, the use of BsMAb leads to decreased background, fine and sharp morphological information, clear and distinct identification of immune reactive sides, and good resolution of ultrastructural details (Milstein and Cuello 1983; Suresh et al. 1986a). Other advantages of using BsMAb in immunodiagnostic assays include improvement in signal to noise ratio and simplification of the assay procedure. They can easily be used as nanoprobes making ultrasensitive detection possible especially in the case of competitive immunoassays. In contrast to BsMAb, chemically conjugated monoclonal and polyclonal antibodies often exhibit higher background (Liu et al. 2003).

Traditionally monoclonal antibodies have been employed in many immunodiagnostic assays. Typically, the use of monoclonal antibodies in diagnostic applications requires chemical manipulation of the antibody such as chemical cross-linking in order to conjugate the detection moiety (such as an enzyme or biotin) onto the antibody. Chemical cross-linking of detection moieties to antibodies has many disadvantages as outlined below (Kricka 1994; Milstein and Cuello 1983). (a) It is a random process where the combination ratio of the antibody and the conjugating chemical is not fixed, generally resulting in batch to batch variations in the detection reagent produced. (b) The size of the resultant conjugated antibodies is not uniform influencing their ability to penetrate, which is especially critical for immunohistochemical applications. (c) Inactivation of active site or binding site of the detection moiety or the antibody and/or aggregation may occur during chemical conjugation. (d) Covalent linkages occur between the detection moiety and the antibody, which is not very preferable for in vivo diagnostic applications. (e) The antibodies produced often have low specific activity and decreased shelf life. The above factors have a significant impact on the performance of the conjugated monoclonal antibodies and therefore the reproducibility of the immunodiagnostic assay method. In contrast, the use of BsMAb completely eliminates the need for chemical conjugation of the antibody with the detection moiety. The bispecific nature of BsMAb ensures that the combination ratio of the antibody and the detection moiety is always 1:1. Further, since BsMAb are generally already tagged with the detection enzyme during the purification step, extra steps of enzyme addition and subsequent washing can be avoided during the immunoassay. BsMAb function as uniform, homogenous immunoconjugates with reproducible high specific activity and therefore serve as an excellent alternative to chemically conjugated monoclonal antibodies.

BsMAb are extremely versatile and function as robust immunoconjugates in the assay. They can be used in a variety of immunodiagnostic assays that are designed in multiple formats including the conventional format such as the microtitre plate assay and as well newer formats such as swab, strip, or filter disk assays. In the newer formats the end point of the assay can be read out visually without the need of any instrument, which makes the immunoassay extremely convenient for use as a point of care detection test in diverse health care settings.

19.3 BsMAb for Diagnosis of Infectious Diseases

One of the major challenges for the global health care system is to control the spread of infectious diseases, which cause 9.5 million deaths annually (WHO (2010), World Health Organization report). Diagnosis of a disease at its early stage is of paramount importance in combating both its progression and spread in the community. In order to develop effective diagnostic immunoassays for various infectious diseases, the high specificity of antigen–antibody interaction can be exploited. Due to their unique characteristics, BsMAb are extremely useful for developing simple, rapid, and point of care immunodiagnostic assays for detection of infective particles (specific antigen of the infective agent). The following sections describe some examples for the use of BsMAb in the diagnosis of infectious diseases caused by both bacteria and viruses.

19.3.1 Diagnosis of Bacterial Infections

19.3.1.1 *Mycobacterium tuberculosis*

According to the World Health Organization (WHO) about 1.3 million deaths were due to tuberculosis (TB) alone in the year 2008 (WHO-TB (2010), World health organization report on TB). Lack of early detection of infected individuals has resulted in the rapid spread of this disease to different parts of the world. The problem is further compounded due to co-infection with human immunodeficiency virus (HIV) that has led to a steep increase in morbidity and mortality. In fact, TB has now become the leading cause of death for HIV-infected patients (WHO-TB (2010), World Health Organisation report on TB). TB is caused by the bacterium *Mycobacterium tuberculosis*, which is highly resilient and rampant worldwide. The current diagnosis for *M. tuberculosis* mainly relies on the result of sputum smear microscopy (SSM) and bacterial culture method, both of which are not useful for early stage detection. SSM is not very sensitive and can only detect if the bacterial load is above 10^4/ml of sputum. Moreover, it is often difficult to collect sputum samples from elderly and pediatric populations who are especially vulnerable to *M. tuberculosis* infection due to their weak immune system. Bacterial culture method, which is still regarded as the gold standard for detecting *M. tuberculosis* takes about 7–10 days to provide results and also requires dedicated culture facilities. In the majority of TB cases, due to the above problems, detection is often performed based on clinical symptoms and results of chest radiograms. However, none of the above methods are specific because many other pulmonary diseases also show similar symptoms and chest radiograms. As a result of this delayed and faulty diagnosis system, patients are often treated incorrectly and in most cases this partly contributes to bacterial evolution towards multi-drug resistant (MDR) and extremely drug resistant (XDR) strains of *M. tuberculosis*.

Recently, sophisticated molecular assays such as interferon gamma release assay (IGRA) and nucleic acid amplification (NAA) assay have become available; however, their performance is also questionable (Campos et al. 2008; Lalvani and Millington 2008; Madariaga et al. 2007). These molecular assays are moreover difficult to perform in resource-constrained countries due to their high cost and requirement of technical sophistication. Currently there is no simple, rapid, and inexpensive point of care immunodiagnostic assay available for the detection of *M. tuberculosis*. Our laboratory therefore has developed a novel and highly sensitive immunodiagnostic assay for the detection of *M. tuberculosis* using BsMAb (Sarkar and Suresh 2010, University of Alberta, Unpublished data). Only a general outline of the developed BsMAb-based immunodiagnostic assay is described here. To design the immunoassay in either strip or swab format, any specific *M. tuberculosis* antigen can be used to generate BsMAb. The unraveling of the complete genome sequence of *M. tuberculosis* (Cole et al. 1998) has made it easier to identify antigens that are highly specific to the bacteria and not other related members of mycobacteria. Some of the important *M. tuberculosis* antigens (Palma-Nicolas and Bocanegra-Garcia 2007) that could be possible candidates for BsMAb generation are mentioned in Table 19.1.

The chosen *M. tuberculosis* antigen is used to first generate a monoclonal antibody and in the subsequent step the hybridoma secreting the monoclonal antibody is fused with another hybridoma secreting monoclonal antibody against any of the detection enzymes like HRPO or alkaline phosphatase to obtain a hybrid-hybridoma or quadroma producing the desirable BsMAb (Kreutz et al. 1998, Suresh et al. 1986b). After isolating a stable clone expressing the BsMAb, it is expanded in a bioreactor and subsequently the BsMAb is purified using affinity chromatography. Our laboratory has designed a novel purification process to isolate BsMAb in which one of the fusion monoclonal antibody partners is directed against HRPO (Bhatnagar et al. 2008). The process allows high efficiency purification of the BsMAb using *m*-aminophenylboronic acid agarose column along with HRPO tagging. The general format of the immunoassay involves coating a solid phase (swab or strip) with the *M. tuberculosis* antigen-specific monoclonal antibody and

Table 19.1 Important *M. tuberculosis* antigens

Protein antigens	ESAT6, CFP10, Mtb48, Mtb81, and 16 kDa-, 30 kDa- and 38 kDa-Protein antigens
Lipid antigens	Diacyl trehalose, triacyl trehalose, trehalose dimycolate and sulfolipids
Polysaccharide antigens	Lipoarabinomannan and arabinomannan

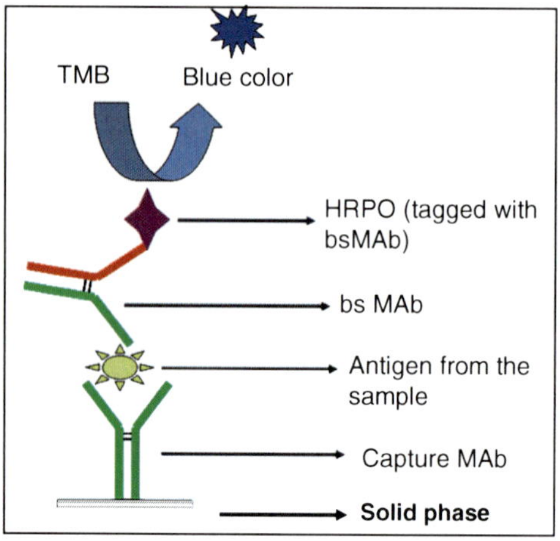

Fig. 19.3 General assay format for BsMAb-based immunoassay. The capture monoclonal antibody (MAb) is first immobilized onto a solid phase. This antibody binds to the respective specific antigen present in the test sample. Upon addition of the corresponding Bispecific monoclonal antibody (bsMAb), one arm of the bsMAb binds to the specific antigen while the other arm bound to HRPO converts the subsequently added TMB substrate to a blue colored product that can be easily detected

then blocking any unbound sites using a suitable buffer. The coated and blocked swab/strip is incubated with the clinical specimen collected from the patient. The *M. tuberculosis* antigen present in the sample would then specifically bind to the monoclonal antibody and other non-specific unbound materials are washed off, after which, the purified BsMAb is used as the detection antibody. As one arm of BsMAb can recognize the antigen, it would bind to the antigen forming a sandwich. Given that BsMAb is already tagged with HRPO, addition of the enzyme substrate 3, 3′, 5, 5′-tetramethylbenzidine (TMB) results in the formation of blue color that can easily be identified. A schematic of the general assay format is shown in Fig. 19.3.

The BsMAb-based immunodiagnostic assay for the detection of *M. tuberculosis* infection has several distinct advantages. The assay is highly specific, sensitive and at the same time easy to perform since it does not require any technical expertise. In addition, results of the assay can be obtained within few hours of sample collection and also the end point can be read out visually. Furthermore, the design of the assay makes it inexpensive so that the test can easily be performed in resource-constrained settings.

19.3.1.2 *Escherichia coli* O157:H7

E. coli O157:H7 belongs to the group of Verocytotoxin or Shiga toxin producing *E. coli* that has surfaced as an important food- and water-borne pathogen worldwide

(Fraser et al. 2004; Law 2000). It causes non-bloody or bloody diarrhea (hemorrhagic colitis) and hemolytic uremic syndrome characterized by hemolytic anemia, thrombocytopenia, and acute renal failure (Griffin and Tauxe 1991). *E. coli* O157: H7 is therefore classified as enterohemorrhagic *E. coli* (EHEC); it is reported to cause about 73,000 cases of infections per year in the United States alone (Mead et al. 1999). Several outbreaks of *E. coli* O157:H7 infection have been associated with recreational and municipal drinking water in North America, United Kingdom, and Japan (Bopp et al. 2003; Hrudey et al. 2003). Current methods to detect *E. coli* O157:H7 in water and food are traditional enrichment and plating on Sorbitol MacConkey agar and Rainbow agar selective media, Polymerase chain reaction (PCR), and other immunological methods (Bennett et al. 1996; Manafi and Kremsmaier 2001). The main drawbacks of these methods include time requirement (24–48 h for culture), difficulty in isolating DNA from sample (for PCR) and low sensitivity. It is important to note that low colony forming units (CFU) of *E. coli* O157:H7 present in contaminated food and water is generally sufficient to cause a severe infection, therefore having a sensitive method for detection during investigations of outbreaks is vital.

A highly sensitive BsMAb-based immunodetection assay was developed by our laboratory for the detection of *E. coli* O157:H7 in water samples (Guttikonda et al. 2007). *E. coli* O157:H7 whole bacteria and *E. coli* O157:H7 lipopolysaccharide (LPS) were used to first generate a monoclonal antibody, the hybridoma of which, was fused with anti-HRPO secreting monoclonal antibody hybridoma (Kreutz et al. 1998; Suresh et al. 1986b) to generate a quadroma producing BsMAb specific for both *E. coli* O157 and HRPO. The BsMAb was subsequently purified using benzhydroxamicacid agarose column (Husereau and Suresh 2001) and used in sandwich ELISA immunoassay. The anti-*E. coli* O157:H7 monoclonal antibody was used as the capture antibody to bind the bacteria from samples and the BsMAb was used as the detection antibody (Guttikonda et al. 2007). Refer to Fig. 19.3 for the general assay format. The detection limits of the assay were found to be 100 and 750 CFU/ml of sample for tap water and lake water, respectively. The immunoassay was further adapted to an immunofilter assay format to suit public health applications such as in water testing laboratories (Guttikonda et al. 2007). As low as 50 CFU of *E. coli* O157:H7/100 ml water were detected indicating that the assay is highly sensitive. Moreover the assay is highly specific since it did not detect a 500-fold excess of other bacterial strains such as *Salmonella*, *Pseudomonas*, and as well non-O157:H7 and non-pathogenic *E. coli*. A higher signal to noise ratio and a clean background was achieved in the assay since BsMAb served as a high specific activity probe. The BsMAb-based immunodetection assay therefore functions as a robust, ultrasensitive, and quick method for the detection of *E. coli* O157:H7. It has distinct advantages as it eliminates the amplification step required by other current assays and could easily be developed as a routine screening assay for detection of *E. coli* O157:H7 in water bodies especially in developing countries.

19.3.1.3 *Bordetella pertussis*

Whooping cough or pertussis caused by the bacterium *Bordetella pertussis* has become a major health concern in recent years especially among the adolescent age group (CDCP 2002, Centers for Disease Control and Prevention). Early and accurate diagnosis is critical in order to prevent the spread of infection and also ensure proper treatment of the affected individual. Conventional diagnosis of *B. pertussis* infection involves identification of the bacterium from culture of clinical nasopharyngeal aspirate samples, which generally takes about 3–7 days (Josephs 2000). Moreover, successful isolation and characterization is in turn dependent on several other factors such as proper specimen collection from the respiratory tract, storage, and culture conditions (Gustafsson et al. 1988). Another method for detection of *B. pertussis* involves the use of direct fluorescent-antibody assay that employs a fluorescence-labeled monoclonal antibody directed against a predominant antigenic LPS molecule present on the outer membrane of the bacterium (McNicol et al. 1995; Peppler 1984). Although detection of LPS is a good strategy, the assay is not sensitive and is also not easy to perform (Tilley et al. 2000).

Our laboratory therefore developed a highly sensitive and simple BsMAb-based immunodiagnostic assay for the detection of whole *B. pertussis* and soluble *B. pertussis* LPS (Tang et al. 2004). The assay is useful not only for analyzing clinical samples but also for immunochemical structural studies and serological characterization of *B. pertussis* LPS. The anti-*B. pertussis* LPS monoclonal antibody secreting hybridoma and the anti-HRPO monoclonal antibody secreting hybridoma were fused to generate a quadroma that produces BsMAb specific to both *B. pertussis* LPS and HRPO. BsMAb was subsequently purified by affinity chromatography using benzhydroxamicacid agarose column (Husereau and Suresh 2001) and used in a homosandwich ELISA immunoassay. Heat-killed *B. pertussis* BP347 by itself or spiked into nasopharyngeal aspirates was first bound by the anti-LPS specific monoclonal antibody, and then the BsMAb was added as the detection antibody (Tang et al. 2004). Refer to Fig. 19.3 for the general assay format. The assay showed high sensitivity with a practical lower limit of detection of ~5 CFU; the extrapolated theoretical lower limit of detection was found to be one bacterium by using the mean + 2 standard deviations of 20 control assays without *B. pertussis*. Interestingly, binding of *B. pertussis* to the anti-*B. pertussis* LPS monoclonal antibody-coated solid phase was found to be irreversible despite extensive washing suggesting a unique molecular velcro effect (Tang et al. 2004). The use of BsMAb in the immunoassay allowed ultrasensitive detection of *B. pertussis* since it avidly captured multiple exposed LPS molecules on the bacterial surface.

The BsMAb-based immunodiagnostic assay was adapted to an immunoswab format to facilitate easy, rapid, and point of care detection of *B. pertussis* in a primary health care setting. The sensitivity of detection was high since as few as ten bacteria could be detected compared to the controls (Tang et al. 2004). The BsMAb-based assay has unique advantages in being highly sensitive and easy to perform. It therefore facilitates early detection of *B. pertussis* infection and accurate monitoring of outbreaks of whooping cough.

19.3.2 *Diagnosis of Viral Infections*

19.3.2.1 Severe Acute Respiratory Syndrome Coronavirus

Severe acute respiratory syndrome (SARS) epidemic affected more than 8,000 people worldwide spreading over 30 countries across 5 continents in the 2002–2003 outbreak (Drosten et al. 2003; Peiris et al. 2003; Poon et al. 2004b). This highly contagious infection claimed over 900 lives in a short period of time with a fatality rate of 9.6%, emerging as a huge global threat to human health. SARS Coronavirus (SARS-CoV) is the etiological agent of SARS, which is believed to have transmitted from wild animals to the human population (Guan et al. 2003; Wang et al. 2006). According to WHO, the absence of a rapid screening test delayed the diagnosis of suspected cases of SARS and resulted in the spread of the disease worldwide. Current methods for the confirmation of SARS infection include detection of viral RNA by reverse transcription-PCR (Jiang et al. 2004; Poon et al. 2004a; Yam et al. 2003), detection of SARS-CoV antibodies in body fluids using indirect fluorescence assay (Chan et al. 2004), and isolation of SARS-CoV from clinical samples (Keyaerts et al. 2005; Yamashita et al. 2005). Since viral culture is a time consuming, tedious, and insensitive method, PCR and antibody detection methods are widely used; however, these two methods are expensive, laborious, and require technical expertise (Wu et al. 2004). A simple, rapid and inexpensive diagnostic test for the specific and early detection of SARS-CoV is therefore extremely important to limit the spread of infection and for proper risk management in the case of a future global outbreak especially since SARS has high rates of transmission and mortality.

Our laboratory has developed a highly sensitive, rapid, and simple BsMAb-based immunoswab assay for early detection of SARS-CoV (Kammila et al. 2008). The assay is designed to detect the most abundant and conserved viral antigen nucleocapsid protein NP (Di et al. 2005; Hiscox et al. 1995; Lau et al. 2004; Rota et al. 2003; Suresh et al. 2008) because its presence in different body fluids (serum, urine, nasopharyngeal aspirate, throat wash samples, and saliva) is suggestive of current infection. Three different monoclonal antibodies that recognize different epitopes on NP antigen and the anti-HRPO monoclonal antibody were employed to generate the BsMAb using hybrid-hybridoma technology. The obtained anti-SARS-CoV × anti-HRPO BsMAb was then purified by a novel, dual sequential affinity chromatography method (Protein-G column followed by *m*-aminophenyleboronic acid agarose column), which was developed by our laboratory (Bhatnagar et al. 2008). In the immunoswab assay, easy-to-use swabs were first coated with anti-SARS-CoV monoclonal antibody to capture NP in the test sample followed by detection with BsMAb (Kammila et al. 2008). Refer to Fig. 19.3 for the general assay format. The assay successfully detected SARS-CoV NP antigen spiked in different matrices such as saline, serum, and pig nasopharyngeal aspirate with high sensitivity. Pig nasopharyngeal aspirate was used as surrogate to human samples to test the ability of the immunoswab assay to detect low concentrations of NP antigen. The immunoswab assay showed NP detection limits of 10 pg/mL

(1 pg/swab) in saline, 20–200 pg/mL (1–10 pg/swab) in pig nasopharyngeal aspirate and 500 pg/mL (25 pg/swab) in rabbit serum. Further, the sensitivity of NP detection was high in BsMAb-based assay compared to a parallel MAb-based assay (20 pg/mL vs. 200 pg/mL in saline; 20–200 pg/mL vs. 200 pg/mL in pig nasopharyngeal aspirate) indicating that the BsMAb-based immunoassay works better than its MAb-based counterpart. Furthermore to determine the robustness of the assay, swabs pre-coated with capture antibody were tested after storage for 2-, 6-, 10-, and 14-week time periods under different conditions (Kammila et al. 2008). The BsMAb-based immunoswabs showed good sensitivity for NP detection even on week 14 at 200 pg/mL (4°C storage) and 500 pg/mL (RT and −20°C storage) indicating a good shelf-life for the immunoswabs.

The immunoswab format has potential advantages particularly in clinical, rural, and primary healthcare settings such as ease of use, early phase detection, and in limiting the spread of infection. It is also extremely useful for SARS screening at all ports of mass human entry (airports, seaports, bus, train, and border stations). Interestingly, the time required to perform the assay was approximately 45 min, which make it a rapid test for SARS diagnosis (Kammila et al. 2008). The assay can easily be used to screen numerous suspected individuals within a short period of time during a future SARS outbreak given the relative ease of accessing nasopharyngeal aspirate with no invasive procedures. The BsMAb-based immunoswab detection of SARS-CoV NP antigen therefore serves as a simple point of care diagnostic test compared to other current methods (Di et al. 2005) and is also highly sensitive, inexpensive, and requires minimally trained personnel.

19.3.2.2 Dengue Virus

Dengue fever caused by Dengue virus is the most prevalent mosquito-borne viral infection in the world as more than 100 million cases are reported each year predominantly in the tropical and sub-tropical regions (Hemungkorn et al. 2007). Dengue virus belongs to the Flaviviridae family and four different serotypes (DENV1, DENV2, DENV3, and DENV4) have been identified so far. In humans, dengue viral infection causes a spectrum of diseases ranging from asymptomatic infection to severe hemorrhagic fever ultimately resulting in death (Hemungkorn et al. 2007). Diagnosis of dengue infection based on clinical symptoms cannot be deemed reliable, and therefore detection of the specific virus, viral antigen, genomic sequence, and/or antibodies is absolutely essential to confirm infection. Current methods to diagnose dengue infection include virus isolation, and other serological and molecular techniques that employ monoclonal or polyclonal antibodies labeled with an enzymatic or fluorometric marker (Shu and Huang 2004). These methods are less reliable, and moreover those that identify the virus or the viral genome are expensive and require specialized laboratories. It is important to diagnose an on-going or recent infection during the acute or early convalescent stages in order to facilitate better treatment, and ensure effective etiological investigation and disease control (Halstead 2007). Currently there is no sensitive, simple, and inexpensive diagnostic assay available for the detection of dengue virus infection at the acute or early stages.

Our laboratory has developed a novel, highly sensitive immunodiagnostic assay for the detection of Dengue virus using BsMAb (Ganguly and Suresh 2010, University of Alberta, Unpublished data). Only a general outline of the developed BsMAb-based immunodiagnostic assay is described here. To design the immunoassay, any specific Dengue virus antigen can be used to generate BsMAb that could be employed in a simple strip or swab assay format. The selected Dengue virus antigen is used to first generate a monoclonal antibody and in the subsequent step, the hybridoma secreting the monoclonal antibody is fused with another hybridoma secreting monoclonal antibody specific for HRPO to produce a quadroma producing the desired BsMAb. The BsMAb is subsequently purified by affinity chromatography using *m*-aminophenylboronic acid agarose column (Bhatnagar et al. 2008). The general format of the immunodiagnostic assay involves coating Dengue virus antigen-specific monoclonal antibody as the capture antibody on calcium alginate tipped swabs with aluminum/or plastic shafts as per previously published protocol (Tang et al. 2004). The swab is then blocked and later incubated with the specific antigen spiked in serum, after which, the purified BsMAb is used as the detection antibody. Addition of the HRPO substrate, TMB, results in the formation of blue color that can be identified visually (Refer to Fig. 19.3 for the general assay format).

The BsMAb-based immunodiagnostic assay for Dengue virus detection has several advantages. The assay is highly specific and easy to perform since it does not require any technical expertise or sophisticated instruments. In addition, the design of the assay makes it rapid and convenient as a point of care detection tool apt for use in primary health care settings and remote villages. These features make the assay very suitable for early detection of dengue infection and thereby control its spread in the community.

19.4 BsMAb for Diagnosis of Cancer

In addition to their un-controlled growth and aberrant cellular physiology, cancerous cells are often characterized by the presence of over-expressed biochemical molecules such as glycoproteins that are present on the cell surface and later shed into body fluids. Such over-expressed tumor-specific antigens have been exploited as markers to enable diagnosis, and in conjunction with a battery of clinical methods can also be of prognostic relevance. Based on specific tumor-associated antigens (TAA), BsMAb can be generated for use in cancer diagnostics (Songsivilai and Lachmann 1990; Souriau and Hudson 2003). Such BsMAb would bind to the corresponding cancer-specific antigen in the test sample and provide an easy read-out regarding the presence of a specific type of cancer. It is becoming more and more apparent that early and confirmative diagnosis of cancer is critical for the control and cure of the disease. Due to their distinct characteristics of being powerful immunoprobes, BsMAb are extremely valuable in developing simple, rapid, and highly sensitive immunodiagnostic assays that can be used for the detection of various types of cancer. The following sections describe the use of BsMAb in detection and diagnostic imaging of cancer.

19.4.1 Prostate Cancer

Prostate cancer is most common in men and is responsible for more deaths than any other cancer, except for lung cancer (NCI 2010, National Cancer Institute Statistics). About 218,890 new cases of prostate cancer were diagnosed in the United States alone during 2007. It is expected that 1 out of 6 men will be diagnosed with prostate cancer during their lifetime; a little over 1.8 million men in the United States are survivors of prostate cancer (NCI 2010, National Cancer Institute Statistics). Prostate-specific antigen (PSA) is a useful tumor-associated marker present in the serum that is widely used for screening and monitoring progression of prostate cancer (Caplan and Kratz 2002). There are several different assays currently available for the measurement of PSA, most of which involve the use of labeled monoclonal or polyclonal antibodies (Armbruster 1993; Cattini et al. 1993; Khosravi et al. 1995; Klee et al. 1994; Leinonen et al. 1993; Oesterling et al. 1995; Vihko et al. 1990; Yu and Diamandis 1993). These assays are technically complex, time consuming, and have lower sensitivity.

Our laboratory therefore developed a novel, highly sensitive, and rapid BsMAb-based immunoassay for the measurement of PSA (Kreutz and Suresh 1997). The anti-PSA monoclonal antibody secreting hybridoma was fused with anti-HRPO secreting hybridoma using the hybrid-hybridoma technology to generate a quadroma producing the anti-PSA $\times$ anti-HRPO BsMAb. The BsMAb was then purified by ammonium sulfate precipitation, anion exchange, and affinity chromatography, after which, it was used in the PSA immunoassay. The anti-PSA monoclonal antibody was used as the capture antibody and the BsMAb was used as the detection antibody to measure PSA levels in the test samples in a 20-min single-step assay (Fig. 19.4). The assay had high analytical sensitivity even with a short substrate incubation time; the detection limit was comparable to most commercially available assays requiring longer incubation times (Kreutz and Suresh 1997). With TMB as the enzyme substrate, a sensitivity of 0.028 mg/L was achieved for 5-min substrate incubation in comparison to the commercially available Hybritech Tandem-EIA^{R} PSA assay that had a detection limit of 0.1 mg/L for 30-min substrate incubation. The developed assay was further evaluated using 138 clinical samples and results were found to correlate with those obtained by an automated Hybritech enzyme PSA immunoassay (Kreutz and Suresh 1997). Overall, the BsMAb-based immunoassay for PSA measurement has rapid kinetics and an excellent detection limit, which makes it a good candidate for use in the development of next-generation automated immunoassays and for rapid screening in health clinics.

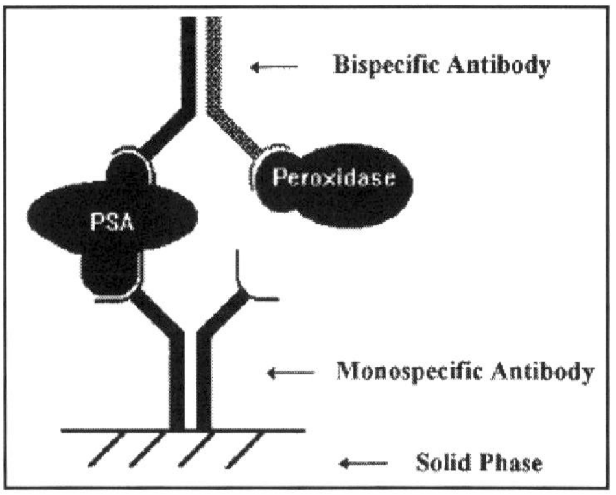

Fig. 19.4 Schematic representation of antibody complex formed during BsMAb-based immunoassay for PSA detection. PSA in the test sample is bound first by the capture monoclonal antibody and then by BsMAb detection antibody tagged to peroxidase (HRPO) to form an antibody sandwich complex. The diagram is adapted from Kreutz and Suresh, Clinical Chemistry, 1997

19.4.2 Cancer Diagnostic Imaging

Molecular imaging has transformed the field of cancer diagnosis with its superior ability to identify and localize cancer based on distinct molecular and functional characteristics of cancerous tissue. ^{18}F-flurodeoxyglucose (^{18}F-FDG)-based imaging using positron emission tomography (PET) or computed tomography (CT) has become popular for the detection and assessment of different types of tumors. Various specific monoclonal antibodies directed against different cancer antigens have also been employed; however, the sensitivity of these techniques is low compared to ^{18}F-FDG-PET/CT. In theory, monoclonal antibodies should have been more ideal imaging candidates owing to their higher degree of specificity for tumor antigens and their ability to discriminate between tumor and inflammation. However, the high molecular weight of the radiolabeled monoclonal antibodies limits tumor uptake and causes insufficient clearance from surrounding tissues resulting in low contrast during imaging. The size also interferes with the ability of these molecules to extravasate from the vascular channels into the extravascular space thus leading to slow accretion of radiolabeled antibodies in tumors (typically 1–3 days to achieve maximum tumor uptake). While tumor uptake of radiolabeled monoclonal antibodies may reach between 10 and 30% of injected dose per gram of tumor in mouse xenograft models, there may only be an accretion of <0.1% injected dose per gram of tumor in humans owing to a larger vascular and extravascular volume of distribution. Although several antibody modifications have been made along with changes to the radionuclide moiety in order to obtain better signal to noise ratios and avoid non-specific tissue uptake by liver, kidneys, and bone marrow, none of them have yielded expected results.

In recent times, BsMAb have proven to be more useful in cancer diagnostic imaging especially using the "pre-targeting" strategy (Cardillo et al. 2004; Gold et al. 2008; Goldenberg et al. 2006, 2007; Rearden et al. 1983, 1985; Sharkey et al. 2007). This strategy aims to deliver high tumor accretion, eliminate background noise due to untargeted radioactivity from blood and normal tissues, and reduce unintended toxicity. It relies on BsMAb's ability to effect tumor localization of the subsequently administered radiolabeled effector molecule. Targeting is achieved with BsMAb that binds to a target antigen with one arm and to a radiometal-chelate or hapten peptide complex (effector molecule) with the other arm. Animal model experiments have revealed an equivalent tumor uptake of BsMAb compared to radiolabeled antibodies and maximum accretion within minutes rather than several hours or days. Even the radionuclide was cleared very rapidly from the body, with more than 80% of the product eliminated in the urine within a few hours, allowing tumor/blood and tissue ratios to often be $\geq$10:1 within 1 h of the radionuclide administration. Tissue retention of the radiolabeled effector molecule was also found to be very low, even in the kidneys and liver that very often show elevated uptake of directly radiolabeled antibodies unless radioiodine is used. The use of BsMAb has significantly enhanced the imaging utility of pre-targeting procedures as compared to using directly radiolabeled antibodies. The de-coupling of the effector molecule from the pre-targeting agent is central to all the variations on this theme that has resulted in enhanced imaging of tumors.

BsMAb can be used in a simple two-step pre-targeting approach without the need for any clearing step or additional agents. Two-step procedures generally involve injection of the pre-targeting agent such as BsMAb and waiting for a day or two before all of the unbound fractions are drained off from the body's circulation by excretion through liver or kidneys (BsMAb are easily cleared from the blood within a day). This is followed by administration of the radiolabeled effector molecule, which would only bind to a specific arm on the pre-targeted BsMAb at the tumor site. Patients would therefore be able to receive BsMAb injection by their oncologist, return a few days later and receive the radiolabeled product in a nuclear medicine facility such that imaging can be accomplished on the same day. Although it may take a few days before an image can be achieved using a pre-targeting approach, experienced investigators and preclinical data have suggested that the images would be superior in terms of specificity, sensitivity, signal intensity, and have minimal background compared to those obtained using radiolabeled monoclonal antibodies. In many cases, BsMAb-based pre-targeting approach has out-performed the results generated by clinically approved radiolabeled monoclonal antibodies, especially in imaging small tumors in colorectal and colonic cancer (Sharkey et al. 2005). In addition to giving better tumor/blood ratios, superior tumor uptake, enhanced tumor/non tumor ratios, the pre-targeted strategy retained tumor-specific signal intensity even 1 day post-injection of the radioactive effector molecule thus providing a 15-fold increase in signal strength at the tumor site. BsMAb are currently being exploited for imaging many CEA-expressing tumors, MUC-1-expressing pancreatic cancer, and CD20-expressing non-Hodgkin's lymphoma (Peltier et al. 1993; Schuhmacher et al. 2001a, b).

BsMAb-based pre-targeting techniques for diagnostic imaging of cancers have taken nuclear imaging to a whole new level and are beginning to complement the present-day gold standard of ^{18}F-FDG-PET. BsMAb are being further engineered to achieve optimal molecular weight and charge for better tumor uptake, rapid blood clearance, higher valency, improved affinity to TAA, enhanced signal intensity, and greater flexibility for binding to a variety of hapten-peptides carrying different radio-isotopes. Improvements in the methods used for the production of BsMAb by "Dock and Lock" molecular engineering strategies and their application in furthering pre-targeting applications in tumor imaging and therapy is described in Chap. 12. Given that highly specific and sensitive images are possible with BsMAb pre-targeting in both the experimental and clinical set up, it is evident that BsMAb-based cancer diagnostic imaging is very promising.

19.5 Conclusion

The use of BsMAb in immunodiagnostic assays has resulted in the development of next generation immunoassays that are highly sensitive, rapid, simple, and cost-effective. The specificity of the assays approaches the theoretical limit of immunodetection making them ultrasensitive. Importantly, these assays are invaluable as a point of care diagnostic tool. Such assays are of immense value especially in some resource-constrained countries to combat the spread of diseases. Rapid detection of communicable diseases has huge impact on the current global health scenario where an infectious disease can make quick transition from an epidemic to a pandemic like the recent outbreak of swine flu. Most of the current diagnostic assays measure specific antibodies to bacterial or viral antigens and therefore may report a previous infection since antibodies circulate in the blood 6 months post-infection. Such assays rarely detect infection at initial stages making early intervention almost impossible. In contrast, BsMAb-based diagnostics measure specific bacterial or viral antigens rather than the antibodies and therefore accurately reflect active diseased state and make early-stage detection possible.

In the coming years, BsMAb will prove valuable for use in diagnosis of different types of cancer. Since BsMAb-based cancer diagnosis has many advantages and is highly sensitive it will be useful not only for early detection of cancer but also convenient for monitoring the progression of the diseased state and assessing response to treatment. Given the immense potential of BsMAb as excellent immunoprobes, there are efforts to engineer newer forms of these antibodies with significantly improved binding specificity and avidity. BsMAb await a plethora of novel applications in the field of diagnostics.

References

Armbruster DA (1993) Prostate-specific antigen: biochemistry, analytical methods, and clinical application. Clin Chem 39(2):181–195

Bennett AR, MacPhee S, Betts RP (1996) The isolation and detection of *Escherichia coli* O157 by use of immunomagnetic separation and immunoassay procedures. Lett Appl Microbiol 22(3):237–243

Bhatnagar PK, Das D, Suresh MR (2008) Sequential affinity purification of peroxidase tagged bispecific anti-SARS-CoV antibodies on phenylboronic acid agarose. J Chromatogr B Analyt Technol Biomed Life Sci 863(2):235–241

Bopp DJ, Sauders BD, Waring AL, Ackelsberg J, Dumas N, Braun-Howland E, Dziewulski D, Wallace BJ, Kelly M, Halse T, Musser KA, Smith PF, Morse DL, Limberger RJ (2003) Detection, isolation, and molecular subtyping of *Escherichia coli* O157:H7 and *Campylobacter jejuni* associated with a large waterborne outbreak. J Clin Microbiol 41(1):174–180

Campos M, Quartin A, Mendes E, Abreu A, Gurevich S, Echarte L, Ferreira T, Cleary T, Hollender E, Ashkin D (2008) Feasibility of shortening respiratory isolation with a single sputum nucleic acid amplification test. Am J Respir Crit Care Med 178(3):300–305

Cao Y, Suresh MR (1998) Bispecific antibodies as novel bioconjugates. Bioconjug Chem 9(6):635–644

Caplan A, Kratz A (2002) Prostate-specific antigen and the early diagnosis of prostate cancer. Am J Clin Pathol 117(Suppl):S104–S108

Cardillo TM, Karacay H, Goldenberg DM, Yeldell D, Chang CH, Modrak DE, Sharkey RM, Gold DV (2004) Improved targeting of pancreatic cancer: experimental studies of a new bispecific antibody, pretargeting enhancement system for immunoscintigraphy. Clin Cancer Res 10(10):3552–3561

Cattini R, Cooksey M, Robinson D, Brett G, Bacarese-Hamilton T, Jolley N (1993) Measurement of alpha-fetoprotein, carcinoembryonic antigen and prostate-specific antigen in serum and heparinised plasma by enzyme immunoassay on the fully automated serono SR1 analyzer. Eur J Clin Chem Clin Biochem 31(8):517–524

CDCP (2002) Centers for Disease Control and Prevention. Pertussis 1997–2000. Morb Mortal Wkly Rep Surveill Summ 51:73–76

Chan PK, Ng KC, Chan RC, Lam RK, Chow VC, Hui M, Wu A, Lee N, Yap FH, Cheng FW, Sung JJ, Tam JS (2004) Immunofluorescence assay for serologic diagnosis of SARS. Emerg Infect Dis 10(3):530–532

Cole ST, Brosch R, Parkhill J, Garnier T, Churcher C, Harris D, Gordon SV, Eiglmeier K, Gas S, Barry CE 3rd, Tekaia F, Badcock K, Basham D, Brown D, Chillingworth T, Connor R, Davies R, Devlin K, Feltwell T, Gentles S, Hamlin N, Holroyd S, Hornsby T, Jagels K, Krogh A, McLean J, Moule S, Murphy L, Oliver K, Osborne J, Quail MA, Rajandream MA, Rogers J, Rutter S, Seeger K, Skelton J, Squares R, Squares S, Sulston JE, Taylor K, Whitehead S, Barrell BG (1998) Deciphering the biology of *Mycobacterium tuberculosis* from the complete genome sequence. Nature 393(6685):537–544

Das D, Suresh MR (2005) Producing bispecific and bifunctional antibodies. Methods Mol Med 109:329–346

Di B, Hao W, Gao Y, Wang M, Wang YD, Qiu LW, Wen K, Zhou DH, Wu XW, Lu EJ, Liao ZY, Mei YB, Zheng BJ, Che XY (2005) Monoclonal antibody-based antigen capture enzyme-linked immunosorbent assay reveals high sensitivity of the nucleocapsid protein in acute-phase sera of severe acute respiratory syndrome patients. Clin Diagn Lab Immunol 12(1):135–140

Drosten C, Gunther S, Preiser W, van der Werf S, Brodt HR, Becker S, Rabenau H, Panning M, Kolesnikova L, Fouchier RA, Berger A, Burguiere AM, Cinatl J, Eickmann M, Escriou N, Grywna K, Kramme S, Manuguerra JC, Muller S, Rickerts V, Sturmer M, Vieth S, Klenk HD, Osterhaus AD, Schmitz H, Doerr HW (2003) Identification of a novel coronavirus in patients with severe acute respiratory syndrome. N Engl J Med 348(20):1967–1976

Fraser ME, Fujinaga M, Cherney MM, Melton-Celsa AR, Twiddy EM, O'Brien AD, James MN (2004) Structure of shiga toxin type 2 (Stx2) from *Escherichia coli* O157:H7. J Biol Chem 279 (26):27511–27517

Gold DV, Goldenberg DM, Karacay H, Rossi EA, Chang CH, Cardillo TM, McBride WJ, Sharkey RM (2008) A novel bispecific, trivalent antibody construct for targeting pancreatic carcinoma. Cancer Res 68(12):4819–4826

Goldenberg DM, Sharkey RM, Paganelli G, Barbet J, Chatal JF (2006) Antibody pretargeting advances cancer radioimmunodetection and radioimmunotherapy. J Clin Oncol 24(5):823–834

Goldenberg DM, Chatal JF, Barbet J, Boerman O, Sharkey RM (2007) Cancer imaging and therapy with bispecific antibody pretargeting. Update Cancer Ther 2(1):19–31

Griffin PM, Tauxe RV (1991) The epidemiology of infections caused by *Escherichia coli* O157:H7, other enterohemorrhagic *E. coli*, and the associated hemolytic uremic syndrome. Epidemiol Rev 13:60–98

Guan Y, Zheng BJ, He YQ, Liu XL, Zhuang ZX, Cheung CL, Luo SW, Li PH, Zhang LJ, Guan YJ, Butt KM, Wong KL, Chan KW, Lim W, Shortridge KF, Yuen KY, Peiris JS, Poon LL (2003) Isolation and characterization of viruses related to the SARS coronavirus from animals in southern China. Science 302(5643):276–278

Gustafsson B, Lindquist U, Andersson M (1988) Production and characterization of monoclonal antibodies directed against *Bordetella pertussis* lipopolysaccharide. J Clin Microbiol 26(2):188–193

Guttikonda S, Tang XL, Yang BM, Armstrong GD, Suresh MR (2007) Monospecific and bispecific antibodies against *E. coli* O157 for diagnostics. J Immunol Methods 327(1–2):1–9

Halstead SB (2007) Dengue. Lancet 370(9599):1644–1652

Hemungkorn M, Thisyakorn U, Thisyakorn C (2007) Dengue infection: a growing global health threat. Biosci Trends 1(2):90–96

Hiscox JA, Cavanagh D, Britton P (1995) Quantification of individual subgenomic mRNA species during replication of the coronavirus transmissible gastroenteritis virus. Virus Res 36(2–3): 119–130

Hrudey SE, Payment P, Huck PM, Gillham RW, Hrudey EJ (2003) A fatal waterborne disease epidemic in Walkerton, Ontario: comparison with other waterborne outbreaks in the developed world. Water Sci Technol 47(3):7–14

Husereau DR, Suresh MR (2001) A general affinity method to purify peroxidase-tagged antibodies. J Immunol Methods 249(1–2):33–41

Jiang SS, Chen TC, Yang JY, Hsiung CA, Su IJ, Liu YL, Chen PC, Juang JL (2004) Sensitive and quantitative detection of severe acute respiratory syndrome coronavirus infection by real-time nested polymerase chain reaction. Clin Infect Dis 38(2):293–296

Josephs JE (2000) Pertussis in the adolescent and adult: a primary care concern. Clin Excell Nurse Pract 4(6):361–365

Kammila S, Das D, Bhatnagar PK, Sunwoo HH, Zayas-Zamora G, King M, Suresh MR (2008) A rapid point of care immunoswab assay for SARS-CoV detection. J Virol Methods 152(1–2):77–84

Keyaerts E, Vijgen L, Maes P, Neyts J, Van Ranst M (2005) Growth kinetics of SARS-coronavirus in Vero E6 cells. Biochem Biophys Res Commun 329(3):1147–1151

Khosravi MJ, Papanastasiou-Diamandi A, Mistry J (1995) An ultrasensitive immunoassay for prostate-specific antigen based on conventional colorimetric detection. Clin Biochem 28(4):407–414

Klee GG, Preissner CM, Oesterling JE (1994) Development of a highly sensitive immunochemiluminometric assay for prostate-specific antigen. Urology 44(1):76–82

Kohler G, Milstein C (1975) Continuous cultures of fused cells secreting antibody of predefined specificity. Nature 256(5517):495–497

Kreutz FT, Suresh MR (1997) Novel bispecific immunoprobe for rapid and sensitive detection of prostate-specific antigen. Clin Chem 43(4):649–656

Kreutz FT, Xu DZ, Suresh MR (1998) A new method to generate quadromas by electrofusion and FACS sorting. Hybridoma 17(3):267–273

Kricka LJ (1994) Selected strategies for improving sensitivity and reliability of immunoassays. Clin Chem 40(3):347–357

Lalvani A, Millington KA (2008) Screening for tuberculosis infection prior to initiation of anti-TNF therapy. Autoimmun Rev 8(2):147–152

Lau SK, Woo PC, Wong BH, Tsoi HW, Woo GK, Poon RW, Chan KH, Wei WI, Peiris JS, Yuen KY (2004) Detection of severe acute respiratory syndrome (SARS) coronavirus nucleocapsid protein in sars patients by enzyme-linked immunosorbent assay. J Clin Microbiol 42(7):2884–2889

Law D (2000) Virulence factors of *Escherichia coli* O157 and other Shiga toxin-producing *E. coli*. J Appl Microbiol 88(5):729–745

Leinonen J, Lovgren T, Vornanen T, Stenman UH (1993) Double-label time-resolved immunofluorometric assay of prostate-specific antigen and of its complex with alpha 1-antichymotrypsin. Clin Chem 39(10):2098–2103

Liu F, Guttikonda S, Suresh MR (2003) Bispecific monoclonal antibodies against a viral and an enzyme: utilities in ultrasensitive virus ELISA and phage display technology. J Immunol Methods 274(1–2):115–127

Madariaga MG, Jalali Z, Swindells S (2007) Clinical utility of interferon gamma assay in the diagnosis of tuberculosis. J Am Board Fam Med 20(6):540–547

Manafi M, Kremsmaier B (2001) Comparative evaluation of different chromogenic/fluorogenic media for detecting *Escherichia coli* O157:H7 in food. Int J Food Microbiol 71(2–3):257–262

McNicol P, Giercke SM, Gray M, Martin D, Brodeur B, Peppler MS, Williams T, Hammond G (1995) Evaluation and validation of a monoclonal immunofluorescent reagent for direct detection of *Bordetella pertussis*. J Clin Microbiol 33(11):2868–2871

Mead PS, Slutsker L, Griffin PM, Tauxe RV (1999) Food-related illness and death in the United States reply to Dr. Hedberg. Emerg Infect Dis 5(6):841–842

Milstein C, Cuello AC (1983) Hybrid hybridomas and their use in immunohistochemistry. Nature 305(5934):537–540

NCI (2010) Key statistics about prostate cancer. http://www.cancer.org/Cancer/ProstateCancer/DetailedGuide/prostate-cancer-key-statistics. Cited on 23 July 2010

Oesterling JE, Moyad MA, Wright GL Jr, Beck GR (1995) An analytical comparison of the three most commonly used prostate-specific antigen assays: Tandem-R, Tandem-E, and IMx. Urology 46(4):524–532

Palma-Nicolas JP, Bocanegra-Garcia V (2007) Innovative strategies to diagnose and monitor tuberculosis patients. Arch Bronconeumol 43(4):225–232

Peiris JS, Lai ST, Poon LL, Guan Y, Yam LY, Lim W, Nicholls J, Yee WK, Yan WW, Cheung MT, Cheng VC, Chan KH, Tsang DN, Yung RW, Ng TK, Yuen KY (2003) Coronavirus as a possible cause of severe acute respiratory syndrome. Lancet 361(9366):1319–1325

Peltier P, Curtet C, Chatal JF, Le Doussal JM, Daniel G, Aillet G, Gruaz-Guyon A, Barbet J, Delaage M (1993) Radioimmunodetection of medullary thyroid cancer using a bispecific anti-CEA/anti-indium-DTPA antibody and an indium-111-labeled DTPA dimer. J Nucl Med 34(8):1267–1273

Peppler MS (1984) Two physically and serologically distinct lipopolysaccharide profiles in strains of *Bordetella pertussis* and their phenotype variants. Infect Immun 43(1):224–232

Poon LL, Chan KH, Wong OK, Cheung TK, Ng I, Zheng B, Seto WH, Yuen KY, Guan Y, Peiris JS (2004a) Detection of SARS coronavirus in patients with severe acute respiratory syndrome by conventional and real-time quantitative reverse transcription-PCR assays. Clin Chem 50(1):67–72

Poon LL, Guan Y, Nicholls JM, Yuen KY, Peiris JS (2004b) The aetiology, origins, and diagnosis of severe acute respiratory syndrome. Lancet Infect Dis 4(11):663–671

Rearden A, Nachtsheim DA, Frisman DM, Chiu P, Elmajian DA, Baird SM (1983) Altered cell surface antigen expression in bladder carcinoma detected by a new hemagglutinating monoclonal antibody. J Immunol 131(6):3073–3077

Rearden A, Taetle R, Elmajian DA, Majda JA, Baird SM (1985) Glycophorin A on normal and leukemia cells detected by monoclonal antibodies, including a new monoclonal antibody reactive with glycophorins A and B. Mol Immunol 22(4):369–378

Rota PA, Oberste MS, Monroe SS, Nix WA, Campagnoli R, Icenogle JP, Penaranda S, Bankamp B, Maher K, Chen MH, Tong S, Tamin A, Lowe L, Frace M, DeRisi JL, Chen Q, Wang D, Erdman DD, Peret TC, Burns C, Ksiazek TG, Rollin PE, Sanchez A, Liffick S, Holloway B, Limor J, McCaustland K, Olsen-Rasmussen M, Fouchier R, Gunther S, Osterhaus AD, Drosten C, Pallansch MA, Anderson LJ, Bellini WJ (2003) Characterization of a novel coronavirus associated with severe acute respiratory syndrome. Science 300(5624):1394–1399

Schuhmacher J, Kaul S, Klivenyi G, Junkermann H, Magener A, Henze M, Doll J, Haberkorn U, Amelung F, Bastert G (2001a) Immunoscintigraphy with positron emission tomography: gallium-68 chelate imaging of breast cancer pretargeted with bispecific anti-MUC1/anti-Ga chelate antibodies. Cancer Res 61(9):3712–3717

Schuhmacher J, Klivenyi G, Kaul S, Henze M, Matys R, Hauser H, Clorius J (2001b) Pretargeting of human mammary carcinoma xenografts with bispecific anti-MUC1/anti-Ga chelate antibodies and immunoscintigraphy with PET. Nucl Med Biol 28(7):821–828

Sharkey RM, Cardillo TM, Rossi EA, Chang CH, Karacay H, McBride WJ, Hansen HJ, Horak ID, Goldenberg DM (2005) Signal amplification in molecular imaging by pretargeting a multivalent, bispecific antibody. Nat Med 11(11):1250–1255

Sharkey RM, Karacay H, McBride WJ, Rossi EA, Chang CH, Goldenberg DM (2007) Bispecific antibody pretargeting of radionuclides for immuno single-photon emission computed tomography and immuno positron emission tomography molecular imaging: an update. Clin Cancer Res 13(18 Pt 2):5577s–5585s

Shu PY, Huang JH (2004) Current advances in dengue diagnosis. Clin Diagn Lab Immunol 11(4):642–650

Songsivilai S, Lachmann PJ (1990) Bispecific antibody: a tool for diagnosis and treatment of disease. Clin Exp Immunol 79(3):315–321

Souriau C, Hudson PJ (2003) Recombinant antibodies for cancer diagnosis and therapy. Expert Opin Biol Ther 3(2):305–318

Suresh MR (1991) Immunoassays for cancer-associated carbohydrate antigens. Semin Cancer Biol 2(6):367–377

Suresh MR, Cuello AC, Milstein C (1986a) Advantages of bispecific hybridomas in one-step immunocytochemistry and immunoassays. Proc Natl Acad Sci USA 83(20):7989–7993

Suresh MR, Cuello AC, Milstein C (1986b) Bispecific monoclonal antibodies from hybrid hybridomas. Methods Enzymol 121:210–228

Suresh MR, Bhatnagar PK, Das D (2008) Molecular targets for diagnostics and therapeutics of severe acute respiratory syndrome (SARS-CoV). J Pharm Pharm Sci 11(2):1s–13s

Tang XL, Peppler MS, Irvin RT, Suresh MR (2004) Use of bispecific antibodies in molecular velcro assays whose specificity approaches the theoretical limit of immunodetection for *Bordetella pertussis*. Clin Diagn Lab Immunol 11(4):752–757

Tilley PA, Kanchana MV, Knight I, Blondeau J, Antonishyn N, Deneer H (2000) Detection of *Bordetella pertussis* in a clinical laboratory by culture, polymerase chain reaction, and direct fluorescent antibody staining; accuracy, and cost. Diagn Microbiol Infect Dis 37(1):17–23

Vihko P, Kurkela R, Ramberg J, Pelkonen I, Vihko R (1990) Time-resolved immunofluorometric assay of human prostate-specific antigen. Clin Chem 36(1):92–95

Wang LF, Shi Z, Zhang S, Field H, Daszak P, Eaton BT (2006) Review of bats and SARS. Emerg Infect Dis 12(12):1834–1840

WHO (2010) World Health Organisation online report on infectious disease. http://www.who.int/healthinfo/global_burden_disease/GBD_report_2004update_AnnexA.pdf. Cited on 22 June 2010

WHO-TB (2010) World Health Organisation online report on tuberculosis. http://www.who.int/mediacentre/factsheets/fs104/en/ Cited on 30 June 2010

Wu HS, Chiu SC, Tseng TC, Lin SF, Lin JH, Hsu YH, Wang MC, Lin TL, Yang WZ, Ferng TL, Huang KH, Hsu LC, Lee LL, Yang JY, Chen HY, Su SP, Yang SY, Lin SY, Lin TH, Su IS (2004) Serologic and molecular biologic methods for SARS-associated coronavirus infection, Taiwan. Emerg Infect Dis 10(2):304–310

Yam WC, Chan KH, Poon LL, Guan Y, Yuen KY, Seto WH, Peiris JS (2003) Evaluation of reverse transcription-PCR assays for rapid diagnosis of severe acute respiratory syndrome associated with a novel coronavirus. J Clin Microbiol 41(10):4521–4524

Yamashita M, Yamate M, Li GM, Ikuta K (2005) Susceptibility of human and rat neural cell lines to infection by SARS-coronavirus. Biochem Biophys Res Commun 334(1):79–85

Yu H, Diamandis EP (1993) Ultrasensitive time-resolved immunofluorometric assay of prostate-specific antigen in serum and preliminary clinical studies. Clin Chem 39(10):2108–2114

Index

R.E. Kontermann (ed.), *Bispecific Antibodies*,
DOI 10.1007/978-3-642-20910-9, © Springer-Verlag Berlin Heidelberg 2011

Printed by Publishers' Graphics LLC